GEOGRAPHY

A MODERN SYNTHESIS

Harper & Row Series in Geography
D. W. Meinig, Advisor

GEOGRAPHY

A MODERN SYNTHESIS

Peter Haggett
University of Bristol

THIRD EDITION

Harper & Row, Publishers
New York   Hagerstown   Philadelphia   San Francisco   London

Credits for Opening Photographs

P. 2 Joel Cordon, DPI; p. 26, Sandy Nixon, DPI; p. 46, Len Rue, Jr., DPI; p. 68, NASA; p. 96, Hal. Fay, DPI; p. 116, E. Johnson, Leo de Wys, Inc.; p. 138, Marc & Evelyne Bernheim, Woodfin Camp & Associates; p. 166, Eric Kroll, Taurus Photos; p. 194, Beckwith Studios; p. 220, The Anaconda Company; p. 242, Ralph Mandol, DPI; p. 268, The Granger Collection; p. 296, Leonard Lee Rue III, DPI; p. 320, De Wys Inc.; p. 350, George Hall, Woodfin Camp & Associates; p. 376, Chris Reeberg, DPI; p. 406, Harold F. Fay, DPI; p. 432, Flying Camera, Inc., DPI; p. 454, Beckwith Studios; p. 476, Alon Reininger, DPI; p. 499, United Nations/Y. Nagata; p. 520, Hap Stewart/Jeroboam; p. 552, NASA; p. 572, NASA; p. 592, Phoebe Dunn, DPI.

The maps in the atlas section and on the cover are courtesy of Rand McNally & Co.

Sponsoring Editor: Bhagan Narine
Project Editor: Molly Scully
Production Manager: Laura Argento
Designer: Janet Bollow
Illustrator: J & R Technical Services
Photo Researcher: Myra Schachne
Cover Designer: Dare Porter
Compositor: Advanced Typesetting Services
Printer and Binder: Kingsport Press

GEOGRAPHY: A MODERN SYNTHESIS, Third Edition

Library of Congress Cataloging in Publication Data

Haggett, Peter.
    Geography: a modern synthesis.

    Includes index.
      1. Geography—Text-books—1945–   I. Title.
G128.H3 1979   910   78–26229
ISBN 0–06–042578–4

Contents

Advisor's Foreword

Geography: A Modern Synthesis is now widely recognized as an accurate title for an important book by an influential author. When it first appeared, it received immediate attention because Peter Haggett was already widely known for a number of major thought-provoking works, and upon examination it was welcomed as an impressively fresh and creative work. Now, seven years later, its use throughout the English-reading world attests to its success in the classroom as a stimulating introduction to a complex and dynamic field.

This success and significance of the book is directly related to major developments within the field of geography. During the past twenty-five years, geography, like many other fields, has undergone a period of intensive re-examination and change. New tools, new methods, new concepts, and new interdisciplinary relationships have led to some major shifts in approaches and emphases and, inevitably, to a good deal of contention and divergence over character, structure, and purpose. Such a period generates a need for new general works that can bring these matters into focus, relate them to the grand themes of geographic inquiry, and give an overarching coherence to the field once more: in short, the need for a new synthesis.

Geography: A Modern Synthesis first appeared just at that critical moment, and it was apparent that the author had directly and imaginatively addressed that great task in a highly original work. It was, at the most fundamental level, an effective synthesis of old and new. It did not call for a basic redefinition or narrowing of the historic field of geography. Rather it recognized the full, rich tradition of geographic work, the essential complementarity and interdependence of its various approaches and emphases, but at the same time, it set a very different framework, opened new perspectives, and applied an array of new concepts and tools to an unorthodox sequence of topics. Thus the book became an important exhibit of a new era in the ongoing development of the field.

The publication of this third edition attests to the fact that the book is more than a landmark; it continues to be a highly effective instrument for introducing students to the fundamentals of the field. And it will be as apparent to the readers of this edition as it was to those of the first and second that much of that success derives from the lively style and individuality of this particular geographic mind at work. *Geography: A Modern Synthesis*, Third Edition, is an invitation to join Professor Haggett in a carefully planned reconnaissance of a great and varied realm. It is a joint undertaking, requiring conscientious work on the part of all participants, but all the special tools are provided, and the results will amply repay the effort. The participants will learn new ways of looking at the world we share and at some of its problems and possibilities, and they will be shown many areas that await further investigation and how they might become explorers themselves. Under the leadership of such a capable and lively guide, this now widely tested tour can be recommended with complete confidence.

D. W. Meinig
Syracuse University

"If I ONLY knew geography!" Chico Pacheco kept repeating the phrase between clenched teeth, lamenting the wasted days of his youth; He had been a notorious cutter of classes. And all the time he had lost during his life, frittering it away on nonsense, when he could have devoted himself, body and soul, to the intensive study of geography, a science whose utility he had only come to realize!

... "I'll have to send to Bahia for some textbooks."

JORGE AMADO

Of the Drawbacks of Not Knowing Geography, and the Deplorable Tendency to Bluff at Poker.

In *OS MARINHEIROS* (1963)

Preface

Geography: A Modern Synthesis, Third Edition, is an attempt to present the whole spectrum of geography in a modern context and within a single volume. It tries to synthesize at two levels: first, by bringing together the the different traditions and themes within the field; second, by stressing the synthesizing role of geography as a whole in relation to neighboring fields. The book is designed to introduce the student with no previous geographic training to a field of rapidly expanding horizons and increasing consequence both as an academic subject and as an applied science. Lying athwart both the physical and social sciences, geography challenges students to abandon familiar and comfortable "straightjackets" and to focus directly on relationships between people and the environment, their spatial consequences, and the resulting regional structures that have emerged on the earth's surface. Geography is uniquely relevant to current concerns both with the environment and ecology and with regional contrasts and imbalances in human welfare.

No single exponent of geography or any other academic field of inquiry can write about the whole of it in detail. Past efforts to do so seem naive in retrospect. The problems that face the beginning student in geography, however, are now so complex that the challenge must be met. It is too easy to take the view that all one can or should do is to have a student take introductory courses in various easily identifiable subfields—physical geography, cultural geography, and so on—and hope that somehow these will produce an integrated view of geography as a whole. The osmosis, however, by which this is supposed to take place is rarely clearly defined.

To begin with the parts of a field of inquiry and go on to the whole seems to me to be a tactic of convenience, forced on us by the rising tide of research and the continuous fission of new subdisciplines. Surely, to confine larger questions to postgraduate seminars is an inversion of the desirable sequence of scholarship. We owe it to those who are starting in a field, and who we

xi

hope will follow us, to look around and ahead just as far as we can. Therefore, in this book I have turned away for a while from my own research patch and forced myself to put the various parts of geography together into what seems to me at this time to be an integrated form.

Third Edition Revisions

In addition to the "normal" updating and elimination of rough areas, this third edition of *Geography: A Modern Synthesis* contains a number of important changes. First, the number of chapters has been increased from 22 to 25. The additional space has been used to give fuller treatment to topics of growing importance in geographic studies: rapid environmental change (now two chapters), the geographer's view of the city (now a separate chapter), offshore areas and ocean resources, the role of the state in shaping world geography, the shift toward a more humanistic geography. New themes are introduced for the first time in several chapters (e.g., multinational corporations and their role in regional economic development in Chapter 22, catastrophe theory and its implications for understanding change in geography in Chapter 25, and so on). An atlas section has also been added to make the book more self-contained.

Second, while the overall strategy of the book has been retained (to help users who have already established patterns of work linked to this structure), there are important changes. Summary sections have been added at the end of each chapter. It is hoped that these will provide a checklist for students on some of the more important ideas presented. In conjunction with the study questions, the summary should enable readers to check out their understanding of a particular chapter independent from a given instructor, and provide a focus for thinking and for reviewing the topics discussed. In formulating the questions, I tried to strike a balance between simple checks on matters in the text and more demanding questions designed to challenge the reader favored with more time or greater interest. Some questions propose simple projects; others raise issues for class debate.

Correspondence suggests that the book is now extensively used not only in the United States and Western Europe, but also in Canada and in Australia and New Zealand. An effort has been made to reflect this wide readership in the choice of regional examples and illustrative material.

In addition to these major changes, there is information added here and there which I hope will prove useful. Some of the overly complicated diagrams have been redrawn, and a high proportion of the photographs has been replaced.

To accompany this third edition the *Instructors' Manual* and *Student Self-Paced Learning Manual* have been revised by Professor Larry K. Stephenson of Arizona State University and Professor John R. Healy of the University of Hawaii at Hilo. These books have been thoroughly reviewed in light of experience gained with the second edition and should help both the instructors using this book and students reading independently.

Revised suggestions for the use of the book in different teaching contexts are given in a new Appendix B.

Geographers are concerned with the structure and interaction of two major systems: the ecological system that links people and their environment, and the spatial system that links one region with another in a complex interchange of flows. These two systems are not separate, but overlapping.

Organization of the
Third Edition

The five parts of this book are like the atoms in a molecular chain or the circles in the Olympic symbol. Each is linked, and the internal structure and linkages between them, and between our two systems, can be explored.

Other geographers might have used different and equally arguable arrangements, but this particular one has two very useful characteristics: enough rigidity to provide a framework on which concepts, methods, and facts can be hung; and enough flexibility to accommodate new discoveries and new perspectives. For the only thing we can predict with confidence is that pressures for expansion and differentiation will continue, albeit perhaps in ways quite unlike those foreseen in this text.

Part One, "The Environmental Challenge," presents a geographer's view of the uncertain planetary environment in which the human population has

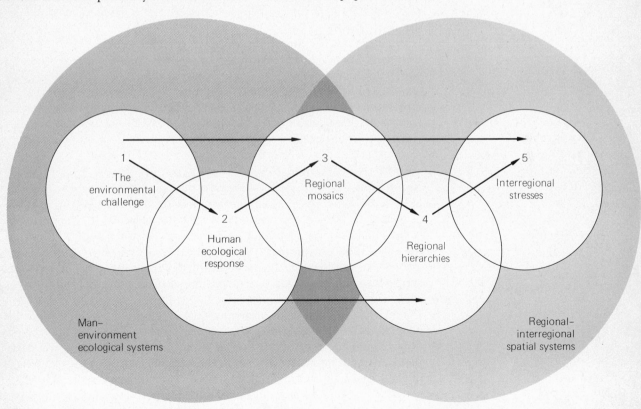

evolved and now lives at ever-increasing densities. Part Two, "Human Ecological Response," takes up the two-pronged response of people to the environmental challenge: adaptation of the environment and human adaptation to the environment. Part Three, "Regional Mosaics," turns from the ecological balance of people and their environment to the cultural fission and divisiveness within the human population responsible for regional contrasts. Part Four, "Regional Hierarchies," shows how the forces of urbanization work to override some of these regional differences and to organize human settlements into chains of city regions. Part Five, "Interregional Stresses," examines the interactions between the regional structures described in Part Four and the problems to which they give rise. Finally, the Epilogue is concerned with the future in two senses: It explores the increasing concern of geographers for the multiple worlds of the future, and it conjectures upon the future of geography itself. A detailed breakdown of the material covered is given in the introductions at the opening of each part.

The organization adopted in this book allows students to look at geography in an integrated way, abandoning the orthodox classification of geography as either physical or human, regional or systematic. Thus, chapters are not titled in the conventional manner or grouped into familiar patterns. Nonetheless, an initial course based on this book will provide the groundwork for full introductory courses in these areas as given in most university departments.

We can show the links between the chapters of *Geography: A Modern Synthesis*, Third Edition, and these courses in terms of a helix. Each complete circuit of the helix represents one level of approach, from the basic approach presented in this volume to advanced postgraduate courses. Topics introduced on one level may be taken up and expanded on others. Concepts, facts, and techniques introduced on one level may be reinforced and integrated on higher levels.

Areas of Emphasis

An attempt has been made to include the majority of concepts that are likely to be of use to the novice geographer. Although I have tried to cover the field in a catholic and impartial way, space is a prime constraint in any volume of this length. Thus, in the last analysis, there must be a heavy personal bias. Although authors cannot eliminate this bias, they should at least identify their own predilections so instructors can correct and modify the text if they so choose. My own partiality, where it presents itself, is fairly evident. I have leaned toward systematic theory and hypothesis rather than toward the elaboration of many regional case studies; I have generally adopted contemporary rather than classical statements of such theory; and I have chosen to restrict the discussion of physical geography to those aspects that contribute most directly to the other subfields of geography. Advanced physical geography is taught in most schools in Britain and the Commonwealth, and students there may find this treatment somewhat oversimplified.

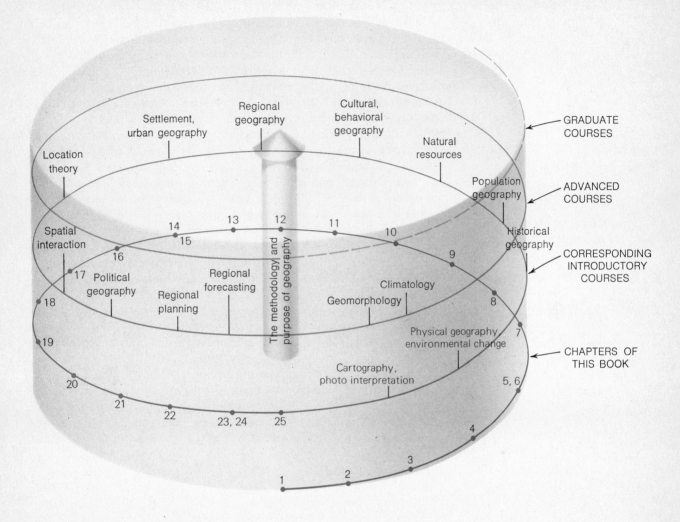

But many students in the United States and elsewhere may be meeting these topics for the first time. Compromise has therefore been inevitable.

Although the emphasis throughout is on concepts and methods, it would be inconceivable to write an introduction to geography without numerous regional case studies. These range widely to stress the global variations in the environment and its exploitation as well as to illustrate differences in the temporal and spatial scales of operation of the forces discussed. Thus, although the bulk of the regional examples is drawn from the world as structured in the third quarter of the twentieth century, some cases are drawn from other times.

Half the figures are concerned with general relationships, and the other half present specific regional cases. Of these, slightly over one-third is drawn from North America, another one-third from Europe, and the remainder from the rest of the world. This balance reflects partly the pattern of research results and partly the likely distribution of those who may use this book. Therefore, even though care has been taken to diversify the regional case studies, a minority of readers may feel themselves deprived of locally relevant examples. As for the scale of examples, one in six is drawn on the world scale, and one in three deals with an area no larger than a single city. Between these extremes, cases are well distributed along a size continuum.

Modern geography lays strong emphasis on the ways of analyzing research problems, and many of these ways are quantitative. Most quantitative methods can be left to later courses, however, and advanced techniques have been eliminated from the text. Those which it seems appropriate to mention are usually described in separate "marginal" discussions for readers to explore or ignore depending on their own inclinations or those of their instructor. Fuller guides to these techniques are listed in the suggestions for further reading.

Each chapter is accompanied by a list of references to guide readers in their further studies. Most are standard and widely available texts found in most college libraries. Journal references, because they change from year to year, are left for more advanced courses.

One's books, like one's children, develop a style and individuality of their own as they grow up. This third edition of *Geography: A Modern Synthesis* reflects two types of maturing influence: first, the many helpful reactions and suggestions of those who have used the book; second, the ongoing trends that continue to work themselves out within geography itself. I hope that readers of this third edition will continue to reinforce both sources through their comments and criticisms.

Peter Haggett
Chew Magna, England

With each succeeding edition a writer's debts grow. In the last three years I've been fortunate to spend time in both Australian and Canadian universities. Friends there, particularly at Monash and Toronto, may see ways in which their shared experience has colored this revision. In Europe, my long-standing debt to Nordic geography has been extended east to Finland, a country whose independent geographic traditions have much to teach its academic visitors.

Again it is to Donald Meinig, advisory editor of this series, that I owe a special debt for encouraging me to complete this revision. He was one of a series of reviewers, including many classroom users, whose critical comments helped to reshape the final book. I am particularly grateful to Professor Louis G. Morton, Department of Political Science, Mesa College; Professor John E. Oliver, Department of Geography, Indiana State University; Professor Christopher Salter, Department of Geography, University of California at Los Angeles; Professor Michael J. Troughton, Department of Geography, University of Western Ontario; and Professor Joseph Velikonja, Department of Geography, University of Washington for their critical reviews.

Students and colleagues at Bristol, Pennsylvania State, Cambridge, and University College, London, where I have taught over the years, have all contributed to the ideas in this volume in ways I hope they will recognize. I am grateful to them and also to the score of universities in North America and Europe that have allowed me, however briefly, to try some of the ideas contained here on their classes. In addition, all general texts draw on such a vast storehouse of published work that to acknowledge the debt in full would mean producing a *Who's Who* of contemporary scholars. The credits given for the figures will, I hope, indicate where my main acknowledgments lie. Books owe more to behind-the-scenes design and production work than authors often concede, and I have been more than fortunate in the talented team at Harper & Row, particularly Bhagan Narine and Molly Scully, that has worked on this volume.

Prefaces are too public a place to express the personal indebtedness of an author to his wife and family—they will know just how much I have to thank them for. Both my father and mother, who played so decisive a part in my interests and education, died while the first edition was being prepared. I dedicate this book to my second child, Timothy, in the trust that he may fulfill his late grandparents' hopes for him and his generation.

Acknowledgments

To teach I would build a trap such that, to escape, my students must learn.

ROBERT M. CHUTE
Environmental Insight (1971)

To the Student

Starting a course in a new subject at college is like driving into an unfamiliar city. We see the sprawling new suburbs, the bustling freeways, the pockets of decay, but find it hard to get an overall impression of the structure or to know where we are. Geography is a Los Angeles among academic cities in that it sprawls over a very large area and merges with its neighbors. It is also hard to be sure which is the central business district.

This book has been written specifically for "newcomers to the city" who have not previously taken courses in geography at college. It attempts to introduce some of the basic concepts geographers use as well as some of the essential environmental facts that form their background. The emphasis of the book is on ideas, or concepts. But these cannot be applied in a vacuum. Certain technical material has therefore been placed in separate discussions that are set outside the main text, and you may skip or explore them depending on the amount of time at your disposal and your taste.

The approach is essentially nonmathematical, and the book can be understood without any training in mathematics. On the other hand, geographers are using mathematics increasingly in their research, and certain aspects of a topic can be stated more explicitly in mathematical terms. These aspects, too, are presented outside the main text in separate discussions. You may wish to return to this material on a second reading.

For those of you who may be going on to further work in geography, each chapter makes some suggestions for further reading in the section entitled "One Step Further . . . ." These suggestions are largely confined to a handful of books that, in turn, open up other aspects of a subject. The final chapter points out some of the areas in which further training can be obtained and the kinds of courses you may wish to take.

Each of you may have your own favorite method for studying a textbook. Certainly no author can tell you which way is best for you, though many students find it useful to skip through a whole chapter quickly to get the general story. Figures have been designed to be self-contained wherever possible and so have been given somewhat fuller captions than normal. When, after a more lengthy reading of the chapter, you feel confident that you've understood it, you can turn to the concepts listed for review in the "Reflections" section to check yourself.

For those of you whose formal study of geography takes you no further than this book, I hope the brief acquaintance will have been a provoking one. If you take with you even some small part of the concern and fascination geographers experience in their exploration of the earth's environment and our place in it, then I will feel that my job is done.

P. H.

GEOGRAPHY

A MODERN SYNTHESIS

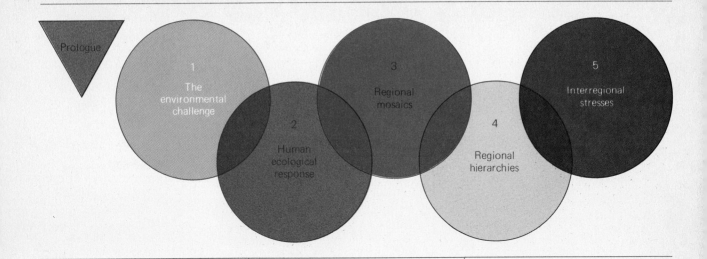

The Prologue introduces the reader to some of the basic concepts geographers use in their study of the planet Earth and its problems. On the Beach (Chapter 1) begins with a familiar scene—people arriving for a few hours of relaxation by the sea—and shows how careful geographic observation mirrors light on human beings and their ecological and spatial behavior. Analyzing this small slice of the human environment allows us to see how geographers approach their subject and provides a launching pad for work on a more vital global scale. The notion of changing geographic scale is a central one in this book, and the transfer of findings on one scale to applications on another remains a major geographic concern. So in *The World Beyond the Beach* (Chapter 2) we look at some of the ways geographers do this through maps. Maps provide the essential spatial language in which many geographic problems are discussed and conclusions and recommendations are given. Ways in which maps are made and used are illustrated, and the increasingly important role of environmental surveillance from space is introduced. Topics in this Prologue are picked up throughout the book, and we shall look back on them again in a fuller context at the end of the book, in the Epilogue.

prologue
Some Basic Concepts

We shall not cease from exploration
And the end of all our exploring
Will be to arrive where we started
And know the place for the first time.

T. S. ELIOT
Little Gidding (1942)

chapter 1
On the Beach

In Nevil Shute's compelling novel *On the Beach* the end of human occupation of the earth is forecast. If you have read the book or seen the old movie, you may recall the small group of survivors in Australia, waiting for radiation clouds to drift over the southern hemisphere—to complete the annihilation already accomplished in the north.

Whatever the merits of Shute's grim forecast, what justifies borrowing his title for the opening chapter of a textbook? Well, there are good reasons for a geographer selecting this title. People were historically creatures of the strandline between water and land. They move like crabs in the denser bottom layer of gas on the surface of the earth, not occupying the water itself, but never far from it. In prehistoric times the beaches were used as a highway; in the Renaissance they were used as a springboard for colonization and conquest. Even in the latter half of the twentieth century, the biggest cities are on the strandline: Three quarters of the world's largest urban centers—those with over 4 million inhabitants—are on the ocean or lake shore. Most of the remainder are on major rivers.

Today's urban dweller remains in an ecological relationship with the earth's resources which is less intimate, but no less fundamental, than that in prehistoric times. This relationship has always been a finely balanced one, in which quite small swings could bring about discomfort or disaster. But the hazards for early peoples were essentially local, and new and empty lands could always be found over the horizon. For today's city dwellers the hazards are regional or global, and most of the new or empty lands have long since been filled or abandoned.

For over 2000 years geographers have been describing and analyzing the ways in which humanity has come to terms—or failed to come to terms—with the planetary environment. In this book we shall try to see what kind of insights into that environment have been achieved.

3

1-1
The Crowded Beach

Our opening photograph shows a small segment of a crowded beach. Figure 1-1 shows a more extensive beach scene. These photographs were taken not from ground level, but from a helicopter, producing maplike pictures of the scene below. Because an overhead shot tends to distort the shape of familiar objects (notably ourselves), you may need a moment or two to sort out what is what.

Beach scenes are ordinary enough events, so what makes them of special interest to geographers? Let us answer that question in a round-about way. If we give three similar rocks to three different people, they may respond in quite different ways: A sculptor may shape the rock into new and more interesting forms, a mineralogist may start to break his up to examine its chemical structure, and a protester may hurl hers through the nearest window. The trains of thought started and the actions taken are determined not by the object at hand (the rock), but by the attitudes of the three individuals toward it. People other than Sir Isaac Newton have been hit on the head by apples falling from apple trees; reacting to the blow by pondering the laws of gravity clearly represents only one of many possible responses!

Figure 1-1. The crowded beach.
A low-level view of a New England beach taken in late summer. [Photograph by Rotkin, P.F.I.]

"SPACE," "LOCATION," AND "PLACE"

Three words which geographers use a lot are "space," "location," and "place." Since they are also words used in everyday language, we need to be sure just how they are being used in this book.

Space means extent or area, usually expressed in terms of the earth's surface. It does not mean space in the sense of outer space (e.g., NASA, the National Aeronautical and Space Administration) or space in the sense of to arrange things in tidy rows.

Location means a particular position within space, usually a position on the earth's surface. Like the word space, it is rather abstract in meaning when compared to the third word of the trio.

Place also means a particular position on the earth's surface: But, in contrast to location, it is not used in an abstract sense but confined to an identifiable location on which we load certain values. So a location becomes a place once it is identified with a certain content of information. Sometimes the content is a physical fact. For example, latitude 27° 59′N, longitude 86° 56′ E is an abstract location which we only recognize as a place once we know it describes the position of Mount Everest, the highest point on the earth's land surface. In other cases, the information content is a human experience. What gives a place its particular identity was a question which occurred to physicists Niels Bohr and Werner Heisenberg when they visited Kronberg Castle in Denmark. Bohr said to Heisenberg:

"Isn't it strange how this castle changes as soon as one imagines that Hamlet lived here? As scientists we believe that a castle consists only of stones, and admire the way the architect put them together. The stones, the green roof with its patina, the wood carvings in the church, constitute the whole castle. None of this should be changed by the fact that Hamlet lived here, and yet it is changed completely. Suddenly the walls and the ramparts speak a quite different language. The courtyard becomes an entire world, a dark corner reminds us of the darkness in the human soul, we hear Hamlet's 'To be or not to be.' Yet all we really know about Hamlet is that his name appears in a thirteenth-century chronicle. No one can prove that he really lived, let alone that he lived here. But everyone knows the questions Shakespeare had him ask, the human depth he was made to reveal, and so he, too, had to be found in a place on earth, here in Kronberg. And once we know that, Kronberg becomes quite a differrent castle for us." [Werner Heisenberg, *Physics and Beyond: Encounters and Conversations* (Harper & Row, New York 1972), p. 51, cited by Yi-Fu Tuan, *Space and Place*, University of Minnesota Press, Minneapolis, 1977), p. 4.]

In the same way, a familiar beach may provoke different reactions even among various types of scientists. Geologists may head for the sand particles and the fluid dynamicist for the breaking waves. Sociologists may study the behavior of the groups using the beach and economists the profits of the hot dog stands. How would a geographer react?

Perhaps the first reaction of a geographer to the beach scene in Figure 1-1 would be to try to pin down exactly where events were occurring in space. A photograph taken from a helicopter allows a much more accurate assessment of the location of people on a beach than a photograph taken on the ground. It is for this reason that most of the photographs you will find in this book are aerial photographs. Concern with locations in space is a characteristic of geographers' curiosity; specifying location accurately is one of the prime rules of the geographic game. An inaccurate description of location causes a geographer to wince in the same way as a linguist would over a mispronunciation, or a historian over an inaccurate date. (See the above discussion of "space," "location," and "place."

From this concern with space come questions of spatial patterns and organization. So a second reaction to the beach scene in Figure 1-1 would be to try explaining the spatial variations which are observed. Why are some parts of the beach packed with people while others are deserted? How much does this circumstance relate to differences in the quality of the beach? Questions of this kind lead to a general concern with the relationship of people to their environment.

A third geographic reaction to the beach scene shown here would be to try to sort the various elements in the photograph into some kind of regional order. Regions are a shorthand way of describing the variable character of an area in an efficient manner. One of the simplest ways of establishing sets of regions is to divide an area into several zones, each of which has certain characteristics that give it a particular character. For example, we could divide the beach into three zones: a swash zone below the regular high water mark, an upper zone above the regular high water mark, and a belt of sand dunes behind. Figure 1-2 shows a cross section of a typical beach zoned in this way. Geographers use this process of division, called dissection, to establish sets of regions. From this geographers go on to try and relate their findings on the beach to others around the world. That is, they try to place their regions in some kind of world focus.

To sum up: A geographer is concerned with three different but interlocking questions: (1) the question of *location*, in which the concern is to establish the precise spatial position of things within a particular area of the earth's surface; (2) the question of *human-environment relations* within the area; (3)

Figure 1-2. Beach environments. Regional division of the beach environment into a series of five distinctive zones, A through E. The principal factors controlling the character of each zone are the tidal range (areas below the high water mark are shaded) and the surface material (sand, mud, rock). The vertical scale has been exaggerated to emphasize the effect of height above sea level.

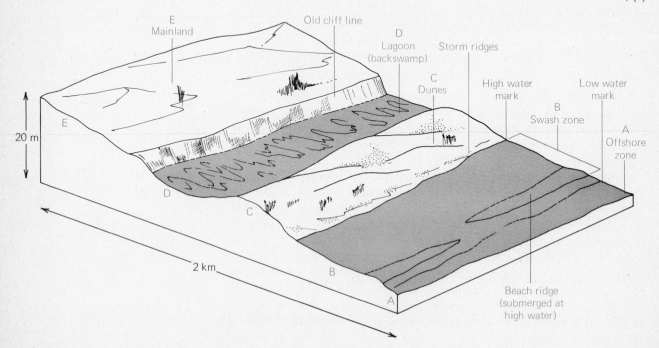

the question of *regions* and the identification of the distinctive character of particular spatial subdivisions of the area.

All three questions focus on the particular character of the beach as a place, and we shall need later to ask a fourth and more general question. How do our findings on the beach relate to other regions, and to the world picture as a whole?

Perhaps at this point we should follow the conventional textbooks and attempt a definition of geography itself based on these central questions. If you like formal definitions, then please turn to the last chapter and look at those given in Table 25-1. These definitions have been deliberately placed near the end of the book to encourage you to draw your *own* portrait of a geographer as you work through the chapters; this should give you a better likeness of geography than would looking up a dictionary entry right away. So let's agree for the moment that "geography is what geographers do" and go on to see them at work on the beach. Later in the book we shall see them working on wider, more important questions and in a global context.

If we wish to answer the geographer's first question, we need to extablish the accurate location of individuals on the beach. The simple question "Where are they?" can, however, be answered in two ways. We can answer it, for example, by analyzing the distribution of individuals in terms of their absolute location, or in terms of their relative location. A person's *absolute location* is his position in terms of an arbitrary grid system. Thus in Figure 1-3(a), the location of individual A is about 9 m east and 6 m north of an arbitrary point of origin at zero. The grid provides a convenient framework on which locations can be fixed. But a person's *relative location* is usually more interesting. For example, individual B is approximately 6 m from A, whereas C and D are almost sharing the same spot. Let us look at the use we can make of both kinds of locational information.

Mapping in Absolute Space   The absolute location of things is important in making accurate maps. Figure 1-3 shows various ways of mapping a population. We begin by assigning each individual on the beach to a corresponding location on the map, representing each person by a single dot. We then place a grid over the dots. By counting the number of dots that falls within each square cell of the grid, we can translate the distribution of dots, or people, into an array of numbers that reflect the density of the population [Figure 1-3(b)]. Crowded parts of the beach are represented by cells with high values, sparser parts by cells with low values, and empty stretches by cells with zero values.

Two kinds of maps can be drawn from these cell values. We can make *choropleth maps* (from the Greek *choros*, area, and *plethos*, fullness or quantity) by assigning different shades of color to each of the cells. By linking cells with similar colors, we can create a general picture of the distribution of

1-2
Space and Time on the Beach

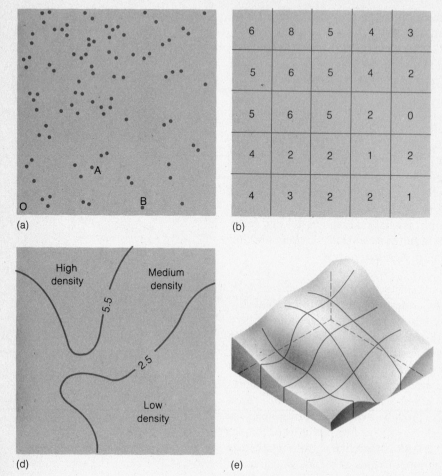

(a)

(b)

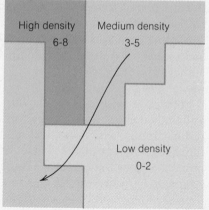

(c)

(d)

(e)

Figure 1-3. Population mapping.
Diagrams (a) through (e) show various ways of mapping population density. (a) is a simple dot map of a 25 m × 25 m segment of the beach. In (b) the area covered by the dot map is divided into 5 m × 5 m cells, and the number of persons in each cell is summed. (c) is a choropleth map of the beach population, and (d) is an isopleth map. The values on each isopleth (2.5 and 5.5) describe the average number of people in each cell. (e) is a three-dimensional isopleth map. All these maps are simply different ways of describing the same spatial distribution of population.

population on the beach [Figure 1-3(c)]. In doing this, we lose some information. Instead of 9 different values (between 0 and 8), we now have only 3. However, we have conveniently simplified our map, or grid, by reducing the spatial pattern from 25 cells to 4 areas. The second and more common way of mapping population distributions using a grid is to draw lines between all points having the same quality or value [Figure 1-3(d)]. Such lines are known as *isopleths* (from the Greek *isos*, equal), and maps of this type are known as *isopleth maps*. These maps also give an accurate picture of variations in population density. A three-dimensional version of an isopleth map would show areas of denser population as peaks and sparser areas as hollows [Figure 1-3(e)].

Organization in Relative Space The relative location of the members of a population is also of great interest to geographers. It helps them to understand why a population is organized, or distributed, in a particular way.

Animal behaviorists like Konrad Lorenz in *On Aggression* and social psychologists like Edward T. Hall in *The Hidden Dimension* have discussed how, like other primates, human beings have a strongly developed sense of territoriality. They surround themselves with visible or invisible "space bubbles" that are sensitive to crowding. Hall distinguishes between four "action zones" based on the distance between people: intimate space, personal space, social space, and public space. *Intimate space* is reserved for physical interactions like loving or fighting, while *personal space* is used for soft talk and friendly interaction. The boundary between these first two zones is about half a meter (1½ ft). Beyond a distance of 1½ meters (about 4 ft), personal space gives way to the *social space* used for formal business and social contacts. At around 4 meters (about 12 ft), the outer zone of public space, in which the preacher and the ice-cream-cone vendor on the beach may operate, begins. Clearly, these zones may vary from person to person and from culture to culture—the personal conversational space of the French may be seen by the more reserved English as an intrusion into their intimate space!

Because of the sense of territoriality, it is possible to argue that individuals arrange themselves on a beach to achieve a certain distance from each other. Their object may be to be as near as possible to those they love, as far as possible from those they hate, and at a convenient distance from those about whom they feel less strongly. When the distribution of people in an area is regarded in this way, it is measured in terms of *interpersonal space*, the linear distance separating an individual from neighbors. To determine interpersonal space, we ask, "How many meters is a given person from his nearest neighbor, his second nearest neighbor, his third nearest neighbor, and so on?" Geographers frequently use this approach in the study of human distributions and settlements. (See Section 15-1.)

The result of analyzing the relative location of a beach population is given in Figure 1-4. The three peaks, near zero, at 2 m, and at 5 m, may be interpreted as related to the space between couples less concerned with the beach (let alone geography!) than with each other, to family groups, and to strangers outside the family groups and keeping a respectful distance from them.

As relative distance increases, so the contacts between human groups change. The limited power of the human voice and eye mean that a large lecture theater is about the limit of direct face-to-face communication. Beyond that the media and the communication system take over. Also as the size of space increases, so the number of people who can be involved gets larger. Thus the implications of the space seen on the beach extend far beyond it, underlying human organization right up to the level of the globe itself. We shall return to it again and again in this book.

Time and Spatial Diffusion The scene in Figure 1-1 is unreal insofar as it freezes at one click of the camera's shutter a constantly changing pattern. The photograph is static; the real beach scene is dynamic. A similar photograph taken at another time would show a different picture. Just how different

ISOPLETH MAPS

Isopleth maps are one of the most common ways in which geographers show spatial distributions. Some of the isopleths are given distinctive names.

Isochrone maps show lines of equal time.

Isohyet maps show lines of equal rainfall.

Isoneph maps show lines of equal cloudiness.

Isophene maps show lines of biological events that occur at the same time (e.g., flowering dates of plants).

Isotherm maps show lines of equal temperature.

Isotim maps show lines of equal transport cost.

One of the commonest isopleth maps you will encounter is the *contour* map, which shows lines of equal height of land above sea level. "Contours" are also used by geographers as a general term for any type of isopleth.

Figure 1-4. Interpersonal distances. This is an idealized profile of the density of population on the beach in terms of the distance between individuals. Note the three characteristic peaks.

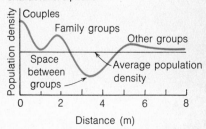

would depend on the timing of the second shot. A picture taken a few seconds later would reveal little change in the population but would catch the movement of the breakers. One taken a few hours later would show an empty beach and a different tide level. One taken some months later in winter might show a change in some of the physical structures. One taken some years later might reveal a substantial change in the form of the beach itself due to erosion.

All geographers work within a specific time context. And as Figure 1-5 makes clear, this context vitally affects the conclusions they draw. In each graph the number of people on a beach is related to the passage of time. Over a period of 100 years (e.g., 1870-1970) the most noticeable change is the increasing use of the beach. On the average, more people were using the beach at the end of the period than at the beginning. This is not surprising when we think of the substantial increases in southern New England's population over this period and the changes in social attitudes toward leisure time.

If we reduce the period of observation to a single year, we find a wavelike trend with a peak in late summer and a trough in late winter. If we reduce it to a single week, the waveform narrows to a sharp peak on the weekend. But over a much shorter period (for example, half an hour), the number of people on the beach remains constant and the trend line is horizontal. These general trends—of accelerating growth, wavelike cycles, or stability—are functions of the period over which observations are made.

Consider what we would see if we observed the beach at regular intervals from daybreak. The first arrivals might well occupy what seemed to be the "best" sites—what is best being determined by the requirements of the group: say, near the surf line for the youngsters, and near the parked car for the old folks. As the best sites were taken, new arrivals would have to occupy less attractive areas or crowd into the already occupied ones, reducing the amount of interpersonal space and producing the population pattern shown in the photograph.

Figure 1-6 traces the evolution of this pattern over three hours. This is a simple example of the spatial *diffusion* of a population, in which the location of an individual is related to the time she arrives. We shall look at more

Figure 1-5.　Changing population densities.

These graphs show the impact of the length of the observation period on the trends detected in population density. (a) shows the historical trend, (b) the seasonal cycle, (c) the weekly fluctuation, and (d) the short-term equilibrium.

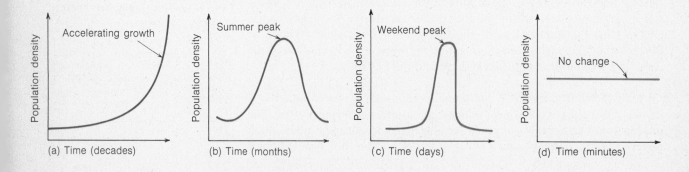

(a) Time (decades)　　　(b) Time (months)　　　(c) Time (days)　　　(d) Time (minutes)

Point of access to the beach

(a) 10 A.M.

(b) 11 A.M.

(c) 12 NOON

(d) 5 P.M.

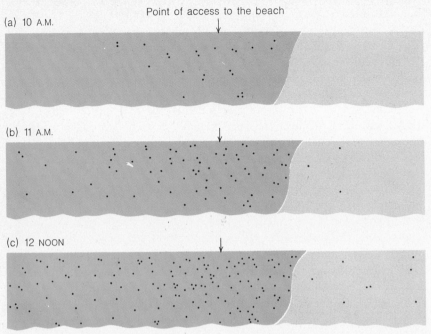

Figure 1-6. Spatial diffusion.
Stages in the diffusion pattern of people arriving on an empty beach at different stages of the morning. Note how the filling-up process is related to the point of access to the beach and to differences in the quality of the beach environment. Emptying the beach is not a simple reversal of filling it up. The homeward drift in late afternoon, shown in (d) below, is less orderly in spatial terms. Note that "environmental quality" can include elements made by people (like lifeguard surveillance) as well as natural features (e.g., sand vs. shingle).

ENVIRONMENTAL QUALITY

High   Low

DIFFUSION

In geography, diffusion is the process of spreading out or scattering over an area of the earth's surface. It should not be confused with the physicist's use of the term to describe the slow mixing of gases or liquids with one another by molecular interpenetration. Different kinds of geographical diffusion are discussed in Section 13-1.

complex examples of spatial diffusion in later chapters. Observation of this process over time not only allows us to consider the present spatial pattern in light of its past development; sometimes, it makes it possible to predict spatial diffusion in the future. Given the first two maps in Figure 1-6 (and knowing what has happened on the beach on similar days in the past), could you make a reasonably accurate prediction of the noontime pattern?

The Tyranny of Time and Space   Together space and time form the framework of the cage within which human life unfolds. Figure 1-7 shows how this cage works. Imagine you are living in a small town near the coast 1-7(a). On a given Saturday in summer, you decide to go to the beach. There is a choice of three: first, the local town beach at A to which you can cycle in half an hour; second, the ocean beach at B which means borrowing the family car and driving for an hour; finally, the best beach of all at C, with fine surf but lying so far to the north that it takes about 5 hours of hard driving to get there. Which are you going to choose?

You can plot your own choice as a path running from home to the beach and back in terms of a space-time box, looking like an aquarium. Figure 1-7(b) shows what happens if you choose B. Notice that the vertical scale is that of the clock running from midnight to midnight over a single day. All three choices are shown in Figure 1-7(c). Notice how your paths through time and space are limited in three different ways. There is the familiar rhythm of night

(a)

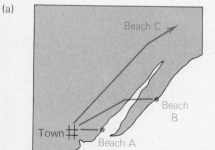

Figure 1-7.   Space-time constraints.

(a) Map of paths taken by an individual family traveling from their home to three local beaches. (b) The route to Beach B plotted in a timespace "aquarium." (c) Cross section to show all three routes in terms of time of day, distance, and the operation of the three fundamental constraints discussed in the text. Much of the work on the geographical impact of these space-time constraints has been by Swedish geographers at Lund University.

(b)

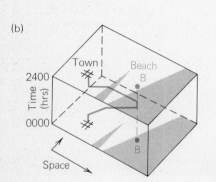

(c)

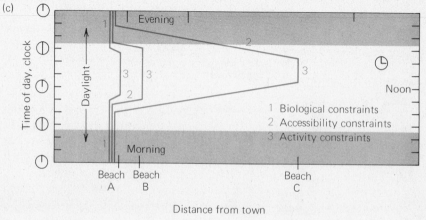

Distance from town

and day, with the biological clock within each of us demanding sleep at more or less regular intervals. So, we want to be home at a certain hour. Then there is the accessibility constraint, as shown by the diagonal lines in the diagram. Can you get hold of the car? Last, there is the activity constraint. You need at least an hour or two on the beach to make the trip worthwhile.

If you continue the three constraints, then you'll see how they box in your set of locational choices. If you can't get the car, then constraint 2 means that only beach A is open to you. Even given the car, beach C lies so far away that it's hardly worth the long drive for the limited time there (constraint 3). So it looks like you'll have to settle for beach B.

Geographers at Lund University in southwest Sweden have made a special study of the way these time and space constraints box in human activity. They have been able to show how the growth of cities, the range of jobs we can take, even the partners we meet and marry are all related to the kind of constraints we've just discussed. If we're very poor or sick, our locational choice may be severely limited to the local area. If we're well off, then Concorde or a private jet may greatly expand our spatial horizons. But, rich or poor, the basic biological constraint of the human being as a circadian (meaning "almost daily") animal remains. None of us can climb out of that cage.

1-3
People and the Beach Environment

The concept of relations between the individual and the environment is basic to geographic thinking and underlies the second of our three basic questions. By an *environment*, geographers mean the sum total of conditions that surround (literally, environ) a person at any one point on the earth's surface. For early people these conditions were largely natural and included such elements as the local climate, terrain, vegetation, and soils. With the rise of civilization people surrounded themselves with artifacts which, because of their sheer scale and longevity, became an integral part of their environment. For today's urban dweller the environment is dominated by the fixed structures of urban life (freeways, city blocks, asphalt surfaces). The natural environment has been either replaced or radically modified.

Human-environment relations have two sides to them. The first side relates to the influence of the environment on human activity. We can express this in symbols as E→H. Second, human activity may alter a given environment. This reverses the order to H→E.

Environmental Impacts on Human Beings (E→H) We can bring the relationship between people and their environment into focus by going back to the beach. The population density on a beach is partly a function of environmental quality. Good beaches (i.e., beaches with fine sand or good surf) tend to attract more users, while poor beaches (those that are, say, polluted by oil or by the local dog population) are shunned. Other things such as weather and accessibility being equal, we can relate the capacity of a beach to attract a population to its environmental quality. Even within the area of our photograph, local variations in environmental quality can lead to variations in population density. Figure 1-8(a) shows the distribution of both the beach population and the beach environment as maps.

The study of two or more geographic distributions varying over the same area is a study of *spatial covariation,* an idea we shall meet repeatedly in this book. When the two maps look alike and the two distributions "fit" one another closely, we say that the two phenomena are associated by area; that is, high values for population density in one area correspond with high values for environmental quality in the same area, and vice versa. Other hypothetical cases with little covariation are also illustrated in Figure 1-8. Comparing pairs of maps in this manner often tells us a great deal about the spatial covariation of different phenomena. Distributions can also be compared by statistical methods, but a discussion of these lies outside the scope of this introductory text.

To view the environmental quality of a beach in terms of, say, its surfing potential is clearly only one possible viewpoint—one strongly influenced by age and nationality of the users. For most of our history our questions would relate more to mundane matters of safe anchorage or shellfish yield than to questions of recreational use. We need then to assume some lens or filter which is placed between people and the environment. This means that what we *see* in a beach may be determined by our age, our interests, our income, our ethnic background, and so on. The environment provides a range of choices, only some of which have ever been seen. For example, a beach backed by very high dunes may suddenly become an attraction for hang-gliding enthusiasts. The environment (i.e., the dunes) has not changed: but what we choose to see and to do in that environment has.

In this book, we shall spend much of Part One describing the earth's environment, and much of Part Two discussing our reaction to it. We shall see there how strongly interpretation of a given environment is influenced by social, cultural, and technological factors.

Figure 1-8. Spatial covariation.
Here we have three hypothetical distributions of population density on a beach. In (a) there is a strong positive correspondence between the distribution of population and the quality of the beach environment. In (b) there is a strong negative correspondence, and in (c) there is little correspondence.

Environmental quality

High   Low

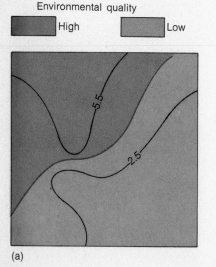

(a)

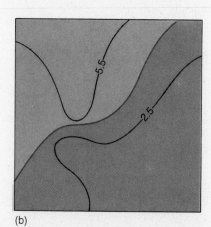

(b)

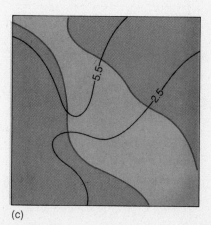

(c)

Human Modifications of Environments (H→E) While people are affected by their environment, they also have some capacity for modifying it. For example, they can alter the form of a beach by erecting defensive walls and change its quality by fouling it with oil and debris. Again time must enter our analysis, for the impact of human action is often *lagged* in time. A lagged impact is one that occurs later at the same place or later and at a different place. An example of the first type of lag is where toxic industrial waste is slowly concentrated in the food tissues of marine animals and the birds that feed on them. (See Figure 8-14.) An example of the second is where protection of a beach on one part of the coast may mean increased wave erosion at another. Likewise, sewage discharged at one point on a river may affect the fish populations downstream.

Beyond the beach most of the world's environments have been strongly influenced by human action. Chapters 8 through 10, later in this book, will describe in detail how massive that change has been.

Human-Environment Systems (E/H) Whether the effect of people on environment, or the effect of environment on people is more important is a chicken-and-egg question. Geographers find it more helpful to think of both relationships as part of a human-environment system. A *system* may be defined as a group of things or parts (called elements) that work together through a regular set of relations (called links) within defined limits (called the system boundary). Thus we can regard the beach as a system in which its various parts—shingle, sand, and mudbanks—are each linked together through a set of relations involving the energy of waves, tides, and winds.

Geographers are particularly interested in systems which link together human beings and environment. Figure 1-9 shows such a system, linking the users of a particular beach with its environment. Here a human system (H/H relations) and an environmental system (E/E relations) are linked into a human-environment system (H/E relations). The links are of various kinds—information, energy, and material. Together they form the flows that bind

SYSTEM

A system is a group of things or parts that work together through a regular set of relations. Thus we can regard the beach as a system in which its various parts—shingle, sand, mudbanks, and the like—are linked together through a set of relations involving the energy of waves, tides, and winds. Geographers are particularly interested in systems which link together people and environment.

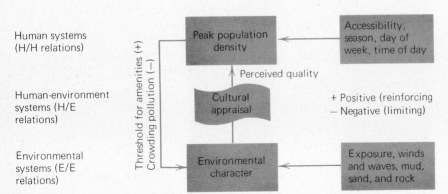

Figure 1-9. **Human-environment relations as a simple feedback system.** Note the loop by which beach character affects the numbers who use it, but this beach-using population itself may feed back on perceived beach quality. Beneficial changes like the installation of amenities will improve the beach character and attract more people (+), while hostile changes will have the reverse effect (−).

the five elements together. Note in particular the circular loop (called a feedback loop) connecting the character of the beach and the density of those using it. Higher levels of use may lead to two effects: first, to growing litter and pollution which may lead people to shun the beach and move elsewhere. Since this effect lessens the use of the beach, this kind of feedback is termed a negative feedback. A second and opposite effect may be for more people to increase the attractiveness of the beach. For example, on a popular beach it may be worthwhile to provide a lifeguard. This second effect is termed a positive feedback, since it tends to increase the level of use.

So our beach forms a very simple example of feedback in a human-environment system. Its study is worthwhile insofar as the same principles are incorporated into much more complex models, such as the world models discussed near the end of the book. (See Section 24-3.)

1-4
The Beach in World Focus

Beaches are fine and pleasant places, and we might all wish to spend more time there. But it would be misleading to suggest that geographers spend more time at the seashore than anyone else. We have chosen to concentrate on it in this chapter because it represents a microcosm, but *only* a microcosm, of the kinds of phenomena that geographers study.

Levels of Resolution In the atlas section in the middle of this book are examples of humanity's relationship with the environment on two quite different scales. In one map we see how the human population is distributed throughout the world, and in another, how the environment of the United States and Canada is distributed throughout the various regions. The world, and the nations of the world, are certainly more usual arenas for the geographer's work than a beach, which is only a tiny section of the human environment. The modern geographer deals with a continuum of environmental regions of increasing size. These range from the microenvironment of the individual in local surroundings to the macroenvironment of humanity as a whole.

Figure 1-10 shows what happens when we change our focus. In 1-10(a) the

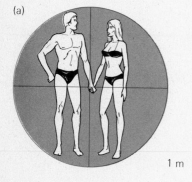

(a)

1 m

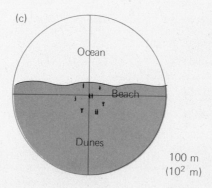

(b)

10 m
(10^1 m)

(c)

Ocean

Beach

Dunes

100 m
(10^2 m)

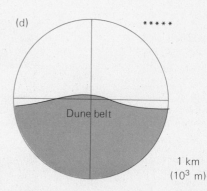

(d)

*****

Dune belt

1 km
(10^3 m)

picture in the camera viewfinder is clearly that of just two individuals on the beach. As we pull our camera back, the picture changes. By (c) the individuals have been lost in the crowd and by (d) even the crowd is no longer visible. But notice that by (e) a new feature, the familiar hook shape of Cape Cod on New England's coastline has come into view. Finally this too disappears (g) and a new feature, the globe itself, comes into sharp focus (h).

So, as geographers stand back from the beach, new spatial realities emerge. Just as with a painting, you have to be the right distance away to see its meaning. For this depends on the level of detail, the spatial resolution, that can be seen. Go too close and the painting becomes a jumble of individual brush marks. Stand too far away and all you can see is a blurred rectangle of canvas on the distant wall.

Orders of Geographic Magnitude   Throughout this text we discuss geographic studies confined to sizes, or *orders of magnitude*, between these two extremes. Although the variations are considerable, the geographer is concerned with a narrow "window" within the scale of scientific inquiry. In Figure 1-11 the main areas of scientific inquiry are plotted along a centimeter scale. Exponential notation (in which 1000 is written as 10^3, 1 as 10^0, 0.001 as 10^{-3}, and so on) is used in order to present a large range of variation on the same diagram.*

*See Appendix A for tables of conversion constants from metric to nonmetric measure.

(i)

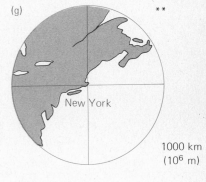

Figure 1-10.   Changing focus on the beach.
The maps show a couple of sunbathers on a Cape Cod beach viewed from higher and higher elevations. The radius of the resulting picture caught in the zoom lens is given in meters (m) or kilometers (km) [n.b. 1000 m = 1 km]. Note that each disk has a radius exactly ten times greater than or smaller than its neighbor, and that each disk *remains* centered on the two sunbathers. Because of the small size of the drawing the human figures quickly disappear and a new feature, Cape Cod, comes into view in (e). By the final drawing (h) this too has disappeared, but we can see one side of the whole earth. The stars, indicating orders of geographic magnitude, are discussed on page 18. (i) shows a Tiros VII satellite photo of Cape Cod. [Photograph courtesy of NASA.]

(h)   *

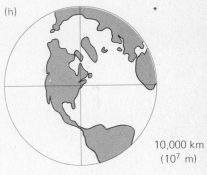

10,000 km
(10^7 m)

(e)   ****

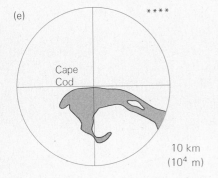

Cape
Cod

10 km
(10^4 m)

(f)   ***

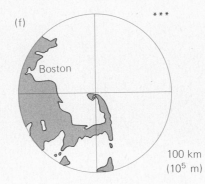

Boston

100 km
(10^5 m)

(g)   **

New York

1000 km
(10^6 m)

ORDERS OF GEOGRAPHIC MAGNITUDE

Geographers have to deal with objects that vary considerably in size. In this book we indicate the size in terms of orders of magnitude:

*First Order of Magnitude. Areas with a range of diameters from that of the surface of the earth itself (with an equatorial circumference of 40,000 km, or 24,860 mi) to 12,500 km (7,700 mi).

**Second Order of Magnitude. Areas with a range of diameters from 12,500 to 1250 km (7,770 to 777 mi). A typical example from the middle of this range is the coterminous United States.

***Third Order of Magnitude. Areas with a range of diameters from 1250 km down to 125 km (777 to 77.7 mi). A typical example from the middle of this range is New York State.

****Fourth Order of Magnitude. Areas with a range of diameters from 125 km to 12.5 km (77.7 to 7.77 mi). A typical example from the middle of this range is New York City.

*****Fifth Order of Magnitude. Areas with a range of diameters from 12.5 km to 1.25 km (7.77 to 0.777 mi). A typical example from the middle of this range is Central Park in New York City.

It would of course be possible, as Figure 1-11 shows, to continue downward to a sixth order and beyond, but the five classes shown cover the main range of geographers' work. Note that differences between the orders of magnitude are not linear: The contrast between the second and fifth order is not a difference of 3 but one of $10 \times 10 \times 10$ (i.e., 10^3, or 1000).

The scale in Figure 1-11 extends from the microworld of the atomic physicist studying cosmic rays with wavelengths of about 10^{-15} cm to the radio astronomer studying galaxies with diameters of 10^{23} cm and more. By comparison, the geographer's world is confined to a narrow band along the middle range of the scale. The smallest objects studied are about the size of a beach or a city block. They are not less than a few hundred meters across, or approximately 10^4 cm. Conversely, the largest object studied is the earth, with an equatorial circumference of around 40,000 km (24, 860 mi) or about 10^9 cm. There is therefore a range of 5, i.e., 10^9 minus 10^4 in the "size" (where size is measured by the distance along the longest axis), of objects studied by geographers. To put this another way, the real world shown in the atlas section is around 100,000 times greater in diameter than the world of the beach.

We use the orders of geographic magnitudes described at the top of this page to keep in focus the areas we are dealing with. These orders of magnitude serve a purpose similar to that of the astronomer's orders of star brightness and are a happy substitute for the jumble of scales usually used in geography books. They remind us that we are dealing not with the real world but with reduced and simplified models of it.

1-5
Models in Geography

The real world is much more complex than a beach. In trying to make sense of the structure of a particular region, geographers often attempt to simulate reality by substituting similar but simpler forms for those they are studying. They do this by constructing models. In everyday language the term *model* is used in at least three different ways. As a noun it signifies a representation; as an adjective it implies an ideal; as a verb, it means to demonstrate. Thus

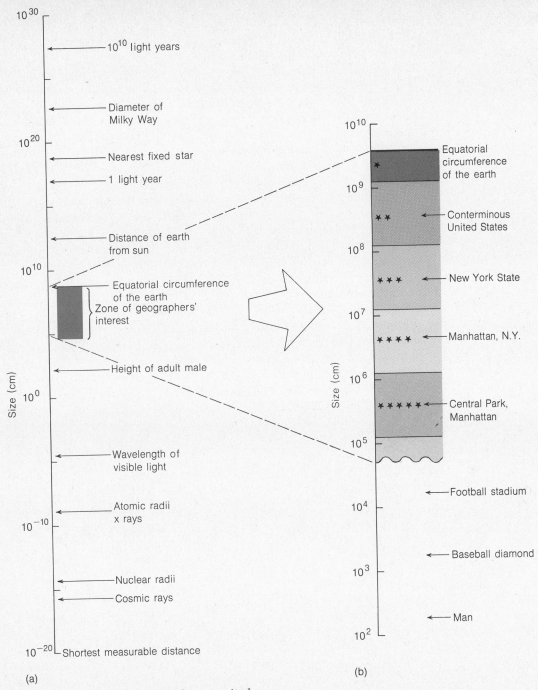

Figure 1-11. Orders of geographic magnitude.
Geographers study a continuum of environments of varying size. However, even this continuum is rather limited when one considers the continuum of environments that comprises the universe. This larger continuum, and the zone of interest to geographers, is represented by the scale in Figure 1-11(a). The enlargement of the zone of interest in Figure 1-11(b) enables us to divide it into zones which are roughly parallel to the orders of geographic magnitude described on p. 18.

MODEL

A model is an idealized representation of the real world built in order to demonstrate certain of its properties.

we are aware that when we refer to a model railway or a model husband we are using the same term in different senses.

In scientific work the term "model" has, to some extent, all three of the meanings. Scientific model-builders create idealized representations of reality in order to demonstrate certain of its properties. Models are made necessary by the complexity of reality. They are a prop to our understanding and a source of working hypotheses for research. They convey not the whole truth, but a useful and apparently comprehensible part of it.

Maps and Other Models We have already seen a simple example of model-building in our study of the crowd on a beach. Our opening aerial photograph, and Figure 1-1, illustrate a first stage of abstraction. They represent the properties of the people on the beach faithfully, but on a different scale. It is common knowledge that most scientists make things larger in order to study them. The optical microscope, the electron microscope, and the radio telescope were scientific breakthroughs because they permitted increasingly powerful magnifications of reality. Geographers are curious folk in that they follow a reverse process. They bring reality down in size until it can be represented by a map. To shrink the universe to manageable size, they use various standard *linear scales*, which determine the ratio of the length of a line segment on a map to the true length of the line on the earth's surface. (See Table 1-1.) For example, 1 cm on a map may represent 1 km (100,000 cm) on the ground. Usually, the scale of a map is given as a representative fraction, as 1/100,000 or 1:100,000. This ratio applies equally

Table 1-1 Standard lengths and areas on ten commonly used map scales

Class and scale	Countries using the scale for major map series	Equivalent on the earth's surface of standard measures on the map	
		1 cm on the map	1 in. on the map
LARGE SCALES			
1:10,000	European	0.100 km	0.158 mi
1:10,560	British and Commonwealth	0.106 km	0.167 mi
1:24,000	United States	0.240 km	0.379 mi
1:25,000	British and Commonwealth	0.250 km	0.395 mi
MEDIUM SCALES			
1:50,000	European	0.500 km	0.789 mi
1:62,500	United States	0.625 km	0.986 mi
1:63,360	British and Commonwealth	0.634 km	1.000 mi
1:100,000	European	1.00 km	1.578 mi
SMALL SCALES			
1:250,000	International	2.50 km	3.946 mi
1:1,000,000	International	10.0 km	15.783 mi

to metric and nonmetric measurement. Thus on a 1:100,000 map, 1 in. is equal to 100,000 in. (about 1.6 mi) on the ground, and 1 cm is equal to 100,000 cm (1km) on the ground.

All the maps in this volume are very selective and partial models of the real world, with all the advantages—and the drawbacks—that simplification brings in its train. Simple scaled-down representations of reality are called *iconic models*. Our chapter-opening photo shows models of sunbathers' bodies at linear scale from 1:30 (foreground) to 1:300 (background). Figure 1-3(a) constitutes the second stage of abstraction—an *analog model*. Here, real people have become points on a map. Clusters of people on the beach have become point clusters on the map. Abstraction is pushed still further in a third type of model, the *symbolic model*, in which real-world phenomena are represented by abstract mathematical expressions, such as that for the population density of a beach. Figure 1-3(d) is a symbolic model. It takes us a step further from reality than either a photograph or a map for a second example of progressive abstraction. See Figure 1-12. Here the subject is part of the Hawaiian island of Oahu (a) as seen from an earth satellite, (b) as shown on a topographic map on the same scale, and (c) as a population density map.

From Models to Paradigms   We may conveniently think of model-building as a three-stage process in which each stage represents a higher degree of abstraction than the last. (See Figure 1-13.) At each stage, information is lost and the model becomes less realistic but more general. Throughout this book we shall return to the idea of models and look at their actual use. We shall encounter many models that are considerably more intricate than the three simple ones in the preceding section, but we will postpone discussion of them until we meet specific examples. It is useful at this point, however, to bring forward the idea of a *paradigm*. A paradigm is a kind of supermodel. It provides intuitive or inductive rules about the kinds of phenomena scientists should investigate and the best methods of investigation. This chapter—like the whole of this book—represents a paradigm of geography.

Research in geography, like research in most fields, is based on a shared paradigm; that is, those who pursue this research are committed to probing the same problems, observing the same rules, and maintaining the same standards. Tedious methodological debates, and concern over what constitutes legitimate research or appropriate methods of analysis, are symptoms of transitional periods in the evolution of a science. Once a paradigm is fully established, debate languishes.

In his provoking history of modern science *The Structure of Scientific Revolutions*, Thomas Kuhn contends that the origin, continuance, and eventual obsolescence of paradigms is the prime factor in the evolution of science. Modern geography has witnessed major shifts in emphasis from descriptive geography toward more analytic work. In the 1960s this focused on mathematical models of how regions grow and interact. In the 1970s there is a

(a)

(b)

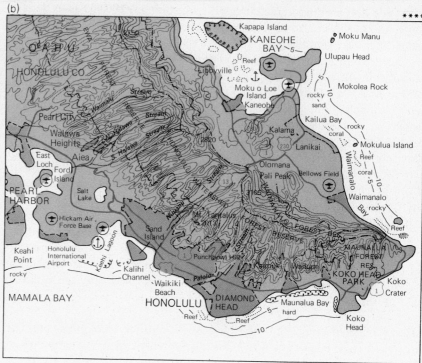

(c)

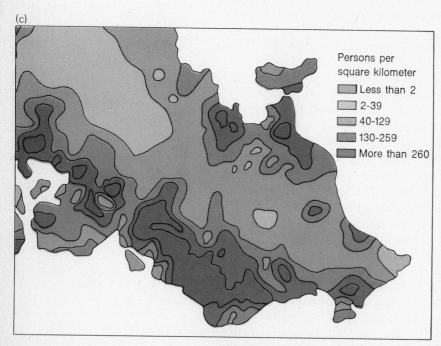

Persons per
square kilometer

	Less than 2
	2-39
	40-129
	130-259
	More than 260

Figure 1-12. Maps as models.
Three views of the southeastern corner of
Oahu, one of the Hawaiian Islands, consisting of the city of Honolulu and surrounding areas. (a) is a satellite photograph of the entire island chain with
superfluous ocean and clouds dropped out
in order to accent the islands' configura-
tions (Oahu is the island arrowed); (b) is a
reproduced section from a U.S. Geological
Survey topographic sheet; and (c) is a
population density map of the area shown
in (b), drawn from 1970 census records.
The scale in (b) and (c) is 1:250,000. The
scale in (a) approaches 1:6,000,000.
[Photograph courtesy of NASA; topographic map courtesy of U.S. Geological
Survey; population map from R. Warwick
Armstrong, Ed., *Atlas of Hawaii* (University
Press of Hawaii, Honolulu, 1973), p. 118.]

renewed stress on human behavior and our response to environmental change. In this book we try to illustrate both the traditional paradigm and the new ones. We regard the current emphases as the most recent phases in a long history of change, in which ways of looking at people on the beach and in the world have been refined, but the essential questions that geographers ask remain unaltered.

From Paradigms to Real-World Problems How can we use the idea of spatial organization and humanity's relations with the environment in ways that are helpful to people? The answer depends partly on the paradigm within which geographers work, and there is a noticeable contrast between the traditional and modern views of the field.

The traditional role of the geographer has been the provision of two types of essential information: locational information on the exact position of events, and environmental information on the quality of particular areas. In response to demands for this kind of information the great geographic works of the Greeks were written, the exploration societies of the early nineteenth century were formed, and the universal geographies of the Victorian period were assembled.

Today, however, geographers are more concerned with optimization— with finding the "best" location for things and making the "best" use of areas. Where should a new model city be located? What is the best site for a hospital within a city? What is the best dividing line between two hostile communities? What is the best use for the more remote Appalachian areas? They are also interested in forecasting projected trends into the future and monitoring the likely effect of policy decisions in a wide variety of situations.

Geographers work in agencies that range from the World Health Organization or the World Bank on the international level to local city halls and county agencies. They dominate the regional planning sections of several countries (notably Britain) and form a significant element in government agencies from Washington to Moscow to Peking. In the world of business or in private agencies, geographers perform essentially the same role: advising on locational priorities, watching for environmental feedbacks, and providing a geographic view of some part of the world.

How important that geographic view is will depend on the problem being examined. In many instances, the importance of spatial and environmental considerations may be negligible. In such cases the contribution of the engineer, the economist, or the educator will surely outweigh that of the geographer. When large-scale environmental issues are discussed, however, a geographer's viewpoint will certainly be needed.

We do not intend to argue for a purely geocentric view of the world's problems. At a time when the walls between academic subjects are crumbling, the isolationist subject makes about as little sense as an isolationist state. Geography has always been heavily dependent on its academic neighbors

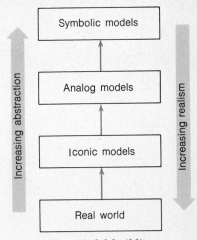

Figure 1-13. Model building.
Model building is shown here as a three-stage process of increasing abstraction.

such as mathematics, the earth sciences, and the behavioral sciences, and it has everything to gain by remaining so.

Geography is interesting today not because of its solution of past problems, but because of its potential contribution to resolving future difficulties. Geographers as a group have been a little embarrassed to discover that locational and environmental questions, so long a part of their classroom discussion, are now a daily topic in the news media, in Senate committee rooms, and on the campus. For over 2000 years, geographers have been studying the world and the place of human beings in it. Suddenly, in the late twentieth century, these seemingly academic preoccupations are being regarded as relevant—to us and to our children. We are taking a new look at ourselves and our world, and, like T. S. Eliot's explorer, "knowing the place for the first time."

Summary

1. Geographic analysis is illustrated by describing a geographer's reaction to a scene common around the shores of most of the more developed nations of the world, a crowded summer beach. These reactions are concerned with questions of (a) precise location and spatial order, (b) human-environment relations, and (c) regional differentiation.

2. Questions of location are answered through the use of an arbitrary grid system showing absolute and relative locations. Grid cell values of population density in absolute space may be used to draw choropleth or isopleth maps. Relation location is another way to deal with the spatial order in populations by measurement of interpersonal space and territoriality.

3. Most geographic distributions are variable over time. Population distributions graphed over time may be broken down to show trends, wavelike curves, or standstills depending on the length of time studied. Spatial diffusion may be studied through the use of such time elements to analyze changing patterns over time.

4. Plotting movements of individuals on space-time diagrams shows the effect of three constraints which "cage in" the spatial range of opportunities. These are (a) biological constraints, (b) accessibility constraints, and (c) activity constraints.

5. All of the conditions around a population at any point on the surface of the earth form its geographic environment. The environment includes both natural elements and elements made by people. Humans may be both influenced by their environment (E→H) and make changes in it (H→E).

6. Human-environment relations may be usefully seen in system terms (H—E). Systems consist of elements, links, and boundaries interacting together. Feedbacks within geographic systems may be both positive and negative.

7. The scale of a geographer's ranges from the microgeographic to the macrogeographic. The five orders of geographic magnitude range from city block size to the entire earth itself. Different features of interest are best studied at particular levels of resolution. These features can be reduced to iconic, analog, or symbolic models, with increasing abstraction from first to last.

8. Geographers use a shared view, a paradigm, as a basis for teaching and research. The approach used in this text illustrates the use of traditional geographic paradigms as well as contemporary ones.

Reflections

1. Note down in a few sentences what you think you are likely to get out of a course in geography. Keep this statement in your file so that you can look back at the end of the term and compare what you expected to gain with what you actually gained.

2. It is sometimes said that disciplines are distinguished from one another by the questions they ask. What special questions do geographers ask? What questions *should* they ask?

3. How much can geographers studying the behavior of man learn from biologists studying the behavior of animals? List (a) some advantages and (b) some disadvantages of such cross-borrowing between disciplines.

4. Consider the graphs of population density over time in Figure 1-5. Try to construct similar graphs for (a) the campus restaurant, (b) the local shopping plaza, and (c) the local airport. Comment on the similarities and differences in these graphs.

5. Use locally available maps with small, medium, and large scales to examine a beach of your choice. (For American students the maps will be at scales of 1:24,000, 1:62,500, and 1:250,000; students in other countries should check with Table 1-1 for guidance.) Familiarize yourself with the conventional symbols used on such maps, and note how the map scales control the richness of the information that can be conveyed.

6. Review your understanding of the following concepts:

space, location, and place
region
absolute and relative
   location
interpersonal space
choropleth and isopleth
   maps

space-time box
feedbacks
linear scales
orders of geographic
   magnitude
models and paradigms

One Step Further . . .

Two paperbacks that provide a lively account of modern geography, with plenty of examples of applied research, are
   Lanegran, D. A., and R. Palm, *An Invitation to Geography*, 2nd ed. (McGraw-Hill, New York, 1978). See chap. 1.
   Taaffe, E. J., *Geography* (Prentice-Hall, Englewood Cliffs, N.J., 1969). See chap. 1.

You may also like to browse through one of the many useful sets of collected readings that give some idea of the range and richness of geographic work: These include
   Dohrs, F. E., L. M. Sommers, and D. R. Petterson, Eds., *Outside Readings in Geography* (Crowell, New York, 1955).
   Detwyler, D. R., Ed., *Man's Impact on Environment* (McGraw-Hill, New York, 1971).
   Blunden, J. *et al.*, Eds., *Fundamentals of Human Geography: A Reader* (Harper & Row, London, 1978).

Ideas on interpersonal space and territoriality are ably treated in
   Yi-Fu Tuan, *Space and Place* (University of Minnesota Press, Minneapolis, 1977).

Beginning students will find it helpful to look through recent numbers of some of the leading geographic serials to see what kinds of work geographers are currently doing. As starters, you might try Geographical Review *(published quarterly) and* Annals of the Association of American Geographers *(also a quarterly). Longer and more substantial research is published in monograph form. Two representative monograph series are* University of Chicago Department of Geography Research Papers *and* Lund University Studies in Geography *(both occasional publications).*

Finally, it will be useful to have at hand a standard world atlas such as Goode's World Atlas *(Rand, McNally, Chicago, 15th. ed., 1978). More specialized maps and atlases will be suggested at appropriate places in the book.*

> Map me no maps, sir, my head is a map, a map of the whole world.

> HENRY FIELDING
> *Rape upon Rape* (1745)

chapter 2
The World Beyond the Beach

The view from the beach is a very limited one. Even if we scramble to the highest point on the sand dunes backing the beach, we can see only a few kilometers out to sea. If, looking inland, the land is also flat, then our total sweep—from horizon to horizon—still covers only a small disk of the earth's vast surface. We are unlikely, even on the clearest of days, to scan more than 0.0008 of 1 percent of the total area of the globe. We could fit over 100,000 disks this size on the earth's surface without them ever touching one another! Even from a high-flying jet the situation is not vastly improved, though we can then see an area half the size of Texas (or about 0.05 of 1 percent of the global area).

It was not until the late 1960s that man first saw the earth from deep space and for most of our history, our visual picture of the world was limited by the local horizon. What lay beyond was a matter for, first, speculation, then calculation, and, finally, confirmation. (This fascinating trail of conjecture and discovery has gone cold as the pieces of the map have fallen into place.) However, the research methods developed along the trail have proved a very useful legacy for modern geography. To record the characteristics of the earth and to preserve and exchange this information, a comprehensive *spatial language* was developed. This spatial language describes the absolute and relative locations of places either in terms of conventional maps and in other, maplike ways. It allows us to unequivocally assign events to locations. To put it simply, by providing "a place for everything" it lets us put everything in its place. In this chapter we present some of the basic grammatical rules of this language and show how geographers use it to describe and record the world beyond the beach.

2-1
Flat World or Round?

It is doubtful if our early ancestors were much concerned with the shape of the earth. As we have seen, their visual horizon, even from the highest peak in the clearest weather, would have been confined to a few hundred kilometers. The facts that observation failed to provide, legend and narrative no doubt found substitutions for. J. R. R. Tolkien's fictional Wilderland (Figure 2-1) very likely recaptures the spirit, if not the substance, of primeval world pictures.

Against this background, the intellectual achievements of the Greek cosmologists are remarkable. Their observations of heavenly bodies led them to deduce a spherical rather than a disklike form for the earth. By 200 B.C., one of the earliest Greek geographers, Eratosthenes of Alexandria, had made rather accurate estimates of the earth's shape. As Figure 2-2 shows, the basic procedures he used in his calculations were simple. Indeed, Eratosthenes' method was (at least until the advent of the satellite) the same in principle as that used to measure the earth in our own time. Assuming that light rays from the distant sun were parallel, as for all practical purposes they are, Eratosthenes calculated the differences in the angle they made with the earth's surface at different points and thereby determined its curvature. Specifically, he compared the angle made by the rays of the noon sun at the summer solstice (June 21) at two places in Egypt: Syene, where the sun is directly overhead and the angle is vertical, and Alexandria, where a shadow of 7° 12′ is cast. Multiplying the known north-south distance between Syene and Alexandria by the difference in the angles gave him a circumference estimate (based on a Greek measure called the *stadium*, which is about one-sixth of a kilometer) for the earth of 46,250 km (28,740 mi). Since current estimates put the figure at 40,000 km (24,860 mi), Eratosthenes' estimate was extraordinarily accurate.

Figure 2-1.   Imagined worlds.
J. R. R. Tolkien's "Wilderland," one segment of his fictional Middle Earth, the scene of *The Lord of the Rings.* Early mapmakers of the real world also used imagination and legend to fill in unexplored areas. Maps of Africa, the "dark continent," contained mythical rivers and lakes as late as the middle of the nineteenth century. [From J. R. R. Tolkien, *The Hobbit* (Allen & Unwin, London, 1937), end map. Copyright 1966 by J. R. R. Tolkien. Reprinted by permission of the publisher, Houghton Mifflin Co., Boston.]

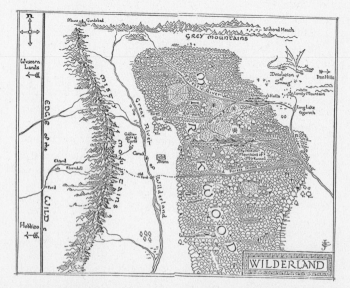

Little improvement was made on Eratosthenes' measurements for another 1800 years. There was an apparent loss of interest in Greek concepts, and their theories about the earth were temporarily replaced, in medieval Europe, by a theological view of the world. Figure 2-3 shows a typical map of that period, with the Holy City of Jerusalem occupying the center of a disk-shaped world. With the Renaissance and the wave of European overseas voyages, interest in the form of the planet was rekindled. Further measurements of the earth were made in the early seventeenth century, when astronomer Willebrord Snell's recalculations at Leiden University triggered a succession of increasingly accurate measurements of the distance over ever-longer sections of the earth's curved surface. From the results stemmed a bitter debate; gross variations in the estimated circumference indicated that the earth could not be regarded simply as a regular sphere. Scholarly opinion divided over whether the earth was flattened at the poles or elongated like a football. By the 1730s, evidence from geodetic expeditions was overwhelmingly in favor of the former view.

Since the eighteenth century, precise measurements have proved that the earth's polar diameter is shorter than the equatorial diameter, and the polar circumference is shorter than the equatorial circumference. (See Table 2-1.) Moreover, the equator does not mark the fattest part of the earth. The distance around the earth is, in fact, greatest along a circle slightly south of the equator. The fact that the earth is an ellipsoid with a somewhat pear-shaped bulge and slightly flattened poles, rather than a perfect sphere, has complicated mapping in many instances.

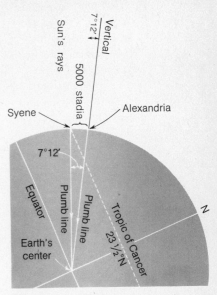

Figure 2-2. The size of the earth.
Eratosthenes, about 200 B.C., estimated the circumference of the earth from the angle at which the rays of the noon sun reached Alexandria and Syene. [From A. N. Strahler, *The Earth Sciences*, 2nd ed. (Harper & Row, New York, 1971), p. 150, Fig. 10-1.]

Figure 2-3. Changing views of the world.
Figures (a) through (e) show representative maps from 1770 years of global exploration. On each map the sites of London (A), Jerusalem (B), and Colombo (C) are indicated.

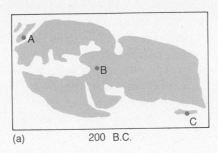

(a)   200 B.C.

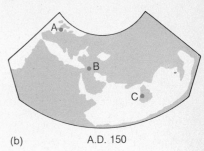

(b)   A.D. 150

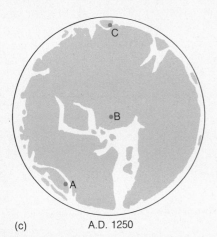

(c)   A.D. 1250

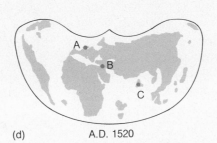

(d)   A.D. 1520

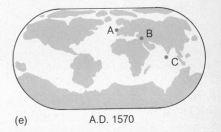

(e)   A.D. 1570

TERMS USED IN MAPPING THE EARTH

Ellipsoids are spheroids (figures like a sphere, but not perfectly spherical) with a regular oval form.

Equator is an imaginary line around the earth which is an equal distance from the North and South poles.

Geodesy is the science that deals with the shape and size of the earth.

Graticules are networks of parallels and meridians drawn on a map.

Grids are arbitrary networks of lines drawn on a map. For example, a network of squares may be used to give a simpler reference system than that of graticules.

Latitude is the distance north or south of the equator measured in degrees.

Longitude is the distance east or west from a prime meridian (usually Greenwich in London) measured in degrees.

Meridians are imaginary half-circles around the earth which pass through a given location and terminate at the North and South Poles.

Parallels of latitude are imaginary lines on the earth which are parallel to the equator and pass through all places the same distance to the north or south of it.

Poles are the two ends (North Pole and South Pole) of the axis around which the earth spins or rotates.

Prime meridians are meridians used as a baseline for the measurement of the east-west position of places on the earth's surface in terms of longitude. The most commonly used prime meridian crosses Greenwich in London.

2-2
Location on the Flat Map

Once we have established the shape and size of the earth, we can go on to specifying locations on its surface. We look first at the devices used on simple maps before considering the more complicated problems posed by the globe.

Place Names   The simplest way of specifying a location somewhere on the earth's surface is to give its name. In everyday language we almost always talk about places in terms of the names attached to them: Chicago, Kansas, or Tibet. This way of specifying locations is termed *nominal specification*. We used it in Chapter 1 to identify particular beaches, and in the preceding section to describe the cities involved in Eratosthenes' calculations. It can describe places that vary in size (from Mount Vernon to Afro-Asia), it can provide distinctive and memorable place names (like the Donner Pass in northern California or Xochimilco in central Mexico), and it can be made as complex or as simple as necessary. The typical address on an envelope exemplifies a hierarchical method of specifying location in which areas of decreasing size are nested within each other. Our resistance to the replacement of familiar addresses by zip codes indicates the innate attractions of a nominal system for everyday use.

Table 2-1   The basic dimensions of the earth

Total surface area	510,056,000 km²	(196,934,000 mi²)
Area of land surface	149,137,000 km²	(57,582,000 mi²)
Circumference measured around the poles	40,003 km	(24,857 mi)
Circumference measured around the equator	40,074 km	(24,901 mi)
Highest point on land surface	+ 8.85 km	(5.45 mi)
Lowest point on ocean floor	− 11.03 km	(− 6.85 mi)

Another advantage of place names is that they may contain a considerable amount of historical or environmental information. Places founded by particular groups tend to have particular types of names. Witness, for example, the many Spanish place names in the American southwest and Texas. These names give us clues to the previous extent of a population. When other information is lacking, such clues may be vital. In a similar manner, names may indicate the kind of environment settlers found on first moving into an area. In southern Brazil, names containing the word "pine" help to depict the original distribution of a type of vegetation now much reduced by deforestation and the clearing of land for agriculture. Work by Zelinsky on the northeastern United States has shown a rich variety in the distribution of stream names. The terms "creek," "brook," "run," and so on each have distinct regional clusters related to the settlement history of the area (Figure 2-4).

If place names are so attractive and useful, why don't geographers use them all the time? Unfortunately, the disadvantages of nominal specification for scientific purposes outweigh the advantages. First, this kind of language is nonunique because several different locations can have the same name. There are scores of São Paulo's in Brazil, and names such as Newport or New Town occur in literally hundreds of forms in the English-speaking world.

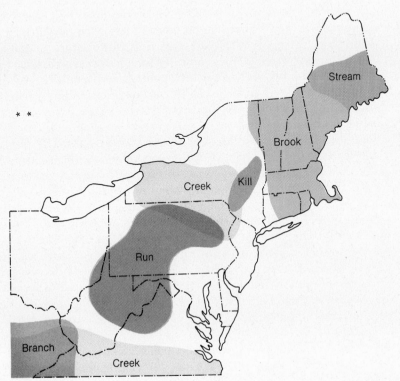

Figure 2-4. Place names as historical evidence.
Stream names in the northeastern United States. The approximate areas within which particular terms are dominant are indicated on the map. "Brook," "run," and "branch" are very common terms and strongly associated with particular locations. "Creek" and "stream" occur less frequently but are also strongly associated with distinct localities. "Kill" is a very uncommon term in the northeast as a whole but very prevalent in areas of early Dutch settlement in New York State. [From W. Zelinsky, *Annals of the Association of American Geographers* **45** (1955), p. 323, Fig. 20.]

**

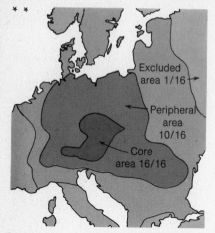

Figure 2-5. Ambiguities in regional names.

The degree of coincidence among 16 geographic definitions of the terms "Central Europe," "Mitteleuropa," or "Europe Centrale." Note the small area of agreement and the very wide area of disagreement. [After K. Sinnhuber, *Transactions of the Institute for British Geographers No. 20* (1956), p. 19, Fig. 2.]

Second, nominal terms are unstable. The same location can have different names at different times (such as Breuckelen and Brooklyn), and in different languages. (The German Konigsberg is the Russian Kaliningrad.) To correct this problem, many countries have established committees to standardize geographic names throughout the world. In 1890 the United States set up a Board on Geographical Names that published authoritative lists for individual countries. Some idea of the immensity of this task can be gained from the size of standard geographic gazetteers. The Merriam–Webster *Geographic Dictionary* contains approximately 40,000 names, and the *Times Atlas of the World* has 315,000 names. The current London telephone directory, which is actually a list of the "locations" of subscribers within the city, contains about half a million names.

We can only guess at the total number of world place names, which probably lies in the trillions. Such a vast array of names creates formidable problems for those concerned with information storage and retrieval, problems which are complicated by the duplication of place names. Various ways of dealing with these problems are being investigated by experimenters at Oxford University, England, who are developing systems for the coding, computer storage, and rapid retrieval of information from geographic gazetteers.

Another serious disadvantage of nominal specification is its spatial imprecision. The boundaries of an area may change while the name remains the same. Turn to a historical atlas and compare the Poland of 1930 with the Poland of 1970. The imprecision on nominal terms for geographic areas is illustrated in Figure 2-5, which shows a series "Mitteleuropa" (Middle Europe) over a 40-year period. Areas included by at least one geographer extend well outside the limits set by other geographers, and the only part of Europe not included in Mitteleuropa by any of the definitions is the Iberian Peninsula. The only area not in dispute, in fact, is the small core area containing Austria and Bohemia–Moravia.

Cartesian Grids If we can't use place names to specify location, how do we proceed? The geographer's answer is to use *reference grids*—mathematical devices that specify the location of a point in relation to a system of coordinates. The type of grid used depends on whether the area involved is small or large. We can treat small areas of the earth's surface as if they were flat planes; in the case of larger areas the curvature of the planet must be taken into account. Reference grids for small areas use a *Cartesian coordinate system*. The location of a place is specified by its distance from two reference lines that intersect at right angles. The horizontal reference line is called the abscissa or *x* axis, and the vertical reference line is called the ordinate or *y* axis. Intersection of the two axes is the point of origin of the system. [See Figure 2-6(a)].

Cartesian coordinate systems are commonly adopted by national mapping agencies, and Britains's National Grid system [Figure 2-6(c)] is a typical example. Its origin lies southwest of the Isles of Scilly; its abscissa runs east and its ordinate north. All geographic locations are measured in kilometers, or subdivisions of kilometers, east and north from these reference lines. In practice, references are given first to the 100-km (62-mi) squares into which the grid is divided, second to distances east of the origin, and then to distances north of it. Thus the precise reference to the United States Embassy in Grosvenor Square, London, is 51 (TQ) 283808; that is, it lies exactly 528.3 km (328.3 mi) east and 180.8 km (112.3 mi) north of the arbitrary origin point.

A modified form of the cartesian system is the *range, township, and section method,* which originated in a land act conceived by Thomas Jefferson in 1784. This rectilinear grid is the basis of surveys in most of the United States

Figure 2-6. Reference grids.
(a) and (b). The location of A with respect to the origin in two alternative spatial reference systems. Note that in the case of Cartesian coordinates the distance east is given first so that the location of A is 20,40 and *not* 40,20. In practice the comma is usually omitted in giving a locational reference so that 20,40 becomes simply 2040.
(c) The 100-km squares of the British National Grid. Each grid square has a distinguishing number or pair of letters. Numbers describe the distance east (the "easting") and distance north (the "northing") of the southwest corner of each square when measured in hundreds of kilometers from the false origin (e.g., the southwest corner of grid square 31 is 300 kilometers east and 100 kilometers north of this origin). The "false origin" is an arbitrary point out at sea chosen for the British map users. It stands in contrast to "true origins" such as the poles or the equator. The letters on the map are part of a wider reference system which ties the British National Grid into other international grids. [Reproduced from *The Projection for Ordinance Survey Maps and Plans and the National Reference System*, with the sanction of the Controller of Her Majesty's Stationery Office. Crown copyright reserved.]

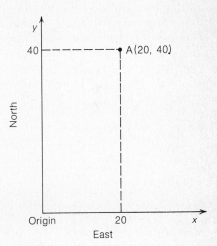

(a) Cartesian coordinates

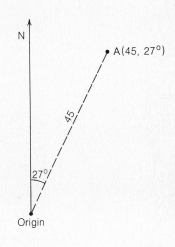

(b) Polar coordinates

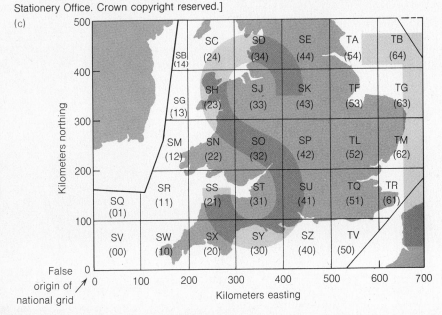

(c)

Figure 2-7. A hierarchic reference grid.
In much of the central and western United States, the location of parcels of land was specified on a rectilinear grid divisible into ranges and townships. Each square of the range and township grid was further divisible into sections and subsections. The "base meridian" and "base parallel" shown in (a) tie this grid system into the worldwide graticules of latitude and longitude described in Figure 2-8. The impact of this reference system is still clearly seen in the pattern of roads and fields: See, for example the landscapes shown in Figure 1-12(b) and Figure 21-5 much later in this volume. [From G. C. Dickinson, *Maps and Air Photographs* (Arnold, London, 1969), p. 124, Fig. 46.]

west of the Appalachians and, in a modified form, much of Canada. It divides land into 6-mile-square townships around a Cartesian axis oriented to the north. (See Figure 2-7.) Each township is subdivided into 1-mile-square sections (numbered 1 to 36 in zigzag fashion). Each section can be further subdivided into quarter-sections and then into 40-acre fields. Even today the imprint of the township and range system on the agricultural landscape of much of the United States is evident.

A less common type of locational reference grid is based on a *polar coordinate system* which specifies the location of a place in terms of an angle, or *azimuth*, and its distance from an origin [Figure 2-7(b)]. In mathematical work, angles are specified counterclockwise from a horizontal; in geographic work, angles are measured clockwise from the north, which is designated 0° or 360°. Thus an azimuth of 225° is a line running southwest from the origin. The polar coordinate system is generally used only when the relation of locations to a single origin is important—for instance, in studying distances a population has migrated from a given place of origin.

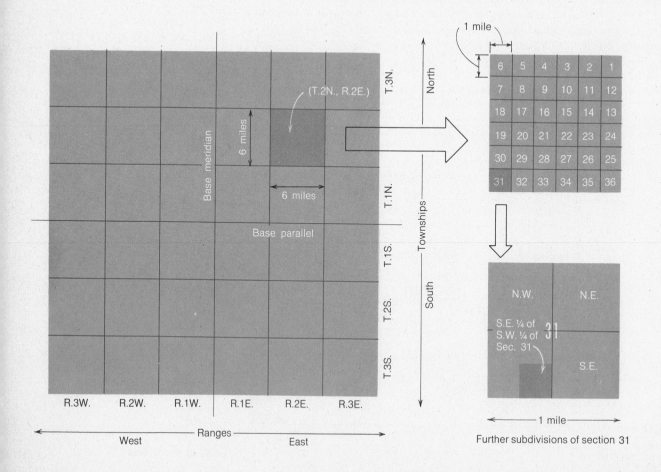

Further subdivisions of section 31

So long as the environment under study is a small one, geographers can use very simple devices for recording location. Once we begin to study very large areas or attempt worldwide studies, then a more sophisticated way of specifying location is needed.

2-3
Location on a Round World

A Spherical Grid The curvature of any portion of the earth's surface depends, of course, on its size. As the area being mapped increases, so does the deviation from a true north-south direction of parallel reference lines on a flat surface. Thus the extension of the range and township system used in the United States as people moved west brought about an increasing divergence of range lines from their true direction. To minimize this divergence, new grids had to be established for each state.

To accommodate the spherical form of the earth, the Cartesian system was adapted to create a geographic grid of meridians and parallels. *Meridians* are imaginary half-circles around the earth which pass through a given location and terminate at the North and South Poles. Thus they form true north–south lines joining two fixed points of reference, the geographic poles, where the earth's axis of rotation intersects the planet's spherical surface. Each meridian is actually a half circle (an arc of 180°). *Parallels* are imaginary lines on the earth that are parallel to the equator and pass through all places the same distance to the north or south of it. They form true east–west lines and are full circles (arcs of 360°). Parallels intersect meridians at right angles. (See Figure 2-8.)

Coordinate positions on the meridian–parallel grid are measured in terms of longitude and latitude. The *longitude* of a place is its distance east or west from a prime meridian measured in degrees. It describes the angle between two planes that intersect each other along the earth's axis and intersect the surface of the earth along, respectively, the prime meridian and the meridian of the place to be located. Prime meridians have a value of 0°. Angles are measured east and west, reaching a maximum of 180° at the meridian opposite the prime meridian. The prime meridian in worldwide use is the Green-

Figure 2-8. The earth's spherical coordinate system.
(a) Parallels of latitude lie in planes oriented at right angles to the earth's axis of rotation. (b) Meridians of longitude lie in planes passing through the earth's axis. (c) Combined parallels and meridians form a spherical geographic grid. [From A. N. Strahler, *The Earth Sciences,* 2nd ed. (Harper & Row, New York, 1971), pp. 13-14, Figs. 1-10, 1-11, 1-13.]

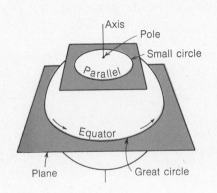

(a) Parallels

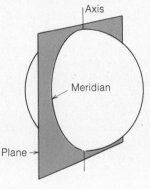

(b) Meridians

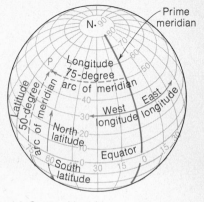

(c) Geographic grid

wich meridian (the former site of the Royal Observatory near London, England), but any meridian can be chosen. Italian topographic series use the meridian that passes through Monte Mario near Rome (12°27'E of Greenwich), and some other European countries use the meridian that passes through Ferro in the Canary Islands (17°14'W of Greenwich).

The *latitude* of a place is its distance north or south of the equator measured in degrees. It is given by the angle between the plane of the equator and the surface of a cone that has its apex at the earth's center and intersects the surface of the sphere along a given parallel. Unlike the prime meridian, the equator is a uniquely determined and natural reference line, rather than an arbitrary one. It also has a value of 0°, and angles are measured north and south, reaching a maximum of 90° at the North and South Poles.

Geographers give the exact location of a point on the earth's surface by specifying its longitude and latitude. In Figure 2-8, the point *P* has the coordinates 50°N, 75°W.

Degrees can be subdivided into 60 minutes of arc (60'), and minutes into 60 seconds of arc (60"). For simplicity, however, latitude and longitude are generally expressed in decimal parts of a degree. Thus, 77°03'41" is rewritten in decimal terms as 77.0614°. Modern worldwide reference systems are sophisticated enough to successfully combine the advantages of both the Cartesian and the spherical systems of reference.

Determining Location on a Spherical Grid   Developing a spherical reference system for the earth is one thing, and determining where a location actually is within that system is another. Finding latitude involves simply measuring the angular elevation of the sun or a star above the horizon, and it was first done with some accuracy more than 1000 years ago. The determination of longitude is more difficult. The first requirement is a reliable clock, so that the local noon time (when the sun is at its zenith) can be compared to noontime at the prime meridian. The difference between the two gives us our correct east–west position. But the clock must be very accurate. One hour on the clock is recording the time during which the noon-time position of the sun has shifted 1/24th part of its east–west track around the earth. At the equator this would mean that an error in time of only 22 seconds on the clock would cause the navigator to be 10 km (6.2 mi) in error in judging his east–west position. It was not until 1761 that this problem was solved when John Harrison's marine chronometer, made in response to a prize offered by Britain's Board of Longitude, provided the breakthrough that allowed position to be determined accurately.

The job of fixing positions with respect to latitude and longitude has become much less tedious in the present electronic era. Figure 2-9 presents one of the electronic systems currently being developed for this purpose. This *Omega system* depends on a pattern of only eight radio stations, the latitude and longitude of which can be given precisely. Each station transmits extremely long radio waves—with wavelengths up to 5 km (3.1 mi)—on exactly

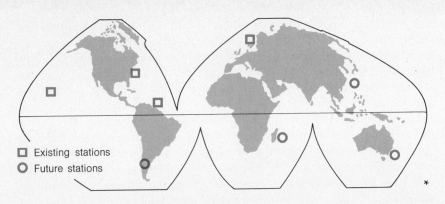

Figure 2-9. Position fixing on the earth's surface.
The U.S. Navy's Omega system uses long radio waves transmitted from a few stations to provide position data. Ships using the system compare signals from a third station as a check. The system represents the latest in a series of moves to reduce the number of fixed points needed to determine position on the globe.

□ Existing stations
○ Future stations

the same frequency in a predetermined sequence. Because these waves can be distinguished by the length of their pulse and the sequence in which they are transmitted, ships and aircraft can plot their progress from a known starting position by comparing signals from two stations and using those from others as a check. When it is completed, positional fixes based on the Omega system should be accurate to about 1 km (.6 mi) anywhere on the global surface. Navigational satellites giving more accurate fixes to a few hundred meters may be used in the future.

The determination of positions on the earth's surface has two interrelated phases. The first consists of fixing primary positions with respect to their horizontal location on the geographers' spherical grid system (and their vertical position with respect to mean sea level). Once a network of primary positions is established, a swarm of secondary positions can be fixed from the horizontal and vertical coordinates of the primary position. The trend in current survey work has been to reduce the need for primary survey positions by extending the area of secondary fixes that can be calculated from the primary points. The Omega system represents an extreme reduction of the number of primary positions needed for global coverage.

Map Projections: A Round World on Flat Paper   Once we have established a system of parallels and meridians and have located places in terms of that system, we have the basic ingredients of a world map. To make one we need only to specify a scale and construct a model globe. If we select a scale of 1:10,000,000, we can represent the real globe by a sphere 127 cm (50 in) in diameter. Making globes like this is not a difficult task; indeed, people have been constructing them for centuries. Martin Behaim made one of the first terrestrial globes at Nuremberg in 1492, and they are still popular for many purposes.

Despite their attractiveness, however, globes are rarely used by geographers in their research. Most globes are less than 1 m (3.3 ft) in diameter and are too small to be of any real value. There are, of course, a few large globes such as the 39-m (128-ft) Langlois globe in France, but larger globes are expensive and

Atlas cross-check.
The world maps in the atlas section give you a good illustration of equal-area projection.

unwieldy to work with. The obvious way around the problem is to substitute flat map sheets for the bulky spheres. How to get a map off the surface of a globe and onto flat paper without tearing or otherwise distorting it is the objective of map projections.

Some projections are very simple to construct. If we place a sheet of flat paper against a transparent globe and shine a light from the opposite side of the globe we can "project" an image of the earth onto the paper. Records of this type of map projection go back to the second century A.D. But the early geographers were to find that there were no easy solutions: Such simple maps had very few useful properties. The puzzling problem of mapping the round world on flat paper attracted some of the best mathematical minds of sixteenth- to nineteenth-century Europe, figures like Gerardus Mercator, Hans Mollweide, and Johann Lambert. Their work is a study in the mathematics of compromise. They showed that, though it is impossible to reproduce, faithfully, in two dimensions all the characteristics of the three-dimensional earth, it is possible to reproduce some of them at the expense of others. We have simply to decide which properties are important and which we are prepared to sacrifice.

In practice, the main decision is whether to select a *conformal projection*, in which the shape of any small area is shown correctly, or an *equal-area projection*, in which a constant areal scale is preserved over the whole map. Although the word "projection" is used here, in a strict sense it should be applied only to maps that are true geometric projections of a sphere. Through usage the word has come to represent any orderly system of parallels and meridians drawn on a map to represent the earth's geographic graticule.

Which projection a geographer chooses will depend on the job the map has to do, and the advantages and drawbacks of the different projections must be carefully balanced. Mapping agencies for individual countries always adopt projection systems that have the most advantages from the point of view of the maps' potential users in that country.

Where maps of the whole world are concerned, the single sheet may be "interrupted" in order to reduce overall shape distortion. If you turn back to Figure 2-9, you'll see an example of this. *Interrupted projections* are usually drawn to place the breaks in the world's oceans, thereby enhancing the shape of the continents. Where the world map is to show features of the world's oceans, then the position is reversed and the breaks are placed in the land areas.

2-4
Location From the Air

The first manned ascent of a balloon in the United States was from Philadelphia on January 9, 1793. George Washington and a large crowd are said to have watched the event. Today's geographers look back on those early ballooning experiments by the Montgolfier brothers and their friends as a watershed (Figure 2-10). From the late eighteenth century onward, people

Figure 2-10.   The beginnings of air photography.
Late-eighteenth-century experiments with ballooning provided the first maplike
perspectives of the earth's surface. [From William L. Marsh, *Aeronautical Prints and
Drawings* (Halton and Truscott Smith, London, 1924).]

SCALE IN AERIAL PHOTOGRAPHS

The nominal or average scale of a vertical aerial photograph can be simply represented by

$$S = \frac{f}{H}$$

where S = the scale,
f = the focal length of the camera,
H = the height of the camera above the ground surface,
and both f and H are measured in the same units of length. Thus, if the camera has a focal length of 20 cm (about 8 in) and the aircraft is flying at a height of 1000 m (i.e., 100,000 cm, or about 40,000 in), the scale at the ground surface will be 20:100,000, or 1:5,000. Areas of higher ground (H_2) are nearer the camera and hence appear larger; conversely, areas of lower ground (H_1) are farther from the camera and appear smaller. Large distortions in scale can be introduced by tilting the camera so the photograph is not truly vertical. This displaces the perspective center of the photograph from the actual center and makes the estimation of both scale and height considerably more complex. [See D. R. Lueder, *Aerial Photographic Interpretation* (McGraw-Hill, New York, 1959). Chap. 1.]

have been able to view increasing portions of the earth's surface from above and to construct maps from direct aerial observation rather than laboriously piecing them together from measurements made on the surface. As the airplane succeeded the balloon, and the space rocket the airplane, the horizon continued to recede until, in the late 1960s, the whole hemisphere of the planet came into view. In this section we shall review this latest phase in environmental reconnaissance and the ways in which it has affected spatial language and the world picture developed by maps.

The Single Aerial Photograph   Photographs of the earth's surface taken from the air present the same problems as those taken from the ground; that is, both give a highly distorted perspective. Each of us, in our first forays with a camera, has come up with prints in which buildings appear to lean drunkenly backwards or portions of the subject's anatomy near the camera are hugely swollen. Identical problems are encountered in photographs taken from the air. Note, for example, the way in which the skyscrapers in midtown Manhattan in Figure 2-11(a) appear to be distorted. The rules that govern such variations in scale in photographs are fairly simple, so long as the camera axis is vertical to the ground surface. (See the marginal discussion of scale in aerial photographs.) As the camera is tilted away from the vertical, more difficult problems occur.

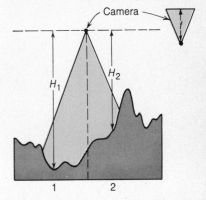

Figure 2-11. Perspective distortion on aerial photographs.

(a) A vertical view of Manhattan from a height of 500 m (1640 ft) shows the extreme distortion caused by difference in height. St. Patrick's Cathedral (in the upper center of the photograph abutting Fifth Avenue) appears much smaller than it really is in comparison to neighboring skyscrapers. [Photo by Lockwood, Kessler, and Bartlett, Inc.] (b), (c) Aerial views taken obliquely are rarely used for mapping. They have the disadvantage that the scale changes greatly from foreground to background. Foreground features may also block those in the background. In compensation the photographs are excellent for visual interpretation. That shown portrays local variations in environmental quality in the North Somerset Plain, southwest England. Note the sharp contrast between the flat, waterlogged valleys of the river Axe and the steep-sloping hills of the Mendip uplands. [Photo by Aerofilms. Map from D. C. Finlay, *Soils of the Mendip District of Somerset* (Soil Survey of England and Wales, Harpenden, 1965), p. 4, Fig. 3.]

(a)

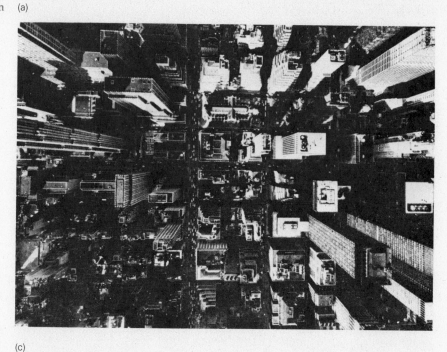

(b)

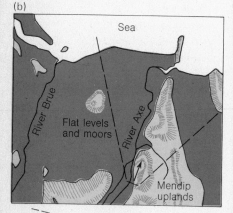

Direction of photographic view in (a).

Upland areas with slopes indicated by hatched lines.

Lowland areas which are flat are subject to flooding.

(c)

* * * *

Because the distortion in aerial photographs varies with the tilt of the camera, it is useful to place aerial photographs in two main classes: first, *vertical photographs*, in which the camera's axis points directly downward to give a plan of the terrain below, and, second, *oblique photographs*, in which the camera's axis points at a low angle to the ground to produce a perspective view. If the camera is tilted enough to include the horizon, the photograph is termed a *high oblique*; pictures that do not include the horizon are termed *low obliques*.

In practice, it is extremely difficult to guarantee an absolutely vertical axis. Thus the first category normally includes photographs whose camera axis is within two or three degrees of the verticle axis. "Verticals" are the most widely used type of photograph because distortions due to scale, tilt, and height can be readily corrected to produce detailed maps. The vertical aerial photograph is not a map, however. Scale can vary not only from one photograph to the next, but also from one part of the same photograph to another as the terrain below varies in height. Oblique photographs cover large areas of ground and present few problems in interpretation because their perspective is more familiar to the viewer. "Obliques" are therefore most popular for illustrative or publicity purposes. They are of limited value for scientific purposes because they contain wide variations in scale, they risk large areas of blocked visibility ("dead ground"), and they possess complex geometric properties from a measurement and mapping (*photogrammetric*) viewpoint. [See Figure 2-11 (b) and (c).]

Stereoscopic Pairs   It is obvious that a flat aerial photograph has only two dimensions, while the terrain it depicts has three dimensions. How is this third dimension, height, shown in a photograph? If we reexamine Figure 2-11, we should find a clue. Not only are the tops of the skyscrapers greatly enlarged, but they also appear to be leaning back from the center of the photograph.

As Figure 2-12 illustrates, differences in the height of objects in a photograph can be determined by observing their horizontal displacement from their true ground position. In a truly vertical photograph this displacement is along lines radial from the center of the photograph. Objects such as hills, which have an elevation above the mean elevation of the photograph, are displaced outward from the center, and low points are displaced inward. The amount of displacement is inversely proportional to the altitude of the camera and directly proportional to the variations in height of the terrain.

Although this horizontal displacement is a nuisance because it prevents the direct use of aerial photographs as maps, it permits direct measurements of relief when photographs are taken in overlapping pairs. Figure 2-13 presents a typical series of photographs in a mapping run. Exposures are timed to allow a 60-percent forward overlap between successive photographs in the same strip. There is also a 25-percent lateral overlap between adjacent strips. Note that the orientation of the photographs is reversed in each successive strip,

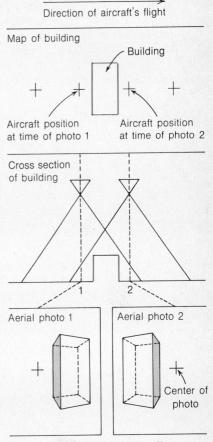

Figure 2-12. Stereoscopic effects. Elevation differences in paired aerial photographs. Successive photographs of the same building from two aircraft positions are shown by horizontal displacements of its image on the two photographs.

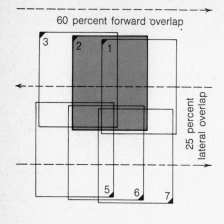

Figure 2-13. Survey flight patterns.
Aerial photographs used for stereoscopic analysis have both a forward and a lateral overlap. This effect is achieved by a flight pattern similar to the track of a lawnmower.

because of the flight pattern. The location of photographs in such a run is facilitated by a *titling strip* automatically included at the top of each print, which gives information on the sortie number, general location, date and time of exposure, focal length of the camera, flying height, and film type.

Pairs of photographs that overlap as in Figure 2-13 are termed *stereoscopic pairs* (from the Greek *stereos*, solid). *Stereoscopes* are devices that allow us to view two overlapping pictures taken from slightly different points of view at the same time. Viewing the two photographs through a pair of binoculars fuses them into a single image having the appearance of solidity or relief. The horizontal displacements are perceived as vertical displacements in a third (vertical) dimension. Advanced stereoscopic equipment permits an observer to convert vertical images into measurements of height and therefore allows the rapid production of contour maps.

Low cost has been a critical factor in the growing use of aerial surveys for environmental surveillance over the last 30 years. In addition to cost advantages aerial photographs may be able to show features which are not visible from the ground and monitor rapid environmental changes. We shall be returning to these increasingly important aspects of this subject toward the end of the book when, in Chapter 23, we discuss remote sensing studies and their role in the geographers' work. So, in this chapter we have moved from a view of primitive people gazing around them on the beach at an unknown world to modern air and satellite surveys. In the next four chapters we shall see what kind of world the centuries of exploration that lay between the two revealed.

Summary

1. Determining the size and shape of the earth has interested geographers since the time of the Greek cosmologist Eratosthenes. Measurements made beginning in the eighteenth century show that the earth is not a sphere, but an ellipsoid; this fact has caused some difficulties in mapping.
2. Specifying location on the earth's surface can be done by giving names to places. This nominal specification has several major drawbacks: It is imprecise spatially, the names are vast in number, nonstandardized, and often nonunique. Reference grids allow the location of points on the earth's surface to be specified by a system of coordinates. Such systems include cartesian coordinate forms and a modified form used in the United States, the range, township, and section method. Special situations sometimes require the use of the polar coordinate system which is based on azimuths and distance from an origin.
3. The problem of locating places on the earth's spherical form using the cartesian system was solved by using the geographic grid of meridians and parallels. Precise location on this grid is indicated by latitude and longitude. Globes are constructed using the spherical grid system plus a scale.
4. Locational information is distorted when transferred from the sphere to a flat map, and map projections have been developed to solve the distortion problem. Projections are only partial solutions, as absolute accuracy can be maintained in shape, or in area, but never in both. Conformal projections emphasize the former, equal-area projections the latter.
5. Geographers today find that aerial photographs are increasingly useful for spatial analysis. After photographic distortion has been corrected, they can be used as substitutes for maps. Stereoscopic pairs of overlapping photos give a three-dimensional view and allow height to be mapped. The relative low cost of aerial surveys makes them very useful in environmental research. Recent extension of aerial surveillance using satellites for remote sensing is discussed in Chapter 23.

Reflections

1. Consult a topographic map of your own locality, and list the names given to streams and water courses. Then do the same for a different part of the country. Are there any variations between the two lists? What do you think might have caused these variations?

2. Have the members of your class check off on a list of "possibles" (a) the countries they consider part of the "Middle East" and (b) the states they consider part of the United States' "Midwest." Can you identify a firm core area on which everyone agrees? Are there any "difficult" borderline cases?

3. Review the section on the earth's spherical coordinate system. How is the location of the earth's (a) equator and (b) poles determined?

4. What effects has the "township and range" system used for land surveys had on (a) the rural and (b) the urban landscapes of the central and western parts of the United States?

5. Consult two atlases in your college's map library which show your own country's outline on different map projections. (The title of the projection is usually given in the lower margin of the atlas page.) Compare these map outlines with the "true" shape of your country on a globe. Which projections appear to produce (a) the most and (b) the least distortion? Does this depend at all on where your country is on the map? Why?

6. Review your understanding of the following terms:

   meridians and parallels   latitude and longitude
   place names   map projections
   reference grids   horizontal displacement
   spherical coordinates   stereoscopic pairs
   township, range, and   stereoscopes
   section

One Step Further . . .

For a basic but brief account of the variety and range of maps used by geographers, see

   Tyner, J., *The World of Maps and Mapping* (McGraw-Hill, New York, 1973)

and browse through one of the two standard works introducing the principles and practice of mapmaking:

   Robinson, A. H., et al., *Elements of Cartography*, 2nd ed. (Wiley, New York, 1978) and

   Thrower, N. J. W., *Maps and Man: An Examination of Cartography in Relation to Culture and Civilization* (Prentice-Hall, Englewood Cliffs, N.J., 1972).

The history of man's efforts to pin down the exact distribution of the world's geographic features is told at length in

   Bagrow, L., in R. A. Skelton, Ed., *History of Cartography* (Harvard University Press, Cambridge, 1964)

and summarized more briefly in

   Abler, R., J. S. Adams, and P. Gould, *Spatial Organization: The Geographer's View of the World* (Prentice-Hall, Englewood Cliffs, N.J., 1971), Chap. 3.

The basic principles of photointerpretation and examples of the use of aerial photographs in research are given in

   Colwell, R. N., Ed., *Manual of Photographic Interpretation* (American Society of Photogrammetry, Washington, D.C., 1960) and

   Leuder, D.R., *Aerial Photographic Interpretation* (McGraw-Hill, New York, 1959).

Its geographic applications are discussed in

   James, P. E. and C. F. Jones, Eds., *American Geography: Inventory and Prospect* (Syracuse University Press, Syracuse, N.Y., 1954), Chaps. 25 and 26, and

   St. Joseph, J., Ed., *The Uses of Air Photography* (John Day, New York, 1966).

Excellent examples of earth photographs from spacecraft are available in a number of NASA publications. For the most recent developments in mapping, see the International Yearbook of Cartography *(published annually) and the* Journal of Cartography *and* Surveying and Mapping *(both quarterlies).*

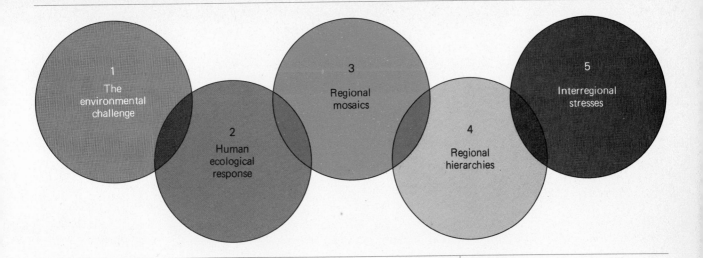

A geographer's view of the uncertain planetary environment in which the human population evolves and lives at ever-increasing densities is presented in Part One. *Environments as Ecosystems* (Chapter 3) provides a context for viewing the earth's environment in terms relevant for the human population. It notes the ecological dependence of human beings on a series of fundamental cycles—such as the carbon cycle or the hydrological cycle—and shows our own species as one of many interdependent organisms each relying on worldwide exchanges of material and energy. *The Global Environment* (Chapter 4) estimates the inherent quality of the earth as a permanent home for humanity. It shows that the fertility of the planet varies enormously from one place to another, and it considers the major factors that underlie this spatial variation at different geographic scales. Geographers find it convenient to look at this variation in terms of a series of environmental regions, each forming a module within which a particular set of ecological relationships is dominant. Both of the last two chapters are concerned with how the environment is changing over time. *Environmental Change* (Chapter 5) concentrates on the slow, long-term, swings of climate and sea level and contrasts these with short-term, season-to-season, and year-to-year variations. *Uncertainties and Hazards* (Chapter 6) picks out the changes of most direct worry to mankind. These are those midterm uncertainties in climate and the sudden and sometimes catastrophic natural events that punctuate the regular cycles of change. Special attention is given to extreme geophysical events—earthquakes, volcanic eruptions, floods, cyclones—that make some environments so hazardous for human settlement. Together, the four chapters of this first part provide an introduction to the "stage" on which the drama of human activity described in the following parts will be played out.

part one
The Environmental Challenge

We will now discuss in a little more detail the struggle for existence.

CHARLES DARWIN
The Origin of Species (1859)

Geographers are sometimes described as those scientists who study "the earth as the home of man." In this first major part of the book we look at the earth from this point of view. In studying the natural environment of the globe, our approach will be somewhat different from that of a "pure" earth scientist such as a geophysicist. Along with the technical difficulties of measuring and mapping environmental characteristics like precipitation or soil types are the socioeconomic ones of estimating which properties are relevant or irrelevant to humans. Different human groups with different technologies view the *same* environment in different ways. Can we identify any common elements that will produce the same reactions among all groups?

People do not lie outside the global ecosystem, but are an intrinsic part of it. How we answer a particular environmental challenge depends largely on our own animal physiology, though our response is modified to some degree by our culture and the economic sys-

tem. Hence we cannot directly equate the utility of an area with its innate environmental qualities.

In this chapter we look first at the nature of the environmental challenge. What do people need from the earth? Second, we see how human beings are inseparable from that environment, linked to it through a complex set of energy and material flows. In this part of the chapter we introduce the concept of systems in general and of ecosystems in particular. This leads to our third section, in which we try to estimate how productive the globe is.

In raising the fundamental questions about the geography of the earth that we have in this chapter, we are entering the field of *physical geography.* This branch of geography analyzes the physical structure of our planetary environment—its land forms, climate, vegetation, soils, and so on. Partly because physical geography is closely linked to other natural sciences (geophysics, geology, meteorology, botany, etc.) and partly because of its longer history, it is currently one of the strongest and most developed branches of

chapter 3
Environments as Ecosystems

geography. (Cf. Section 25-2, especially Figure 25-8.) It has the most highly developed theoretical models, and its predictive capability has already reached levels that are unlikely to be matched in the rest of geography for some decades, if at all. Anyone attempting, therefore, to summarize its concepts within a few chapters must do so with a broad brush. Students who have already taken introductory earth-science courses or an elementary course in physical geography will wish to move quickly through the next chapters or concentrate on the readings suggested in the section "One step further. . . ." If the concepts discussed there are new to you, remember that they are the first steps in a field of geography you will be able to study in much greater depth in advanced courses.

3-1
The Nature of the Challenge

The main challenge to the human occupation of the earth's surface is posed by its environment. We have already defined that environment in our opening chapter as the sum total of conditions which surround (literally, environ) human beings on the surface of the earth. But this broad definition now needs to be refined. Let us look at that environment in more detail.

The Structure of the Environment   The most fundamental divisions of the environment are shown in Figure 3-1. In 3-1(a) we note the fundamental difference between the nonliving world and the living world of which man forms a part. In environmental terms, the former is the abiotic environment and the latter is the biotic environment. The abiotic environment can be subdivided in terms of its physical state as a solid, liquid, or gas. The solid earth is termed the *lithosphere* (from the Greek word, *lithos* meaning rock). This is surrounded by two shells each up to 11 km (7 mi) thick: a discontinuous layer of liquid termed the *hydrosphere*, and a continuous layer of gas termed the *atmosphere*. In comparison with the size of the planet earth with its diameter of 12,700 km (7,900 mi), then these two layers are very thin indeed, much thinner in proportion than the skin on a good apple.

The living world of the *biosphere* [See Figure 3-1 (d)] is even thinner. If we define the biosphere as the biotic environment in which living things are found, then it extends through the full depth of the oceans but is confined to the lower layers of the atmosphere (birds, flying insects, micro-organisms). On the land surface, the biosphere extends upward to over 110 m (360 ft) (the giant redwood trees of northwest California) and down many metres in the micro-organisms in the soil or in deep caves and rock fractures.

Within the shallow layer of the biosphere is an immense variety of living organisms. Already over a million animal species, and a quarter of a million plant species have been described by biologists. The total that exists is certainly far larger than that. The 4.3 billion human beings at present inhabiting the earth's surface belong to a single species, *Homo sapiens*.

Man's Environmental Requirements   The long evolution of *Homo sapiens* as an animal species has given us some highly specific environmental requirements. Like an atmospheric fish we swim through an oxygen-rich gas found only near the surface of one of the minor planets. In view of the multiplicity of physical and chemical conditions in the known universe, humans are a remarkably specialized creation, with only a tenuous toehold on survival. In a reduced oxygen level of our surroundings, we start to pant; in an increased proportion of hydrocarbons, we start to cough; immersed in water, we drown within a few seconds; deprived of water, we atrophy and die within a few days. But as denizens of the planet Earth we appear to be much more robust. Given an unpolluted atmosphere, our tolerance of climatic conditions (precipitation, wind, and solar radiation) actually found on earth is reasonably good.

The climatic element to which we show the greatest sensitivity is probably temperature. We are warm-blooded mammals with an average body temperature around 37°C (98.6°F). Prolonged exposure to conditions that raise or lower this normal body temperature more than a few degrees leads to permanent tissue damage and death. The extreme ranges of body temperatures ever recorded for a living person are 44°C (111°F) and 16°C (61°F). Let us compare these with the actual variations in the surface air temperatures on the earth given in Table 3-1. This shows that our extreme tolerance is only one-fifth of the earth's temperature range. Note also that we are much better suited to the hotter ranges of the planet's surface than to the colder. Extreme cold is a much more significant limit to human existence on the earth than extreme heat.

How can we measure whether an environment is tolerable for human life? Many attempts have been made to assess climatic environments using simple combinations of temperature, humidity, radiation, and wind-speed measurements. One simple index used by the U.S. National Weather Service is the temperature–humidity index (THI). This index is based on temperature readings on two Fahrenheit thermometers, the bulb of one of which is kept

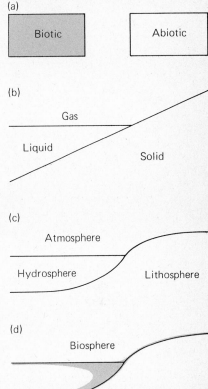

Figure 3-1.   Main elements in the structure of the environment.
Note the way in which the biosphere is concentrated along the interfaces between the three abiotic environments—the atmosphere, hydrosphere, and lithosphere.

Table 3-1   Global extremes of temperature

Thermal characteristic	Temperature		Location
	(°C)	(°F)	
INDIVIDUAL READINGS			
Highest	58.0	136.4	San Luis Potosí, Mexico (1933)
Lowest	−88.3	−126.9	Vostok, Antarctica (1960)
RANGE OF READINGS			
Widest	88.9	160.0	Verkhoyansk, Siberia
Narrowest	13.4	24.1	Fernando de Noronha, South Atlantic
ANNUAL AVERAGES			
Hottest	31.1	88.0	Lugh Ganane, Somalia
Coldest	−57.8	−72.0	78°S, 96°E, Antarctica

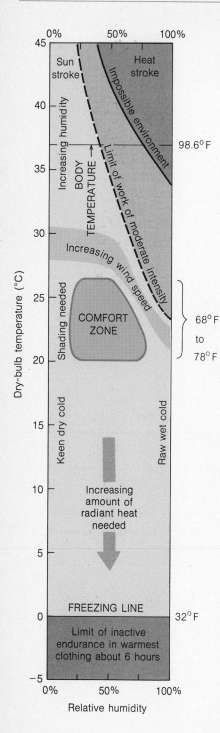

Figure 3-2. Human tolerance of climatic ranges.
The graph shows comfort, discomfort, and danger zones for inhabitants of temperate climatic zones. Dry-bulb temperatures in °F are shown to the right of the diagram. [After V. Olgay, from R. G. Barry and R. J. Chorley, Atmosphere, Weather, and Climate (Methuen, London, 1968), p. 251, Fig. 7-1.]

permanently damp. Evaporation will make the temperature recorded on the wet-bulb thermometer lower than that recorded by the normal dry—bulb thermometer. When the air is humid, there is little evaporation and the readings on the two thermometers are similar. The THI is the sum of the two readings multiplied by a constant (0.40) and added to another constant (15). If the two thermometers record temperatures of 70°F and 65°F, the THI will be 0.40 × (70 + 65) + 15, or 69. At a THI of 75 in still air, about half the people in an office feel discomfort; at 80 few remain comfortable; and at 86 regulations suggest that all workers (at least in federal buildings) be sent home.

These comfort levels pertain to still air. As Figure 3-2 shows, winds reduce the effect of high temperatures and humidity and make conditions more bearable. On the other hand, when temperatures are low, strong winds considerably increase the level of discomfort. Thus, any index of human comfort must clearly include a chilling factor related to air speeds.

However sophisticated an index is, it can only describe average reactions. Individuals vary considerably in their ability to withstand stress. Our sex, body characteristics, genetic heritage, degree of acclimatization, and cultural background all affect our environmental tolerance. Most indexes have been tested on urbanized North Americans, and we would expect the reactions of Nepalese, Kikuyu, or Eskimo groups to be somewhat different.

Environmental Resources as Population Regulators   Human life is maintained by an inward flow of nutrients and an outward flow of wastes. Figure 3-3 shows the estimated daily inputs and outputs for a medium-sized male of 70 kg (154 lb). Inputs in the form of water, food, and oxygen allow the renewal and growth of body tissue, and provide the energy for breathing, blood circulation, and movement. The energy consumed is transformed to heat and cycled back to the atmosphere. Excretory matter from the metabolic process forms the outputs, partly solids, partly liquids, and partly gases which complete the input-output cycle.

The energy needed to maintain human life is measured in terms of calories. Such daily needs are related to age, sex, body weight, and the amount of work to be done. Whereas a two-year-old child can subsist on 1000 calories a day, a pregnant woman would need twice that amount. A man carrying out very

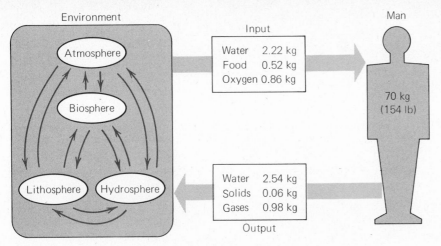

Environment

Man

Figure 3-3. Man's dependence on the environmental support system.
Human life depends on a continuous exchange of materials with the earth's environment. Figures given in the chart are the estimated daily inputs and outputs for an adult male.

heavy labor will require between 3000 and 4000 calories a day. Energy needs are simply a quantitative measure, whereas the food inputs must have certain qualitative requirements. For example, our diet needs to contain around 10 percent of its content as protein.

Food supply is one of the critical input mechanisms by which the environment controls all animal numbers (including human numbers too, in the long run). Consider Figure 3-4, which shows two ways in which a population may grow. In the first, changes in numbers are constrained by the food supply. Increased numbers born in one season mean less food for each member of the population, give a reduced chance of survival, and return the population to its original position. If you follow the sequence of arrows, you'll see how population tends to return to an equilibrium level. Loops of this kind are described as *negative* (or self-regulating) feedbacks and are typical of stable populations. Figure 3-4(b) shows by way of contrast an uncontrolled situation. In this case, food supply is abundant, there are no checks on growth enforced by starvation, and so the population grows rapidly. These unstable situations are characterized by *positive* (or self-reinforcing) feedbacks.

Figure 3-4. Feedbacks and population control.
The diagrams show how feedback mechanisms regulate population size. In (a), negative feedbacks lead to self-regulation. In (b), positive feedbacks lead to changes that are self-sustaining. [After W. B. Clapham, Jr., *Natural Ecosystems* (Macmillan, New York, 1973), Fig. 104, p. 12.]

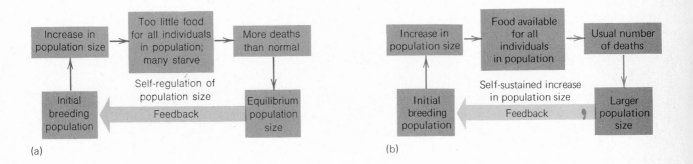

KEY TERMS IN THE STUDY OF ECOSYSTEMS

Biomes are the major environmental zones of the earth marked by a distinctive plant cover (e.g., the subarctic tundra biome).

Carrying capacity is the largest number of a population that the environment of a particular area can carry or support.

Climax is the state of equilibrium reached by the vegetation of an area when it is left undisturbed for a long period of time.

Communities are groups of animals and plants that live in the same environment and depend on each other in some way.

Ecological efficiency measures the ability of organisms in a food chain to convert the energy received into living matter.

Ecology is the study of plants and animals in relation to their environment.

Ecosystems are ecological systems in which plants and animals are linked to their environment through a series of feedback loops.

Food chains describe the series of stages that energy goes through in the form of food within an ecosystem.

Food webs are complex networks of food chains.

Predator-prey relations describe the links between the population of

one set of animals (the prey) that are hunted for food by another set (the predator).

Seres are the transitional stages in plant succession.

Succession describes the orderly sequence of change in the vegetation of an area over time as it passes through transition stages (seres) toward an equilibrium (or climax).

Trophic levels are the main stages in the food chain, green plants occupying the first level, plant-eaters the second, animal eaters the third.

3-2
Systems and Feedbacks

The relationship between a population (like that of human beings) and its environment via food supply is an example of an ecosystem. The word *ecosystem* is a shorthand term for "ecological system." So, to understand something about these systems, we must try to define clearly what we mean by a system and by ecology. This will allow us to see something about the nature of ecosystems, their structure, and why their understanding is so important in geography.

The Nature of Systems   A system is defined here as "a set of components and the relationships between them." As Figure 3-5 shows, it consists of three

Figure 3-5.   Cycles in the ecosystem. (a) The two major energy and material exchanges between the biosphere and the abiotic environment. (b) Internal movements of energy and nutrient materials within the biosphere. Note the arrangement of plants, animals, and decomposers into a food chain.

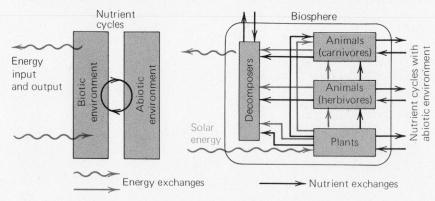

(a)   (b)

essential ingredients, the components, the links between them, and the boundary which separates the system from the rest of the world.

Geographers are usually concerned with three main types of system:

Morphologic Systems These are systems in which relations between individual components are built up by observed association to produce positive or negative bonds. Changes in the level of one component cause associated changes in other components. Such systems vary in the number of components they have, the strength of the links between them, and the arrangement of the links into positive or negative feedback loops.

We can illustrate the links and loops in a morphologic system by considering the way in which coral reefs are found. (See Figure 3-6.) *Corals* are minute marine animals that live in huge colonies in shallow tropical seas where their combined limy skeletons form reefs and atolls. They have attracted great interest from geographers since Charles Darwin's Pacific voyages in the *Beagle* in the 1830s, and they are still one of the most fascinating of marine ecosystems. Many of the organisms that secrete calcium carbonate to make coral reefs are sensitive to the depth of water. As the depth decreases, sunlight becomes more abundant and the rate of growth of the reefs increases. This accelerated growth further decreases the depth of the water, increases the light, accelerates the growth of algae, and so on, in a positive feedback relationship. However, the inability of the organisms that make up the reefs

Figure 3-6. The coral reef puzzle.
Apollo 7 photo of the Tuamotu archipelago in the southern Pacific Ocean. *Atolls* are elliptically shaped reefs of coral enclosing a lagoon—those shown on this photo vary from between 15 and 40 km (10–25 mi) across their longest axis. The reefs are composed of coral limestone, the accumulated skeletons of corals and associated organisms. Exactly why corals form an atoll shape has intrigued scientists over the last century. Charles Darwin suggested that atolls began as reefs fringing an island which sank over a long geologic time. Others thought that coral reefs could build up on their own debris and would form atolls by growing most actively on the outside of the reef. The balance between the building activity of the corals and the destructive energy of the surf breaking on the reefs is shown in Figure 3-7. [Photograph courtesy of NASA.]

Figure 3-7. Feedbacks in the coral reef system.
The role of the sea level in regulating the growth of algae on a hypothetical coral reef is shown via a flow diagram. Notice that in this example each component (box) in the system contains a physical or biological process. [From J. W. Harbaugh and G. Bonham-Carter, *Computer Simulation in Geology* (Wiley-Interscience, New York, 1970), p. 268, Fig. 7-4.]

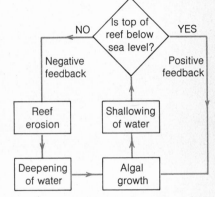

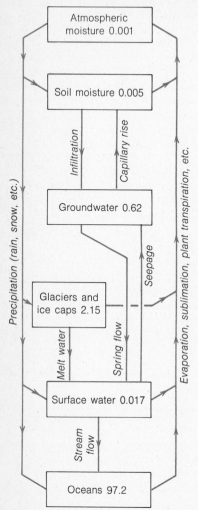

Figure 3-8. Earth's hydrologic cycle.
The diagram shows the main components in the global hydrologic cycle. The figures refer to the percentage of all terrestrial water in each of the six main stores. [After R. J. More, in R. J. Chorley and P. Haggett, Eds., *Models in Geography* (Methuen, London, 1967), p. 146, Fig. 5-1.]

to grow above sea level and the breakup of the reefs by pounding waves introduces an effective negative feedback that limits growth. Figure 3-7 presents a flow chart in which the elevation of the reef with respect to the sea level acts to regulate its growth, initiating either positive or negative feedback loops.

Cascading Systems These are systems in which relations between individual components involve transfers of mass or energy. The *output* from one component becomes the *input* for another. Inputs and outputs can be controlled by *regulators*. Feedbacks between components occur from the sequence of inputs and outputs, which may be lagged in time.

Figure 3-8 shows as an example of a cascading system the earth's hydrological cycle. This is made up of the oceans (which contain 97 percent of all the world's water) plus five other stores or regulators—the atmosphere, the soil, the rocks, glaciers, and lakes and rivers. Note how the water circulates as inputs and outputs from one store to another, sometimes moving as a liquid and sometimes as water particles in a gas.

Control Systems Morphologic and cascade systems can be modified by human intervention. This intervention may take the form of restricting the levels of individual components or governing the flows of inputs and outputs.

Bushfires provide a simple example system which may be controlled, at least in part. Consider Figure 3-9 (a) which shows a fire as a cascade-type system. Each season litter accumulates from the trees to build up combustible material. In dry summer conditions a fire trigger (say, a lightning strike) may start up a fire. This spreads to provide burnt-over areas. Since the fire burns up the litter, this reduces the chance of another fire until the vegetation has grown and litter again accumulates. The pattern of litter accumulation under natural conditions is given in Figure 3-9 (b). Six natural sources of fire are assumed over a time period of about 50 years. Fires vary in intensity, depending on the length of interval since the last fire.

Two strategies for controlling this natural system are considered. In Figure 3-9 (c), there is regular preventive burning every five years: This keeps the fires at low intensities and under some measure of control. In Figure 3-9 (d), a complete prevention system is attempted. The natural fires that start are extinguished wherever possible. Note, however, that since there are so few fires, litter goes on accumulating, so the chance of a large uncontrolled fire goes up. This is shown for the third fire in Figure 3-9 (d). In this case, the fire could not be put out and a massive burnout resulted. Clearly natural systems can be only partly controlled and, without full understanding, attempts at control may lead to making the problem even more severe than under a wholly natural system.

The essential idea of a system is therefore a very simple one. It focuses attention on the behavior and interlinking of several components of the environment, each working together.

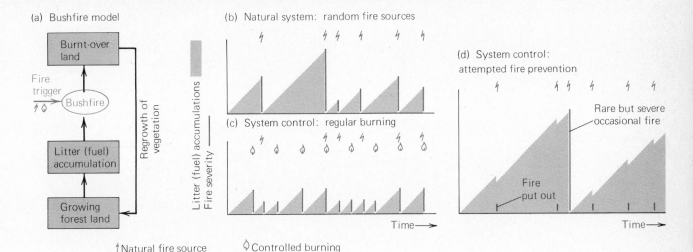

(a) Bushfire model

(b) Natural system: random fire sources

(c) System control: regular burning

(d) System control: attempted fire prevention

Rare but severe occasional fire

Fire put out

↯ Natural fire source   ◊ Controlled burning

The Nature of Ecosystems As we noted earlier, ecosystems are ecological systems in which plants and animals are linked to their environment through a series of links, some of which form feedback loops. The term "ecological" dates back to 1868, when the German biologist Ernst Haeckel used it in discussing his studies of plants in relation to their environment. It stems from the simple Greek word *oikos*, meaning "a house" or "a place to live in" and serves as a direct link to the geographer's concern with the earth as the home of humanity. You will note close links between geography and ecology at many points throughout this book.

Ecosystems: A Small-Scale Example The easiest way to unravel the structure of an ecosystem is to take a small-scale, familiar example. We could stay on the beach and look at the reactions of plant and animal life to the twice-daily changes of environment in the tidal zone. A still better example can be found by moving inland to a small lake like that in Figure 3-10.

A lake is a body of standing fresh water. What physical inputs and outputs does it receive? As a participant in the hydrologic cycle, it derives inputs of fresh water from stream inlets and from rainfall and loses water through stream outlets and evaporation. Its most important input is, however, the solar energy it receives from direct sunlight. This will warm the upper layers of the lake very strongly in summer and set up important vertical differences in water temperatures over the seasons.

In addition to the physical processes by which water flows, sediment is deposited, temperatures change, and so on, there are also far more complex biological processes going on in and about the lake. Sunlight furnishes energy used by microscopic green plants in the lake (the *phytoplankton*) to convert inert chemicals in the water into food. These organisms provide food for small larvae and crustaceans (the *zooplankton*), which are eaten by small fish.

Figure 3-9. Bushfires as uncontrolled and controlled systems.

A bushfire model (a) with outbreak patterns under natural conditions (b). Human interference by controlled burning (c) may reduce the severity of fires. Attempts at complete protection (d) may increase fire severity because of litter accumulation. This may be compensated by the relative infrequency of fires. Note that the model shown is greatly simplified.

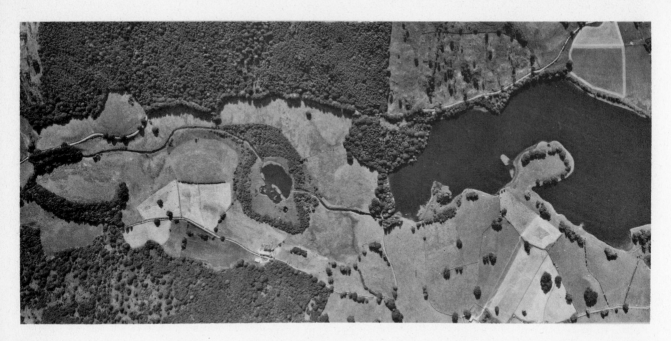

Figure 3-10.   Lakes as environmental features.

Outdubs Tarn (center) in the Lake District of northwest England is one example of the lake ecosystems discussed in this chapter. This small lake (about 50 m across) is gradually being filled in by sediment from the stream flowing into it (left) and by encroaching vegetation. Mats of reed and mosses are shown as a light ring around the dark waters of the lake itself; scrub timber (darker gray in tone) forms the outer ring, encroaching in turn on the reed-moss layer. The final infilling of this small lake is a slow process in human terms, perhaps a few centuries, but rapid in terms of geological time. Note also the outflowing stream draining to the much larger and deeper lake, Esthwaite Water (right). [Aerofilms photo.]

[See Figure 3-11(a) and (b).] These small fish are eaten by larger fish, which may eventually provide food for animals and for people themselves. Plants and animals die and decay, releasing chemicals back into the lake waters. We show these links in the lake ecosystem in Figure 3-11(c). Of course, this is a highly simplified picture of a process that may involve hundreds of living species and very complex chemical chains. Figure 3-11(b) shows some of the inhabitants of a typical lake.

Not all ecosystems are as clearly defined as lakes. Many have boundaries which are hard to establish and which conceal very important internal variations. We shall look at some of the major worldwide ecosystems in Section 4-4, on "Major Environmental Regions." The simple lake ecosystem does, however, illustrate three elements which are critical in all ecosystems—from the smallest to the largest. These are the cycling of energy chemicals (especially carbon) through biological populations, and the linking of these biological populations into food chains. We now look at each element in turn.

Energy in the Ecosystem   Tracking flows of energy through an ecosystem has proved difficult. Figure 3-12 shows one attempt to measure energy flows in an oak–pine forest on Long Island, in New York state. The energy being stored in this New England type of mixed woodland is estimated to be 2650 g of living matter per m^2 each year, but this is only a rough guess.

If you follow the flows around, you'll see that 19 percent of the energy received is stored by the trees, a further 2 percent in the humus of the upper layer of the soil under the trees. Thus the total storage is about one-fifth of

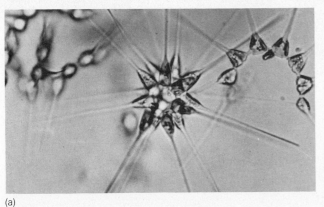
(a)

(b)

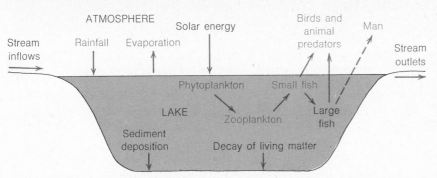

ATMOSPHERE

Stream
inflows
Rainfall   Evaporation   Solar energy

Birds and
animal
predators

Man

Stream
outlets

Phytoplankton   Small fish

LAKE

Zooplankton   Large
fish

Sediment
deposition   Decay of living matter

(c)

Figure 3-11. Lake ecosystems.
The critical lower links in aquatic food chains
are provided by (a) microscopic plant life
(phytoplankton) and (b) animal life
(zooplankton), magnified here to many times
their actual size. (c) In this simplified flow
diagram of the main links in a lake
ecosystem, different colors are used to
indicate physical and biological processes.
(d) A more detailed diagram of a food web
shows the many organisms that take part in
just one segment of the system. Ecologists
find that the more complex the food web is,
the more stable the *ecosystem* is. [Photo (a)
by Walter Dawn and (b) by Hugh Spencer,
National Audubon Society. Diagram (d) from
W. B. Chapham, Jr., *Natural Ecosystems*
(Macmillan, New York, 1973), p. 113, Fig.
4-7.]

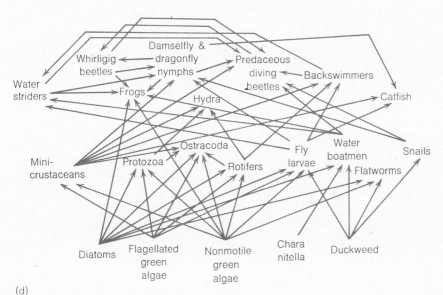

(d)

Figure 3-12. Energy flow in a woodland ecosystem.

The quantities were calculated from a study of oak-pine forest on Long Island, New York. The total energy input of 2650 g living matter/m² year is set equal to 100. Of this 79 percent is returned as respiration, while the remainder is stored in the increased biomass of the forest. [After G. M. Woodwell, *Scientific American,* No. 3. (Sept., 1970), p. 64. Copyright 1970 by Scientific American, Inc. All rights reserved.]

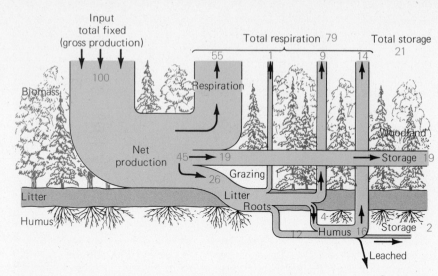

the total energy received. The remainder is returned to the atmosphere through respiration. Comparison of different stands of vegetation shows that the stored share falls as the stands become more mature. So, if you abandon a field in New England, it will be shrub-covered within about 15 years, covered with pine forest within 50 years. If you leave it for more than a century, then oak and hickory will take over from the pines. The bushes and trees simply represent stored energy and, in the early phases of regrowth, the storage is rapid but decreases with increasing maturity. Ecologists believe that very old forests would form a stable plant community, called a *climax*. Under climax conditions the input and output of energy flows would be balanced with no accumulated storage. In other words, the forests would show little or no change in mass from decade to decade.

Nutrient Cycles in Ecosystems   One essential element in both the lake ecosystem and the woodland ecosystem is the conversion of solar energy into living matter. How is this piece of alchemy achieved? Let us use the *carbon cycle* to illustrate one of the most important aspects of this conversion process.

We have already seen that carbon is available in the earth's lower atmosphere as carbon dioxide (CO_2). This gas forms a small but vital 0.033 percent of the total volume of air. It is important climatically as a heat-absorbing blanket, helping to regulate air temperatures near the earth's surface. Biologically, carbon dioxide is essential to plant growth: Green plants with the pigment chlorophyll combine carbon dioxide and water through *photosynthesis* (from the Greek, "putting together with light") to produce all the food materials necessary for life. (See the marginal discussion of photosynthesis in the carbon cycle.) Photosynthesis is actually a cluster of interrelated chemical reactions activated by solar radiation at the wavelength of visible light. Green plants may be regarded as the basic *producers* in the carbon cycle, because they manufacture consumable energy (food in the form of carbohydrates) from atmospheric carbon and solar energy.

The carbon cycle is completed and carbon dioxide returned to the atmosphere by the processes summarized in Figure 3-13. Consider the food produced by land plants. These are eaten by animals (here termed *consumers*), and the energy stored as food sustains activity at high rates. Some of the carbon from carbohydrates is stored in the body, and the rest is excreted by respiration as carbon dioxide. Consumers can be divided into herbivores, carnivores, and omnivores, depending on whether they eat only plants, only

PHOTOSYNTHESIS IN THE CARBON CYCLE

We can describe the overall process of photosynthesis in the carbon cycle by a simple chemical equation,

$$\text{Light} + n\,CO_2 + n\,H_2O \xrightarrow[\text{of chlorophyll}]{\text{in the presence}}$$

$$(CH_2O)_n + n\,O_2$$

In other words, green plants extract carbon dioxide (CO_2) and water (H_2O) from their environment, return the oxygen (O_2) to the environment, and incorporate the remaining substances into carbohydrates (represented here by CH_2O). These carbohydrates are decomposed to provide energy or passed on to other parts of the food chain. (See Figure 3-13.) Rates of photosynthesis are critically related to the intensity of light. At low light intensities, the rate of photosynthesis is slower than the rate of plant respiration; respiration involves the oxidation of the carbohydrates and the release of carbon dioxide and water. At a slightly higher light intensity, the two rates are equal. Above this point, the rate of photosynthesis surpasses the plant respiration rate and carbohydrate products accumulate. The greatest, or saturation, rate of photosynthesis is reached in full sunlight. In addition to light, photosynthesis requires adequate moisture and proceeds most rapidly at temperatures between 10°C and 50°C (50°F and 122°F).

Figure 3-13.   Carbon cycles and the world carbon balance.
The figures indicate the estimated stores (boxes) and annual flows (arrows) of carbon in units of 10⁹ tons. *Carbonification* is the conversion of dead plant and animal remains into coal, oil, and similar fossil fuels. *Diffusion* refers here to the interchange of carbon dioxide gas between the atmosphere and the oceans by molecular mixing. (Note the different use of the term *diffusion* in Chapter 13.) The carbon cycle shown is only one of the major cycles of important chemical elements through the environment. Similar flows occur in the *nitrogen cycle* and the *potassium cycle*. [After J. McHale, *The Ecological Context* (George Braziller, New York, and Studio Vista, London, 1971), p. 52, Fig. 21. Copyright 1971 by J. McHale. Reprinted with permission.]

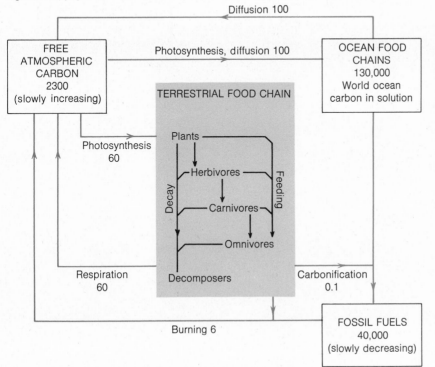

animals, or a mixture of the two (as people do). The final role in the carbon cycle is played by *decomposers*. These are bacteria and fungi which break down the carbon stored in the tissues of dead plants and animals. In the decomposition process, carbon is again returned to the atmosphere or to soil water.

Not all producers and consumers are decomposed as soon as they die. Organic matter is stored and concentrated geologically for millions or billions of years as peat, lignite, coal, petroleum, and natural gas. (See Section 9-2 on fossil fuels.) Plants are also burned as fuel by people. Like eating, burning separates the elements in carbohydrates and returns carbon to the atmosphere as either carbon monoxide or carbon dioxide.

In any event, carbon from the atmosphere is circulated through a chain of living organisms to return eventually to the atmosphere. At each stage in this process it combines with different elements in various chemical forms, and each of these combinations is accompanied by energy transfers. Rearrangements of molecules and energy transfers (by photosynthesis in plants, and by metabolic synthesis in animals) are the essential processes that allow human life on the earth to continue. We have selected the carbon cycle, as an illustration of how energy transfers are accomplished, but we would need to supplement our description of it by a description of other cycles, like the nitrogen cycle, to fully explain the exchange processes involved. (See Table 3-2.) Each cycle represents an essential link in the ecosystem, for it includes biological elements (producers, consumers, and decomposers) as well as inorganic elements (e.g., carbon dioxide gas in the atmosphere and the carbons stored as fossil fuels).

Table 3-2   Major geochemical cycles in the ecosystem

Group	Cycled compound or element	Role in the biosphere
COMPOUND	Water (H_2O)	Major component of biosphere. Wood is 50% water, many mammals are 85% water. Water acts as solvent for other mineral nutrients.
MAJOR ELEMENTS	Oxygen (O)	Major constituent of living matter (70%). Basic building block. Oxidation important component in growth processes.
	Carbon (C)	Major constituent of living matter (18%).
	Hydrogen (H)	Major constituent of living matter (11%).
MINOR ELEMENTS	Nitrogen (N)	Plentiful in atmosphere but scarce in biosphere.
	Sulphur (S)	Bacteria play major part in releasing sulphur compounds for recycling.
	Phosphorus (P)	Important in photosynthesis. Limiting element in ecosystem growth.

Food Chains in Ecosystems   All animals get their food from plants, either directly or indirectly by feeding on other animals that feed on plants. Thus, the process of photosynthesis and mineral cycles like the carbon cycle provide the basis of lengthy *food chains*. We have already encountered a simple example of a food chain, stretching from the millions of microscopic plants [the phytoplankton of Figure 3-11(a)] on a lake surface to human fishermen.

In the world's oceans, fish like tuna that are caught and consumed by man are directly dependent on a three- or four-link chain. Phytoplankton are consumed by larvae and shrimps, which are in turn eaten by squids and small fish, which form part of the food consumed by tuna. However, in each case it takes from 5 to 10 food units (calories) of the prey to produce one unit of the predator; this difference is termed a *food-conversion ratio*. One unit of tuna consumed by a human being represents an estimated 5000 units of phytoplankton.

It is useful to represent the levels in a food chain as a series of food pyramids, as in Figure 3-14. Each step of the pyramid is termed a *trophic* level (from the Greek *trophe*, food). The first level (T_1) at the base of the pyramid is composed of green vegetation with energy contained in the plant tissues. The second level (T_2) consists of herbivorous animals that feed on the plants; the third level (T_3), of carnivorous animals that feed on herbivorous animals; the fourth level (T_4), of carnivorous animals like man that feed on other carnivorous animals and all the lower tiers. The fifth and final level (T_5) is made up of decomposers that break down the dead tissues of organisms at all the other levels of the food chain.

Biologists have shown us the exact structure of trophic levels for individual communities. For example, they have analyzed the food chains and conversion ratios of over 200 species of fish in coral reefs in the Marshall Islands in the Pacific Ocean. By estimating the dry mass of organisms, from plankton and algae to sharks, they showed that the base of the pyramid (T_1) consisted of producers with a weight of 703 grams (g) per m². Above these organisms were herbivores (132 g) and finally carnivores (11 g). Other researchers have tried to estimate the actual energy flows between the different species in a community.

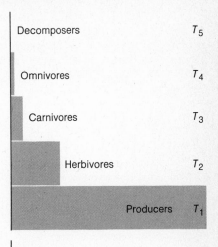

Mass of living materials per unit of area

Figure 3-14.   Trophic levels.
The pyramid shows the relative dry weights of living materials typical on the five main trophic levels of an ecosystem.

Selection of Ecosystem Units   Because environmental variation occurs on many geographic scales, geographers have developed systems of regions that can be modified to these scales. One of the most versatile regional units identified so far has been the *watershed*, or *catchment area*, of a stream. (See Figure 3-15.) Stream watersheds form a convenient unit because they can be simply and unambiguously defined from a topographic map. They are independent of scale, in that large river basins like the Amazon can be broken into a hierarchic system of smaller basins; like a toy Polish doll within larger dolls, each smaller basin fits exactly within the next larger one. Each subdivided basin can be identified and its streams numbered in a way which provides a measure of relative size.

Figure 3-15.   Watershed hierarchies.
Pictured are three examples of the
hierarchical breakdown of large watersheds
into smaller units. Figure (a) shows the
Arroyo de los Frijoles basin near Santa Fe,
New Mexico. Figure (b) shows a small
segment of this watershed, the Arroyo
Caliente basin, in greater detail. Figure (c),
a further enlargement of a single catchment, illustrates *stream ordering*. A hierarchy of stream segments can be ordered
in several ways. One of the commonest
ordering systems (Strahler ordering) desig-
nates the fingertip tributaries as order 1
channels. Order 2 channels are formed by
the junction of two first-order channels;
order 3 channels by the junction of two
second-order channels; and so on. The
watershed in (c) is therefore a third-order
unit. [From L. B. Leopold *et al., Fluvial
Processes in Geomorphology* (Freeman,
San Francisco, 1964), p. 139, Fig 5-4.
Copyright 1964.]

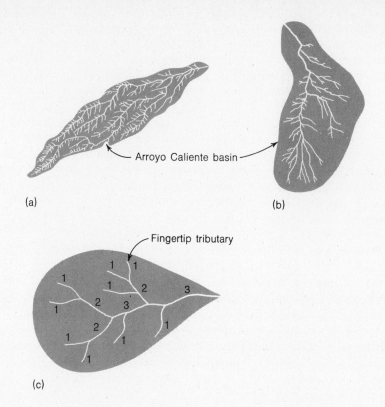

Watersheds also have other advantages as a basis for regional divisions.
Soil-related changes in vegetation reflect location within the watershed. Also,
the physical features of a basin directly affect the hydrologic characteristics of
the streams draining it. A rainstorm falling on a long, narrow basin is likely to
produce a lower peak in the water level of the stream draining it than a similar
storm falling on a rather broad, almost circular basin. Watersheds therefore
form useful ecosystem units for agencies interested in flood control, navigation, hydroelectric power production, or soil conservation.

The founding of the Tennessee Valley Authority in 1933 established a
trend in using river basins as planning units that has spread around the world.
Schemes for the São Francisco River in Brazil, the Snowy River in Southeast
Australia, or the lower Mekong Delta all involved a combination of planning
measures keyed to the use of water resources. Their adoption allowed the
competing demands for water—for irrigation, flood control, hydroelectric
power production, and navigation—to be dealt with through a single controlling authority. Watershed units are employed in heavily urban areas as
water pollution control problems worsen. For all these purposes, the watershed provides a convenient and natural spatial unit.

In Australia, the government's Division of Land Use Research has developed a *land unit* system to aid in the ecological mapping of that country.

TERMS USED IN THE STUDY
OF SOILS

A-horizons are the upper layers of a soil, often rich in organic material and subject to leaching as moisture seeps downward.

B-horizons are the next layers of a soil below the A-horizon where some of the chemical elements (notably iron) leached from the top soil accumulate.

Brown earths are rich brown soils formed in the middle latitudes where the prevailing natural vegetation is deciduous woodland.

C-horizons are the lowest layer immediately below the B horizon. They are made up of decomposed rock from which soil has not yet begun to form.

Catenas are sequences of soils which vary with relief and drainage though normally derived from the same parent materials.

Chernozems are rich, black soils formed in middle-latitudes where the prevailing natural vegetation is grassland.

Gley soils are waterlogged soils.

Horizons are the main layers or strata within a soil.

Laterites are red-colored soils formed in tropical regions. They are heavily leached and consist largely of aluminum and iron oxides.

Leaching is the removal of soluble chemical compounds from the upper layers of a soil by moisture seeping downward.

Pedalfers is a general term for soils formed in humid regions where some compounds (notably calciums) have been removed by leaching, leaving aluminum and iron as the main constituents.

Pedocals is a general term for soils formed in dry areas where there is little leaching and the soils remain rich in calcium carbonates.

Pedology is the scientific study of soils, including their characteristics, their origins, and their use.

Podzols are the soils formed under cool, moist conditions where the natural vegetation is coniferous forest or heath. They are poor and very heavily leached where the the soils are sandy.

(See Figure 3-16.) Each land unit describes a local environment that arises from the variation in four elements (climate, geology, soil, and vegetation) and the interactions between them. Land units are mapped from aerial photographs at scales of 1:10,000 to 1:25,000, and combined into larger units called *land systems* which are shown on maps with a scale of 1:1,000,000.

Recognition of land units usually proceeds in the way outlined in Figure 3-16. An initial division is based on the gross geology and land forms. This is broken down into finer terrain types. At a still finer level, the position on a slope is important in determining the type of drainage, the soils, and the vegetation. (See the above discussion of soils). However, vegetation, and to some extent soils, also reflect the influence of climate conditions. An estimate of the capability of each land unit for agricultural production is made, and units are combined into land systems.

The earth we have studied in the last two sections is an ecosystem. That is, it contains an intricate and delicate network of cycles and feedbacks that have both nonliving elements (the atmosphere, hydrosphere, and lithosphere of Figure 3-1) and living elements. In this third section, we turn to the question of the productivity of those cycles. How far do they provide the basic foodstuffs on which the human population is dependent?

3-3
Ecosystem Productivity

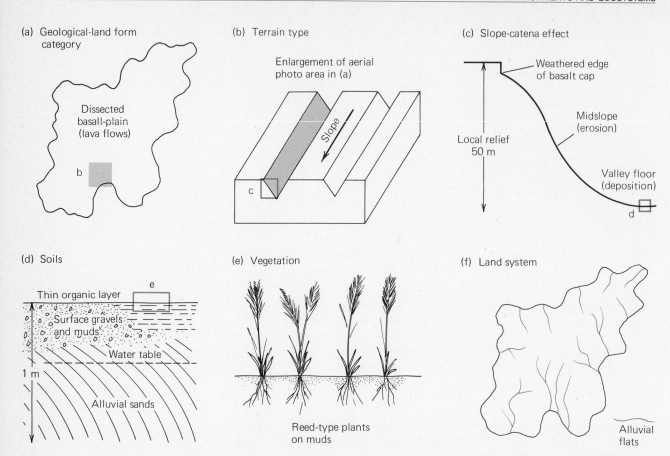

(a) Geological-land form category

Dissected basalt-plain (lava flows)

b

(b) Terrain type

Enlargement of aerial photo area in (a)

Slope

c

(c) Slope-catena effect

Weathered edge of basalt cap

Local relief 50 m

Midslope (erosion)

Valley floor (deposition)

d

(d) Soils

Thin organic layer

Surface gravels and muds

Water table

1 m

Alluvial sands

e

(e) Vegetation

Reed-type plants on muds

(f) Land system

Alluvial flats

Figure 3-16.   Ecological land units.
Assessment of land potential in Australia is often carried out in terms of land units. Environmental attributes are measured from aerial photographs and at sample sites on the ground to show individual *land units* to be recognized. These combine distinctive vegetation, soil, slope, and terrain conditions within a basic geologic-landform category. Land units are combined to give more generalized *land systems* such as the "alluvial flats land system" shown in (f). Maps of Australia at the scale of 1:1,000,000 showing land systems with their resource appraisal are produced by the Division of Land Use Research of CSIRO, Canberra.

Global Productivity   Forecasters at Resources for the Future (RFF) predict that, given the present levels of solar energy, at the present the most food that could be produced by photosynthesis is about 10^{11} tons per year.

Let us see how the RFF estimate was produced. The forecasters reasoned as follows: The prime source of energy on our planet is the sun, which radiates electromagnetic energy waves and high-speed particles into space. Since this constant emission represents almost all the energy available to the earth (except for a small proportion from the decay of radioactive minerals), it can be used to estimate the total amount of energy available to human beings. Green plants store solar energy through photosynthesis, and thus we can estimate the theoretical totals of dry organic matter (i.e., plants minus moisture) that the earth could produce. But dry organic matter is not always edible. Even in the case of croplands, well below half of the gross product may be edible. As for grazing lands, an energy-conversion factor of 12 to 1 must be used to convert the energy consumed by stock (i.e. the grain) to the food value of the stock to humans (i.e. animal products). Thus, the original figures must be revised downward to a maximum of around 10^9 tons per year. This gives a rough order of magnitude to overall global productivity.

Table 3-3 provides a breakdown of this total productivity between different global environments. It separates out the production of organic matter (at the trophic level T_1 in terms of Figure 3-14) from the matter edible to humans (at level T_4). Note that more organic matter is created by photosynthesis in the tropical rain forests than in any other environment on earth. Forest regions as a whole account for 40 percent of the total plant productivity of the globe, and the oceans for another 20 percent.

However, the figures in the table are only estimates, and they refer only to the production of organic matter on the first trophic level. Most of the organic products of forests are forms of wood, and, with our present technology, very little of these products can be converted to an edible form. As for the organic largesse of the oceans, it is of little benefit to man because we lack

Table 3-3 The productivity of global environments[a]

| | | Dry organic matter | |
		Maximal production per year	Maximal *edible* production per year
Environment	Area		
FOREST			
Tropical rain forest	3.9	30.2	6.1
Temperate deciduous forest	1.0	2.4	1.0
Temperate coniferous forest	2.9	12.7	3.1
Taiga	0.8	1.5	0.2
Total	8.6	46.8	10.4
GRASSLAND			
Humid grasslands	2.9	11.2	10.2
Arid grasslands	4.3	7.5	5.1
Total	7.2	18.7	15.3
CULTIVATED LAND	2.0	7.8	71.1
OTHER LAND			
Wetlands and swamp	0.8	3.0	—
Tundra	1.8	0.9	—
Hot desert	4.3	0.6	—
Cold desert	3.5	—	—
Total	10.4	4.5	
OCEAN AND LAKES			
Deep sea	65.5	19.9	—
Shelf, lagoon	5.1	2.7	2.5
Fresh water	0.8	0.3	—
Total	71.4	22.9	2.5

[a]Theoretical estimates of maximal production by photosynthesis in different types of environments. All figures in the table are percentages of the world total. Because of rounding, the total for each column may not equal 100%.

SOURCE: Data from R. U. Ayres, *Science Journal* **3**, No. 10 (1967), p. 102.

the technology to harvest it properly. The long food chains in the sea, the poor food-conversion ratios at each point, and the present wasteful fishing methods mean that we actually derive little food from the oceans. In practice, about 70 percent of the food we eat comes from cultivated land, and the world outlook for food production in the immediate future is going to depend essentially on improving the yield of land already under cultivation. The likely contribution of new virgin lands is actually quite marginal. In the longer term (A.D. 2000 and beyond), the vital areas for increasing man's food supply are the tropical forests and ocean shelves. We return to the issue of food production in the discussion of population in Chapter 7.

Interregional Analogs   Interregional analogs use environmental information from one region in predicting the potential use of another. For example, the American Institute of Crop Ecology has established crop analogs to pinpoint climatically similar areas in the United States and the Soviet Union. These climatic analogs have two functions. First, the climatic requirements of a particular U.S. crop variety can be established, and foreign areas can be screened to determine zones where it might grow. Second, the climatic needs of foreign crops can be identified and used to determine suitable growing ranges in the United States. Either way, the analog method identifies areas where new crops can be planted with some hope of success. Of course, the fact that areas are climatically analogous does not imply that they are the same in other ways. The soil in one area may be unsuitable for crops that would grow in the other, or the land may be more valuable for other purposes.

Like land units, interregional analogs do provide a useful first approximation of potential crop areas. As the pressures on the global environment increase, so the need to map and match the differences from one place on that globe to another increase. In the next chapter we move on to look at these variations and ask what causes them.

Summary

1. The fundamental division of the earth's environment is into the biotic environment (the biosphere) and the nonliving or abiotic environment. The latter is divided into the atmosphere, the hydrosphere, and the lithosphere.

2. Identification of the properties of the abiotic physical environments which are important to human life shows that physiological constraints such as oxygen level and thermal conditions are basic. Man is also directly connected to his physical environment through a series of nutrient cycles.

3. The atmosphere, hydrosphere, lithosphere, and biosphere are closely related through a network of links and feedbacks to form ecosystems. A very important part of ecosystem structure is the process of conversion of solar energy into living matter through the carbon cycle. In this, chains of living organisms circulate carbon from the atmosphere through various chemical forms, allowing life to be maintained. The fact that all animals receive their food directly from plants, or indirectly from animals that feed on plants, means that all are part of food chains. Various levels in food chains can be thought of as steps on a pyramid. Each such step is a trophic level.

4. Ecosystems contain feedbacks, both positive and negative. Positive feedback increases change in the ecosystem while negative feedback suppresses

change. The latter type acts as a stabilizer in an ecosystem, while the former breeds unstable conditions.

5. Ecosystem appraisal leads to estimation of the earth's environment for food crop production. Terrain types plus climate, soil, and vegetation information are used to produce ecological types. Site study of each type reveals their agricultural and settlement potential. Such studies can be used to develop inter-regional analogs, useful in assessing potential productivity between different regions around the globe.

Reflections

1. Consider the reactions of human beings to variations in temperature indicated in Figure 3-2. What do you think would be the limits of human settlement on your own continent if people were unable to construct artificially heated shelters? For how many months of the year would your own college town be habitable?

2. What do you understand by the term *ecosystem*? Using the diagram in Figure 3-11 as a guide, trace the main links in one other typical ecosystem.

3. Do you think that the term ecosystem should be used only for natural plant and animal communities? List the (a) advantages and (b) dangers of viewing human communities as ecosystems. What would be the closest parallel to phytoplankton in a collegiate ecosystem?

4. If you had to divide a local agricultural area into environmental zones, what criteria would you use? How far would terrain differences provide a main key to contrasts in land use and farm productivity?

5. Review your understanding of the following concepts:

biosphere	positive and negative
comfort zones	feedbacks
food chains	watershed hierarchies
trophic levels	climatic analogs
carbon cycles	

One Step Further . . .

The recent widespread interest in ecology has led to the publication of a large number of excellent brief introductions to the field. See, for example,

Clapham, W. B., Jr., *Natural Ecosystems* (Macmillan, New York, 1973), Chaps. 1 and 2, and

Chute, R. M., Ed., *Environmental Insight* (Harper & Row, New York, 1971), Part 2.

Geographic perspectives on ecological systems are provided by Simmons, I. G. *The Ecology of Natural Resources*, 2nd ed. (Arnold, London, 1980).

while a more advanced approach to ecology, stressing the quantitive aspect and its direct relevance to people, is given in

Watt, K. E. F., *Ecology and Resource Mnaagement: A Quantitative Approach* (McGraw-Hill, New York, 1968), Chaps. 4 and 5.

More detailed climatic and environmental analyses of all the main ecological zones are available in most standard physical geographies. A good reference is

Trewartha, G. T., *An Introduction to Climate* (McGraw-Hill, New York, 4th ed., 1968).

Current research is regularly reported in the leading geographic journals. You might also like to look through some of the increasing number of serials devoted to ecological topics, such as Ecology (quarterly) or the more popular Your Environment (monthly).

Behold, a sower went forth to sow; and when he sowed, some seeds fell by the way side . . . some fell among stony places, where they had not much earth . . . but others fell into good ground, and brought forth fruit, some an hundredfold, some sixtyfold, some thirtyfold.

The Gospel According to
St. Matthew, XIII, 3-8

chapter 4

The Global Environment

As we move outward into deep space, the intimate world of the beach and our small-scale ecosystems disappears. Viewed from this remote distance, the earth is simply a pale blue planet, much of it wreathed in clouds, with the outlines of tawny continents dimly visible. But even though the detail is invisible, relations between people and the environment are still observable. The scale of observable relations is vastly different though; our range of vision now includes the whole of man's 4.3 billions on the earth's 510 million square kilometers of surface. In place of individual sunbathers on the beach, we now have great clusters and nebulae of population set against a continental background. In place of wet sand and dry soil, we have an environmental range that runs from the Amazonian forest to Arctic ice caps.

In this chapter we shall look at the earth from a global perspective and assess some of its qualities as an environment for humanity. We shall follow the mapmakers of the past in their search to fill in the missing details of the global picture. In the millions of years in which people have been present on the earth, they have moved far beyond the beach, spreading over continent and island. But, like the sower's seed in Christ's parable, in some areas they have found the living hard and the land unfriendly and their stay has been brief; in others there has been a steady growth of population to massive levels.

How do the different parts of the earth vary, like the sower's field, from good to bad as a home for people—and why do these variations occur? If we can answer these questions, then we shall be able, later in the book, to go on to see what part they play in shaping the distribution and activity of the human population. For much has been written about overpopulation and the ability of the human race to feed itself in the centuries ahead. But before we can work out our planet's food production, we need first to know something about the way in which its different environments support all life. We shall turn first, therefore, to the living things at the bottom of the food chain, the plants, and see how their growth varying from location to location provides the key to unlock many of the doors through which we will later need to pass.

69

4-1
A World Productivity Map

If we look at the world around us, the visual evidence of strong environmental contrasts is so compelling that we should begin by reminding ourselves that, compared to other wanderers in the solar system, the earth is a uniform planet. It is almost spherical, with a radius of 6365 km (3955 mi) that varies by less than 0.02 percent. In height, no two places on its surface vary by more than 20 km (about 12 mi), or the length of Manhattan Island in their vertical distance.

Despite this uniformity compared to harsher planets, the earth presents considerable environmental variety in terms of man's specialized needs. In this section we look at two critical questions: How shall we measure the environment, and what kind of pattern do we find in it?

Measures of Environmental Contrast The geographer's approach to measuring environmental diversity will be somewhat different from that of a "pure" earth scientist such as a geophysicist. Like our questions about the beach in Chapter 1, our questions here relate not to the abstract physical properties of the environment, but to its interpretation in human terms. It is a simple matter to show, on a map, the differences between the tropical equatorial zone and the polar zones. (See Figure 4-1.) We can, for example, show the constancy of warmth in one, and the constancy of cold in the other, in terms of the amount of *permafrost*, subsoil that remains permanently frozen all year round. Turning the pages in the environmental section of an atlas will show you dozens of ways of indicating evident contrasts in heat and cold, wetness and dryness, constant and fluctuating climatic conditions, and the like.

Many geographers have wrestled with the problem of devising a single measure of environmental differences. Some have concentrated on the lithosphere and developed terrain-related schemes; more commonly, climate or climate-plus-vegetation has been used as an indicator of environmental variations. Vegetation has attracted special interest, since it may be seen as (a) a biotic response to variations in the abiotic environment and (b) an indicator of the potential productive uses of an area for human settlement.

One crude measure of variations in vegetation is the sheer mass of plants that grow in a particular area. Suppose we could bulldoze all the vegetation growing in sample kilometer-square patches throughout the globe and weigh the resulting piles of trunks, stems, and leaves. The pile that was once a tropical rain forest would weigh many thousands of tons; the pile gathered in the open woodland of the savannah would weigh between one-tenth and one-hundredth as much. The muskeg swamps of northern Canada would provide still smaller piles, and the Arctic ice caps and the most arid parts of the deserts would provide nothing.

Using data of this kind, plant physiologist Helmut Lieth was able to match up the productivity of natural vegetation with prevailing climate. Note that the curves in Figure 4-2 represent *average* relationships of plant productivity with heat and moisture; the actual data show a scatter of points around the

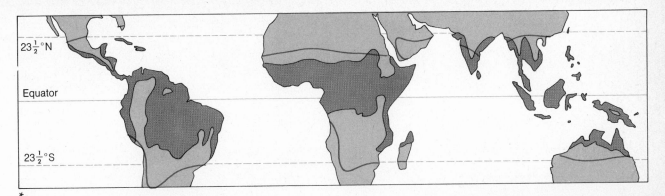

*

▨ Temperature above 21°C (70° F) all year
___ Average daily range of temperature greater than average annual range

Figure 4-1. Global contrasts in warmth.
The *tropics* are formally defined as the areas of the earth between the Tropic of Cancer
(latitude 23½°N) and the Tropic of Capricorn (latitude 23½°S). Thus the tropics lie in
those latitudes where the sun is vertically overhead at some time of the year. These
limits correspond very roughly with the two measures of warmth shown. At the other
extreme the *polar* zones lie north of the Arctic Circle (66½°N) and south of the Antarctic
Circle (66½°S). Two thermal measures of the Arctic polar zone, *permafrost* (subsoil that
remains frozen all the year round) and *pack ice* (large areas of ice covering the entire
sea surface) are shown. Note that these extend south of the Arctic Circle in Siberia
but lie well to the north over the sea areas north of Scandinavia. [Based on R. Common,
in G. H. Dury, Ed., *Essays in Geomorphology* (Heinemann, London, 1966), p. 68, Fig. 7.]

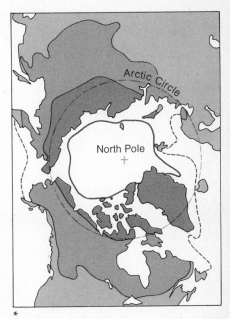

*

▨ Continuous permafrost
▨ Sporadic permafrost
___ Limits of year-round pack ice
- - - Average spring maximum of pack ice

line. Nonetheless, the story they show is a very clear one. Hot, wet areas
produce a very dense vegetation cover like that of the equatorial rain forest
shown in Figure 4-2(b). As we move toward drier or cooler conditions, so plant
growth slows down and the vegetation cover becomes sparser.

Of course, plant growth is not just a matter of average conditions. Some
geographers have found that the relationship between plant productivity and
climate is a rather complex one. For example, the Swedish geographer Sten
Paterson has determined that productivity increases with the length of the
growing season, the average temperature of the warmest month, the annual
precipitation, and the amount of solar radiation and decreases with the annual range of temperatures in an area. By combining values for these factors
into a single index, he was able to give numbers to different places on the
surface of the earth indicating their potential for plant growth. (For a more
detailed description of Paterson's work, see the discussion on page 74.) Consider an example: Portland, Maine, has an index of 3.1. By contrast, Miami,
Florida, has an index of 22.3; and Belem, at the mouth of the Amazon River
in Brazil, has one of 118.0. Many measures similar to Paterson's are available,

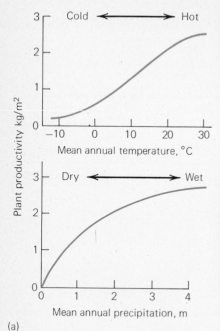

(a)

(b)

Figure 4-2. The productivity of natural vegetation.

(a) Using sample data from 53 sites around the world, Helmut Lieth was able to match up plant productivity with variations in their mean annual temperature and mean annual precipitation. Only the generalized curves are shown here. (b) A combination of high temperature and high rainfall gives the very high levels of plant growth typical of the equatorial rain forests of the world. That shown is from central Amazonia in Brazil. [(a) From H. Lieth, *Human Ecology,* Vol. 1 (1973), p. 304. (b) Photo by author.]

and differences between the world maps based on these various indexes are relatively minor.

A Six-Zone World Map   Figure 4-3 presents a world map whose contours are based on Paterson's index of productivity. It has six arbitrary divisions, and we label them, like examination grades, from A to F. F is a nonproductive zone with index values up to 0.25. Areas in this grade are readily identifiable as the very cold and very dry areas of the world: ice caps, the fringes of northern tundras, midlatitude deserts. Zone E, a band of very low productivity with index values between 0.26 and 1.00, borders on Zone F. It is a relatively narrow band around the deserts but covers significant sections of North America and Soviet Asia. A low-productivity band (D) with index values of 1.01 to 3.00 is confined mainly to cool, temperate climates having the most rain in the summer. This band includes central and eastern Europe and northern states of the Great Plains region in North America. The areas of the tropics in this band, too, are relatively small.

The fourth band (C), with index values of 3.01 to 10.00, is one of medium productivity and covers some of the most densely populated areas in the world. (Cf. the map in the atlas section.) It includes the eastern United States, Western Europe, and much of India and South China, as well as the savannah areas of Africa. A higher-productivity band (B), with index values of 10.01 to 50.00, is restricted to the tropics; most of this area is in South America. Finally, there is a zone (A) of very high productivity (index values greater than 50.00) in the equatorial belt: The Amazon basin in South America, the Congo basin in Africa, and the Indonesian archipelago in southern Asia lie in this zone.

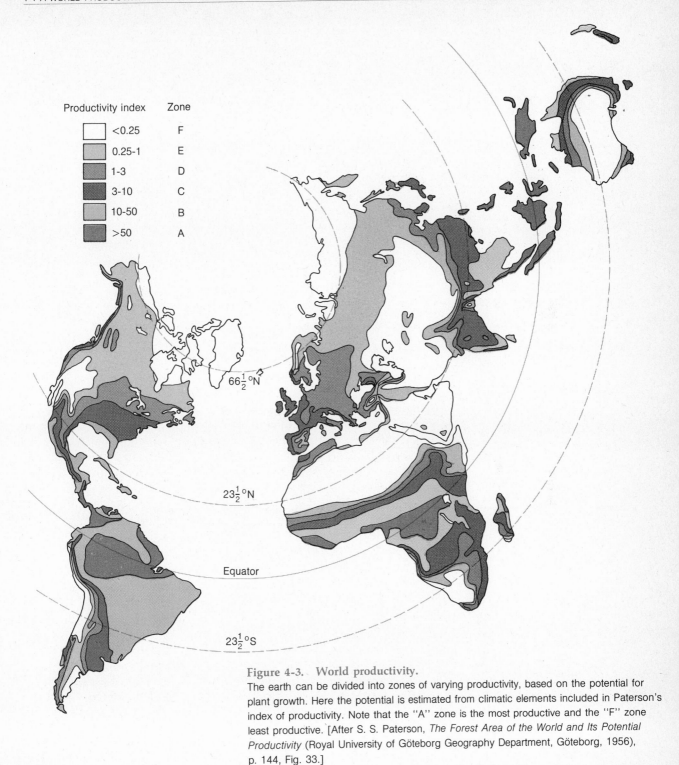

Productivity index   Zone

<blank>   <0.25   F

0.25-1   E

1-3   D

3-10   C

10-50   B

>50   A

$66\frac{1}{2}°$N

$23\frac{1}{2}°$N

Equator

$23\frac{1}{2}°$S

Figure 4-3. World productivity.
The earth can be divided into zones of varying productivity, based on the potential for plant growth. Here the potential is estimated from climatic elements included in Paterson's index of productivity. Note that the "A" zone is the most productive and the "F" zone least productive. [After S. S. Paterson, *The Forest Area of the World and Its Potential Productivity* (Royal University of Göteborg Geography Department, Göteborg, 1956), p. 144, Fig. 33.]

PATERSON'S INDEX OF PLANT PRODUCTIVITY

The index of plant productivity on which the map in Figure 4-3 is based is calculated using the following equation, the righthand side of which is a combination of basic climatic elements:

$$I = \frac{T_m PGS}{120(T_r)}$$

where I = an index of plant productivity,

T_m = the average temperature of the warmest month in degrees centigrade,

T_r = the annual range in average temperature between the coldest and warmest months in degrees centigrade,

P = the precipitation in centimeters,

G = the growing season in months, and

S = the amount of solar radiation expressed as a proportion of the radiation at the poles.

The growing season is calculated by counting the number of months in which the average monthly temperature reaches or exceeds the threshold needed for plants to grow (assumed to be at 3°C). Thus, Portland, Maine, with an average temperature in the warmest month of 19.7°C, a temperature range of 24.9°C, a rainfall of 106 cm, a growing season of 8 months, and a radiation value of 0.56, has an index of 3.13. For a full discussion see S. S. Paterson, *The Forest Area of the World and Its Potential Productivity* (Geography Department of the Royal University of Göteborg, Göteborg, Sweden, 1956). Note that this index is only one of many ways of estimating the potential plant growth in an area from climatic data. All give a similar picture of global productivity patterns but differ in various details.

The photographs in Figure 4-4 show how dramatic zonal differences in vegetation can be. However, it is important to remember that we are talking about zones of *potential* productivity; Paterson's index measures the plant life that the climate of a particular environment *could* support. The actual plant cover of an area will reflect other factors such as its vegetational history or the degree of human interference with the ecosystem. People may reduce plant growth (e.g., by pollution) or increase it (e.g., by irrigation). Furthermore, broad zonal maps such as Figure 4-3 do not reveal local areas whose productivity is affected by factors other than the ones included in an index.

4-2
Global Keys to the Map

What factors determine the productivity patterns in Paterson's maps? Like the population patterns on the beach, the *productivity contours* of Figure 4-3 can be unraveled by locational analysis. Geographers have found that the natural fertility of any location on the earth's surface is governed by four main factors: (1) its solar climate (corresponding to its latitude), (2) its location relative to the general atmospheric circulation of the earth, (3) its location relative to the continents, the oceans, or other major terrain features, and (4) local environmental factors. We shall begin by looking at the first two of these factors, which operate at the global level, before turning to the part played by the other, smaller-scale elements in the puzzle.

Latitudinal Factors Geographers in ancient Greece regarded the climatic environment of any part of the earth's surface as largely a result of its latitude.

(a)

(b)

The Earth was thought to slope (the Greek word *klima*, used in climatology, means slope) *away* from the sun north of the latitude of Greece and the Mediterranean Sea, making the northerly climate increasingly colder. The part of the planet to the south was thought to slope *toward* the sun, eventually becoming a torrid zone too hot for human life.

Figure 4-5(a) is a simplified picture of the Greek world view. The earth lies in a beam of solar radiation whose angle of incidence is directly related to latitude. As we move from the equator toward the poles, equivalent amounts of energy are spread over ever wider areas of the globe, producing an equatorial *torrid* zone, a midlatitude *temperate* zone, and a high-latitude *frigid* zone.

To understand this low-latitude to high-latitude gradient we must look at the energy relations between the earth and the sun. Each day the earth intercepts a massive beam of solar energy (estimated at 17×10^{13} kilowatts). This energy is emitted on different wavelengths; the peak intensity is the visible (daylight) part of the spectrum, but the beam also includes important shortwave (ultraviolet) and longer wave (infrared) emissions. If the earth were like the moon and had no atmosphere, the impact of these latitudinal variations in energy would be catastrophic for biotic life. Daytime temperatures near the equator would rise to some hundreds of degrees centigrade, while in the winter night at the poles, temperatures would fall almost to the absolute zero of outer space. Actually, as we saw in Section 3-1, the most extreme shade temperatures ever recorded at the earth's surface (in 1933 and 1960) differ by

Figure 4-4 Barren environments. Extreme cold and extreme dryness lead to environments with no plant cover. Both the icefield and the desert sand dunes shown receive an F grade in Paterson's productivity rating. [Photograph (a) by De Wys Inc. and (b) courtesy of the U.S. Department of the Interior, National Park Service Photo, George A. Grant.]

North Pole

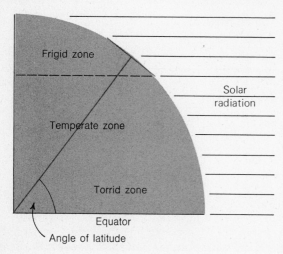

(a) Greek model with latitude effects

North Pole

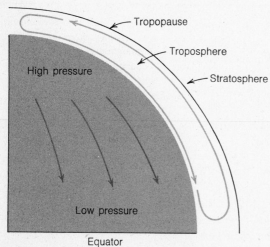

(b) Single-cell atmospheric model with no rotation

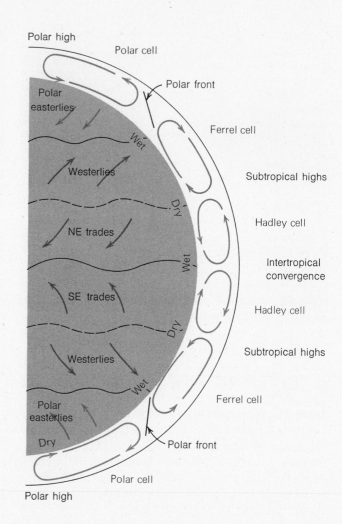

(c) Triple-cell (Hadley-Ferrel model) with rotation

Figure 4-5. General circulation of the earth's atmosphere.
Here we see three stages in the historical evolution of concepts of atmospheric circulation and climatic zones. The Hadley-Ferrel model (c) names the three main wind belts of the world—the *polar easterlies*, the *westerlies* of the middle latitudes, and the *trades* of the tropical zone. (The word "trades" comes from a Latin word indicating that the winds blow in a constant direction, and has nothing to do with trade in the sense of commerce.) Two important lines of air convergence are also shown. The *polar front* marks the junction of the polar easterlies and the westerlies in both the northern and southern hemispheres. Storms marked by low pressure move west along this front, playing a major part in the weather of the United States and Western Europe. The *intertropical convergence* is a zone of low atmospheric pressure separating the two trade wind systems. It lies more or less along the equator but moves slightly north and south with the seasons. Areas of air divergence with high pressures occur at both poles (the *polar highs*) and again at about 30°N and 30°S (the *subtropical highs*).

less than 150°C (270°F), and the average temperatures of the hottest and coldest locations on the earth vary by less than 90°C (160°F). Moreover, these are extremes. The average contrasts between any pair of locations are much more subdued.

Clearly, latitudinal variations in temperature are dampened and modified in some way. To explain this softening of thermal contrasts on the earth's surface, we must bring a thin coating of air into our model of the earth's relationship with the sun.

Earth's Atmospheric Filter   The planet earth is surrounded by a thin but critically important shell of gases—its *atmosphere*. This shell is held to the earth's surface by gravitational attraction and, as we should expect, is densest at the bottom, thinning rapidly as we move upward. The zone of greatest importance to man is the *troposphere* [see Figure 4-5(b)], which gets its name from the Greek phrase for a "turbulent realm." The upper boundary of the troposphere, the *tropopause*, varies seasonally in height but on the average is 9 km (5½ mi) above the poles and 17 km (10½ mi) above the equator. Above the tropopause lie the stratified layers of the *stratosphere*.

The troposphere is so thin, relative to the size of the earth, that it would be barely detectable on the regular classroom globe. Why then is it so important in any analysis of our planet as a home for man? There are four overriding reasons: (a) It contains the invisible and odorless gas, oxygen (about 20 percent of its volume), which is essential to human life. Climbers above about 6 km (about 20,000 ft) find that they need additional oxygen as the concentration of this element in the air diminishes. High-flying aircraft (and of course spacecraft) must carry with them their own oxygenated atmospheres. (b) Carbon dioxide, which is critical to plant life, is also present in the atmosphere, although in minute quantities (less than half of 1 percent). The role of carbon dioxide in plant growth and the food chains dependent

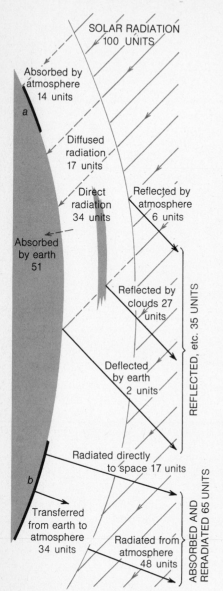

Figure 4–6. The earth's solar radiation budget.

The diagram shows how 100 units of solar radiation are affected by the thin layer of atmosphere surrounding the planet. (For demonstration purposes, the height of the atmosphere in relation to the earth's curvature is exaggerated.)

on it was discussed in Section 3-2. (c) Water vapor is drawn up from ocean and sea surfaces into the troposphere and is circulated and redistributed over the earth's surface as precipitation. Like oxygen and carbon dioxide, this water vapor is essential to the life and growth of organisms. (d) The gases in the troposphere and the layers above it act as a filter and a blanket. Harmful short-wave radiation from the sun is absorbed and reflected, while long-wave radiation from the earth itself is retained.

A detailed account of what happens to solar energy when it strikes the earth's atmospheric shell is given in Figure 4-6. A small amount (6 percent) is reflected by the atmosphere and a larger amount (27 percent) by the world's cloud cover. The remaining two-thirds is absorbed by the earth and its atmosphere, and reradiated. If you follow the arrows around the diagram you will find that the amount of energy received (100 units) is exactly balanced by the amount of energy reflected and reradiated. This *global energy balance* is the engine that powers not only the circulation of air and water (ocean currents), but the food chains of which humanity itself is a part.

The variation between the torrid, temperate, and frigid zones recognized by the Greeks (and shown in Figure 4-5) is also related to this cycle of energy. Latitudinal variations in temperature indicate the average angle of incidence of the sun's rays, which is at a maximum in the tropics but decreases toward the North and South Poles. The energy beam from the sun must pass obliquely through the earth's atmosphere and falls on a wider surface area. Compare areas *a* and *b* in Figure 4-6 to see the effect of these combined factors on the amount of energy received in the second area.

The General Circulation of the Atmosphere Like any gas or fluid, the atmosphere responds to temperature changes—increasing in density when it is cooled, decreasing in density when it is warmed. As we might expect, therefore, the warmer (and lighter) air near the equator rises and flows poleward, to be replaced by cold (and dense) air near the poles moving equatorward along the surface. These two flows together make up a *convective circuit*, illustrated in Figure 4-5(b). The unequal heating of the air by the sun sets up compensating currents that act like escalators, redistributing heat across various latitudes, and tempering the simple Greek pattern of hot and cold zones.

The Hadley-Ferrel model As early as 1686, English astronomers had outlined a circulational model to explain why the lower latitudes, despite receiving more heat, did not become progressively hotter. By incorporating the effects of the earth's rotation into the model, it was possible to give a reasonable explanation of the trade winds, those steady air currents blowing toward the equator from the northeast in the northern hemisphere and from the southeast in the southern hemisphere.

Figure 4-7. Earth's cloud cover.
Major climatic zones are suggested in this picture of the entire earth's cloud cover on a single day (February 13, 1965). It is made up from 450 photographs received from the weather satellite Tiros 9 on its pole-to-pole orbit of the earth. Note the relative absence of cloud cover from the desert areas of North Africa and Arabia (a), central Australia (b), and the middle section of the west coast of South America (c). [NASA photograph.]

These early models were far from satisfactory, however. If we look at the motions of the earth's atmosphere as revealed by telltale cloud patterns in satellite photos (Figure 4-7), we can see two important features left out. First, there are extensive cloudless areas of very dry descending air at latitudes of about 30°N and 30°S (above the Sahara Desert, for example.) Second, there are belts of strong westerly winds in the middle latitudes of both hemispheres. The westerly drift of clouds across the North Atlantic illustrates this circulatory force.

The missing elements in the explanation of atmospheric movements were provided by English scientist George Hadley in 1735 and American meteorologist William Ferrel in the following century. Their model [Figure 4-5(c)] replaced the single pole-to-equator convective circuit by a series of *three* convective circuits in each hemisphere. The Hadley-Ferrel model—although much modified in detail—still forms the basis of modern concepts of atmospheric circulation. It not only accounts for the puzzling feature of the westerlies (now seen as caused by the westward rotational deflection of poleward-moving air), but throws light on global patterns of precipitation. Each of the main belts of winds is defined in the caption of Figure 4-5.

Global Precipitation Patterns Though a number of factors affect precipitation, its general cause is the cooling of air that contains water vapor. Since water vapor is present throughout the troposphere, it is the location of cooling and drying processes at the global level which explain the wet and dry areas of the earth. Such processes are indicated in the Hadley-Ferrel model by

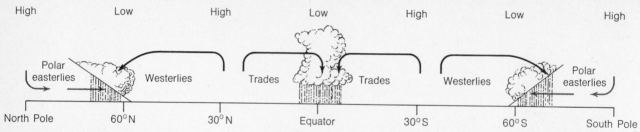

Figure 4-8. Circulation and precipitation patterns.

A simplified cross section of the earth's atmosphere shows the general pattern of circulation. Note the association of the three zones of upward air motion (low pressures) with belts of rain. The four high-pressure areas are marked by descending air motion and very low precipitation. The sharp discontinuities at latitudes of 60°N and S mark the polar fronts.

areas of upward and downward motion in the general atmosphere. As Figure 4-8 shows, there are three main zones of upward motion: (a) an *equatorial convergence* zone, where the two trade-wind systems meet, and (b) two *polar front* zones at around 60°N and 60°S, where warm tropical westerly air overrides the colder air flowing outward from the two poles. A map of world precipitation (Figure 4-9) confirms that these are precisely the locations where high values are observed.

Using the reverse argument, we should expect zones of low precipitation in areas of downward-moving air. As Figure 4-8 shows, such zones coincide with

Figure 4-9. The world's wettest and driest areas.

The world's wettest areas straddle the equator and the windward margins of continents. The location of the driest areas is related to tropical high-pressure cells. [After *The Times Atlas* (The Times, London, 1958), Vol. 1, Plate 3.]

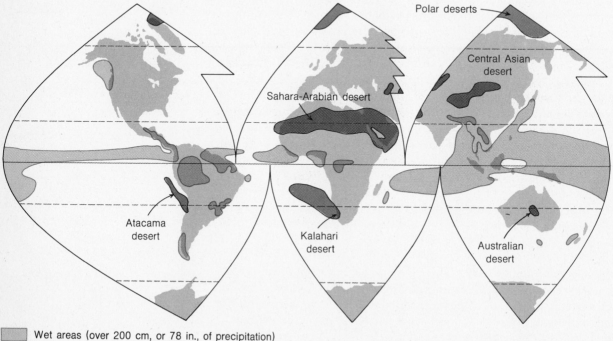

Wet areas (over 200 cm, or 78 in., of precipitation)
Dry areas (below 10 cm, or 3.9 in., of precipitation)

the two poles and with two zones of divergence at 30°N and 30°S; again, we find that these are zones of low precipitation on the world map. Note on Figure 4-9 the location of the world's main desert zones in relation to the Hadley-Ferrel model.

We began our locational analysis of global productivity patterns in Figure 4-3 by assuming a simple, airless planet. We have seen how adding a thin but critical layer of atmosphere to the harsh Greek model of torrid, temperate, and frigid zones produced a more realistic model that went a long way toward explaining the overall pattern of warmth and wetness. We now add another thin but critical layer—the oceans—to our world picture.

4-3
Continental Keys to the Map

Oceans and the World Water Cycle   Water vapor was noted earlier as one of the critical components of the earth's atmosphere. Water as a gas or small droplets in the air forms only 1 part in 100,000 of the planet's overall water resources. Over 97 percent is concentrated in the great water sheets of the world's oceans. Not only do the oceans cover over 70 percent of the earth's surface, but their lowest point (Marianas Trench in the western Pacific, 11.03 km, or 6.85 mi deep) greatly exceeds the highest point on the land surface (Mount Everest in the Asian Himalayas, 8.85 km, or 5.5 mi high). Perhaps the simplest way to remember the oceans' size is to recall that if the earth had a smooth surface it would be covered everywhere, to a depth of about 3 km (1.86 mi), by water.

Water vapor is carried and distributed over the earth's sea and land surface by the global wind belts described in the Hadley-Ferrel model of the atmosphere. Each year an estimated 33,600 km³ (8,100 mi³) of water evaporates from the ocean surface. Figure 4-10 shows what happens to that water. About 89 percent of it is returned directly to the ocean by precipitation. The remaining 11 percent moves over the earth's land surfaces before precipitating out. Precipitation falling on land may be either returned directly to the

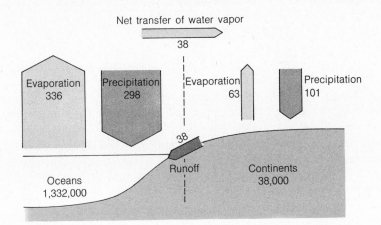

Net transfer of water vapor

38

Evaporation
336

Precipitation
298

Evaporation
63

Precipitation
101

38

Runoff

Oceans
1,332,000

Continents
38,000

Figure 4-10.   The world water balance.
The diagram shows major water flows between the oceans and the continents. The figures indicate flows in 100 ton³ of water. [After A. N. Strahler, *The Earth Sciences,* 2nd ed., (Harper & Row, New York, 1971), p. 586, Fig. 33-2.]

atmosphere by evaporation and transpiration, or temporarily stored (in lakes, ice caps, in the upper soil layers, or more deeply as groundwater). Eventually all the water is returned to the oceans by flowing streams and melting glaciers. Hence, the balance between the moisture leaving the oceans as water vapor and returning as a liquid is maintained.

This global circulation of water is termed the *global hydrologic cycle*. We saw earlier in Figure 3-12 the earth's main stores of water, as well as the direction of transfers. Outside the world's oceans most of the water is locked up in glaciers and ice caps. Only 0.001 percent of the total is in the atmosphere. To understand the varying pattern of precipitation around the world, we must understand something of how that small atmospheric fraction is shunted from one location to another by the general circulation of the atmosphere.

Continents and Islands   Temperatures vary more on land than they do over the sea. But less than a third of the earth's surface is land, and this land is distributed in an asymmetric way in relation to latitude. Most of it is in the northern hemisphere. The climatic implications of the division of the planet into areas of land and water are dramatic. The thermal conductivity of the surface layers of land and sea is quite different. Land heats and cools much more quickly than the sea; bodies of water act as heat stores whose temperatures fluctuate much less than those of adjacent bodies of land.

Thus, there are considerable global variations in the seasonal range of temperatures. The *average* annual range is lowest at the equator and increases with latitude, from about 3°C (5°F) to 60°C (110°F) at the South Pole. The smallest ranges are in oceanic islands near the equator; the largest ranges are in midcontinental locations in high latitudes. On Saipan in the Mariana Islands of the western Pacific, the highest and lowest temperatures ever recorded, the *extreme* range, differ by only 12°C (22°F). By contrast, Olekminsk in central Siberia has an extreme range ten times longer, from −60°C (−76°F) to 45°C (113°F). The conventional distinction between continents and smaller land masses is an arbitrary one. All can be regarded as "islands" of different sizes, and within each an increased range is detectable, though it weakens as the area becomes smaller.

A broad distinction can be drawn between the climates of continental interiors and areas near the sea. *Continental* climates are characterized by great ranges of temperature (both between day and night and between winter and summer), low humidities, and very variable precipitation. This variability in precipitation shows as a strong seasonal contrast, but also in year-to-year irregularity. *Maritime* climates have the reverse characteristics: smaller temperature ranges, higher humidities, and more uniform precipitation. These contrasts are not symmetrical on a continent, but are related to the latitudinal location of the land area with respect to the circulation pattern of both the atmosphere and the oceans. We can combine the effects of latitude and continentality and superimpose them on an idealized continent of low and uniform elevation. The resulting distribution of precipitation on the conti-

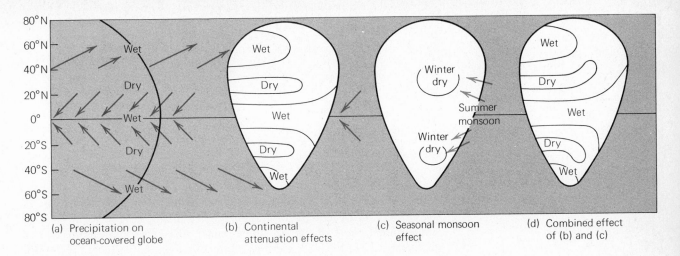

(a) Precipitation on
    ocean-covered globe

(b) Continental
    attenuation effects

(c) Seasonal monsoon
    effect

(d) Combined effect
    of (b) and (c)

nent [Figure 4-11 (b)] should be interpreted in terms of the airflows shown in
Figure 4-8. The dry areas coincide with the subtropical high-pressure areas,
and the wet areas with the storms in the tropical zones. The westerly circulation of air brings a sequence of midlatitude storms over the western margins
of the continent. If you compare the distribution of wet and dry zones with
the direction of air movements worldwide, you will see why precipitation
decreases away from the continent's oceanic boundary.

Because of the great variations in air temperature and pressure over continents, we must add a further seasonal effect to our model of precipitation
patterns [Figure 4-11(c)]. The differential heating of the continental air
masses causes maritime air to be drawn inward in summer to replace the
warm, light, and rising air over the continents. In winter, the colder, heavier,
and descending air above the world's land masses flows outward toward the
seas. The impact of these flows on the world's largest land mass, where the
most striking seasonal reversals occur, is discussed in Chapter 6. (See Section
6-1 on monsoon India.)

Mountains and Plains   Within the overall climatic framework formed by
the effects of latitude and continentality, we can detect a hierarchy of
smaller-scale factors. The geographic scale of their impact is minor, but they
may play a decisive part in determining the use that can be made of an area.

We have seen that the earth is almost a perfect sphere and that variations
in its surface elevation are equal to only a small fraction of its radius. On a
classroom globe, these variations would appear as almost imperceptible
bumps and hollows. Table 4-1 shows the proportion of surface area in each
altitude zone on the earth. We shall ignore for the present the variations in
areas below sea level (although we shall return to them in Section 20-2 when
we consider territorial claims on the shallow waters of the continental
shelves). Land above sea level ranges up to 8.9 km (5.53 mi) in height, but over
two-thirds of it is at elevations below 1 km (.62 mi), and less than one-tenth is

Figure 4-11. Continentality.
The impact of a hypothetical continent on
global patterns of precipitation. Note how
the regular bands of wet and dry are
progressively distorted. The continent is
assumed to have low, uniform elevation.
Compare with the actual distribution of wet
and dry areas as shown in Figure 4-9.

Table 4-1 Distribution of the earth's surface area by height

CONTINENTS		
Over +4 km	(+2.5 mi)	0.5 percent
+2 to +4 km	(+1.2 to +2.5 mi)	3.3 percent
0 to +2 km	(0 to +1.2 mi)	25.3 percent
OCEANS		
−2 to 0 km	(−1.2 to 0 mi)	11.5 percent
−4 to −2 km	(−2.5 to −1.2 mi)	18.7 percent
−6 to −4 km	(−3.7 to 2.5 mi)	39.7 percent
Below −6 km	(−3.7 mi)	1.0 percent

above 2 km (1.24 mi). The city of Denver, the highest located major city in the United States, is at an elevation of 1.6 km, or almost 1 mile.

The direct effect of these relatively small differences in elevation on the characteristics of lowland and highland environments is striking. At an altitude of 8 km (roughly 5 mi), the density of the atmosphere is less than one-half its density at sea level. High elevations have a thinner shell of atmosphere above them and receive considerably more direct solar radiation than sea-level locations, but they lose much more heat by radiation from the ground surface. Within the lower layers of the atmosphere, temperature decreases with elevation at an average rate of about 6.4°C per kilometer. This rate of loss in temperature with height is termed the *thermal lapse rate*. It is generally the same throughout the troposphere.

The effects of elevation on temperature are twofold: As elevation increases, the average temperature of an area decreases, and the daily range in temperature increases. Both effects are due to clearer, more rarefied air, which allows more solar radiation to reach the ground surface (raising the midday temperature) and also permits a more rapid heat loss from the ground at night. The net effect is to reproduce the changes in temperature we may encounter as we move from one latitude to another over a short vertical distance. If we live on the equator and wish to see some snow, we can either travel 8000 km (5000 mi) poleward to find snowbanks at sea level or climb 4.5 km (15,000 ft) up!

The symmetry between changes in climate due to latitude and changes due to elevation is shown in Figure 4-12. Because snowlines change with the seasons, the upper limits of plant growth have been used as a substitute measure of climatic differences. Timberlines reach their greatest sea-level extent at a latitude of about 72° in the northern hemisphere and 56° in the southern hemisphere. They are at higher altitudes as one moves from these latitudes toward the tropics. A mean temperature of 10°C (50°F) for the warmest part of the year appears to be a prominent factor limiting forest growth, and this limit is met at varying elevations in different parts of the world.

We can spot the effect of altitude on plant cover within small areas as well as on the world scale. Figure 4-13 shows the variations in zones of vegetation on a small tropical island. Note that the effect of elevation is complicated by the direction in which moisture-bearing air is traveling and by variations in

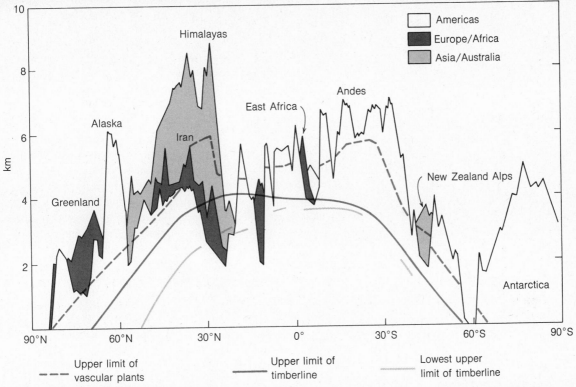

Figure 4-12. World timberline variations.
The altitude at which plants will grow in mountain areas changes at different latitudes, as also does the timberline. [After L. W. Swan, in W. H. Osburn and H. E. Wright, Jr., Eds., *Arctic and Alpine Environments* (Indiana University Press, Bloomington, Ind., 1968), p. 32, Fig. 1.]

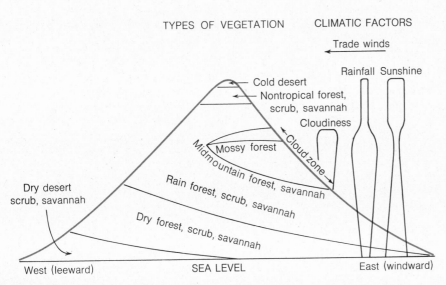

Figure 4-13. Timberline variations and ecological zoning.
This idealized cross section of a typical high oceanic island in the tropics shows the changing zones of vegetation caused by differences in the elevation of various parts of the island and by the location of the island with respect to the rain-bearing trade winds. Note the contrast between the height of vegetation zones on the wet eastern (windward) side and the drier western (leeward) side. *Savannah* is open forest with scattered trees mixed with scrub and grassland to give a parklike pattern. *Mossy forest* is a belt of damp forest in the cloudy zone on the windward side with mosses on the living trees and ground surface. The changing width of the columns at the right of the diagram indicates the relative amount of cloudiness, rainfall, and sunshine at each attitude. [From S. Haden-Guest, *et al.,* Eds., *World Geography of Forest Resources* (American Geographical Society, New York, 1956), p. 620, Fig. 58.]

cloud cover with elevation. The simple relationship between height and variations in vegetation is disturbed on the windward side of the island by a characteristic "cloud forest" belt. The leeward side of the island is much drier, and the forest belts start at different elevations. Similar rain-shadow effects, related to the prevailing moisture-laden winds from the Pacific, are visible in the forest levels in the southwestern United States (e.g., in the Sierra Nevada or White Mountains).

Minor Environmental Contrasts   Latitude, continentality, and altitude combine to account for the largest share of systematic environmental variation on the global and continental levels. On local levels, broad patterns are broken up by the effects of another interlocking set of variables: terrain, slope, drainage, surface geology, and soils. These are too small to show up on Paterson's world map (Figure 4-3), but we shall examine their combined effect when we look at variations within the major environmental zones discussed in the next section.

4-4
Major Environmental Regions

Geographers have long been intrigued with the possibility of reducing the spatial variety of the globe to a single, comprehensive scheme of regions. We look with some envy at the relative order the labeling of species brought to the previously jumbled world of botany and the periodic table brought to chemistry. Can any similar way be found to classify the mosaic of different environments and ecosystems geographers study?

Nine Basic Zones   Earlier in the chapter, we discussed the varying productivity of the global environment in terms of a six-point scale. This ranged, like school grades, from A to F. Areas in zone A were highly productive; areas in F had almost no potential. (See Figure 4-3.) We can now take this idea somewhat further by distinguishing three main divisions within the biosphere in terms of types—forested, intermediate, and barren.

The three maps in Figure 4-14 show the land areas of the globe divided into nine basic environmental zones, or *biomes,* ranging from the *polar zone* at high latitudes to the *equatorial zone* at low latitudes. Note that the boundaries shown are only approximate and give a broad-brush picture. The actual boundaries may be very irregular and much modified by human interference. Rather than review each zone's characteristics at length, we have summarized them in Table 4-2. You may find it useful to match up these zones with the productivity ratings given in the extreme righthand column of the table and already encountered in Section 4-1.

The table indicates each zone's share of the total land area, but this is not necessarily an indication of its importance to man. For example, the *Mediterranean zone* (only 1 percent of the earth's land surface) has played a part in human development out of all proportion to its small size; whereas the largest zone, the *savannah zone* (24 percent of the land surface), has played a minor role.

Each of the nine regions is designed to relate in one simple scheme as many different physical conditions as possible. Thus, the savannah zone has the environment it has because of a mixture of climatic, vegetational, hydrologic, and soil factors. It lies within about 30° of the equator. Its largest single area is a horseshoe-shaped belt in Africa, about one-half of that continent, but it also covers much of southern Asia. While temperatures are high around the year, there is a pronounced summer rainfall and a dry winter season. Annual rainfall totals range from 25 cm (10 in) to 200 cm (79 in) and vary moderately from year to year. In southern Asia, West Africa, and northern Australia, the rainfall is associated with disturbances in the monsoon flow south of the equatorial low-pressure trough. Late summer hurricanes add significantly to rainfall totals in various parts of this zone.

The vegetation of the savannah zone is highly differentiated. It has heavy forests near its boundary with the equatorial zone and sparse shrubs and grasses near its arid border. Savannah vegetation consists of an open patchwork of drought-resistant shrubs and trees with expanses of tall, coarse grasses. Variations on this theme include the thorn forest of East Africa and the dense, semideciduous jungle of Thailand and western Burma. Variations in the length and intensity of the rainy season relate both to the variety of vegetation and to soil and hydrologic conditions. Alternating precipitation leads to seasonal variations in soil characteristics and intense differences in the flow levels of rivers draining the savannah and monsoon areas.

Boundaries Between Zones   The zoning system outlined in Table 4-2 and mapped in Figure 4-14 is a useful capsule guide to the world environmental mosaic. Terms like "equatorial," "savannah," and "Mediterranean" serve as a way of summarizing the dominant characteristics of regional climates and soil-vegetation complexes. Unfortunately, there is not complete harmony on how many zones there should be or what criteria should be used to separate one zone from another.

We can see the reasons for this lack of harmony by looking more closely at another of the major zones, the *boreal* zone. Generalizing, we can say that the boreal zone is a climatically determined ecological unit covered (except in settled areas) by forests dominated by coniferous growth. The boundary of this zone is not clearly marked by any abrupt change in the environment. Coniferous forests are widely dispersed in regions outside the boreal zone, in the Mediterranean area and parts of Central America, for example, so a definition of the zone based on vegetation alone is difficult. The poleward limit is conventionally marked by the Arctic tree line. In practice, however, this "line" is really a belt where trees grow only in the most favorable sites; *muskeg* (waterlogged depressions filled with sphagnum moss) occupies the hollows, while *tundra* (dwarf shrubs, herbs, lichens, and mosses) occupy the more exposed ridges.

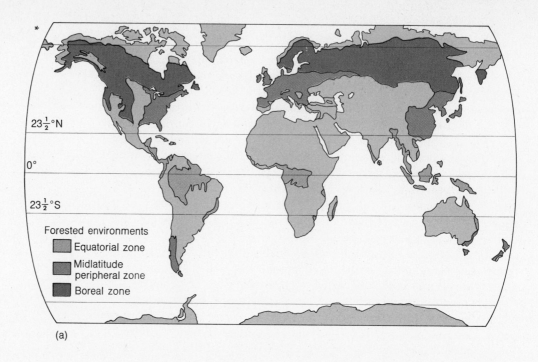

(a)

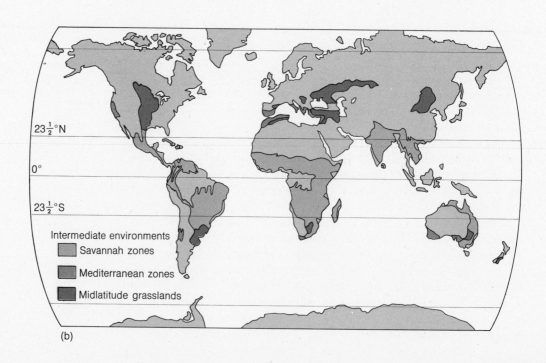

(b)

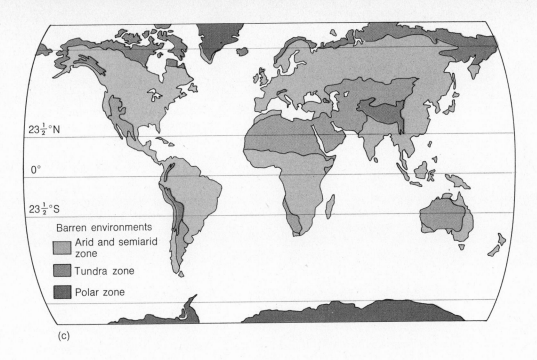

23½°N

0°

23½°S

Barren environments

Arid and semiarid zone

Tundra zone

Polar zone

(c)

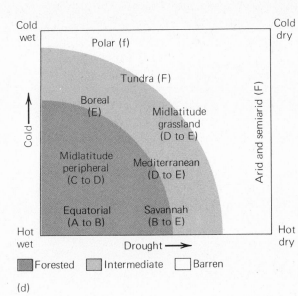

Cold wet

Polar (f)

Tundra (F)

Boreal (E)

Midlatitude grassland (D to E)

Cold

Midlatitude peripheral (C to D)

Mediterranean (D to E)

Equatorial (A to B)

Savannah (B to E)

Hot wet

Drought →

Cold dry

Arid and semiarid (F)

Hot dry

Forested   Intermediate   Barren

(d)

Figure 4-14.   Major land biomes.
(a) to (c) Biomes are distinguished by their distinctive climates and vegetation. (See Table 4-1 for details.) This scheme is one of many geographers have proposed in an attempt to break down the complex ecologic mosaic into simple patterns. Note that it is highly generalized; sharp local differences occur, notably in highland areas. (d) Generalized relationship of the biomes to climatic variation in cold and drought. The letters A through F refer to Paterson's productivity grades as given in Figure 4-3. [After M. Vahl and J. Humlum, *Acta Jutlandica* **21** (2-6) (1949), p. 28, Fig. 1.]

Table 4-2 Major land biomes

Type	Zone name	Percentage of land surfaces[a]	Main areas[b]	Dominant vegetation cover
Forested	Equatorial	8	Amazon Basin, S. America (51) Indonesia, S. Asian peninsulas (21) Congo Basin, Africa (19)	Natural broadleaved evergreen forest; wide variety of species; swamp forests on floodplains and coast
	Midlatitude peripheral	7	Europe (38) E. China (30) E. United States (26)	Broadleaved, deciduous, and mixed forests, merging with warm, temperate, evergreen forests on eastern periphery
	Boreal	14	Russia and Scandinavia (62) Canada, Alaska, N.W. United States (37)	Needleleaf forests; relatively uniform stands with small number of species (e.g., spruce, fir, pine, larch)
Intermediate	Savannah	24	African Tropics (48) S. America (21) S.E. Asia (20)	Ranges from open, tall-grass savannah to deciduous monsoon forest; gallery forest along stream systems
	Mediterranean	1	Lands around Mediterranean Sea (49) S. Australia (31)	Evergreen drought-resistant hardwoods and shrubs
	Midlatitude grasslands	9	C. Asia and E. Europe (42) C. North America (23) E. Australia (15)	Grasslands varying from tall-grass prairie to short-grass steppe with decreasing humidity
Barren	Arid and semiarid	21	C. Asia (42) Sahara and S. W. Asia (30) C. and W. Australia (10)	Widely dispersed, drought-resistant shrubs; salt flats, plantless sand, and rock deserts
	Tundra	5	N. Canada and Alaska (53) Russia and N. Scandinavia (42)	Low herbaceous plants, mosses, and lichens
	Polar	11	Antarctica (87) Arctic (13)	Ice caps; no plant life

[a]Computed on the basis of boundaries shown in Figure 4-14. They do not correspond exactly with the proportions given in Table 3-3, which were computed from boundaries drawn on a slightly different basis. Note that characteristics of all the zones are highly general; for examples of internal variations within zones, see the text.
[b]The figures in parentheses give the percentage of the total land surface of each zone in each area.

Main man-made changes in vegetation	Main precipitation characteristics	Main thermal characteristics	Productivity rating on Paterson's map[c]
Very variable ranges from extensive clearing and cultivation (e.g., in Java) to little impact (e.g., in Amazonia); low to very high population densities	High rainfall (over 100 cm, or 39 in) throughout year; heaviest at equinoxes	Uniformly high temperatures, little seasonal variations	A to B
Very extensive clearing and cultivation; medium-to-high population densities throughout	Moderate precipitation (75–100 cm, or 30–39 in) all year; heaviest in winter or autumn on western periphery, in summer maximum in eastern warm temperate area	Cool to warm temperate; seasonal range increases with continentality	C to D
Limited clearing on equatorward fringes; very low population densities	Light precipitation (25–50 cm, or 10–20 in) mainly in summer	Short, cool summers; very large annual temperature range	E
Burning, grazing, variable clearing, and cultivation; high population densities on flood plains in monsoon Asia, otherwise low	Variable rainfall (25–200 cm, or 10–79 in) with pronounced spring or summer maximum	Warm; small seasonal variations	B to E
Extensive clearing and cultivation, especially in lands around Mediterranean Sea; variable population density	Low to moderate rainfall (50–75 cm, or 20–30 in) with pronounced summer drought	Warm temperate; moderate annual range	D to E
Hunting and grazing; main settlement and cultivation in last 150 years; low population density	Low to moderate rainfall (30–60 cm, or 12–24 in), mainly in spring and summer; considerable year-to-year variation	Very strong seasonal variations; cold winters dominated by invasions of polar air	D to E
Little impact outside small irrigated areas	Very low rainfall (0–25 cm, or 10 in) with considerable year-to-year variation	Very high summer temperatures; seasonal variations range from moderate in tropics to very high in midlatitudes	F
Little impact	Low annual totals (10–40 cm, or 4–16 in) with late summer or autumn maximum; light winter snowfall	Severe cold; short, cool summers	F
No impact	Low annual totals; little detailed precipitation data	Extreme cold; no months with above-freezing average temperatures	F

[c]See Figure 4–3.

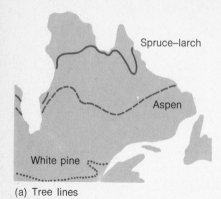

(a) Tree lines
* *

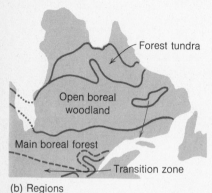

(b) Regions

Figure 4-15. Variations within major ecological zones.

Tree lines correspond roughly with subdivisions of the boreal zone in eastern Canada. [After F. K. Hare, *Geographical Review* **40** (1950), p. 617, Fig. 4. Reprinted with permission.]

In the early part of this century, geographers Alexander Supan and Vladimir Koppen equated this transitional belt with an isotherm of 10°C (50°F) for the mean daily temperature for the warmest month. Later geographers confirmed that the northern forest boundary follows essentially a thermal limit, though they found a growth threshold of 6°C (42°F). Similar constraints apply to the equatorward limits of the zone, at least in the humid sections. For example, the southern boundary roughly approximates the line along which there is a mean daily temperature of 6°C in six months of the year. It is here that the broad-leaved forests of the midlatitude zone begin. However, the boundary between the boreal zone and the midlatitude grasslands in Siberia and the Canadian prairie areas is controlled by the dryness of the environment rather than by its temperature.

Local Contrasts Within Zones   Even if geographers were in complete agreement on how to draw biome boundaries, another difficulty would still remain. The zones we have described give only a broad-brush picture of an immensely detailed mosaic of environments. Environmental variations occur on many scales, and these variations tend to break up and differentiate the major, subcontinental zones. For instance, within the boreal zone itself we can distinguish sharp contrasts. Canadian geographer Kenneth Hare recognizes three subzones within the boreal zone in the northern hemisphere. (See Figure 4-15.) The first subzone is formed by the close main boreal forest, where the crowns of the trees touch. This *closed-crown forest* occupies at least half the area of the zone, except in the driest sites. In the drier areas such as the Mackenzie Basin of Canada, the species of trees found change, grassy openings occur, and soil tends to be alkaline rather than acid. The second subzone is the *woodland* subzone, where lichen breaks up the closed-crown forest. The open, almost savannahlike woodlands are sometimes called *taiga*, though this word is used by Russian authorities to designate the whole boreal zone. The third subzone is the *forest-tundra* subzone, a mixture of tundra on the drier ridges and woodlands in the valleys. This area is a classic example of an *ecotone*, or transitional belt, where two major environmental zones—the tundra and the boreal—interpenetrate and blend into each other.

Another example of local contrasts within a major land biome is provided by Figure 4-16. This shows how one midlatitude grassland—the Great Plains—has a major break along the hundredth meridian.

The division of the global land surface into the nine major zones listed in Table 4-2 has three limitations. First, strong internal variations (related to elevation, geology, or groundwater) may occur within zones, and boundaries between zones may be fuzzy. Second, the boundaries of zones are slowly but continuously changing because of long-term alterations in the climate of the present postglacial period. Third, the zonal vegetation described in Table 4-2 describes only the undisturbed plant cover that is either known to exist or assumed to be able to grow should human intervention in the ecosystem

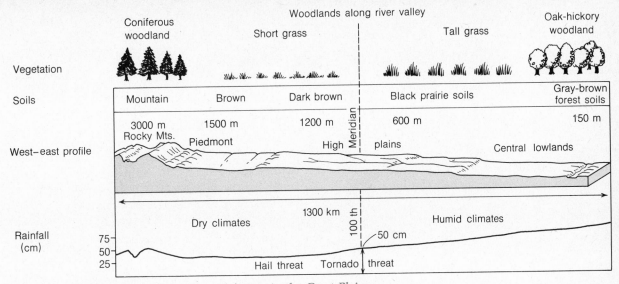

Figure 4-16. **Interlocking environmental factors in the Great Plains.**
This idealized cross profile of the Great Plains of North America shows related changes
in climatic, vegetational, and soil conditions. Note that as the moisture situation changes
from humid to dry, notable changes occur in soils and vegetation. Soils change from
black in the humid east to dark brown and brown on the dry western margins of the
Plains. Very critical changes are caused by a layer of salts in the subsoil at the lowest
limit to which seasonal rainfall penetrates. This layer is as deep as 1.2 m (4 ft) in the
east but rises to only 0.2 m (8 in.) in the west. An important dividing line occurs between
the 98th and 100th meridians of west longitude, where the layer of salts rises above the
0.8 m (2.6 ft) mark. Along this line black soils give way to brown, and tall prairie grass
gives way to short bunch grass. As a result of grazing and wheat farming, much of the
original vegetation in the west has disappeared. But the dividing line created by the
changing depth of the salt layer remains, separating the productive ex-prairies from
the western areas where cultivation is still risky without specialized irrigation or dry-
farming techniques. [After W. R. Mead and E. H. Brown, *The United States and Canada*
(Hutchinson, London, 1962), p. 216, Fig. 41.]

cease. In some zones, like the midlatitude woodland zone, little or no natural
plant cover remains; in others, like the savannah zone, the exact role of
human beings is difficult to assess.

   To sum up, in this chapter we have taken a measure of our planet's productivity and used it to look at variations in the geographic environment. We
have illustrated how the intricate and complex spatial pattern of our environment can be broken down into a series of patterns of different sizes. Within
each fragment a different set of local environmental factors comes into play.

   The environment described is not, however, a fixed one. In the next
chapter we shall go on to see how it changes, and what special challenges these
changes pose for human occupation and modification of the planet earth.

Summary

1. Geographers are interested in measuring the total productivity of the earth from the viewpoint of human occupation. To do this, they frequently use plant productivity in terms of vegetation growth and describe this through climatic indices. The Paterson index produces a six-grade world map, with equatorial rain forests in grade A and deserts and ice caps in grade F.

2. The factors controlling spatial variations in productivity are at three levels: global, continental, and local.

3. At the global level, productivity patterns are related to the energy relations between the sun and the earth. Latitudinal variations in climate due to the angle of incidence would make much of the earth uninhabitable were it not for the atmosphere. The atmosphere distributes the solar energy it receives such that a global energy balance is maintained which is important to both biotic and abiotic parts of the environment. Convective flows between cold and warm air masses are created which redistribute heat. The Hadley-Ferrel model of this redistribution plus the rotation of the earth help explain the existence of the world's major wind systems: the trades, the westerlies, and the polar easterlies.

4. The oceans contain most of the earth's total moisture, and, as shown by the Hadley-Ferrel model, a great deal of water vapor is transported from them to adjoining continents where much of it falls as precipitation. Runoff from the continents returns the water to the oceans to form part of the global hydrological cycle.

5. At the continental level, differential heating between land and sea gives rise to distinctive continental climates with a wider range of temperature, lower humidities, and more variable precipitation than maritime climates. Seasonal changes in these climatic types cause reversals of air movements commonly known as monsoons.

6. At the local level, the continental patterns in productivity are broken up by smaller-scale factors. These include terrain, slope, geology, and soils.

7. The combined effect of environmental variations at the different spatial levels lead to a series of nine land biomes. These are grouped into three main classes, the forested (equatorial, midlatitude peripheral, and boreal), the intermediate (savannah, mediterranean, and midlatitude grasslands), and the barren environments (arid, tundra, and polar).

Reflections

1. Consider Paterson's map of potential productivity (Figure 4-3). List (a) the advantages and (b) the disadvantages of measuring potential fertility in this way. What additional information would you like to have before deciding on the productivity of a particular environment?

2. What are the main wet and dry areas of the continent in which your country is located? Why are these areas located where they are?

3. What kind of climate would the United States have if it were shifted several thousand miles south, so that the equator passed through Washington, D.C.? Would the country be more or less productive, assuming that productivity is measured adequately by Paterson's index?

4. Look at some atlas maps of January and June temperatures for North America. Why do the thermal contours bend equatorward in winter and poleward in summer?

5. Use Figure 4-10 to trace what happens to rain falling on the earth. How long do you think it takes rain falling in your own locality to get back to the ocean? How might you figure this out?

6. How are the timberline limits on high mountains like those one sees as one moves poleward? List the ways in which the climates of poleward and high-mountain regions are (a) similar and (b) dissimilar.

7. Review your understanding of the following terms:
permafrost
troposphere
tropopause
westerlies
trade winds
intertropical convergence
The Hadley-Ferrel model
solar-radiation budget
hydrologic cycle
biomes

One Step Further . . .

A useful nontechnical introduction to some of the issues missed in this chapter is given in

Oliver, J. E. *Perspectives on Applied Physical Geography* (Duxbury, North Scituate, Mass., 1977).

Fuller descriptions of the globe's physical environment and the basic findings of the earth sciences are given in numerous sources. Among the best introductions to this subject are

Strahler, A. N., *Principles of Earth Science* (Harper & Row, New York, 1976).

Earth Science Curriculum Project, *Investigating the Earth* (Houghton Mifflin, Boston, 1967).

Butzar, Karl W., *Geomorphology from the Earth* (Harper & Row, New York, 1976), Chap. 16.

Major zonal variations around the world and the climatic factors that lie behind them are discussed in an interesting way in

Trewartha, G. T., *Earth's Problem Climates* (University of Wisconsin Press, Madison, Wis., 1961).

There is a long list of descriptive works about each of the major zones. The interlinking of the major physical systems at the surface of the earth is described in

Miller, D. H., *The Energy and Mass Budget at the Surface of the Earth* (Association of American Geographers, Washington, D.C., 1968).

Manners, I. R. and M. W. Mikesell, Eds., *Perspective on Environment* (Association of American Geographers, Washington, D.C., 1974).

Research on the global environment is reported in all the major geographical journals. Look also at Scientific American *(a monthly) which carries regular reports on current work by environmental scientists.*

The world's a scene of changes, and to be Constant, in Nature were inconstancy.

ABRAHAM COWLEY
Inconstancy (1647)

chapter 5
Environmental Change

The environmental patterns described in Chapter 4 are but a frame in a feature-length film. The earth is probably over 4.5 billion years old, and in that time all the major environmental boundaries—of land and sea, mountain and lowland, tropic and pole—have never ceased to change. Environmental change, like death and taxes, is one of the few certainties in life. Many of these changes occurred so far back in time that they are of interest only to geologists. But others, which have been taking place during the last million or so years, are of direct interest to us all. They continue, slowly but perceptibly, to transform our environment.

In this chapter we look at these slow, long term changes and contrast them with rapid, short-term changes that form the familiar calendar of environmental change. The daily sequence of dark and light over most of the globe (note that conditions near the two poles are different) is one such cycle; the sequence of the seasons from high summer through fall, winter, and spring is another. Again, as we shall see, the seasonal pattern varies in strength around the globe but is everywhere present in some degree.

5-1
Long-Term Swings

How do we know what environmental changes have been taking place since man's emergence on this planet? Dim memories of change have been passed down through the ages in oral traditions and written records. The account of the flood in *Genesis* may not be literally true, but there is no doubt from related evidence that major floods did occur in appropriate parts of Asia Minor in Biblical times. However, very precise evidence of climatic change, in the form of written records, is available only for a very short period. By including early occasional records of heavy storms or extended cold periods, we can construct long runs of meteorological observations for a few parts of Europe. Figure 5-1 presents a remarkable 280-year record of winter temperatures for central England compiled by climatologist Gordon Manley. But even with a record of this length, finding a pattern is difficult. It is not unlike trying to make sense of changes in Wall Street share prices.

Reading the environmental record   In practice, therefore, geographers find that they have to build up a picture of environmental change from a wide variety of indirect sources. Let us examine some of these sources. Early researchers were aware of significant climatic changes primarily from macroscopic organic remains. All sorts of evidence, from the excavation of elephant and rhinoceros skeletons on the edge of the tundra to the discovery of warm-water shells in cold-water streams, pointed to a substantial climatic change in the recent past. In the early nineteenth century, Victorian scientists reported the growing size of the Alpine glaciers in Central Europe. They also described small streams meandering through great valleys that, judging by their cross sections, must once have carried much larger flows of water. Similarly, old beach levels were mapped many meters above existing lake levels.

Disputes occurred not over the degree of change but the order and time of changes. One of the most important sources of evidence on the order of climatic changes is provided by *pollen analysis*. As all hay fever sufferers know, plants that depend on the wind for pollination produce many thousands of microscopic pollen grains. (See Figure 5-2.) We can detect a changing climatic pattern from statistical analysis of the relative abundance of different types of grains preserved in lakes, peats, and muds. Table 5-1 summarizes the main sequence of climatic and vegetational conditions in Western Europe, indicated by pollen analysis, since the end of the last major expansion of the

Figure 5-1. Climatic change.
This graph of winter mean temperatures in central England from 1680 to 1960 serves as a historical record of climatic change. Temperatures have been averaged over ten-year periods to produce a smoother curve. (See the discussion on page 121 for an explanation of averages.)
[From G. Manley, *Archiv for Meteorologie, Geophysik und Bioklimatologie* **9** (1959).]

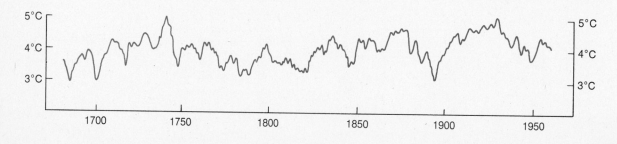

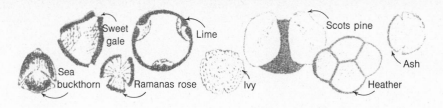

Figure 5-2. Pollen types.
The analysis of different types of pollen provides vegetational evidence of recent climatic shifts. The statistical frequency of pollen grains (magnified about 200 times in the diagram) preserved in peat deposits allows us to reconstruct the probable plant cover during different time periods. [From H. Vedal and J. Lange, *Traer og Buske* (Politikens Forlag, Copenhagen, 1958), p. 208.]

ice caps. The present cool, rainy climate, which began about 400 B.C., is the ninth phase in a series of postglacial oscillations. About 500 B.C., this part of Europe had a rather warm continental climate conducive to extensive pine and hazel forests.

Pollen evidence of recent shifts in belts of vegetation has been studied mainly since the 1920s. It helps us to sort out the order of environmental changes, but it leaves unsolved the problem of exactly when they occurred. That we can give actual dates to within a few decades is due to the remarkable advances made in the early 1950s by physicist F. Willard Libby at the Institute for Nuclear Studies in the University of Chicago. By 1947, carbon-14, a radioactive form of carbon that loses half its radioactivity in the first 5750 years of its existence and half the remainder in each 5750 years that follow, had been discovered in nature. Its constant rate of decay enabled Libby to devise a method of dating organic material. This technique, called *carbon dating,* lets scientists correlate organic evidence of how the world looked in the past with other biological, geologic, or archeological evidence. Carbon dating has proved to be extremely accurate for a period of 1000 years, but

Table 5-1   Main postglacial ecological changes[a]

Period	Time (B.C.)	Climate	Dominant Cover (Main Species)
Sub-Atlantic	Since 400	Cool coastal	Woodland: beech and hornbeam Cultivation and clearings
Subboreal	2500–400	Continental (cold winters, warm summers)	Woodland: oak and ash Cultivation and clearings
Atlantic	5500–2500	Warm coastal	Woodland: oak and elm
Boreal	8000–5500	Warm continental	Woodland: pine and hazel
Later Dryas	9000–8000	Arctic	Tundra
Allerod	10,000–9000	Cool subarctic	Scrub: birch and aspen
Early Dryas	15,000–10,000	Arctic	Tundra: large barren areas

[a]General sequence typical of lowland areas of northwest Europe.

Figure 5-3. Dating world climatic changes.
The bristlecone pine (Pinus longaevia) is the earth's most long-lived inhabitant. In the White Mountains of California, a count of the annual tree rings suggests these trees may reach ages of nearly 5000 years. Study of dendrochronology (from the Greek words for "tree" and "time") allows estimates of climatic conditions to be gauged from the width of the annual growth rings. The evidence of the bristlecone pine has proved particularly useful since it inhabits a semiarid area and its growth is very sensitive to rainfall amounts: Rings are wide in wetter years, narrow in drier years. Sometimes parts of individual rings are missing, but multiple borings around the tree circumference allow a complete picture to be built up. By matching the rings of living and dead trees, the record has been extended back to 6200 B.C. This long record has helped to check the dates derived from radio-carbon dating. [Photo by John Bonnell, De Wys Inc.]

evidence from tree-ring counts from the very old bristlecone pine suggests that corrections are needed for periods beyond 2000 years. (See Figure 5-3.)

Carbon dating has now been supplemented by other radioactive dating methods. Most new dating methods have been used to study the top few centimeters of deep-sea sediments because they hold the greatest promise as an archive of environmental change. Microscopic analysis of these sediments confirms the postglacial warming indicated by pollen analysis: a rise of about 8°C in mean ocean temperatures in the North Atlantic during the last 15,000 years, and an even greater warming (12°C) over a similar period in the Mediterranean. Most recent work is on the magnetic orientation of old sediments, which shows the position of the earth's magnetic field at the time they were laid down and allows us to date material as much as 20,000 years old.

Patterns of Change: The Pleistocene Epoch What kind of environmental changes have these techniques revealed? In discussing this, it is helpful to use geologists' terms and confine ourselves to the Quaternary Period, the fourth and last of the four major geologic divisions of earth history. This period is conventionally divided into two epochs: the longer Pleistocene epoch of some 3.5 million years and the Recent epoch, covering the last 25,000 years. Human beings probably developed from primate forebears sometime in the second half of the Pleistocene epoch. The human species emerged in a period that, from the standpoint of previous geologic periods, was one of intense environmental contrasts and rapid changes. Global differences in elevation, climate, and vegetation, for example, were sharper than they had been for the last 250 million years.

The earth's climate, which had been cooling slowly for the last 65 million years, grew much colder about 2 million years ago. The effect of this cooling was to lock up more of the world's water in the form of ice. In North America ice caps formed over central Canada and Labrador and spread as far south as Missouri and southern Illinois. In Europe ice caps formed over Scandinavia and advanced south into England and east almost to Moscow. In the southern hemisphere and in the tropical highlands the evidence of glaciers is less clear, but a large expansion of the ice fields is indicated. This expansion did not occur in a single surge, but entailed several slow advances and retreats, separated by mild and sometimes long *interglacial periods*. [See Figure 5-4(c).] In Europe there were four main periods of glacial expansion: the Günz, Mindel, Riss, and Würm glaciations. These correspond timewise to the Nebraskan, Kansan, Illinoian, and Wisconsinan glaciations in North America.

The impact of the ice sheets on the planet was threefold. The first effect we can predict from our knowledge of the earth's hydrologic cycle. (Refer to Figure 3-8 for an outline of this cycle.) As more water was stored as ice, the return flow to the oceans lessened and sea levels fell. At their largest the expanded ice caps lowered the ocean levels by approximately 100–125 m (328–410 ft). Although the vertical drop in water levels was relatively small, the horizontal effects were striking. The shallow continental shelf fringing the main land masses was exposed. For example, the shoreline of the northeast United States was extended by 100–200 km (62–124 mi). As a result, new routes between continents and islands were opened. Although the archeological evidence on this point is conflicting, it seems probable that man's entry into the New World by way of a land bridge with East Asia (now the Bering Straits) was in this period.

A second environmental change produced by the ice sheets was the compression of broad climatic and vegetational belts toward the equator. The productivity zones mapped in Chapter 3 were sharpened and realigned. For example, the Sahara Desert in North Africa (zone F) may have moved as far south as latitude 10°–15°N, compressing the savannah and equatorial zones into a narrow area. Figure 5-5 shows the sand dunes of this Pleistocene desert, now covered with vegetation and lying well south of the present desert margin.

Thirdly, expanding ice also bulldozed and realigned the river systems of much of North America and Europe. The Great Lakes system emerged and was shaped as a frieze of vast seas on the edge of the retreating ice field. Northern Canada and Finland have a landscape fretted with millions of small lakes and pools left among the jumbled debris at the base of the ice sheets.

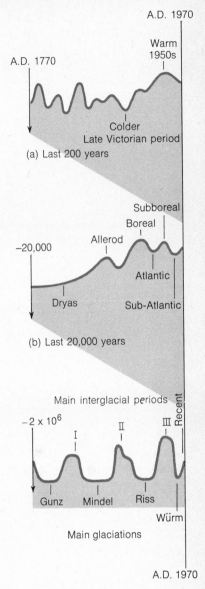

(a) Last 200 years

(b) Last 20,000 years

(c) Last 2,000,000 years

Figure 5-4. Continuities in climatic change.
The graphs show the general pattern of temperature changes over three time periods: (a) the last 200 years, (b) the Recent epoch, and (c) the Pleistocene epoch. Note that each period is 100 times longer than the period immediately above it.

Figure 5-5. Changing desert margins. Ancient longitudinal dunes, now under cultivation in the savannah belt of northern Nigeria, were formed under desert conditions and provide evidence of major climatic shifts in the Pleistocene epoch. Note that the dunes form the parallel narrow strips; the large square patterns are simply the individual air photographs in the mosaic. The prolonged droughts in the Sahel region (in the early 1970s) have led some scientists to conclude that the desert margin is continuing to shift south. [From A. T. Grove and A. Warren, *Geographical Journal* **134** (1968), p. 244, Plate 1. Crown copyright photos by Directorate of Overseas Surveys.]

Debris, ranging in size from a few centimeters to over 50 m (164 ft), was scraped and gouged from one area to be smoothed and plastered on another. It reshaped the terrain of the northern halves of both Europe and North America.

Patterns of Change: The Recent Epoch   Almost 10,000 years ago (about 8000 B.C.), the latest of the poleward shifts of climatic zones began. The continental ice caps contracted, and the glacial tundra climates of lowland North America and Europe gave way slowly to the present middle-latitude climates. The general warming continued until about 6000 B.C., when the *Atlantic* climatic stage, characterized by temperatures 2.5°C higher than the temperatures we experience today, began. This stage lasted until about 3000 B.C. [See Figure 5-4(b).] From this climatic optimum there has been a general but irregular deterioration. The *Subboreal* stage (2500 B.C. to 400 B.C.) was colder than now, and sea levels were generally what they are today. The evidence of climatic changes in the last 2000 years is more detailed, and we can detect continuing swings of temperature. A low point in the record was reached in the northern hemisphere in the middle of the eighteenth century,

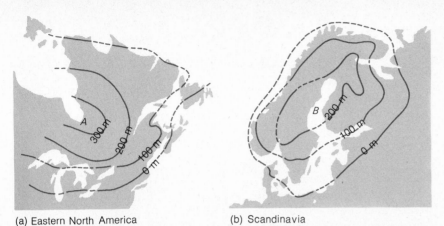

(a) Eastern North America   (b) Scandinavia

Figure 5-6. Changing beach levels. Contour lines show the height of old beach lines above the *present* sea level. They indicate how the melting of two huge ice caps (centered over points A and B) was followed by an upward warping of the land surface after the weight of ice had been released. Since the greatest weight was near the centers of the caps, it is here that the upward adjustment of the land surface has been greatest. [From B. W. Sparks, *Geomorphology* (Longmans, London, 1960), pp. 329, 332, Figs. 190, 192.]

and temperatures remained low into the nineteenth century. Whether the Recent epoch is a separate and warmer phase of earth history is a matter of dispute. The present warmer conditions may be merely a prolonged interglacial stage.

These later stages of retreating ice caps have encompassed two opposing phenomena: landward and seaward movements of shorelines. As the ice caps have melted, there has been a general worldwide rise in ocean levels of about 30 cm (1 ft) a century. This has resulted in a considerable net loss of land, particularly because the coastal plains that emerged during the glacial maxima (the most intense periods of glaciation when the greatest volume of water was backed up in ice) provided some of the most attractive sites for early human settlement and communication. The rate of loss is trivial, however. For example, human populations in the Gulf of Mexico have been driven inland only 15 km (9.3 mi) every 1000 years.

Around centers of former ice caps, shorelines have moved seaward and land has risen. At their maximum, the centers of the Labrador and Scandinavian ice caps may have been up to 3 km (1.87 mi) thick. This enormous weight caused a compensating downward displacement of the earth's crust. As the ice has melted, so the crust has recovered, but slowly and haltingly. Today the land around the former ice cap areas (e.g., the Hudson Bay and Scandinavia) is still slowly rising, and the sea is still retreating, leaving lines of old marine beaches inland from the present coast. (See Figure 5-6.)

One cannot be aware of the massive long-term changes in our global environment without wishing to know their cause. However, to answer all our questions on this subject would take us well outside the bounds of geography into the realm of sunspots, mountain-building cycles, and other geophysical events. (See "One step further..." at the end of the chapter.) We can, however, give two brief illustrations of the ways in which change continues to occur today. Both involve the notion of *cycles* and *succession*.

5-2
Cycles and Successions

Cycles in the Lithosphere We noted in the last chapter how millions of km³ of water evaporates from the oceans, moves over the continents as water vapor, precipitates out, and returns to the seas as rivers flow and ice melts. (Refer to Figure 4-10 for the details of this transfer.) But the rivers do not return unloaded to the sea. On the average, the Mississippi River brings back 1 ton of sediment in every 1200 tons of water; in floods this figure may rise to 1 ton in 400 tons of water. We can see one dramatic result of this process in the Mississippi delta, which looks like a growing bird's foot. Much less dramatic, but equally inexorable, is the general wearing away of the land surface within the area drained by the river (i.e., within its *catchment*, or *watershed*). In one person's lifetime the effect is miniscule, perhaps 3.6 mm is worn away in a 70-year span. But over a million years this adds up to an average lowering of the land of over 51 m (167 ft).

We should be careful not to extend our calculations too far. The slow reduction of the continental surfaces by this return flow of the hydrologic cycle does not necessarily mean that the elevation of the continents is decreasing. There are two reasons for this. To understand the first, we must look more carefully at the earth's crust. This crust is made up of two layers. The first is an upper layer of granitelike rocks called *sial*; below this is a heavier layer of rocks, called *sima*, which extends under the continents and the ocean basins. Thus, the lighter continents appear to float on the heavier rocks of the earth's mantle. Slow reductions in the continental mass by erosion are compensated for by upward movements of the land surface. But the balance is not achieved instantaneously, and the stresses caused by these adjustments are among the many reasons for earth tremors and earthquakes.

A second reason the elevation of the continents tends to remain the same is that the erosional forces set in motion by the hydrologic cycle are self-limiting. High regions are more easily eroded than lower ones, so as the elevation of the land is reduced the rate of lowering also declines. The interaction among the processes of erosion, sedimentation, and compensating uplift can be seen as part of yet another basic environmental cycle (Figure 5-7). Physical geog-

Figure 5-7. Erosion cycles.
This idealized cross section shows major elements in the cycle of erosion, sedimentation, and uplift. Geographers are concerned with the surface phases of the cycle; geologists and geophysicists are mainly interested in the subsurface phases.

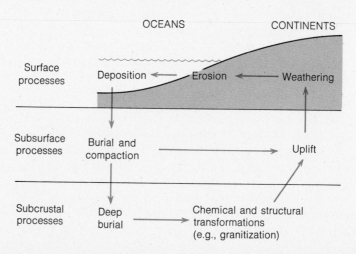

raphers have been particularly interested in the erosion phases of the cycle. The Davisian school, named after the American geomorphologist W. M. Davis, stressed the role of declining rates of erosion over time and developed a series of *geomorphic cycles*, each related to the shaping of land masses under different climatic conditions. (See the discussion of the Davis cycle on pages 106 and 107.)

Most erosion was found to be due to the action of water rather than glacial action or wind. (See Table 5-2.) Although these processes are, in the main, very slow, they can be punctuated by rapid changes that severely affect us. The shifting of the Chinese Hwang Ho River from an outlet north of Shantung in the years 1192 and 1938, and its reversal in 1852, resulted not only in many deaths but in long-term changes in patterns of settlement and land use. Similarly, compensating movements of the earth's crust may cause dramatic surface effects such as earthquakes and volcanic activity. (See Section 6-3).

Succession in the Biosphere A second type of slow change that is directly observable occurs among plants. When a field is first abandoned, it is bare. But it does not remain that way very long. Pioneer plants rapidly take root and flourish, establishing a simple community of weeds. Gradually, more brushy plants invade the community; a few quick-growing trees emerge, and some of the earlier plants are forced out. Over the decades a mature woodland may develop, but slow changes in the plant community continue. In the very long run (which may be hundreds or thousands of years) an equilibrium state may be achieved in which no further change occurs unless it is externally induced by a climatic change. As we saw in chapter 3, this vegetational equilibrium is termed a *plant climax*. The stages by which plants colonize an area and replace one another is termed a *plant succession*.

Frequently, the very long-term changes of the erosional cycle and the shorter-term changes in vegetation may be intertwined. This is very clearly shown in the history of many lakes. Lakes may begin as very deep bodies of clear water with few plant nutrients. As sediments are carried into the lake by

Transfers	Millions of Metric Tons per Year
ERODED FROM CONTINENTS	
By streams	9.3
By wind	0.06 to 0.36
By glaciers	0.1
Total	9.46 to 9.76
DEPOSITED IN OCEANS	
Shallow waters (less than 3 km)	5 to 10
Deep waters (more than 3 km)	1.2
Total	6.2 to 11.2

Table 5-2 Estimated transfers of mineral matter from continents to oceans

SOURCE: Data from S. Judson, *American Scientist* **56**, No. 4 (1968), p. 371.

THE DAVISIAN CYCLE

William Morris Davis was probably the world's most influential physical geographer. In his lifetime he published over 600 papers and books which contained his evolving ideas about the way the world's terrain had developed. His voluminous writings included *The Rivers and Valleys of Pennsylvania* (1889), *Physical Geography* (1898), *Geographical Essays* (1909), and *The Coral Reef Problem* (1928). Born in Philadelphia in 1850, he spent most of his academic life at Harvard, continuing very active work after his retirement in 1912. He died in California in 1934 at the age of 84.

The Davisian model of the way in which landforms change can be simply stated as an equation:

$$Landforms = function\ (Structure + Process + Stage)$$

By *landforms* Davis meant the physical shape of the earth's surface terrain; *structure* was the geologic composition of the land, and its original elevation above sea level as the result of

mountain-building forces. By *process* he covered all the decay and removal processes—such as chemical rotting, river erosion, downhill shuffling of material—by which the earth's surface was slowly reduced in height. *Stage*, the most important element in the Davisian equation, recognized that landforms evolve over time.

Davis found it was useful to divide the time periods into three distinct and recognizable stages—youth, maturity, and old age. Together these formed a life cycle of landforms, generally known as the Davisian cycle.

Evolution of landforms under the action of running water finding its way back to the sea as part of the hydrologic cycle was regarded by Davis as the norm. It formed a standard cycle which could then be modified in many ways. First, the cycle could be modified by climatic change (what Davis called *climatic accidents*). For instance, the sequence of landform stages under a very dry climate would give more importance to windblown material movements. The arid cycle also had to recognize that streams from occasional storms did not reach the sea but petered out in debris fans or salt lakes. In contrast, under glacial conditions ice takes the place of water as the major erosive element in the cycle. Second, the normal cycle could be modified by changes in the relative postion of sea level. This could be due to an absolute

inflowing rivers, the water becomes chemically enriched and the depth of the lake lessens. Plant productivity may build up in the lake itself, and vegetation may begin to encroach on the lake margin, with rooted plants near the shore acting as a sediment trap. Mosses and sedges build up, and great floating rafts of vegetation extend into the lake itself. [See Figure 5-8(a).] In the final stages of the lake's history, it may slowly fill with vegetation and sediment until eventually a forest is established and land plants take over from aquatic ones. Figure 5-8(b) shows just such a generalized succession of vegetation.

Although many environmental changes are very slow, we can clearly see them in operation in the contemporary world if we look carefully enough. We now turn to a different set of phenomena, to the rapid, short-term changes that demand our attention and of which we can never be unaware.

drop or rise in its level as happened during the main ice ages when more or less volumes of ocean water were locked up as ice. It could also be caused by shifts due to earth-building movements. These changes in elevation were termed by Davis *interruptions* to the cycle. Such interruptions meant that a landscape which had reached an old-age stage could be rejuvenated and go back to a youthful stage.

William Morris Davis traveled very widely, wrote profusely, and fitted ever more complex landforms into his scheme. Although modern work has given different interpretations to some of his findings, notably by emphasizing the greater mobility of the earth's crust, Davis's contribution to understanding the evolution of the earth's surface form remains monumental. He was the first geographer to develop a cohesive scientific theory which allowed the natural world of

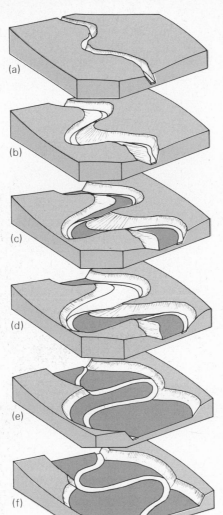

(a)
(b)
(c)
(d)
(e)
(f)

terrain to be seen as an evolving, living landscape. Throughout his life Davis, a geologist by training, stressed the importance of the physical earth in geographic studies. He played a leading part in the development of the Association of American Geographers (founded in 1905) and has left a permanent mark on the form and function of American geography.

For a sympathetic but critical biography of Davis with fascinating insights into both his academic and his private life see *The Life and Work of William Morris Davis* by Richard Chorley, R. P. Beckinsale, and A. J. Dunn (Methuen, London, 1973). [Davis's photo is from p. 438 of that volume; the diagram, from W. M. Davis *Physical Geography* (Ginn, Boston, 1898), Fig. 152.]

Cambridge University philosopher Ludwig Wittgenstein liked to illustrate the motions of the earth by spinning himself around, while at the same time circling around one of his friends. In the meantime, the friend was supposed to walk, following a leisurely, curving path, across the lawn. His biographers don't record how long this giddy game was kept up, but it would be as relevant today.

The earth has three motions. First, it moves with the sun as it orbits the center of the Milky Way once every 200 million years. Second, it travels around the sun once every 365.26 days. Third, it spins like a top around its own axis once every 23.94 hours. The planet's motion along the solar orbit is largely of astronomical interest, but its second and third motions are of direct and vital significance to us.

5-3
Short-Term Roundabouts

Figure 5-8. Plant succession as a cause of environmental change. Lakes are often good examples of the intertwining of long-term erosional changes and short-term vegetational changes. Pictured are four stages in an idealized sequence of plant succession in a lake environment. These stages are not of equal length: Commonly, the later phases of a succession are slower than the earlier ones. In a geologic sense, all lakes are ephemeral and have a finite life span. [Photo by Beckwith Studios.]

(a)

(b)

STAGE I

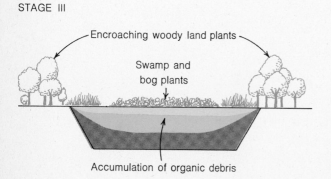

Marginal stream

Lake

STAGE II

Sedges and reeds

Mosses and floating plants

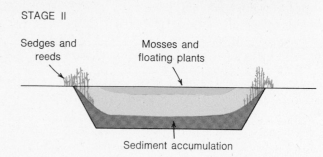

Sediment accumulation

STAGE III

Encroaching woody land plants

Swamp and bog plants

Accumulation of organic debris

STAGE IV

Complete colonization by shrubs and trees

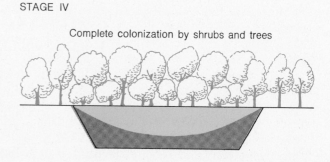

The Daily Round   The regular sequence of darkness and light that accompanies the daily rotation of the earth is so familiar that we ignore it. Yet we have evolved biologically in phase with this regular cycle, and our heartbeats, our blood pressure, our urine flow, even our sexual awareness, all have a distinct daily rhythm. Recall that in Chapter 1 a person on the beach was affected by this cycle; in Chapter 14 we go on to look at the basic rhythms of human communal activity and find that all our settlements are adapted to this same 24-hour beat.

From a strictly environmental viewpoint, the main effect of the earth's turning away from the sun is to cut off its darkened areas from massive inputs of solar energy. Thus the night is a period of energy loss by radiation from the land surface and falling temperatures. From dawn onward, the average amount of incoming solar energy increases; it reaches a noontime peak, then declines again as evening approaches. Average air temperatures follow the same pattern, but peak in the early afternoon.

On some warm summer days we can observe this daily buildup and decline of temperatures by watching cycles of cloud formation. Figure 5-9, a Gemini V satellite photograph, shows the view looking south along Florida's Atlantic shoreline. Note especially (a) that the clouds consist of thousands of isolated cells, giving the whole expanse a mottled appearance, and (b) that the clouds stop at the ocean's edge and are apparently absent from the main lake areas within the peninsula. This cloud pattern illustrates the noontime stage in the development of *tower* clouds, deep, rapidly developing clouds with small bases but considerable vertical depth. Such clouds are formed from the cooling of vertical columns of moist air that develop as the land surface heats up rapidly on a summer day. As some of the vertical columns grow stronger, larger and taller clouds predominate, some giving heavy rain showers. As the land cools toward evening, the tower clouds flatten and decay, so that the sky is mostly clear by nightfall. At dawn a new cycle of cloud formation will begin. Clear areas between the clouds over land are related to countervailing downward movements of air that compensate for the rising cloud columns. Clear areas over the sea and lakes are related to the different rates of warming of the land and sea.

The Seasonal Round: Temperature   Night and day are related to the rotation of the earth; winter and summer are related to the earth's revolution around the sun. Figure 5-10 shows that the planet's path around the sun lies on an imaginary flat surface (termed the *orbital plane*) that cuts through the sun. The earth's axis, around which it spins (shown as a line connecting the two poles), does not stick up vertically into the orbital plane but is tilted at an angle of 23½°. It is this tilt, together with the earth's motion around the sun, that produces our seasons. In late December the northern hemisphere is tilting away from the sun, so on each revolution it receives solar radiation for less than half a day. On December 21 and 22 the noon sun is vertically overhead at a latitude of 23½°S (the Tropic of Capricorn), but locations

Figure 5-9.   Daily cycles of environmental change.
This view looking south over Florida's Atlantic coast with Cape Canaveral in midphoto was taken from Gemini V. The mottling over the land occurs as tower clouds form in response to rising warm-air currents. When the temperature falls and these thermal currents die away, the clouds will dissipate. Note that clouds are absent both from the sea and from the main lake areas. Onshore and offshore winds are generated by the relative differences in the temperatures of land and sea areas. [Photograph courtesy of NASA.]

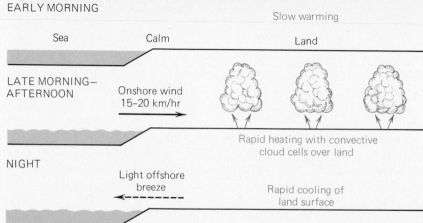

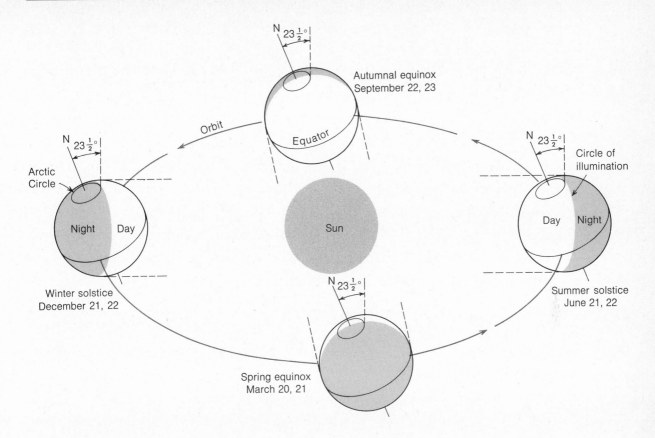

north of the Arctic Circle (latitude 66½°N, i.e., 90°−23½°) receive no direct sunlight. If you follow the diagram around, you can work out the seasonal alteration through the northern spring, summer, and autumn, and trace the reverse patterns in the southern hemisphere. Thus the traditional spring months of March, April, and May in the northern hemisphere herald colder weather in the southern hemisphere and form its autumn.

Outside the tropics the essential characteristic of the seasonal cycle is a swing in temperatures. Thus for crops winter is the dormant period, spring that of sowing and germination, summer that of growth and maturity, and autumn that of harvest. As Figure 5-11 shows, this familiar cycle is related to changes in solar radiation. How much radiation is received at any point on the earth's atmospheric surface is related to its location in terms of latitude. Note the shifting position of the latitude where the sun is vertically overhead in the diagram. The amount of solar energy received is not greatest at this latitude because in summer areas in higher latitudes have a longer day (i.e., more hours when they fall within the illuminated area shown in Figure 5-10). The annual flux of temperatures over the earth's surface lags behind that of solar radiation. As Figure 5-11(b) shows, this lag is caused by air temperatures

Figure 5-10.   Seasonal rhythms.
This simplified diagram of the earth's annual orbit around the sun shows how the fixed orientation of the earth's axis in relation to its orbital plane gives rise to the familiar sequence of seasons. [From A. N. Strahler, *The Earth Sciences,* 2nd ed. (Harper & Row, New York, 1971), p. 50, Fig. 4-1.]

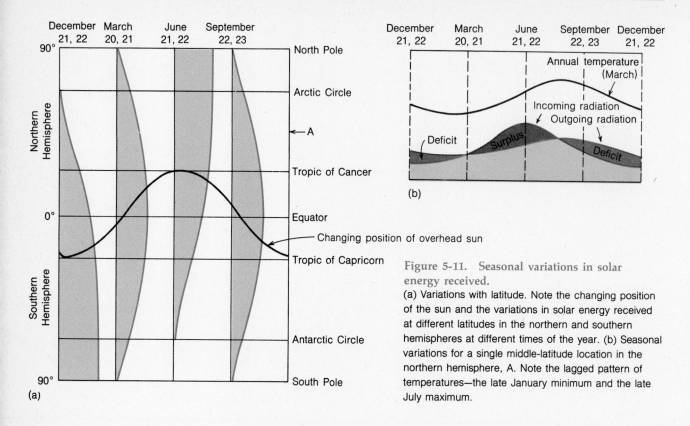

Figure 5-11. Seasonal variations in solar energy received.
(a) Variations with latitude. Note the changing position of the sun and the variations in solar energy received at different latitudes in the northern and southern hemispheres at different times of the year. (b) Seasonal variations for a single middle-latitude location in the northern hemisphere, A. Note the lagged pattern of temperatures—the late January minimum and the late July maximum.

that continue to rise so long as the incoming energy from solar radiation exceeds the energy reradiated from the earth. Thus, New York has its highest average air temperatures not late in June but in the middle of August. Offshore, the lag in sea temperatures is still longer.

The Seasonal Round: Water Deficits The regular swings of peak solar radiation cause a continuous north-to-south, south-to-north shift of the atmospheric circulatory system described in Figure 4-8. In late June, the northern summer, the whole zonal sequence is shifted north by up to 20° of latitude. This shift brings subtropical high-pressure areas with dry, warm, descending air over areas like California and the Mediterranean, but sends the trade wind belt, with damp, unstable air, into northern Nigeria and Venezuela. By late December the system has been shifted 40° of latitude southward, bringing winter rainfall to California but a winter drought to northern Nigeria. We must therefore modify our general continental pattern of precipitation to include areas of seasonal deficits arranged symmetrically in the northern and southern hemispheres.

Because of its significance for vegetation and crop production, geographers are interested in measuring the seasonal variation in these environments. Let us consider the seasonal balance in Berkeley, California (Figure 5-12), as typical of the summer-deficit areas. Berkeley receives about 63 cm (25 in) of precipitation in an average year, over half of it, as Figure 5-12(a) shows, in the three winter months. If we compute the area's potential water loss from evaporation over the year, we find that it is only slightly more, 70 cm (about 28 in); hence, the moisture deficit appears to be 7 cm (close to 3 in). However, most of the evaporation loss [Figure 5-12(b)] comes in the hot, summer months when rainfall is at its lowest. Some of the precipitation that falls in winter can be stored in the soil as soil moisture. Thus the soil serves as a short-term reservoir, and the moisture can be drawn on by a growing crop to compensate for a lack of rainfall. Still, there is not enough water from this source for plant growth to reach its full potential. From April on growth is inhibited, until by August the monthly deficit reaches 5 cm (2 in), as shown in Figure 5-12(d). These monthly deficits add to a yearly sum of 18 cm (17 in). This large moisture deficit is caused not by a lack of rainfall, but by its seasonal concentration. Excess winter rain cannot all be stored. Once the soil is sodden, additional rain runs off into streams.

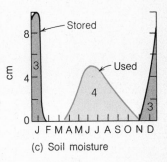

(c) Soil moisture

Figure 5-12.   Seasonal moisture balances.
This series of graphs shows monthly changes in the relationships between precipitation, potential evapotranspiration, and moisture stored in the soil for Berkeley, California. "Evapotranspiration" is the term used to describe the return of water to the atmosphere from the soil surface and from transpiration from plants. If you look closely at (e) you will see that it's made up from all the information shown in graphs (a) through (d). The same colors and numbers are used in all five charts. The most important chart is (d), which shows the severe late-summer drought in Berkeley in an average year.

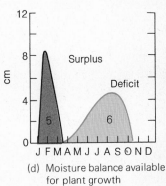

(d) Moisture balance available
     for plant growth

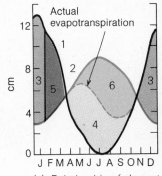

(e) Relationship of elements
     shown in (a) through (d)

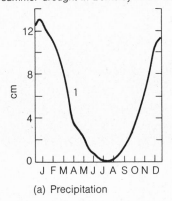

(a) Precipitation

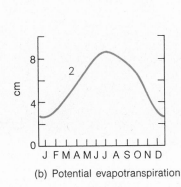

(b) Potential evapotranspiration

Information on the moisture deficit of a particular environment is useful in assessing its irrigation needs. If we extend our calculations, we can show other significant deficit areas over the western part of the United States.

In tropical latitudes more complex patterns of seasonal moisture deficits are encountered. Annual swings of temperature are less significant, and the daily range in values is often greater than the seasonal range. Important seasonal changes are related more closely to periods of rain and drought than to variations in temperature. Rainy seasons are directly related to the weather at the zone of convergence between the trade winds of the two hemispheres as it swings first northward and then southward in its annual cycle. Figure 5-13 shows an ideal cycle of shifts with distinctive two-season peaks of precipitation at the equator, the first in March-April and a second in October-November. North and south of the equator the two peaks merge into a single rainy season.

In interpreting Figure 5-13 we should recall that it shows an ideal situation. In reality irregularities of air flow may blur the picture and convert a regular seasonal cycle into a tragically uncertain pattern of precipitation. This has been seen most recently in the prolonged drought in the Sahel region of Africa. This lies between latitude 10° and 15°N on the southern border of the Sahara Desert. As Figure 5-13(c) shows, rains should fall during the June-July hot season. Their failure has led to the famine conditions on an unprecedented scale. We turn to the reasons for such failures in the next chapter.

Figure 5-13.   Seasonal patterns of precipitation in the tropics.
Tropical rainy seasons (c) are associated with the rhythmic northward and southward oscillation (b) of the high-rainfall belt associated with the intertropical zone of convergence where the trade winds from the two hemispheres meet (a). However, the uneven distribution of land and sea in the tropical zone, as well as monsoon effects, tend to blur the simple seasonal patterns shown here.

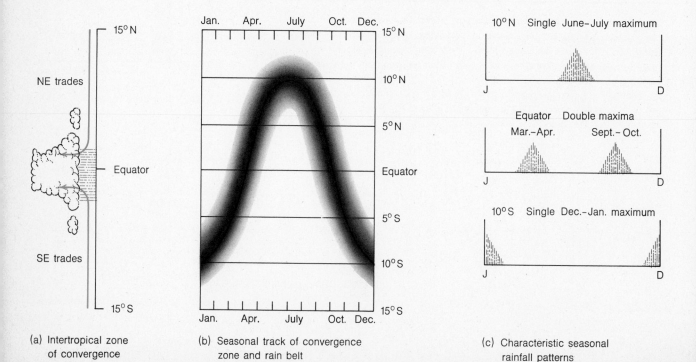

(a) Intertropical zone of convergence

(b) Seasonal track of convergence zone and rain belt

(c) Characteristic seasonal rainfall patterns

Summary

1. Environmental patterns are subject to changes over varying lengths of time. Knowledge of such environmental instability is important to us in our attempts to come to terms with our surroundings.

2. In terms of long-term changes the time period of greatest interest to the geographer is the Quaternary. The earlier epoch in this period is the Pleistocene, a time of environmental instability and rapid change caused by expansion and contraction of polar and alpine glaciers. Long-term storage of moisture alternately lowered and raised the ocean level, thus exposing the continental shelf. In addition, climatic zones were compressed southward and glacial debris was moved over a large region, radically changing the physical landscapes of North America and Europe. Evidence of continuing environmental change in the second part of the Quaternary, the Recent epoch, is found in both landward and seaward movement of shorelines and in climatic fluctuations.

3. Some of the forces of environmental change can be viewed as *cycles* or *successions*. Cycles in the lithosphere are shown by the effect of river erosion in the removal of continental surface material, which is balanced by upward land movements based on different rock densities and the reduction of erosive power through lowering of sea level. Succession in the biosphere can be seen in the response of vegetation to environmental changes. Different plants follow each other in a regular progression until the vegetation no longer shows systematic change. When this state of equilibrium is established in an area, the resulting vegetation is termed a plant climax.

4. Short-term changes are illustrated by daily and seasonal rhythms. The diurnal variation in temperature on the earth's surface can be seen through observation of cycles of cloud formations over coastal areas. Seasonal variation is based on the earth's revolution around the sun. The existence of seasons is due to the 23½° tilt of the earth from the vertical to the orbital plane. Latitude affects the amount of solar energy received at any point on the earth's surface.

Reflections

1. Set up a debate on whether very slow and long-term environmental changes are (a) too remote to concern modern man or (b) give us an essential perspective on our tenure on the planet Earth. Which side would you support? What arguments would you use?

2. Consider what is happening in Figure 5-10. If the earth's axis were tilted further from the orbital plane (say 33½° rather than 23½°), what effect would this have on the planet's climate? Where would the Tropics of Cancer and Capricorn and the two polar circles be? What changes would you expect in the climate of your own college town?

3. Use an atlas map to look at the seasonal pattern of precipitation in the tropics between latitudes 10°N and 10° S. To what extent do the maps resemble the idealized model in Figure 5-13? How would you explain the discrepancies?

4. Review your understanding of the following terms:

pollen analysis   erosion cycles
carbon dating   plant succession
Pleistocene epoch   seasonal moisture
Recent epoch   balance

One Step Further . . .

The long-term environmental swings in the later Pleistocene period are summarized in
    Strahler, A. N., *The Principles of Earth Science* (Harper & Row, New York, 1976) Chap. 16.

and treated in detail in
    Lamb, H. H., *Climatic History and the Future* (Methuen, London, 1977) and
    Gribbin, J., Ed., *Climatic Change* (Cambridge University Press, Cambridge, 1977).

How environmental variability is measured, its global patterns, its effect on human use of the earth are joined in
    Barry, R. G. and R. J. Chorley, *Atmosphere, Weather, and Climate* (Methuen, London, 3rd edit., 1977) and
    Maunder, W. J., *The Value of Weather* (Methuen, London, 1970).

Research in the areas treated in this paper is summarized in the regular geographic journals. Research on climatic topics is summarized monthly in Weatherwise *and* Weather.

> When men lack a sense of awe, there will be a disaster.
>
> LAO TSU
> *Tao Te Ching*, Ch. 72.

chapter 6

Uncertainties and Hazards

Once every few months the newspaper headlines stop concentrating on the usual round of politics and diplomacy to report a natural disaster. It may be a drought in Bihar, India, a cyclone in Bangladesh, an earthquake in Turkey, a flood in Missouri. Readers are jolted by the reminder that someone else's environment is not so house-trained as we assume ours to be, but the disaster rarely stays in the headlines for more than two days. It slips back into small type on inside pages, and finally drops out of the news altogether.

Such cataclysmic reminders of the instability of the environment have greater impact on slow variations in the levels of land and sea, the silting of a reservoir or the cycle of the seasons that we met in the last chapter. In occupying the earth, we were forced to learn to cope with the uncertainty of nature, with a multitude of possible risks and changes rather than a consistent opponent. In this chapter we view some examples of these more extreme environmental instabilities and try to piece them into a broad pattern of change.

6-1
Mid-term Mysteries

Long environmental swings like postglacial cooling are too remote to worry us; short daily and seasonal rhythms are so repetitive that we have learned to adapt to them. Even year-to-year fluctuations can be coped with if the surplus of one year can be carried over to the next. Irregular and uncertain fluctuations are what hit us hardest. But they are also one of the most difficult things for geographers to interpret. The records are too short for statistical patterns to appear, and the links to physical theory are too tenuous to provide reliable guides for forecasting. Here we shall illustrate with two regional examples the difficulties caused by irregular geographic phenomena.

The Great Plains: Mid-latitude Uncertainties   In the middle latitudes the boundaries between the major wind systems are continually shifting. Thus the polar front we described in Section 4-2 may wander considerably about its average location at about 60°N and 60°S. Fluctuations in the westerly circulation of the atmosphere in the middle latitudes are associated with waveforms that follow a four- to six-week cycle. The cycle begins with a zonal latitudinal flow in which waves of increasing amplitude build up to produce poleward and equatorward movements of air. Circulation then breaks these air movements into cellular patterns before the zonal flow is slowly reestablished (Figure 6-1). During the wave maximum [Figure 6-1(c)], strong incursions of freezing air from the north and warm, tropical air from the south may greatly distort "normal" climatic conditions. Past records of the world's climates reveal that these cycles are part of much larger swings which can last for several years. These cycles, in turn, are part of longer-term climatic shifts in the Recent epoch. Thus, the drought typical of arid zones may extend well outside the normal desert boundaries in one year, and precipitation characteristic of wetter regions may make incursions into an arid zone in the next year. A map of the world climate based on the meteorological records for 1981 would not be the same as one based on similar records for 1980, even if exactly the same classification criteria were used.

These shifts in climate become critical in areas where agriculture is carried on at the margins of humidity. For example, we saw in Chapter 4 that rainfall

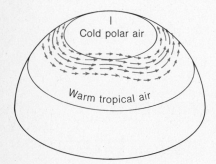

(a) The air stream begins to undulate

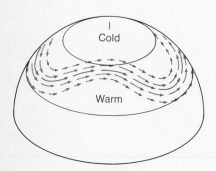

(b) Waves begin to form

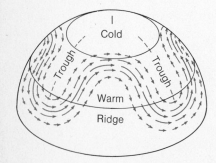

(c) Waves are strongly developed

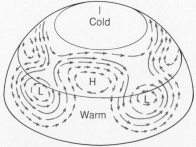

(d) Cells of cold and warm air are formed

Figure 6-1.   Changes in the westerlies. This six-week cycle of wave formation and dissipation in westerly air flow (Rossby waves) is one of the more regular of the short- and long-term fluctuations that lead to the characteristic instability of mid-latitude climates. The general pattern of the westerlies was shown in Figure 4-6. Longer-term patterns of change are discussed later in this chapter. [After J. Namias, from A. N. Strahler, *The Earth Sciences,* 2nd ed. (Harper & Row, New York, 1971), p. 247, Fig. 15-26.]

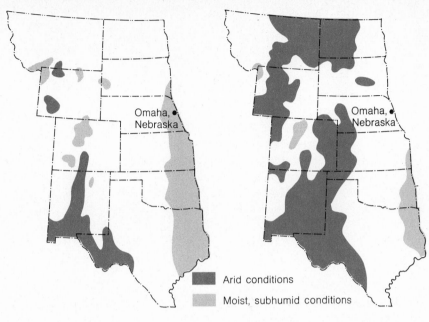

(a) "Normal" climate

(b) 1934 climate

Arid conditions

Moist, subhumid conditions

Figure 6-2. Climatic variations in the Great Plains.

The maps show the stark contrasts between (a) the "normal" climatic pattern with that experienced in (b) the drought year of 1934. The second map shows the kind of conditions that were described so vividly in the opening chapters of John Steinbeck's *Grapes of Wrath*. Note that arid conditions extended as far north as the Canadian border and covered an area five times greater than normal. The photo shows dust clouds over Lamar, Colorado, during the height of the great drought of 1934. [From C. W. Thornthwaite, in the U.S. Department of Agriculture's *Climate and Man* (Government Printing Office, Washington, D.C., 1941), Fig. 2, p. 182. Photo from the Granger Collection.]

in the Great Plains of North America decreases from around 125 cm (49 in) in the humid east to 25 cm (10 in) in the dry west. (See the map of the region's "normal" climate in Figure 6-2.) But rainfall may fluctuate not only from year to year but from decade to decade. The 1930s saw a disastrous run of dry years in the plains; conversely, the 1940s were generally wetter there than average. In the 1950s the regional pattern varied still further, with little rainfall in the southern plains but average conditions in the north.

What do we mean when we talk of "average" conditions? One way of interpreting records is to filter out small variations and leave only the prin-

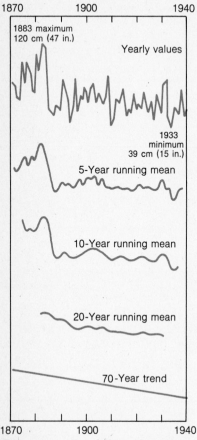

1870   1900   1940

1883 maximum
120 cm (47 in.)

Yearly values

1933
minimum
39 cm (15 in.)

5-Year running mean

10-Year running mean

20-Year running mean

70-Year trend

1870   1900   1940

Figure 6-3. Rainfall trends in the Great Plains.

The graph shows the smoothing effect of running means on the record of annual precipitation in Omaha, Nebraska, from 1871 to 1940. [From E. E. Foster, *Rainfall and Runoff* (Macmillan, New York, 1949).]

cipal swings. Figure 6-3 displays a 70-year record of precipitation for a Great Plains location—Omaha, Nebraska. Graphing the original yearly values produces an irregular pattern, but by selecting 5-, 10-, or 20-year means we can smooth it out to emphasize the general decline in rainfall over the period. Smoothing out data can be useful in checking trends; but, as investors in stocks often find, the curves that result are of limited value in predicting what will happen next. Even wholly random data can have deceptively plausible rhythms and trends. (See the discussion of averages and trends on the facing page for more comments on this subject.)

The problem of variability in rainfall in the Great Plains has its counterpart in the other midlatitude grasslands of the world—the South American Pampas, the South African Veld, the Australian Murray-Darling Plains, and so on. In humid regions the annual range of precipitation is small and poses few problems for agriculturalists; in the desert the drought is expected and plans are made accordingly. It is in semiarid areas like the Great Plains that settlers have frequently been fooled, because these locations are sometimes desert, sometimes humid, and sometimes a hybrid of the two. Good years draw optimistic individuals into very marginal areas where a run of bad years has caused failure and tragedy in the past. A knowledge of the climatic environment's intrinsic variability may prevent future misfortunes.

Monsoon India: Tropical Uncertainties Figure 6-4(a) is a conventional rainfall map of the Indian subcontinent. It shows the average amount of rain that may be expected to fall in any one year and distinguishes between very wet areas (southwest India, the eastern Himalayas and Assam, and the Burma coast) and very dry ones such as the Thar Desert. Like all maps based on averages, it should be viewed with caution until we know how much variability is concealed by the averages. We can illustrate the difficulty of working with averages by looking at maps for the two months of January and July [Figures 6-4(b) and (c)]. The January map shows the situation during the *winter monsoon* period, when India is dominated by dry, colder air moving from the cold high-pressure cell over central Asia. (See the discussion of "continentality" in Section 4-3.) The June map shows the contrasting period at the height of the *summer monsoon*, with moist, warm, tropical air moving from the Indian Ocean as southwesterly winds in response to the low-pressure cell developing over central Asia.

This regular seasonal reversal of wind directions and precipitation patterns lies at the heart of the Indian agricultural system. The rainy summer season from June to September provides some 90 percent of the annual water supply of the subcontinent and is especially critical for crops like rice, which depend on waterlogged conditions. The end of the dry season and the sudden "burst" of the summer monsoon is awaited with anxiety. Figure 6-4(d) shows the average dates for the onset of the monsoon rains. Ceylon in the south begins its wet season some two months later than the Indus Valley in the northwest.

AVERAGES AND TRENDS

We have already seen in Figure 6-2 how misleading maps of *average* environmental conditions can be. Their unreliability is directly related to the way averages are calculated. When we wish to determine the average of a distribution of values, we usually find the *arithmetic mean.* If we have a string of five rainfall values of 57 cm, 69 cm, 85 cm, 96 cm, and 116 cm, we obtain the average value by summing all the values and dividing the result by the number of observations. In this case the sum of the values is 423 cm and the number of observations is 5, so the arithmetic mean, or average, is 84.6 cm. This is a reasonably satisfactory result, as it is very close to the *median* of the distribution (the middle value in the series when the observations are ranked from lowest to highest) of 85 cm.

In studying Great Plains or monsoon rainfall we characteristically get less well-behaved sets of observations (i.e., sets in which there are some exceptionally high values due to exceptionally wet years). The same thing happens when we measure flood levels in rivers or, for that matter, the income of individuals. What happens when we try and use averages to describe these "skewed" distributions? We can illustrate the problem by going back to our rainfall records and making the last value much bigger— say, 196 cm rather than 96 cm. If we now recalculate our mean, we find it has risen to 104.6 cm (i.e., the new sum of 523 cm divided by 5). In this case we are less happy with the mean. Four of the five observed rainfall values lie *below* this average; 104.6 cm seems unrepresentative either of the four "normal" years or of the one "abnormal." Note, however, that the median of the distribution is unaffected by the change in the final value; it remains at 85 cm. Clearly, the arithmetic mean is poorly representative of the many natural events that tend to have a few very high values and a large number of low ones. In these conditions the median is probably a preferable proxy.

Another type of average, one we met in Figure 5-1, is the *moving average.* Moving averages are used in the study of environmental trends and may be calculated from either arithmetic means or medians, depending on the skewness of the observations.

MOVING AVERAGES

Moving averages are a simple means of smoothing time series by adding the values at regular intervals over a period and dividing the result by the number of observations. If we have a set of yearly values for rainfalls (y), the 5-year moving average for a particular year (t) is

$$\frac{y_{t-2} + y_{t-1} + y_t + y_{t+1} + y_{t+2}}{5}.$$

Thus, if our rainfall values in cm for the first seven years are 57, 69, 85, 96, 116, 141, and 124, the corresponding 5-year moving averages are —, —, 84.6, 101.4, 112.4, —, and —. Note that moving averages cannot be computed for the end values of the series. They can, however, be calculated for any length of time, depending on the length of the series available and the degree of smoothing required. As Figure 6-3 shows, the longer the period of the moving average, the greater the amount of smoothing. It is preferable to use an odd number of years in calculating this kind of average, so that the midpoint of the period to which the average refers will be an actual year. Moving averages can also be extended to two dimensions to smooth map series.

The special anxiety over the monsoon relates to (a) its timing and (b) its character. Delays in its onset affect planting conditions, jeopardize irrigation regimes, and may—if followed by poor rains—lead to famine and millions of deaths through starvation. On the other hand, exceptionally heavy rainfalls may lead to flooding, wash seeds from the soil, cause landslips, and so on.

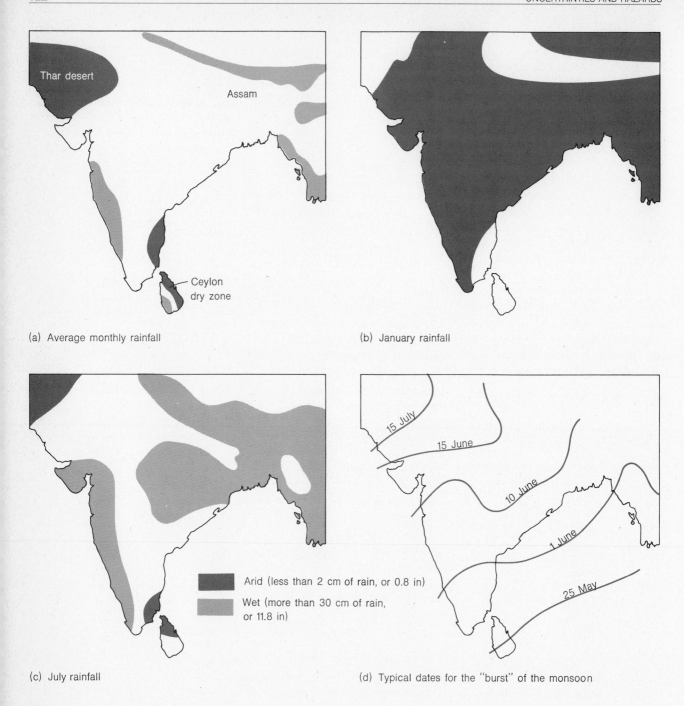

Figure 6-4.   The Indian monsoon.
The map shows contrasts between (a) average monthly conditions, (b) dry-season conditions, and (c) wet-season conditions. The dates in the fourth map (d) are averages; major delays in the "burst" occur in some years.

Figure 6-5. Variability in monsoon climates.
The graph shows year-to-year variations in July rainfall recorded at Anuradhapura in the dry zone of Sri Lanka between 1906 and 1945. Note that the average value seriously overestimates the rainfall likely to occur in a typical year. Each point indicates the July rainfall for one year. [After B. H. Farmer, in R. W. Steel and C. A. Fisher, Eds., *Geographical Essays on British Tropical Lands* (George Philip, London, 1958), p. 238, Fig. 4.]

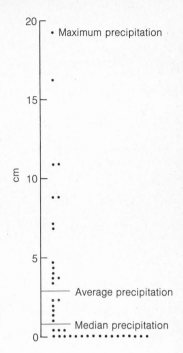

The intense variability of the Asian monsoon environment is illustrated by Figure 6-5, which shows 40 years of July rainfall records for Anuradhapura in the dry zone of Sri Lanka. (The location of Anuradhapura is in north-central Ceylon.) Note how misleading is the average July rainfall of 3 cm (1.9 in). In 15 years no rainfall at all was recorded in this month, while in one year nearly 20 cm (7.9 in) fell. The high values for a few years distort the average upward, so that the midpoint of the distribution, the median, is a better indicator of the probable rainfall in future Julys.

Both the examples we have chosen, the Great Plains and the Indian subcontinent, illustrate the puzzling nature of middle-term environmental changes. In both cases the causes of change are complex and are now beginning to be unraveled. But the human reactions and consequences are clear. Food production depends on three factors—the area planted and harvested, the level of husbandry (in terms of selection, fertilizer, and care), and the weather during the planting, growing, and harvesting periods. Climatic fluctuations of the kind examined may disastrously upset the food-producing ecosystems. Both examples also underline the care needed in the interpretation of maps and help to explain why modern geographers are so interested in probability theory. In the kind of environments we have just described, a knowledge of the odds is the first step toward wise resource planning.

Both the Great Plains and Monsoon India are regions of extreme climatic uncertainty. But it would be wrong to think of this uncertainty as simply variation about a "normal" climate for each area. The normal may itself show sudden changes, often with severe implications for the peoples whose food supply may be dependent on it. Wisconsin climatologist Reid Bryson argues that such flips in climatic conditions can be found in the past. Let us take two of his examples.

6-2
Abrupt Climatic Changes

Regional Examples More than three thousand years ago a distinctive and great civilization, that of the Myceneans, was thriving on a sunny plain in southern Greece. The capital city of Mycenaea, some 95 km (60 mi) southwest of Athens, was the trading center for the Aegean Sea and much of the eastern Mediterranean. Excavations of the city revealed walls ten meters thick and a kilometer long and the signs of a warlike and sophisticated people whose exploits live on in Greek literature (Figure 6-6).

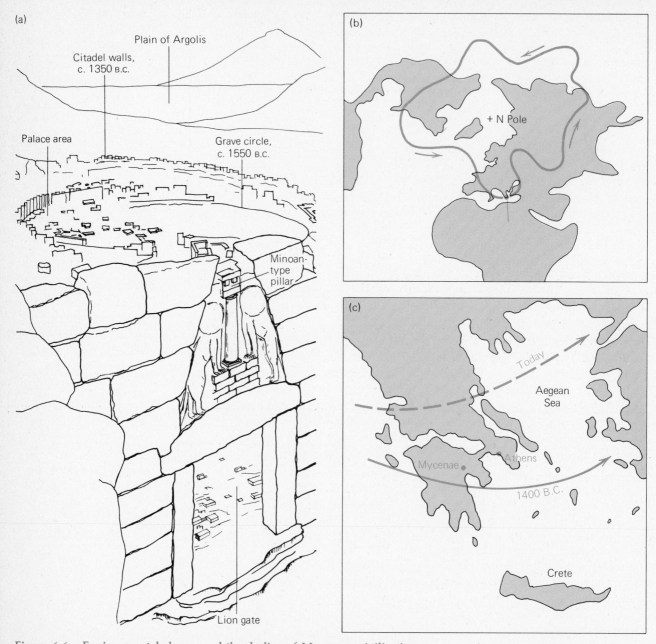

Figure 6-6. Environmental change and the decline of Mycenean civilization.
(a) Mycenae was the leading political and cultural center of mainland Greece from about 1450 to 1200 B.C. Thus Homer in the *Iliad* describes Agamemnon, the king, as the most powerful of Greek rulers. The sudden decline of Mycenae has been conventionally ascribed to invasion from outside, but recent research suggests that environmental change may have been a critical factor, with food shortages leading to internal overthrow. (b) One possible hypothesis put forward by Bryson links the decline to changes in westerlies over the northern hemisphere (see also Figure 6-1). Map (c) indicates possible switches in winter storm tracks over the eastern Mediterranean. Such switches are known to have occurred within the last few decades. For a full discussion see the argument put forward by Reid Bryson in *Climates of Hunger*, from which the maps were drawn (details given in "One Step Further . . . ," p. 135.)

Abruptly the power of Mycenaea began to decline. In 1230 B.C. the palace and main granaries were attacked and burned while other tributary cities began to decay. Early archaeologists excavating the sites in the late nineteenth century thought the answer to this decline lay in invasion by other Greeks, the Dorians, coming from the north. But as more research has gone on and more sites have been excavated, this now appears less and less likely. About a decade ago the classical scholar Rhys Carpenter, in a book called *Discontinuity in Greek Civilization*, suggested that an abrupt climatic change leading to loss of crops, famine, and civil disorder was a more likely cause. It was, he argued, the Mycenaeans who burned their own cities, and not outside invaders.

Could climatic change be the answer to the riddle? Before answering this, let's look at a second example, this time from North America.

Northwestern Iowa has today a yearly precipitation of 63 cm (25 in.) and is a rich producing area for corn and soybeans. About A.D. 1200 it was the center of a thriving Indian culture, the Mill Creek people. Excavation of settlements along this branch of the Little Sioux River shows that the Indians grew corn and ate bison and deer along with whatever smaller game they could catch. Figure 6-7 is based on the dating of the piles of debris containing bones and potsherds (broken pieces of pottery) that were found around the settlements. Note the curious way in which the number of bones and potsherds fall off quickly after A.D. 1200. By the time Columbus was crossing the Atlantic there were no bones, no potsherds, and no signs of the Indians. All the evidence suggests that the Mill Creek people had abandoned their villages and moved on.

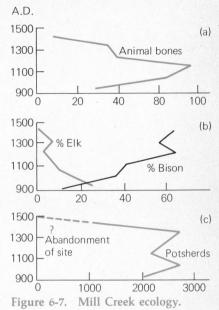

Figure 6-7. Mill Creek ecology.
Changes in the ecological conditions at Mill Creek in northwest Iowa between A.D. 900 and 1400 as shown by archaeological excavations. (a) Number of animal bones showing at peak about 1100. (b) Changing percentage share of animal bones showing elk (a woodland species) decreasing and bison (a grassland species) increasing. (c) Number of potsherds— pieces of broken pottery—with an abrupt decline after 1300. [From R. A. Bryson and D. A. Barreis, *Journal of the Iowa Archaeological Society*, **15** (1960), pp. 290–291.]

Flips in the Westerlies The two examples given above are from several cited by Bryson in his fascinating book, *Climates of Hunger*. The title suggests climates that changed so quickly that they were no longer able to support the plants and animals they once did, including those plants and animals (wild or domesticated) with which particular human cultures had built up their numbers. If climates change slowly and progressively, cultures may be able to adjust, but sudden change may be too sharp for the ecological or social system to cope with.

But what links these two peoples, separated in space and time, both with each other and with modern problem regions in the Sahel or India? Bryson sees a vital link in the changing circulation of the westerlies in the northern hemisphere. Look back at Figure 6-1 and note the way in which the westerlies loop and wave their way in a midlatitude circuit around the pole. Notice that since the waves must "catch their own tail," they must make a *whole* number of loops—say, three, four, or five—and cannot make any in-between numbers (e.g., 3.12 or 4.76 waves). As slow environmental changes warm or cool the atmosphere, so the area covered by the westerlies contracts or expands and its configuration may change. But the crucial implication of Figure 6-1 is that the pattern can only change by one whole number.

Bryson argues that it is precisely these jumps from one pattern to another that may abruptly change the climate of some midlatitude areas. Since we don't have good climatic records for these early time periods, it's impossible to be absolutely sure, whether this was what happened to the Myceneans or the Mill Creek peoples. But if we look at the modern record and argue that what *has* happened *can* happen, then it's possible to show that just the kind of flips required have occurred over the period for which records are available. Sudden shifts in the westerlies can move around the storm tracks along which the rain clouds move and sharply alter the pattern of drought or flood.

The patterns, once established, may last for long periods, and there is evidence that there has been a 200-year drought in the United States corn and spring wheat belt within the last 1000 years. The only certainty is that climate will change again one day. Ironically the climate in the Iowa area went back to a moister pattern soon after the Mill Creek people pulled out.

But what is the significance of these studies of past climatic change to our modern world? Bryson argues that the present pattern of crops has been developed very efficiently for what we consider to be "normal" climate today. Given increasing world population (see Chapter 7), the effect of an "abnormal" year like 1972 on world food production is enormous. But Bryson goes on to show that our present climate is *not* normal. If we take the period from 1931 to 1960, then since 1880 three out of four decades in the northern hemisphere were colder. The chances of more variable weather around the globe, with shorter growing seasons in the main food producing areas, are high for the coming century, and as we shall discover in Part Two of this book, part of this change may be self-induced by a growing pollution and other atmosphere-affecting human activity.

6-3
Extreme Geophysical Events

A basic distinction can be drawn between changes like drought which, even if abrupt, come on over a period of weeks and months and those in which the speed of onset is very fast. There are a number of climatic conditions—blizzards, hurricanes, floods—that are felt by the human population within hours. Similarly there may be events within the slow evolution of the lithosphere or the oceans that are of short duration but of high impact (e.g., earthquakes, volcanic eruptions, avalanches, or tidal surges). These sudden changes in the environment form a class of *extreme* geophysical events; we can see some typical examples in Table 6-1. This table shows extremes in the atmosphere (hurricanes), the hydrosphere (floods), and the lithosphere (earthquakes and volcanic eruptions). Note that all these are part of humankind's abiotic environment as defined in Chapter 3. Extreme events may also occur in the biotic environment (see the cholera pandemic shown in Figure 13-1).

Hurricanes Geographers estimate that about one in five of the world's natural disasters is caused by very severe storms in the atmosphere. Their frequency and sudden onset put them near the top of the hazards league in terms of loss of life. Tropical storm Agnes killed 118 people along its path up the eastern seaboard of the United States in June, 1972.

As far as hurricanes and tornadoes are concerned, the United States' losses are dwarfed by those of Asian countries. For example, on November 13, 1970, the greatest natural disaster of the century struck the low-lying delta areas of the Ganges and Brahmaputra rivers at the head of the Bay of Bengal. A tropical cyclone with a storm surge and winds of over 160 km (100 mi) per hour destroyed 235,000 houses and 265,000 head of cattle, and led to over 500,000 human deaths.

Violent vortices of this type are termed *tropical cyclones,* or *hurricanes.* They form in moist tropical air between 5° and 15° from the equator and move poleward along characteristically sickle-shaped paths. (See Figure 6-8.) For example, in the North Atlantic region hurricanes form between Africa and South America and move west and north into the Caribbean, the Gulf of Mexico, and the offshore areas of the eastern United States before curving back toward the northeast. The most commonly affected areas in this region are the Caribbean islands, but Florida may also be hit. Very occasionally, a hurricane will not cross the land until as far north as New England. Considerable research is being conducted on ways of controlling hurricanes by seeding them at an early stage of development to trigger precipitation before they reach critical land areas. At present, however, we can only reduce the damage by more accurate tracking and forecasting of approaching storms. This will allow more time for danger areas to be cleared of people and for at least some vulnerable property to be protected.

Table 6-1 Types of extreme environmental events

Abiotic environment (geophysical)		
Atmosphere	Hydrosphere	Lithosphere
Storms (tornado, hurricane, typhoon) (6)[a]	Floods, river (6)	Earthquake (6)
	Floods, coastal (6)	Volcanic eruptions (6)
Snow blizzard (6)	Avalanches, snow	Avalanches, rock
Droughts (6)		Shifting sand migration
Biotic environment		
Floral population	Faunal population	
Dutch elm disease in tree populations	Cholera pandemic in humans (13)	
Wheat stem rust	Foot-and-mouth disease in cattle	
Algal blooms in eutrophic lakes (22)	Locust plagues (23)	
Water hyacinth infestation in waterways		

[a]The number in parentheses refers to the chapter in which the event is discussed.

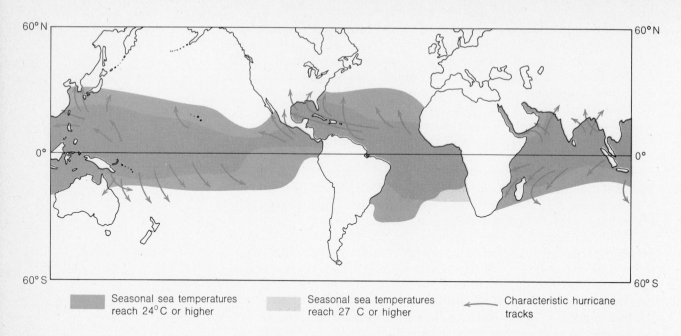

Seasonal sea temperatures reach 24°C or higher	Seasonal sea temperatures reach 27 C or higher	Characteristic hurricane tracks

Figure 6-8. Hurricane danger zones. Tropical cyclones (hurricanes) form in areas where the sea surface temperatures are high. Most originate in latitudes between 5° and 15° where the sea temperature in the hurricane season is 27°C (81°F) or higher. Characteristic tracks of hurricanes are shown on the map by arrows.

Floods Flood hazards occur in two distinct zones: coastal areas and areas bisected by rivers. *Coastal flooding* follows above-average sea levels caused by (a) unusual atmospheric conditions (e.g., the high seas created by onshore hurricane or tornado winds) or (b) earth tremors or volcanic eruptions that set up huge tidal surges. *River flooding*, a more frequent hazard, is related to heavy precipitation, rapidly melting snow, and—very rarely—the collapse of natural or manmade dams and the release of impounded waters.

In both coastal and river flooding the hazards are made greater by the attractiveness of such locations as places for human settlement. Some 12 percent of the United States' population elects to live in areas where there are periodic floods. The fact that flood losses have topped $1 billion in recent years must be weighed against the advantages—fertility, flatness, etc.—that make areas near the water so attractive. The *floodplain* of a river is created by water spilling over the normal channel limits (often banks or levees) and depositing sediment over the surrounding plain. Under natural conditions, this floodplain, or "spillplain," will be covered by water for a small but rather regular number of days each year. Where there are human settlements, this natural overspill and sedimentation process is interrupted by building artificially high levees and protective dykes. This has the desired effect of keeping a stream within its main channel, but it means that additional debris is deposited on the stream bed, raising the height of the channel and creating a need for still higher artificial levees. Many of the world's major streams now run across their lower floodplains in artificially constrained channels some meters above the level of the surrounding densely settled land. When floods occur under these conditions, the depth of the flooding and its impact on the human population are immense. The effects are particularly severe in densely used floodplains.

Earthquakes   However difficult hurricanes are to forecast, they appear relatively predictable compared with the abrupt and cataclysmic environmental changes that occur with earthquakes and volcanoes. Although only one or two earthquakes causing much damage occur each year, about 150,000 earth tremors are detected every year. If we measure tremors by the area over which their effects were felt, then the largest in recent times was probably the Assam earthquake of 1897, which affected an area of 4.2 million km^2 (over 1.6 million mi^2, half the continental United States!). The largest earthquake in the conterminous United States was the San Francisco earthquake of 1906. The Tangshan quake in northern China in 1976 may well have killed 400,000 people. In addition to the loss of life, we must add damage by fire and destruction of buildings caused by such events. There are also longer-term environmental changes resulting from the displacement of sediments and the redirection of river courses that may cause secondary hazards.

Although the specific timing of earthquakes is unpredictable and no effective countermeasures (other than appropriate building restrictions) are possible, the location of earthquake-prone areas is well established. Figure 6-9 shows that earthquakes occur principally in two elongated belts. The first passes around the Pacific Ocean and includes in its North American segment the Aleutian Islands, southern Alaska, and the Pacific coast of Canada and the United States. This circum-Pacific belt is estimated to account for some 80 percent of all the earthquake energy released on the planet. A second major belt runs from Portugal through the Mediterranean, the Middle East, and the Himalayas and meets the circum-Pacific belt in the Indonesian islands.

As more oceanographic exploration is completed, the presence of other earthquake belts associated with midoceanic ridges is being established. These areas form the boundaries between a series of great structural plates rather like the sutures of a human skull. Movements and adjustments in pressure appear to take place in these critical tension zones within the earth's crust.

Volcanic Eruptions   More lasting environmental changes than those produced by earthquakes probably stem from volcanic activity. The largest such explosion in historic times was probably the Krakatau explosion of 1883, which blew away two-thirds of an island and triggered a tidal wave estimated at 45 m in height (almost 150 ft) that broke upon the adjacent coast of Java with great destructive force. However, volcanic activity can also have beneficial effects. Slow accumulations of volcanic lava and dust may lead to the creation of entirely new land areas. The Hawaiian Islands, for example, were formed in this way. And while some ejected materials remain rocklike and sterile for centuries, others decompose into exceptionally fertile soils. Parts of

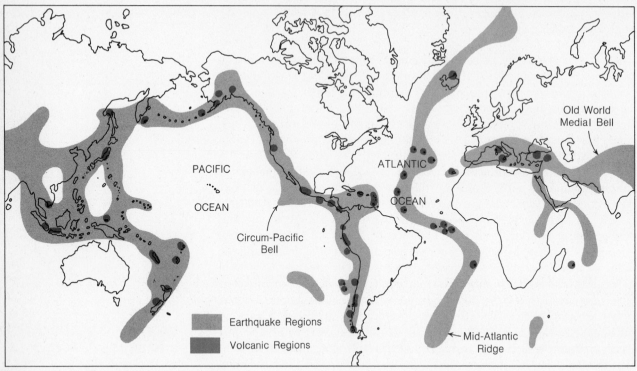

Figure 6-9.   Earthquake-volcanic hazard zones.
Schematic map of main areas of earthquakes and volcanic activity during the most recent geological period (i.e., that most relevant to human beings). Note the concentration into three main zones. The most important is the circum-Pacific zone or girdle which accounts for about 80 percent of all earthquake activity. The Old World medial belt running from the Mediterranean to Indonesia accounts for most of the remaining activity with the mid-Atlantic ridge forming a third and less active region.

Java, the Japanese island of Kyushu, and south India typify volcanic areas with fertile soil structures that support high populations.

The location of volcanic activity broadly follows that of earthquakes. The specific locations of areas of active volcanoes are aligned on the earthquake zones in Figure 6-9.

6-4
The Environment As Hazard

Each of the situations described in the preceding sections of this chapter represents some degree of hazard to the human population. It's clear, however, that the degree of hazard is not simply a function of the natural event itself. A severe earthquake in the unpopulated Arctic may represent a much less severe hazard than a mild earthquake in a densely packed city. A severe storm that occurs at harvest time may be a disaster; an equivalent storm on the bare fields of winter may pass without comment. The notion of a hazard is, like that of resources, one that only makes sense in terms of an evaluation of the earth as a home for the human population. (Figure 6-10).

(a)

Figure 6-10. Uncertainties in the natural environment.
Extreme and irregular natural events such as (a) hurricanes and (b) earthquakes provoke a wide variety of reactions related to the perception of the groups affected and the magnitude, frequency, duration, and spacing of the events. (See Table 10-4.) [Photographs (a) from United Press International, (b) by J. Eyerman, Black Star.]

(b)

Pervasive and Intensive Hazards   How then can we evaluate the environment as a hazard? In a book entitled *The Environment as Hazard* by geographers Ian Burton, Robert Kates, and Gilbert White, seven measures found to be significant in human terms are proposed. The first is the *magnitude* of an event, say, the height of a flood or the intensity of an earthquake tremor. The next four all relate to time: the *frequency* of occurrence, the *duration* of the event, the *speed of onset* from first warning signs to peak, and the *temporal spacing* in terms of randomness or regularity. The last two are more specifically geographical: the *areal extent* over the earth's surface and the degree of *spatial concentration* within that area. Figure 6-11 summarizes the last six characteristics for a major drought, a severe blizzard, and an earthquake. You may like to draw profiles for some other events, say, a major flood, on the same diagram. Note that a basic difference emerges between *pervasive* hazards, such as drought, and *intensive* hazards, such as tornadoes. Thus tropical storm Agnes, which moved up the east coast of the United States from June 19 to 23, 1972, brought intense hazards along a narrow track for a short duration. But by the time the storm had passed, 118 people had lost their lives and $3.5 billion worth of property damage had occurred—in monetary terms the greatest material disaster from natural causes suffered by the United States to date.

Return Periods   But how frequently is such a disaster to occur? One approach is to look back over the records of a particular type of event and see what the figures say. Thus an agriculturalist may phrase his question in terms of averages. How likely is a crop failure in any 10-year period? But to the settler on the coast or floodplain, the strength and frequency of extremes may

Figure 6-11.   Hazard profiles.
Ways in which it is possible to draw a profile for natural hazard events in terms independent of magnitude. Sample curves for a drought, a blizzard, and an earthquake are shown. [From I. Burton, R. W. Kates, and G. F. White, *The Environment as Hazard* (Oxford University Press, New York, 1978), Fig. 2.4, p. 29.]

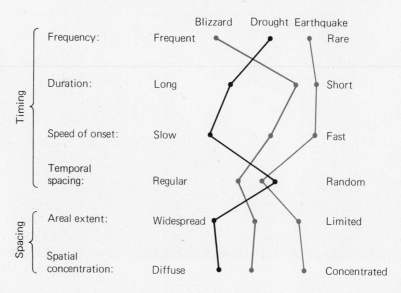

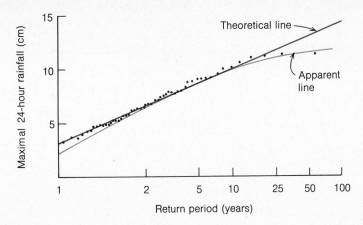

Figure 6-12. Predicting extreme values for natural hazards.
The graph shows the frequency with which the greatest amount of rainfall in a 24-hour period reached particular levels in any one year during the period 1950-1956 in Nantucket, Massachusetts. Each year's record is shown by a point on the graph with the lowest value at lower left and the highest part at upper right. Note that the numbers on the horizontal axis of the graph are not regularly spaced. By drawing the axis in this way the sequence of years is found to fall along a line (labeled the "apparent line"). This apparent line can be approximated by a straight line (the "theoretical line"). Extending the straight line allows an estimate to be made of the greatest rainfall to be expected once in a century—about 14 cm (5.51 in.). Further extensions beyond the 100-year mark on the graph are possible but are likely to be of uncertain accuracy. [From D. M. Hershfield and M. A. Kohler, *Journal of Geophysical Research* **75** (1960), p. 1728, Fig. 1. Published by the American Geophysical Union.]

be more important: What will be the highest flood or the strongest wind in any given period?

One way to answer both these questions is to try to calculate a *return period*, the average interval within which one event of a specified size can be expected to occur. To do this, we first rank all observations of a particular phenomenon according to their magnitude, from the largest (1) to the smallest (n). The return period is then equal to $(n + 1)/r$, where r is the rank of a particular observation. Suppose we have a 49-year flood record which gives us the highest flood level each year. Then the average return period of the tenth-largest flood will be $(49 + 1)$ divided by 10; a flood as large or larger should recur on average once every 5 years. Because the timing of floods is irregular, there may be several large floods in our 49-year record. One way around this problem is to plot the magnitude of an event against its return period on a graph. Thus we can average the recorded observations by drawing a straight line through them, as in Figure 6-12. This method allows us to estimate the most likely return ratios on the basis of all the records available. Figure 6-12 shows that we can expect Nantucket to have a heavy rainfall of 10 cm/day once in a decade and a 14 cm/day rainfall once every century. Of course, these estimates are only averages. The rainfall that comes once every 1000 years may still fall next week!

This type of frequency analysis depends on rather simple assumptions. It provides a useful first approximation of the size of the risks a given environment is likely to pose. We assume, for example, that the pattern of floods or heavy rainfalls is not undergoing cyclic changes. If floods of a certain river are getting steadily worse, possibly because of deforestation, then this method will underestimate the size of a 100-year flood. Like rainfall, floods may tend to occur in clusters. The great floods on the Ohio and Mississippi rivers in 1936 and 1937 were paralleled three decades later by the disastrous floods of 1964 and 1965.

The Human Dimension   An approach to hazards in geophysical terms is only part of the story. Some hazards may become more frequent because of human action. A good example is floods in urban areas; storm water now flashes off acres of concrete and asphalt straight into streams where once it filtered more slowly through soil and vegetation. For others the geophysics remains the same, but the hazard grows. This is true for coastal floods where an inundation that affected few and was a local agony only a century ago may now affect many and be of international concern.

As the human population has grown, so it has tended to move into areas of high attraction but high environmental danger. We saw in Chapter 1 that many cities are located in coastal areas and many more are in vulnerable flood plains. The property values in floodable areas of such cities are very high. The world as a whole has become increasingly dependent on efficient food production from a few regions which, in some cases, notably the midlatitude wheatlands, may show major swings in their productivity from one harvest to the next.

Hazard research is now a major field in geography, and we shall be returning to it in later chapters (see especially Section 10-4). Here we note simply that it must involve study not only of the range of environmental experience (floods, droughts, cyclones, etc.) but also human reactions.

How do we cope with hazards, both as individuals and as societies? We may simply accept losses in a resigned way or learn to live with them by sharing risks (say, by insurance or internationalizing disaster relief). Very extreme situations may lead to complete reappraisal of the environment with radical changes in land use or relocation. The stream of "Okies" moving west to California from the drought-stricken Great Plains was vividly portrayed in John Steinbeck's *Grapes of Wrath*. Relocation is the most extreme reaction, and most societies seem to "stay put and do something different" rather than move on. Study of different reactions may help to evolve new and better ways of coping with hazards.

Geographers argue that to understand the environment we need to see it in human as well as physical terms. It is, therefore, at this point in the book that we turn from study of the planet itself to consider the fast-growing species that increasingly populates the "natural" environment. From here on we shall be concerned with men and women and the way their growing numbers have intruded on the global scene. We begin Part Two by looking at the human response to the environmental challenges we have just described.

Summary

1. Long- or short-term environmental variations may be easier for humankind to cope with than those of middle length. Two major food-producing regions affected by year-to-year uncertainties in rainfall are the midcontinental grasslands (illustrated by the Great Plains of North America) and the tropical monsoon regions (illustrated by India).

2. Abrupt climatic changes can be shown to have occurred frequently in middle-latitudes over the last

few thousand years. Such abrupt changes appear to be linked to "flips" in the pattern of the westerlies. Such flips alter the average location of the storm tracks along which rainclouds move and may sharply reduce or increase the precipitation over a given area. Once established, a new average track may persist for several decades (possibly, up to 200 years).

3. Extreme geophysical events are occasionally superimposed on the regular cycles of environmental change. These may affect the atmosphere (hurricanes), the hydrosphere (floods), or the lithosphere (volcanic eruptions and earthquakes). When such events affect human population, they are termed natural hazards.

4. Basic distinctions can be drawn between pervasive and intensive hazards. Pervasive hazards (typified by drought) build up over a long time period and affect a wide area. Intensive hazards (typified by tornadoes) have a short duration and affect a sharply defined area. For some natural hazards, the likely future danger can be estimated in terms of return periods.

5. Environmental evaluation demands measurement in human as well as physical terms. The *same* geophysical event can be shown to have a radically different impact when it occurs in areas of different human occupation.

Reflections

1. Using newspapers and recent magazines, make a list of any extreme geophysical events (floods, earthquakes, eruptions, etc.) reported. Then look at the spatial distribution of these events. Is there any pattern to them? How would you explain such patterns?

2. Do you consider that you live in a "very safe" or "very unsafe" environment so far as natural hazards go? Explain your reasoning. Can you suggest any measures that individuals, corporations, or government might take to make conditions safer? Would you support these measures?

3. Collect some figures for monthly rainfall in your own locality and plot them on a dispersion graph, using Figure 6-5 as a guide. Comment on the distribution of values. Is the mean a good description of the usual monthly rainfall?

4. Review your understanding of the following terms:

westerlies   pervasive hazards
monsoon   intensive hazards
hurricanes   return periods
circum-Pacific belt

One Step Further . . .

An outstanding study of mankind and the world's changing weather is

Bryson, R. A., and T. J. Murray, *Climates of Hunger* (University of Wisconsin Press, Madison, 1977).

This traces how environmental trends are linked to world food supplies and to human institutions. A closeup of our attempts to cope with a changeable geographic environment is given in a classic study of human settlement of the North American grasslands:

Webb, W. P., *The Great Plains* (Ginn College, Waltham, Mass., 1959),

and is supplemented by a modern account of similar problems in an Australian context:

Meinig, D. W., *On the Margins of the Good Earth* (Rand McNally, Chicago, 1963).

For studies of human reactions to environmental risks (a topic expanded in Chapter 10), see

White, G. F., Ed., *Natural Hazards* (Oxford University Press, New York, 1974).

Burton, I., R. W. Kates, and G. F. White, *The Environment as Hazard* (Oxford University Press, New York, 1978).

The former contains a wealth of regional examples while the latter is more systematic in approach. Research in the areas treated in this chapter are summarized in the regular geographical journals and in general scientific journals such as Science *(weekly).*

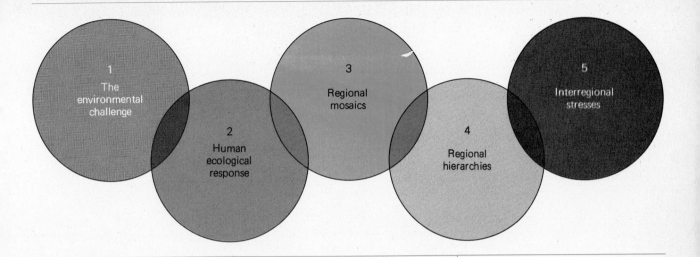

The human side of the human-environment system is considered in Part Two by viewing people in an ecological context. Our response to the environment in terms of population growth is analyzed in *The Human Population* (Chapter 7). We examine there alternative models of population growth in the face of different environmental constraints, and we note the explosive consumption of material resources that accompanies accelerating population growth. *Pressures on the Ecosystem* (Chapter 8) looks at humanity's place in the natural world and the effects of increasing numbers of people on the natural environment, both direct and indirect. These effects range from minor intervention through population densities to the firing of grasslands by hunting societies to the far-reaching pollution caused by our urban, industrial society. Special consideration is given

to the geographer's view of the many-sided pollution problem. *Resources and Conservation* (Chapter 9) defines natural resources and shows how we estimate the size and probable duration of reserves. Both optimistic and pessimistic views of the energy crisis are presented and the role of conservation in planning the future use of resources is explored. Finally, in *Our Role in Changing the Face of the Earth* (Chapter 9) we look at the total effect of human ecological dominance in terms of the changing landscape. As our numbers have increased, so the environment itself has been modified. But finding out how much it has been changed, or in what ways, is a more complex problem than it appears at first sight. We look at the ways geographers measure and estimate changes in human use of the land and at ongoing trends in land use. We close Part Two by looking at the philosophical frameworks within which geographers view people in relation to the planet Earth.

part two

Human Ecological Response

> A finite world can support only a finite population; therefore population growth must eventually equal zero.
>
> GARRETT HARDIN
> *The Tragedy of the Commons* (1968)

> The optimist proclaims that we live in the best of all possible worlds; and the pessimist fears this is true.
>
> JAMES BRANCH CABELL
> *The Silver Stallion* (1926)

chapter 7
The Human Population

Just when man first appeared on the earth is a matter of speculation. We know that several humanoid primates emerged during the last 3.5 million years of the earth's most recent geological period—the Quaternary period—but much of our theorizing about human origins depends on the interpretation of a small number of critical skeletal remains. *Homo sapiens,* with a larger brain (the dividing line is conventionally drawn at 1000 cc), can be traced back to an interglacial period about two and a half million years ago.

Though archaeologists may argue over the exact date of evolutionary advances, two broad ecological generalizations can be made. *First,* human beings are very recent arrivals on the biological scene. The earth itself is about 4.5 billion years old. The first living forms, algae and bacteria, originated about 2.2 billion years ago, and the first primitive mammals about 0.22 billion years ago. By these standards, members of the species of mammal we call *Homo sapiens* are the new kids on the block! To use a familiar analogy, their arrival occurred in the last second of an hour-long global history.

Second, the human population of the earth has grown in numbers since its emergence to a staggering level of 4.3 billion, and —what is more important— half that growth has come in the last 35 years. It is not difficult to see that the present explosion of the human population can be only a short-lived ecological phenomenon. Stanford biologist Paul Ehrlich has shown that if the world's present population continued to grow at its present rate (i.e., doubling every 35 years), by A.D. 3000 there would be 2000 people piled on every square meter of the earth's surface—land, sea, and ice included. If we extended this nightmare projection far enough, the universe would eventually consist of a ball of closely packed people expanding outward at the speed of light!

In this chapter we look at the critical facts behind the immense growth of the human species on the earth. First, we review the process by which growth occurs. How is it affected by the relationships between births and deaths? How do we measure population change? Second, we look at the checks on population growth. Here the questions center on how other species control their numbers, and whether human beings are in any sense a special or an exceptional case. Third, we look at the facts of world population growth. What happened in the past? What is happening now, and what are its long-term implications? Is zero population growth possible—or even desirable?

This chapter is a necessary forerunner to the remaining three in Part Two, which look at the impact of increasing numbers of people on the world's ecosystems (Chapter 8), on resources (Chapter 9), and on the landscape (Chapter 10). An understanding of population and its growth are so important to our understanding of the human place on the earth that we shall be returning to this theme constantly in the remainder of the book.

7-1
Dynamics of Population Growth

While the facts of birth and death at an individual level are clear, their effect on the growth and decline of a *population* (that is, a collection of individuals) is more opaque. Here we look at the processes which shape population growth and the kind of yardsticks we use to measure it. In this chapter we shall be concerned mainly with the human population. Much of the reasoning we shall use could, however, be applied to animal populations as well.

Births, Deaths, and Growth   The total population of any area of the earth's surface represents a balance between two forces. One is *natural change*, caused by the difference between the number of births and deaths. If births are more numerous than deaths in any period, the total population will increase. If they are less numerous, it will decrease. This simple relationship is modified by a second force, *migration*. When immigrants are more numerous than emigrants, there will be a population increase. (This assumes, of course, that we are ignoring natural change for the moment). When emigrants are more numerous, there will be a population decline.

As Figure 7-1 shows, net changes in population totals are caused by the interaction of four elements: Births and immigrants tend to push the total up; deaths and emigrants tend to bring the total down. Although migration may be the most important factor in small areas (for example, in a small village or a city block), it is less significant on the national level. For the world as a whole, migration is irrelevant because all movements take place within the limits of the recording area. In other words, until interplanetary travel comes along, the planet Earth can be safely treated as a *closed* system for demographic purposes. We shall therefore concern ourselves in this chapter with the natural change component, leaving the more complicated analysis of *open* population systems (in which migration is significant) until later in the book. (See especially Sections 12-2, 14-2 and 18-1.)

IMPORTANT TERMS IN THE STUDY OF POPULATION

Birth rates measure the proportionate number of births in a population.

Carrying capacity is the largest number of a population that the environment of a particular area can carry or support.

Census is an official counting of the population.

Crude rates are vital rates which are not adjusted for the age and sex structure of a population.

Death rates (or mortality rates) measure the proportionate number of deaths in a population.

Fecundity rates measure the biological capacity of females in a population to produce offspring.

Fertility rates measure the actual production of offspring by females in a population.

Migration is the movement of people from one area to another.

Migration change is the net change in the total population of an area due to migration.

Morbidity rates measure the amount of illness in a population.

Natality rates (see Birth rates).

Natural change is the net change in the total population of an area due to the balance of births and deaths.

Population pyramids show the age and sex distribution of a population.

Replacement rates are estimates of the extent to which a given population is producing enough offspring to replace itself.

Reproduction rates measure the number of girls born to females in the childbearing age groups (roughly 15 to 45 years) in a population.

Saturation is the level at which the population of an area exactly equals its carrying capacity.

Standardized rates are vital rates which are adjusted for such factors as the age and sex structure of a population.

Survivorship curves show the proportion of a given population surviving to a particular age.

Vital rates are the measures which describe changes in the size and structure of a population.

We can illustrate the effects of natural change on a small scale by looking at a single island population. We choose an island that, over the period studied, has a very small migration and the population approximates a closed system. Figure 7-2 shows the effects of births and deaths on the total population of the small island of Mauritius in the Indian Ocean. In 1900 the island had a total population of approximately 0.3 million, which increased rather slowly to around 0.4 million by 1950; since then it has increased sharply to nearly 0.9

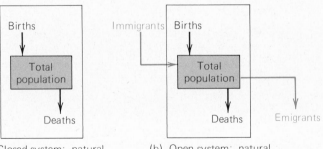

(a) Closed system: natural change only

(b) Open system: natural plus migration change

Figure 7-1. Population change. The total size of the population of any part of the earth's surface may be thought of as like the water level in a bath. It is the result of inflows (shown at the top of the diagram) and outflows (shown at the bottom).

Figure 7-2. **Natural components in population change.**
Changing patterns of fertility and mortality are shown for the Indian Ocean Island of Mauritius. Related changes in social and economic conditions and natural hazards are indicated. Since the last date shown on the graph total population has continued to grow and is now 0.9 million. By the mid 1970s the birth rate had fallen below 30 but the death rate remained at around 7. [From Population Reference Bureau, *Population Bulletin* **18,** 5 (1962), Fig. 1.]

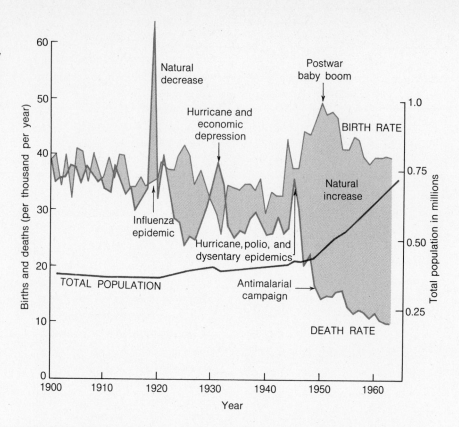

million. The diagram shows the changes in births and deaths over the period, expressed as birth and death rates per thousand inhabitants per year. Births and deaths were almost in balance until about 1920, when general medical improvements began to lower the death rate. Individual peaks in the two rates are associated with both natural disasters, such as hurricanes and epidemics, and economic fluctuations such as the 1929 depression and the postwar boom of the late 1940s. The 1970s have seen a fall in the birth rate, and the island's population is now increasing at an annual rate of 1.3 per cent.

In describing change on the island we have been talking in terms of "rates." How are these rates measured, and how should we interpret them? There are several ways of measuring the vital rates of a population, of which the crude rates of birth, death, and growth are the simplest. The *crude birth rate* is defined as the number of births over a unit of time divided by the average population. The *crude death rate* describes the number of deaths per unit of time; *crude growth rate* describes the difference between the number of births and deaths per unit of time, each divided by the average population in the time interval. Thus if there were 25 births and 18 deaths in a year on an island whose average population during that year was 500, the crude birth rate would be 50 per thousand; the crude death rate, 36 per thousand; and the crude growth rate, 14 per thousand.

EXPONENTIAL POPULATION GROWTH

Exponential models of population growth describe a simplified situation in which growth (or a decline) is unchecked and the rate of change is constant. We express this simply as

$$\frac{dN}{dt} = rN$$

where N = the number of people,

r = the rate of natural increase (a constant), and

$\frac{d}{dt}$ = the rate of change per unit of time.

The expression states that the amount of growth is related to the size of the population; the larger a population is, the faster it grows.

To simplify the computation, we can rewrite this as

$$N_t = N_0 e^{rt}$$

where N_t = the number of people at time t,

N_0 = the number of people at time 0, and

e = 2.71828

The constant e is the base of Napierian, or natural, logarithms and is the sum of the infinite series

$$1 + \frac{1}{1} + \frac{1}{2 \times 1} + \frac{1}{3 \times 2 \times 1}$$

$$+ \frac{1}{4 \times 3 \times 2 \times 1} + \cdots$$

If we start with 1000 people (N_0) and assume a growth rate of 1 percent per annum ($r = 0.01$), then we can show, by substituting these values in the equation, that after 70 years ($t = 70$) the original population will have doubled ($N_t = 2000$). In another 70 years the population will have doubled again. Exponential models show the critical importance of small changes in the rate of natural increase. If we halve this rate ($r = 0.005$), the population doubles only every 140 years. Despite their simple structure, exponential models are a useful way of describing recent phases of human population growth. [See A. S. Boughey, *Ecology of Populations* (Macmillan, New York, 1968), Chap. 2.]

These rates are described as *crude* because they fail to take into account such factors as the age and sex of the members of the population, or migration. We should certainly expect an island with a large number of young adults to have a higher birth rate and a lower death rate than an island inhabited only by octogenarians! Hence, demographers have defined and developed much more sophisticated measures of change called *net* rates, which take into account the structure of the population. They are somewhat complex to go into here, but interested students will find references to discussions of these more refined measures in "One Step Further . . ." (p. 165).

Growth Rates and Doubling Times   The uncertainties involved in estimating future survival and marriage rates, together with the scarcity of data for much of the world, often force us back to a simpler view of population growth. For example, between October 1, 1967, and September 30, 1968, there were 3,453,000 live births and 1,906,000 deaths in the United States. If we take the estimated midyear population of the country on March 31, 1968, as 198,400,000, simple arithmetic indicates that for every 1000 Americans there were 17.4 births and 9.6 deaths. The excess of births over deaths was 7.8 per thousand, and the annual rate of natural increase was less than 0.8 percent.

If we were simply to add 8 new individuals each year to the 1000 still living, it would take 125 years for the population to double (125 × 8 = 1000). But this is not what happens. The people added to the population also increase at a rate of 8 per thousand. Population grows exponentially, just like money earning compound interest in a bank. (See the discussion of exponential population growth on the previous page.) The doubling time is, therefore, shortened from 125 years to only 87 years. As the rate of natural increase rises, the doubling time decreases sharply. When it is 2 percent (the current world rate), the doubling time is 35 years. For some of the populations of tropical Latin American countries with rates over 3.25 percent, the doubling time is only a little over 20 years!

Survivorship Curves and Age Pyramids   In order to understand the rate of growth of a population, we must know something about its age and sex structure. By studying the ages at which the members of a population die, we can establish *survivorship curves*. As Figure 7-3 shows, these tell us the number of survivors of an original group (say all those born in a given year) according to their age at death. If this were a perfect world from which all accidents and infections had been eliminated, so that we all lived into our 80th year, the curve would have an abrupt right angle [as in Figure 7-3 (a)]. If all members of a given population have exactly the same capacity for survival, their survivorship curve has this shape. In practice, the curves for real populations have complex forms, but there is a tendency for the populations of advanced countries to have curves closer to the hypothetical right-angled one than primitive populations or those of underdeveloped countries. Figures 7-3(b) and (c) go on to illustrate the estimated survival curves for three populations with low survival rates (those of the Stone and Bronze Ages and of China in 1930) and others with medium or high survival rates (like New Zealand and the Netherlands in the 1950s).

Another useful and often-used method of portraying the structure of a population is the *population pyramid*. This is a vertical bar graph showing the

Figure 7-3.   Survivorship curves. Here, idealized survivorship curves (a) are contrasted with actual examples of low survival curves (b) and medium and high survival curves (c) from different cultures and from different time periods. The vertical axis measures the number of surviving members of a population of 1000 births in relation to their age. Age is measured on the horizontal axis in years. [From C. Clark, *Population Growth and Land Use* (Macmillan, London, and St. Martin's Press, New York, 1967), pp. 39–40, Figs. 11A, 11B.]

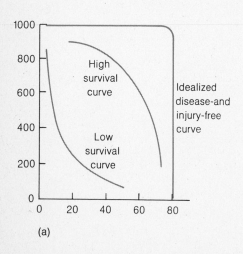
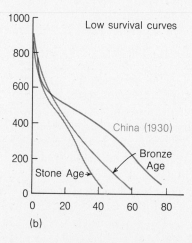
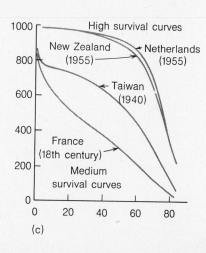

proportion of individuals in various age ranges. Figure 7-4 shows some idealized age pyramids for various human populations. The number of males is measured left of the axis and the number of females right of the axis. We shall come back to population pyramids later in this chapter and also in Chapter 20, where we look at the economic implications of highly skewed population-age distributions for developing countries.

To sum up: The population size of any part of the earth's surface is determined by two forces, natural change (births and deaths) and migration. Generally, the larger the population of an area, the more important natural change becomes, as opposed to migration. At the world level, only natural change is important. Rates of natural change can be measured in several ways and the resulting distribution of population can be shown by survivorship curves and population pyramids.

The compound course of population growth we have described is representative only of the *biotic potential* of a population, that is, of its theoretical rate of growth when it is allowed to develop in an optimal environment of unlimited size. This is rather like measuring the growth of a favorite indoor plant that we regularly feed, water, and fuss over. Under natural conditions,

7-2
Ecological Checks on Growth

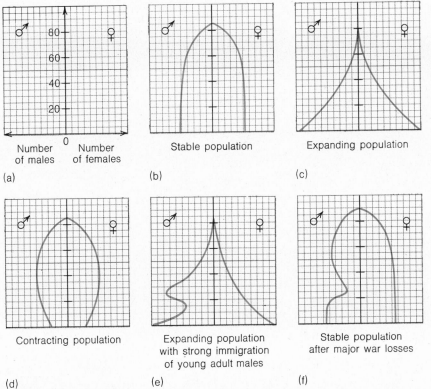

(a) Number of males 0 Number of females

(b) Stable population

(c) Expanding population

(d) Contracting population

(e) Expanding population with strong immigration of young adult males

(f) Stable population after major war losses

Figure 7-4. Population pyramids. Characteristic pyramids for different populations at different stages of growth. These give only the bare bones of the situation and should be compared with population pyramids of actual countries later in the book (See Figure 21-5). Females tend to survive into old age in greater numbers than males and so the curves in the figures shown here are less symmetrical than drawn.

we would expect this potential for growth to be checked either by some natural limits or, as seems likely in the case of both animal and human populations, by cultural, social, and economic constraints. Here we look at some of the models used to understand these checks and how they operate.

Ecological Feedbacks and the Malthusian Hypothesis   In an earlier chapter we came across an example of an animal population that increased very greatly in numbers, the crown of thorn starfish. If detailed figures are available, we can chart these kinds of eruptions in animal numbers. Figure 7-5 shows one example of two animals whose numbers follow a cyclic pattern. Look first at the curve for the hares. Note how it rises explosively upward, reaches a peak, and then falls. If you look at the second curve (the lynx) you'll see why! Lynx become more abundant as their prey—the hares—grow in numbers. Eventually both populations collapse and a new cycle starts again.

We have already met feedback models that show the kind of forces which bring numbers back into line with the capacity of the local environment to

Figure 7-5.   Fluctuations in animal populations.

(a) Relations between two Canadian wildlife populations, one a predator (the lynx) and another its prey (the snowshoe hare). The hare population is large when lynx numbers are low (1) and small when lynx numbers are high (3). Conversely the lynx population is large when there is a plentiful supply of hares (2) but low when hares are scarce (4). (b) The sequence of situations, (1) through (4), gives rise to cycles in the populations of both animals over time. The lynx cycle lags slightly behind the hare cycle. (c) The graph shows changes in the abundance of lynx (predators) and snowshoe hares (prey) over a 90-year period, determined from the number of pelts received by the Hudson's Bay Company. [Data from MacLulich; reprinted, with permission, from E. P. Odum, *Fundamentals of Ecology*, 3rd ed. Copyright 1971 by the W. B. Saunders Company, Philadelphia.]

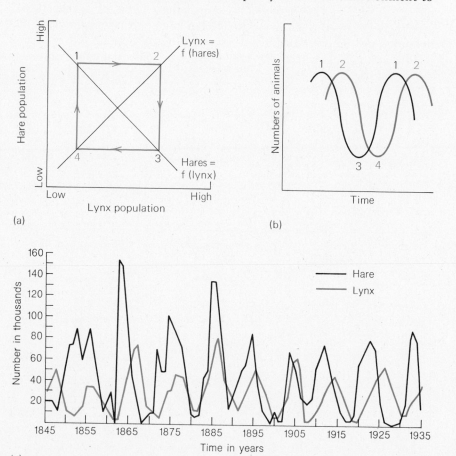

LOGISTIC POPULATION GROWTH

When a population is allowed to develop in an optimal environment of unlimited size, its growth follows an *exponential* curve. If we now introduce a fixed *carrying capacity*, or *saturation level* (K), the potential for biological growth, or the biotic potential, will be modified by environmental pressures.

We can introduce this environmental pressure into the exponential growth model

$$\frac{dN}{dt} = rN$$

described earlier by subtracting

$$\frac{K - N}{K}$$

from rN. We then have

$$\frac{dN}{dt} = rN \cdot \frac{K - N}{K}$$

where

N = the number of individuals in the population.

K = the maximal number of individuals allowed by the carrying capacity.

r = the rate of growth per individual, and

$\frac{d}{dt}$ = the rate of change per unit of time.

Both N and K also can be expressed as population densities. This modified

growth curve is termed a *logistic* growth curve and has a characteristic S-shape. Its calculation is described in Section 13-2. [See also A. S. Boughey, *Ecology of Populations* (Macmillan, New York, 1968), Chap. 2.]

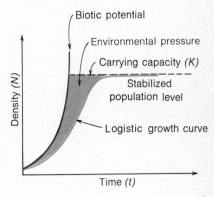

support them. As Figure 3-4(b) (p. 51) showed, an increase in population size may mean less food per individual, more deaths than normal, and a subsequent reduction in numbers. Conversely, a decrease in population may set in motion a chain of events that raises the number in a group to the original level. Clearly the model is a very rough one, but it broadly fits the observed facts for many animal species.

But what about human beings? The world population has been uniformly increasing over at least the last 500 years. Does this mean simply that the wavelength of the human population is a very, very long one—that it will hit a peak at some point in the next few centuries and then stabilize or decline? Such questions are impossible to answer with confidence without going into questions of human culture, human economics, and human politics—the substance of Parts Three, Four, and Five of this book. We should, however, at this point, note the analogies between human and other animal populations.

These analogies troubled Thomas Robert Malthus, the English demographer, in writing his now-famous *Principles of Population*, published in 1798 [Figure 7-6(a)]. Malthus saw dire ecological consequences in the continuing growth of the human population. He claimed that population has a tendency to increase geometrically (by increasing amounts, as does the series 1, 2, 4, 8, 16 . . .) while the food sources for that population, even with improving agricultural methods, increase arithmetically (by a constant

Figure 7-6.   The malthusian equation.
(a) Modern study of population geography can be traced back to the
English demographer, Thomas Malthus (1766–1834), whose book on
the principles of population growth was published in 1798. (b) He
proposed that a population would always outrun its food supply in the
long run since population grows geometrically while food supplies grow
arithmetically. In the diagram, the food supply is originally at a level
of 10 units and increases by 3 units in each time period; the
population is originally at a level of 0.1 but doubles in each time
period. Whatever figures are chosen, in the long run the exponential
curve will eventually intersect the arithmetic curve. (c) Increasing
crowding of population would put such pressures on the food supply
that, under the malthusian equation, only hunger, disease, and war
could bring numbers back to supportable levels. [(a) From the
Granger Collection. (c) Photo by Werner Bishof, Magnum.]

(a)

(c)

(b)

amount, like the series 10, 20, 30, 40, 50 . . .). As Figure 7-6(b) shows, given
these assumptions Malthus was able to demonstrate that any rate of population increase (however small) would eventually exceed any conceivable food
supply. When growth reached that point, it could be kept in check, according
to Malthus, only by "war, vice, and misery." [See Figure 7-6(c).] Yet the basis
for the arithmetic growth of agriculture was never made clear. Moreover, in
the 1817 edition of his book, Malthus paid considerably more attention to

the curtailment of population increases through birth control than to the gloomy devices of war, vice, and increasing human misery.

The Carrying Capacity of Environments We can explain a simple Malthusian check on population growth by imagining a fixed level above which numbers cannot expand. At this saturation level, the population exactly equals the *carrying capacity* of the local environment (i.e., the number of members of a given species it has the biological capacity to provide food for). This is represented in Figure 7-7 by a population ceiling.

What will happen as population growth approaches this ceiling? Three situations are conceivable. First, the rate of increase may continue unchanged until the ceiling is reached, and then abruptly drop to zero. Second, the rate of increase may decline as it approaches the ceiling, eventually falling to zero. Third, the population may overshoot the ceiling periodically, only to be reduced by food shortages, and oscillate above and below the carrying capacity (as in Figure 7-5).

The instantaneous adjustment implied in the first solution seems rather improbable, not least because the mechanism by which such a sudden change might be achieved is unclear. It is unsupported either by empirical evidence on human numbers or by the growing number of studies of other animal populations. The second solution, in which the rate of increase tapers off as numbers approach the critical level, is more plausible. (See the discussion of logistic population growth on page 147.) Such a solution does, however, imply more knowledge about environmental limits and more social control over births than we presently have.

The third possibility as population approaches the critical level is illustrated in Figure 7-7(c). Here, the relationship between the population and the carrying capacity of the environment is reflected in changes in both birth and death rates. Too many people (i.e., a population above the carrying capacity) leads to deaths through starvation and fewer births; this brings the population down. This overshooting and undershooting of the saturation level, as population fluctuates above and below the carrying capacity of the environment, is commonly encountered in animal populations. Periods when a species is abundant follow periods when it is scarce, in a rather regular rhythm.

Because the first accurate census data for human populations became available only in the late eighteenth century, it is difficult to determine which, if any, of these simple models are appropriate to the human situation. Historical trends in world population reveal that the current exponential pattern of growth is relatively recent. The early period of man's tenure on the earth was one in which the Malthusian constraint of hunger played a key role, and the first two graphs in Figure 7-7 may be more relevant to that time.

Malthusian Checks on Population: Famine Here we look at how the food supply acts as a check on the human population in two contexts: during specific local famines and in the longer run on a global scale.

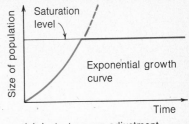

(a) Instantaneous adjustment

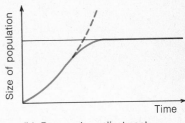

(b) Progressive adjustment

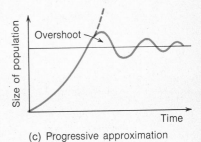

(c) Progressive approximation

Figure 7-7. Environmental constraints on growth. These graphs illustrate three hypothetical relations between a population growing exponentially and an environment with a limited carrying capacity (saturation level).

Table 7-1 Population losses from major famines

Location	Dates	Estimated deaths[a] (millions)
India	1837	0.8
Ireland	1845	0.75
India	1863	1.0
India	1876– 1878	5.0
East China	1877– 1879	9.0
China	1902	1.0
China	1928– 1929	3.0
USSR	1932– 1934	4.0

[a]Famines before the nineteenth century are poorly documented, and estimates of the number of deaths vary widely.

Famines and Local Food Shortages   We can find some rough indications of how environmental checks operate by piecing together historical records of local breakdowns of food-population balances. Famines may be closely related to environmental events (as in the case of a drought) or largely unrelated (as was the extensive famine in the refugee population of central and eastern Europe at the close of World War II). We can argue, however, that a poor region with limited food stockpiles whose population has a food intake only slightly above the starvation line and whose climate varies greatly from year to year is more likely to experience famine than other, more fortunate regions.

Although records of famines are difficult to assemble, early historical accounts support the view that famines were once much more pervasive and extensive. The combination of high population densities in rural areas and low caloric intake, in addition to occasional failures of monsoon rains to arrive on schedule, make south and east Asia the world's main disaster areas for famines. The world's worst recorded famine occurred between 1877 and 1879, when an estimated 9 million people died in China. (See Table 7-1.) In Europe, the worst disaster of this type was the Irish potato famine of 1845. Large areas with high populations were supported at subsistence levels by a single plant species, the potato. Blight followed by crop failures in 1845 and again in 1846 brought famines of Malthusian dimensions; deaths, massive emigration (800,000 people moved out of Ireland in the next 5 years), plus a sharply reduced birth rate lowered the population of 8 million (based on the 1841 census) to around half that figure by the end of the century.

When the forces that trigger famine are environmental, as when the monsoon rains fail, we can regard them as fluctuations in the carrying capacity of the environment itself. Thus, we can discard the idea of a fixed limit on population growth (used in Figure 7-7) and replace it with a variable limit. Figure 7-8 shows a series of changes over time in the carrying capacity of an area. Environmental changes are of three kinds. First, there are *nonrecurrent changes* that may be (1) abrupt [Figure 7-8(a)], such as those following the overrunning of fertile fields by a flow of lava, or (2) more gradual, such as those caused by a deteriorating climate or eroding topsoil. Second, there are *periodic regular changes* [Figure 7-8(b) and (c)], including annual variations in productivity connected with seasonal variations in growing conditions. Changes of this type are caused by, for example, low winter temperatures in the boreal zone or summer droughts in the Mediterranean zone. Third, there are *periodic but irregular changes* [Figure 7-8(d)]. That is, environments may have irregular periods of low productivity induced by irregular natural events like river plain flooding. We have already encountered examples of all three types of environmental instability in Chapter 6.

Famine and Migration   Local human populations respond to changes in the carrying capacity of their environment in different ways. Regular seasonal changes may be coped with either by storing food for use during the season of low productivity or by regular migrations to other areas. (Livestock, for

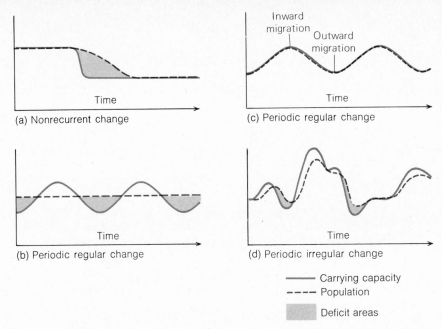

(a) Nonrecurrent change

(b) Periodic regular change

Inward migration

Outward migration

(c) Periodic regular change

(d) Periodic irregular change

——— Carrying capacity
– – – – Population

Deficit areas

Figure 7-8. Environmental change and population size.
The graphs show hypothetical responses of population numbers to changes in the carrying capacity of a given region. In (a) the population adjusts slowly to a permanent reduction in carrying capacity over a long time. In (b) the timespan is shorter and the population copes with periodic regular changes by storing food during good years or seasons. In (c) inward and outward migration keep the population in line with the food supply. Where the changes are seasonal regular migration movements of animal stock (termed *transhumance*) may occur. In (d) both strategies are used to cope with irregular fluctuations in the food supply.

example, are moved from low to high pastures in the European Alps as the seasons change.) Periodic but irregular changes pose more severe problems. If the change is for a relatively short period (as in the case of river flooding), temporary abandonment of the area may solve the problem. More serious climatic changes may be both too lasting and too widespread for evacuation to provide a solution. Here the classic famine symptoms set in. Often, the pattern is reinforced by the consumption of the next season's seed corn for food and the resulting loss of productive capacity in the following season. Longer-term declines usually lead to steady emigration and a cumulative fall in population.

All these responses to environmental change involve some spatial (migratory) movements. There may be outward movements (seasonal, periodic, or permanent) of population from areas where there is a food deficit or inward movements of food from areas of surplus. These strategies obviously apply only in the case of local famines. Such spatial reshuffling of population and resources would not help in the event of a global famine.

Global Food Shortages Setting aside for a moment critical local shortages, how far is the earth as a whole able to support the increasing demands for food that a growing population makes? We can gain some rough idea of the present situation by taking current demands and comparing them with estimates of the earth's ultimate food-producing capacity. The United Nations World Health Organization (WHO) has estimated that the world's people today consume around 10^7 tons of food per year. As we saw in Section 3-3, forecasters at Resources for the Future (RFF) predict that, given the present levels of solar energy and the present distribution of world climate, the maximum amount of organic matter that could be produced by photosynthesis is about 10^{11} tons per year. A comparison of these two estimates suggests that only a trivial portion (about one-hundredth of 1 percent) of the earth's ultimate food-production capability is being used.

But how relevant are these estimates really? Organic matter is not always edible. Even in the case of croplands, well below half of the gross product may be edible. As for grazing lands, an energy-conversion factor of 12 to 1 must be used to convert the energy consumed by stock to the food value of the stock to man. Thus, the original figures must be revised downward to a maximum of around 10^9 tons per year. If we also raise the WHO estimate of food consumption to allow for such things as preharvest losses (30 percent), postharvest losses (30 percent), edibility, conversion factors, and noncropland, then the totals converge rapidly. It would be more realistic to say that the farming industry is operating at less than 15 percent of its maximal productive potential.

But the load on the earth's food-producing area is not evenly spread. Most food is produced from a small part of the global surface. The intensive pressure on the cultivated land of the world was indicated in Table 3-3. Cultivated land constitutes only 2 percent of the earth's total area (land and sea) but it is capable of producing nearly three-quarters of the world's potential output of edible matter. The grasslands are second in the productivity of edible material. The greatest gap between total productivity from photosynthesis and productivity of edible material is in forest areas. The edible fraction from the oceans and inland waters is quite small; there is little evidence that the world's oceans (despite their enormous area) will make anything but a marginal contribution to world food production in the next generation or two.

Using the most conservative estimates, we can say that the food-producing capacity of this planet is vast, even with our existing technology. Given improved standards of production, its capacity to feed a population much larger than 4.1 billion is not in doubt. Unfortunately, calculations of global demand conceal immense local differences in food consumption. (See Figure 7-9.) Broadly speaking, the ratio of the well-fed to the undernourished is about 1:6. About 20 percent of the people in the underdeveloped countries are undernourished (i.e., receive less than the minimal number of calories they need per day), and some 60 percent lack one or more of the essential nutrients, commonly protein.

Malthusian Checks on Population: Crowding and Conflict   In our models so far, we have envisaged a simple situation in which a single, homogeneous world population monopolized the exploitation of the available resources. But what happens if we partition the population (into "haves" and "have nots," for instance) and allow different groups to compete for resources? We

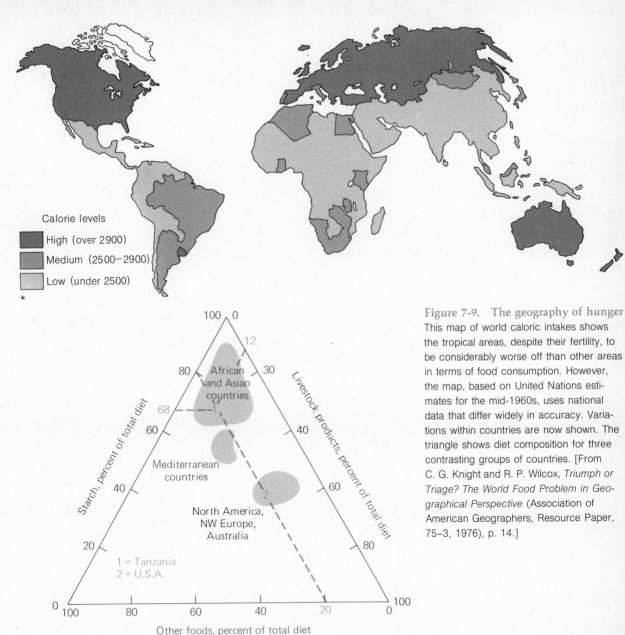

Figure 7-9. The geography of hunger
This map of world caloric intakes shows
the tropical areas, despite their fertility, to
be considerably worse off than other areas
in terms of food consumption. However,
the map, based on United Nations estimates for the mid-1960s, uses national
data that differ widely in accuracy. Variations within countries are now shown. The
triangle shows diet composition for three
contrasting groups of countries. [From
C. G. Knight and R. P. Wilcox, *Triumph or
Triage? The World Food Problem in Geographical Perspective* (Association of
American Geographers, Resource Paper,
75–3, 1976), p. 14.]

can, of course, follow Malthus and assume that competition will lead to conflict, conflict to wars, and wars to a reduction in population. But despite the
immensity of war losses in the last 250 years, there is no evidence that they
have checked the exponential increase in population. To be sure, individual
countries, like France after World War I, experienced severe checks on
growth, but these lasted little more than a single generation. Wholesale

Table 7-2 Population losses from major conflicts

Conflict	Dates	Estimated deaths (millions)
World War II	1939–1945	7.3
World War I	1914–1918	7.2
Taiping Rebellion	1851–1864	6.3
Spanish Civil War	1936–1939	6.3
1st Chinese Communist War	1927–1936	6.1
La Plata War	1865–1870	6.0
Indian Communal Riots	1946–1948	5.9
Russian Revolution	1918–1920	5.7
Crimean War	1853–1856	5.4
Franco-Prussian War	1870–1871	5.4
Mexican Revolution	1910–1920	5.4

SOURCE: Data from L. F. Richardson, *Statistics of Deadly Quarrels* (Boxwood Press, Pittsburgh, Pa., 1960). Richardson gives a figure of 5.8 million dollars for the U.S. Civil War (1861–1865), but this is almost certainly an overestimate. Figures generally refer to direct losses of military personnel; civilian losses would inflate those given here.

checks on population growth from international and other conflicts would demand far more deaths than the biggest conflicts have yet produced. Note, however, that Table 7-2 includes only estimated deaths and does not allow for low birth rates due to a reduction in the male population and separation of families.

Because the historical record is unhelpful in assessing the losses from any future nuclear conflict, we can only turn to more speculative evidence. Ecologist L. B. Slobodkin has considered theoretical patterns of competition between two populations living in the same area and having different rates of growth and saturation levels. Each populations's logistic growth curve flattens as its density reaches the carrying capacity of the environment. Figure 7-10 shows representative density curves over time for the two populations. An equilibrium is established when both cease to grow. But both populations compete for the same resource, so the growth of one is dependent on the growth of the other. Slobodkin's aim was to specify, given these conditions,

Figure 7-10. Competition and population growth.
The graphs show two possible outcomes when exponentially growing populations (N₁ and N₂) with fixed saturation levels (S₁ and S₂) must depend on the same resources to grow.

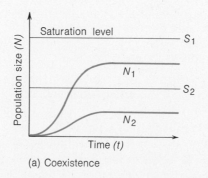

(a) Coexistence

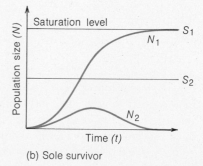

(b) Sole survivor

whether the two populations could coexist in a state of equilibrium or whether one would progressively dominate the available resources to the exclusion of the other. With his theoretical model, two outcomes are possible. Either both populations coexist, but their numbers remain below the saturation levels that would otherwise prevail, or only one population survives and reaches its relevant saturation level. In other, and more complex cases, both populations may fluctuate cyclically. Note the example of the predator (lynx) and prey (snowshoe hare) populations in Figure 7-5.

Although the Slobodkin model refers only to two populations in highly simplified conditions, its ecological implication for the competitive relations between different populations and between subgroups within the same populations is important. It illustrates how we can simulate future conditions of conflict and thus try to avoid them. Furthermore, it underscores the fact that it is not so much the level of resources as our ability to share and distribute them that lies at the heart of the population–resources dilemma.

Non-Malthusian models   In this chapter we have looked at the most important of the ecological models of human population, the Malthusian. Other views of human population density, notably those of Boserup, will be taken up in the next chapter. (See Section 8-3.)

How far do actual patterns of population growth follow the abstract models discussed above? We shall continue to consider the situation at the world level because this is the only level at which we can legitimately ignore migratory transfers of population. In this sense, then, despite its size, the world is one of the simplest population systems.

7-3
The History of World Population Growth

Past Patterns of Growth   Conclusions about past population growth might usefully begin with Oxford University demographer Colin Clark's axiom that most of the historical (still more the prehistorical) evidence on population is not very accurate and that, as we go back in time, its accuracy generally diminishes further. It is therefore extremely difficult to estimate the size of early populations. Archeologists suggest that, at the beginning of agriculture, the population of the world was not more than 10 million. For the beginning of the Christian Era the estimate is 250 million, and for A.D. 1650 it is double that figure, 500 million. As we noted earlier, for periods from the late eighteenth century on, the increasing number of national censuses makes the estimating process somewhat easier. One of the articles of the United States Constitution required that "enumeration shall be made within three years after the first meeting of the Congress . . . and within every subsequent Term of ten Years." The actual census (Figure 7-11) conducted in 1790 represents a milestone in demographic history: Only Sweden had collected accurate information on such a scale before, and most countries were

Atlas cross-check.
At this point in your reading, you may find it useful to look at the World Population Density map in the atlas section in the middle of this book.

SCHEDULE *of the whole Number of* PERSONS *within the several Diftricts of the* UNITED STATES, *taken according to* " An Act providing for the Enumeration of the Inhabitants of the United States;" *paffed March the 1ft, 1790.*

DISTRICTS.	Free white Males of fixteen years and upwards, including heads of families.	Free white Males under fixteen years.	Free white Females including heads of families.	All other free perfons.	Slaves.	Total.
* Vermont	22,135	22,328	40,505	255	16	85,539
New-Hampfhire	36,086	34,851	70,160	630	158	141,885
Maine	24,384	24,748	46,870	538	NONE	96,540
Maffachufetts	95,453	87,289	190,582	5,463	NONE	378,787
Rhode-Ifland	16,019	15,799	32,652	3,407	948	68,825
Connecticut	60,523	54,403	117,448	2,808	2,764	237,946
New-York	83,700	78,122	152,320	4,654	21,324	340,120
New-Jerfey	45,251	41,416	83,287	2,762	11,423	184,139
Pennfylvania	110,788	106,948	206,363	6,537	3,787	434,373
Delaware	11,783	12,143	22,384	3,899	8,887	59,094
Maryland	55,915	51,339	101,395	8,043	103,036	319,728
Virginia	110,936	116,135	215,046	12,866	292,627	747,610
Kentucky	15,154	17,057	28,922	114	12,430	73,677
North-Carolina	69,988	77,506	140,710	4,975	100,572	393,751
South-Carolina	-	-	-	-	-	-
Georgia	13,103	14,044	25,739	398	29,264	82,548

	Free white Males of twenty-one years and upwards, including heads of families.	Free Males under twenty-one years of age.	Free white Females, including heads of families.	All other Perfons.	Slaves.	Total.
S. Weftern Territory	6,271	10,277	15,365	361	3,417	35,691
N. Do.	-	-	-	-	-	-

Truly ftated from the original Returns depofited in the Office of the Secretary of State.

TH: JEFFERSON.

October 24, 1791.

* This return was not figned by the marfhal, but was enclofed and referred to in a letter written and figned by him.

Figure 7-11. America's first census.

A page from the modest 56-page pamphlet giving the results of the first United States census, taken in 1790. Thomas Jefferson was required under the Constitution to conduct a census to insure a fair base for tax apportionment between the states. It has been repeated every subsequent ten years. [Courtesy of the U.S. Bureau of the Census. From L. Broom and P. Selznick, *Sociology*, 6th ed. (Harper & Row, New York, 1977), p. 262.]

to lag behind for some further decades. Even for the middle 1970s, the United Nations' estimate of world population of around 4.3 billion is very approximate. Figure 7-12 presents a general picture of the increase from the beginning of the Christian Era projected to A.D. 2000.

Viewed in a historical context, the present expansion of the world population by about 2 percent per year must be extremely rapid. If the expansion is projected backward in time, the population reduces to a single human couple by only 500 B.C. In fact, we know that human populations were inhabiting the earth in 500,000 B.C. Thus, the average rate of population increase in this early period must have been extremely slow. Rates of increase as low as 0.01 percent per year have been proposed, but any model assuming continuous growth is probably unrealistic and unhelpful. By comparing the human population with other mammalian populations, we can conjecture that primitive populations underwent major fluctuations, and a fluctuating model on the lines of Figure 7-8(d) may be more appropriate.

The Demographic Transition   In the last two hundred years, industrialization and urbanization appear to have caused a transition in the ways in which population grows. This *demographic transition* can be represented as a sequence of changes over time in vital rates. (See Figure 7-13.) We can recognize four connected phases in this sequence. In the first, or *high-stationary*, phase, both the birth and death rates are high. Although both rates vary, we can assume that the greatest variation is caused by deaths stemming from famines, wars, and diseases. Because the gains in population during a period when death rates are high are canceled by the losses when death rates are low, the population remains at a low but fluctuating level.

The second, or *early-expanding*, phase is characterized by a continuing high birth rate but a fall in death rates. As a result, the life expectancy increases and the population begins to expand. The fall in the death rate is ushered in

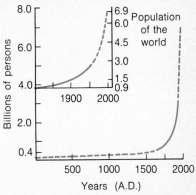

Figure 7-12.   World population. The graphs show the general pattern of increase in the estimated world population over the last 2000 years and (inset) the last 200 years. The solid line indicates periods for which reasonably good census material is available; the dashed line indicated estimates. [From H. F. Dorn, in P. M. Hauser, Ed., *The Population Dilemma* (Prentice-Hall, Englewood Cliffs, N. J., 1963), p. 10, Fig. 1. Copyright © 1963 by The American Assembly, Columbia University. Reproduced by permission of Prentice-Hall, Inc.]

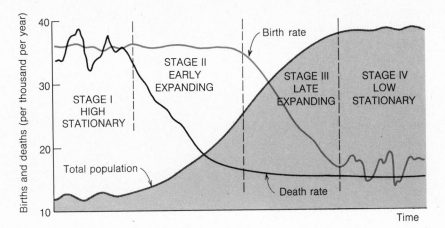

Figure 7-13.   The demographic transition. The graph shows four stages in a demographic sequence in which industrialization and urbanization are important factors. Note the variability of death rates (due mainly to famines and epidemic diseases) in Stage I and the variability of birth rates (due mainly to cycles in economic prosperity and attitudes to family size) in Stage IV. Compare the general curves shown with the actual pattern of change for the island of Mauritius in Figure 7-2 and the sequence for Sweden in Table 7-3.

Table 7-3   Evolution of the Swedish population in terms of the demographic transition

Stage	Dates	Total population (millions)	Crude rates per 000		Age structure (percent)		Job structure (percent in agric.)	Urban structure (percent in Stockholm area)	Notes
			Birth	Death	Children (0–14)	Old (65–)			
I High stationary	–1750	1.8	36	27	33	6	n.a.	8	High birth and death rates, stable age structure. Average of 8 children per marriage (4 or 5 gain adulthood)
II Early expanding	1810	2.5	33	26	32	5	n.a.	7	Death rates start to fall. Sharp increase in total population. Delayed marriage. Rural over-population and overseas emigration.
III Late expanding	1870	4.4	30	18	34	5	72	6	Birth rates start to fall. Slow change in age structure. Fall in farm jobs. Urbanization.
IV Low stationary	1930	6.3	14	12	25	9	39	13	Fluctuating birth rates. Small families dominate, marriages less stable. Sharp increase in elderly. Continued urbanization. Social welfare state.
	1975–	8.2	13	11	20	15	7	19	

SOURCE: Data from *The biography of a people: past and future population changes in Sweden* (Royal Ministry for Foreign Affairs, Stockholm, 1974).
n.a. Not available.

by improvements in nutrition, in sanitation, in the stability of the government (which means fewer wars), in medical technology, and so on. The third, or *late-expanding*, phase is characterized by a stabilization of the death rate at a low level and a reduction in the birth rate. As a result, the rate of expansion slows down. The fall in the birth rate is associated with the growth of an urban–industrial society in which the economic burden of rearing and educating children tempers the desire for large families, and birth control techniques make family planning easier.

The fourth, or *low-stationary*, phase is a period when birth and death rates have stabilized at a low level; consequently, the population is stationary. This period is unlike the high-stationary phase in that the death rate is more stable than the birth rate.

Because of its unique population records which begin in 1750, Sweden's growth is of special interest. Table 7-3 summarizes the historical record over a 225-year period. Note the way in which demographic rates, age structure, job structure, and geographical distribution are all related.

To what extent do other countries today fit this pattern? Table 7-4 summarizes the present position. Note that it is confined to large population aggregates (countries with 10 million people or more) and that there may therefore be considerable local variation. Let us take Australia as an example.

The country as a whole falls in Stage IV, with its birth rate at 19 and death rate at 8 per thousand. But Canberra, its national capital, is expanding fast with a young population moving into new government-related jobs. Its birth rate of 24 and death rate of 8 per thousand would put it into a Stage III category.

But for the world picture, the position is clear. Because of the rapid spread of medical technology, no major countries are found in Stage I. The high-stationary phase characterizes countries with an uncertain and low level of food production; most of the population in these countries engages in agriculture. This phase was universal among the human population during most of its early history, but now it is restricted to the more isolated and primitive groups. Most of the populations of the Third World of Latin America, Africa, and South Asia fall into the early expanding phase. Population is expanding rapidly, as environmental and medical technology have brought substantial improvements in life expectancy. Some of the countries

Table 7-4

Table 7-4 Demographic stage reached by large countries[a]

Stage II Early expanding	Stage III Late expanding	Stage IV Low stationary
Birth rate over 35 Death rate over 15	Birth rate over 20 Death rate 15 and below	Birth rate 20 and below Death rate 15 and below
AFRICA Algeria 39/17 Ethiopia 46/25 Ghana 47/18 Kenya 48/18 Morocco 50/17 Nigeria 50/25 South Africa 40/16 Sudan 49/18 Tanzania 47/22 Uganda 43/18 Zaire 44/23 ASIA Afghanistan 51/27 Burma 40/17 India 43/18 Iran 45/17 Iraq 49/16 Indonesia 48/19 Nepal 45/23	AFRICA Egypt 35/13 ASIA China 33/15 Korea, North 39/11 Korea, South 36/11 Malaysia 38/11 Pakistan 36/12 Philippines 45/12 Sri Lanka 30/8 Thailand 43/10 Turkey 40/15 LATIN AMERICA Argentina 23/9 Brazil 38/10 Chile 28/9 Colombia 45/11 Mexico 46/8 Peru 42/11 Venezuela 41/8	ASIA Japan 19/7 AUSTRALASIA Australia 19/8 EUROPE Belgium 13/12 Czechoslovakia 20/12 France 15/10 Germany, East 10/13 Germany, West 10/12 Hungary 18/12 Netherlands 14/8 Poland 18/8 Spain 19/8 United Kingdom 14/12 USSR 18/9 Yugoslavia 18/9 NORTH AMERICA Canada 15/7 United 15/9

[a]Figures following each country refer to the crude birth rate and death rates (per thousand per year). Thus Algeria has a birth rate of 39 per thousand and a death rate of 17 per thousand. Only countries which had a population of 10 millions or more in the mid 1970s are included.

in that group have already experienced substantial drops in the birth rate. This decrease is related to socioeconomic changes, where people work in their occupations, and is reinforced by family planning techniques; thus the populations of some of these countries appear to have moved into the late-expanding phase. Western Europe, the United States, Canada, and Australia are all examples of countries whose populations appear to be moving into the fourth stage. In Section 19-2 we review the global variation in the stages now reached by different countries and relate these to levels of economic development.

Current Trends   The accelerated increase in population in the twentieth century suggests that the exponential model of population growth discussed earler in this chapter may not be wholly inappropriate for the modern period. It is extremely difficult, however, to proceed from a general recognition of the type of expansion that is occurring to a precise forecast. For example, Figure 7-14 presents six estimates made back in 1965 of world population changes before A.D. 2000 that give end-of-the-century populations varying from 4.5 to 7.6 billion. The considerable range of these estimates underlines the immense variation in logical projections of the long-term trend, especially when we allow for possible innovations in birth control techniques and attitudes toward family size over the remaining years of the century. Each projection tends to reflect the trends existing when the forecast was made. There is evidence that improvements in sanitation and medicine, which allowed a decline in death rates, are now leveling off. Birth rates are likely to be more volatile in the future as both fertility drugs and contraceptive devices permit greater control over family size.

Current demographic projections are concerned largely with the next 20 years, the period up to A.D. 2000. This range is justifiable for two reasons. First, it approximates the extreme limits to which current trends extrapolated. Beyond that date, the assumptions, let alone the calculations, become so fraught with error that longer-term forecasts become somewhat absurd. Second, 25–30 years represents one biological generation for a slow-breeding species like us. We hope that our children can improve the range of their own forecasts.

But the world should not end in A.D. 2000, and we can speculate, although in a negative way, on some characteristics of the period beyond it. It seems certain that, in a historical context, the period from 1700 to 2000 is one of exceptionally high increases in population (and high rates of natural resource extraction). To extrapolate present rates of increase into the future is dangerous, but it is instructive. For the United States, a continuation of the rates of population increase prevailing in the mid-twentieth century would result in a density of one person per acre of total land surface—cropland, desert, land of all kinds—by A.D. 2100. The density would rise to around one person per square meter by A.D. 2600. Of course, such projections represent only statistical mumbo jumbo because of the extremely high probability of an inter-

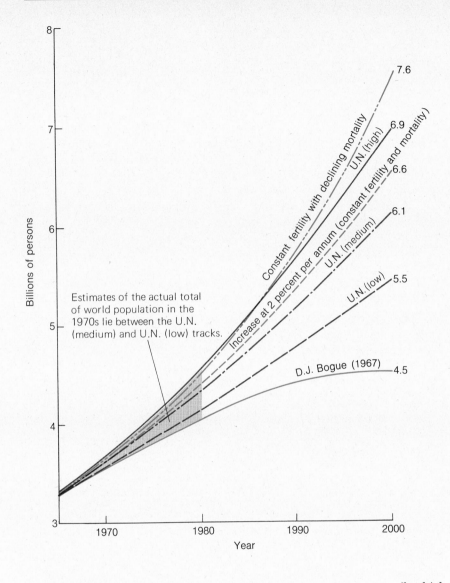

Figure 7-14. World populations projected to A. D. 2000.

These six estimates of the total number of people likely to be around at the end of the century are based on evidence available in the 1960s. The highest estimate assumes that fertility will remain constant as mortality declines. The United Nations provides three separate estimates (high, medium, and low), based on varying assumptions. Most recent information available to the United Nations is now causing demographers to lower their estimates slightly. [From N. Keyfitz, in *Resources and Man: A Study and Recommendations* by the Committee on Resources and Man of the Division of Earth Sciences, National Academy of Science-National Research Council, with the cooperation of the Division of Biology and Agriculture. (Freeman, San Francisco, copyright © 1969).]

vening decrease in population growth. Indeed, we can now see the high growth rates of the 1950s and 1960s as a short-term occurrence. Thus, although the precise mechanisms that will be operating are not clear, there are some grounds for thinking that the next century will be one of considerable limitations on population growth.

Zero Population Growth?   As more and more people realize the serious consequences of continued exponential growth, public opinion is shifting rapidly toward a concern for putting on the brakes. ZPG (zero population growth) is becoming an increasingly popular goal, particularly among the

REPLACEMENT REPRODUCTION

To calculate the replacement rate, we have to take into account several factors. First, we know that the human species produces about 106 male births for every 100 female births. Second, we have to account for the proportion of female offspring who may be expected to die before they themselves reach the reproductive age (say, 15 to 49 years). Next, since the basic data used by demographers usually relates to married women, an allowance has to be made both for those who do not marry, and for those who have offspring outside marriage. So, taking all these factors into account, a family size greater than 2 is needed for replacement reproduction. The actual figure will vary slightly from one country to another, but the figure of 2.3 for the United States is reasonably representative. Once that standard is set up, we can compare it with the actual family size and see whether a population is replacing itself. This is commonly expressed as a percentage: A country with a 100 percent replacement rate is just in balance. By dividing the actual family size by the replacement family size and multiplying by 100, the actual replacement rate can be computed. A family size of 2 would mean a replacement ratio of $(2 \div 2.3) \times 100$, or 87 percent. A single-child family would give 43 percent; 3 children, 130 percent; 4 children, 174 percent; and so on. Currently, none of the countries of the western world is reproducing at replacement rate. In 1977, the estimated rates were 78 percent for the United States, 84 percent for England and Wales, and 88 percent for Australia. Note, however, that these rates can change very sharply. As recently as 1970, the United States had a replacement rate of 115 percent. [See C. Clark, *Population Growth and Land Use* (Macmillan, London, 2nd ed., 1977).]

younger folk of the developed nations. But is zero growth possible? And if so, how soon can it be achieved?

In the strictly mathematical sense, the achievement of a stable, nongrowing population is a straightforward matter. If all the breeding couples in a population together produce just as many children as are needed to replace the present generation, we have what the demographers term *replacement reproduction*. (See the above discussion.) In current terms, this means that the average number of children per married couple should be 2.3. Two children per couple would presumably not be enough to replace the current population because not everyone gets married, not everyone has children, and not all children live to reproductive age.

Even if replacement reproduction rates were acheived immediately, it would take many years for population growth to slow down. The present shape of the population pyramid for the whole world shows very large numbers of children below the reproductive age who will eventually move into the critical childbearing age-band (15–45 years). If all were to adopt a family size of 2.3 as a target, the population would still continue to rise to a peak 1.6 times its present level. If the present programs for birth control through family planning and socioeconomic change were successful in achieving a slow reduction to the replacement rate by A.D. 2000—a *very* optimistic assumption—then the world population would rise to a peak at about 2.5 times its present size sometime in the next century. Thus, there is a momentum about population growth that is very hard to change quickly.

Does this mean that an immediate reduction in population growth is impossible? Research by demographer Tomas Frejka has shown that, for the United States, this is only possible if the target family size is first set well *below* the replacement rate—as low as 1.2 children per family. If this rate were to prevail for a couple of decades, the total population would decline, and the average family size would then need to be raised to above the replacement level. Figure 7-15 shows, on the left, the actual variations in the United States' birth rate over the last sixty years and, on the right, Frejka's calculation of the seesaw birth rates necessary to keep the population at the 1970 level over a 400-year period. According to Frejka, this long period is necessary for the violent cycles of population increases and declines to be smoothed out.

In the first half of this period, the remedy is as unwelcome as the illness. The effects of attempting to stabilize the population at present levels are scarcely less awful than those of unbridled growth. If Frejka's calculations are correct, the United States would go through alternating phases in which its age pyramids fluctuated violently from a dominance of old folk to a dominance of the young. As Figure 7-15 suggests, these phrases would have a period of about 80 years and would become less pronounced as the centuries passed. Enormous economic and social problems would be created in the meantime. This line of alarmist argument is supported by the French historian Chaunu who regards the births lost in the present downturn in western Europe as creating in the next century "a demographic disaster comparable only with that caused by the Black Death."

It would appear, then, that despite an increasing awareness that people cannot go on reproducing at present rates for many more years, any substantial reduction in the rate of population growth is likely to be a slow process if the present cultural barriers to birth control are maintained. Changes in birth rates to a replacement level will not produce zero population growth im-

Figure 7-15. Zero population growth. On the left are the *actual* birth rates in the United States over the 60 years since 1910. On the right are Frejka's estimates of the fluctuating birth rates needed to maintain the U. S. population at its *present* level (i.e., for ZPG) for the next 400 years. Note that the time scales for the curves on the left and the right are different. [Data from Population Reference Bureau and from A. Frejka, *Population Studies* **22** (1968), p. 383, Fig. 1.]

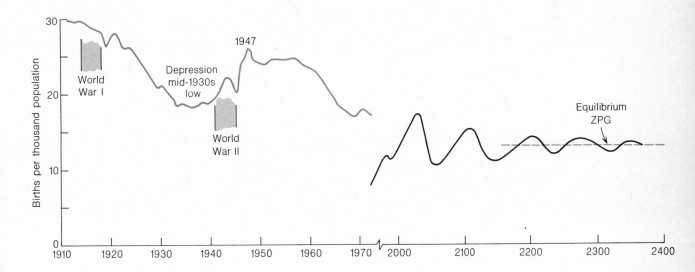

mediately; moreover, an immediate reduction of the growth rate could generate a series of alternating increases and decreases which would last for the next ten or twenty breeding generations. The most likely outcome, from the viewpoint of the late 1970s, is that the world population will continue to grow for the next few generations of *Homo sapiens*, but there will be a slowdown in the rate of growth.

Summary

1. The distribution of the human population around the globe forms the single most important focus in human geography. In one form or another, population concepts underlie all the remaining chapters of this book.

2. Births and deaths (natural change) and movement of people (migration change) are the two forces determining the population size of any area on the earth's surface. At the world level, only natural change is important since the population system is *closed*. At regional and local levels, migration change may be important if the system is *open*.

3. Rates of natural change can be measured in several ways: crude birth rate, crude death rate, or crude growth rate. Such rates can be refined by measurements of change called net rates. Study of such rates shows world population growing at an exponential rate over recent centuries.

4. Survivorship curves show the number of survivors of some original population according to their age at death. Such curves take different forms for different historical periods and different areas, such variations being generally dependent on levels of technological development. Population pyramids are used to show the distribution of age–sex groups in a population.

5. Exponential growth of a population is a useful concept only if we consider it in relation to biotic potential or carrying capacity. Natural, cultural, or socio-economic constraints result in negative feedbacks to check growth. In the Malthusian hypothesis, continuing human population growth will eventually be greater than the growth of food supply. This can be discussed in terms of a Malthusian check, a point above which population cannot grow. This establishes a saturation level at which the population equals the environment's carrying capacity. Environmental checks on human food supplies may result in famine. Such checks are considered to be fluctuations in the carrying capacity of a local area and may be nonrecurrent, periodic regular, or periodic but irregular. Spatial or migratory movements may be a compensatory result of such fluctuations.

6. Population crowding leads to another Malthusian check, conflict, and the problem of competitive growth. Slobodkin's model of this problem using two populations indicates that only two outcomes are possible: subsaturation level of coexistence, or single survivor at saturation level.

7. Currently rates of population increase are falling in the western world and a few countries are now below the replacement reproduction level. Achievement of a long-term balance of births and deaths in terms of zero population growth (ZPG) is likely to be difficult since cyclic oscillations in age structure are likely to be set up.

Reflections

1. Using the latest national census figures for your own country or state, map the pattern of birth rates and death rates for each province, state, or county. Attempt to explain the spatial pattern which results.

2. Construct hypothetical population pyramids for populations which are (a) expanding rapidly, (b) declining sharply, (c) static, or (d) recovering from major wars. Can you match these populations with populations of real countries today?

3. How would you define the term "overpopulation"? Do you think your own country or state is overpopulated? Why? Suggest and defend an optimum population for your area. How does it compare with the optimum populations suggested by others in your class?

4. Set up a debate in the class on the motion that: "The ideas of Malthus on the balance of population and food supply are no longer relevant." On which side would you prefer to speak? Why?

5. Do you think the United States will achieve zero population growth in your lifetime? Which other countries seem able to achieve this goal?

6. Look closely at Figure 7-13 and compare it with Table 7-3. Which phase in the demographic sequence best describes the situation in your own country today? Which phase describes the situation 100 years ago? What factor might account for a change in the pattern of population growth?

7. Review your understanding of the following concepts:

    crude birth rates   carrying capacities
    crude death rates   saturation levels
    crude growth rate   the demographic
    survivorship curves   transition
    population pyramids   zero population growth
    the Malthusian
       hypothesis

One Step Further . . .

Two basic texts by geographers that outline the main concepts of population geography in a systematic manner are
  Zelinsky, W., *Prologue to Population Geography* (Prentice-Hall, Englewood Cliffs, N.J., 1966) *and*
  Clarke, J. I., *Population Geography* (Pergamon, Elmsford N.Y., 1966).

The dynamics of population growth, ways of describing population statistics, and ways of making demographic projections are described in
  Hauser, P. M. and O. D. Duncan, Eds., *The Study of Population* (University of Chicago Press, Chicago, 1959) *and*
  Bogue, D. J., *Principles of Demography* (Wiley, New York, 1969).

For a classic survey of population trends, now somewhat outdated but still a basic reference text for historical trends, see
  Carr-Saunders, A. M., *World Population: Past Growth and Present Trends* (Barnes & Noble, New York, 1965), *first published in 1936.*

The relationship of population to food supply is ably, if controversially, argued in
  Ehrlich, P. R. and A. H., *Population Resources and Environment: Issues in Human Ecology* (Freeman, San Francisco, 1970).

Problems of population control and the issue of zero population growth are well treated in
  Westoff, L. A. and C. F. Westoff, *From Now to Zero: Fertility, Contraception, and Abortion in America* (Little Brown, Boston, 1971).

In addition to the regular geographic journals, look at Demography *(published semiannually) and* Population Studies *(a quarterly) for substantive reports on current research. The* Population Reference Bureau *(Washington, D.C.) publishes very useful bulletins and annual data sheets. The United Nations Statistical Office publishes an annual* Demographic Yearbook, *an indispensable guide to world data on population.*

Minimata before 1953 was a tiny, unimportant fishing village off the coast of Japan. In that year, it began to gain worldwide notoriety as a fearful symbol of the consequences of human intervention in natural ecosystems. In that year, many residents of the village went down with a mysterious and deforming disease of the nervous system. Termed the *Minimata disease,* it was later traced to concentrations of a deadly mercury compound—methyl mercury—in human body tissues. Altogether, 900 people in the neighborhood of the village were affected by mercury poisoning; of these, 52 died and nearly twice that number were crippled beyond recovery.

Once the cause of the disease had been identified, it was not hard to trace the souce of the mercury to the wastes discharged into Minimata Bay by a giant chemical plant. But Minimata was not an isolated case. The same problem of mercury poisoning was later to force a 1967 ban of fishing in 40 Swedish rivers and lakes. In 1970, a scare developed in North America when a graduate student at the University of Western Ontario turned up dangerously high mercury levels in his research on the fish populations of Lake St. Clair. (Lake St. Clair lies on the Canadian–U.S. border northeast of Detroit.) Elevated mercury levels were subsequently found in 30 other states in the United States.

Minimata and mercury provide only one example of the problems that can arise because of human intervention in the tangled web of ecosystems, on which the survival of human populations on the earth ultimately depends. In this chapter we take a longer view and try to place recent problems in a historical context. We shall look first at the scale and pattern of human intervention and then at the increasing degree of intervention as human population densities were built up. Finally, we shall look at the most recent pollution problems to examine why the Minimata disease was at first so puzzling and to speculate on what further environmental backlashes may lie ahead of us.

chapter 8
Pressures on the Ecosystem

8-1

Intervention: Benign or
Malignant?

Before considering the ways in which humanity intervenes in natural ecosystems, readers may find it useful to have a recap of some of the points made
in Chapter 3. (See Section 3-2, "Food Chains.") There we noted (a) that
ecosystems were structured webs connecting the material environment and
its biological population, (b) that the main linkages in the ecosystem were
food chains running from simple phytoplankton to higher animals, and (c)
that the size of biological populations appeared to be controlled by complex
feedback mechanisms related to the food supply.

As one of the higher animals, *Homo sapiens* comes at the end of both
terrestrial and marine food chains. As both herbivores and carnivores, people
are consuming predators of both plant and animal products. Though once a
prey to a few other higher animals, humans are now almost entirely removed
from this role if we discount the few deaths due to sharks, tigers, and the like.
We remain vulnerable, however, to a host of microorganisms—most notably
the disease-carrying virus and bacillus populations. This position of the
human species in the ecosystem has been reinforced by two further critical
factors: first, the dramatic exponential increase in numbers of the species,
and second, its growing power to modify food chains through technology. To
put it simply, our "natural" position in the ecosystem gave us a potential for
dominance; our subsequent technological development and increase in
numbers enabled us to capitalize on this potential.

Designs for an Improved Ecosystem   Most human changes in ecological
systems have had a benevolent purpose. For in most cases, intervention has
been directed at improving the productivity or the habitability of a given
environment.

Consider the example of the lake ecosystem discussed in Chapter 3. (See
Figure 3-11.) How could people modify the lake so as to "improve" it for their
own purposes? If we assume the objective of the improvement to be the basic
one of increased food production, then we can easily draw up a list of means.
At the beginning of the list will come simple schemes like that of the selective
killing of fish or animals that prey on species with a food value to humans.
Thus, we might try to eliminate species like pike in order to increase the
numbers of trout or carp. At the end of the list will come major schemes for
environmental reorganization like draining the lake and using the fertile
lake-bed soils for direct crop production. The first type of intervention
demands only primitive resources (plus some basic understanding of the
ecosystem); the last type demands an advanced technology and a high input
of resources, for it entails sweeping a whole ecosystem away to replace it with
another.

All such processes result in one or more of three main categories of change.
First, there is the impact on other animal populations. Thus, the expansion in
the number of *Homo sapiens* has been accompanied by a great expansion of
some animal species (usually domesticated ones such as the horse, the cow,
and the chicken) which are of direct use to human beings. At the same time,

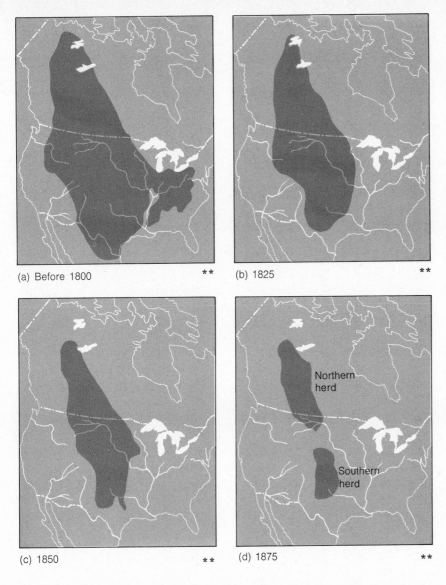

(a) Before 1800   ✱✱   (b) 1825   ✱✱

(c) 1850   ✱✱   (d) 1875   ✱✱

Northern herd

Southern herd

Figure 8-1 Human impact on animal populations.

The maps show the diminishing range of the North American bison over a 75-year period of European settlement, from (a) before 1800 to (d) 1875. A conservative estimate of the numbers of bison when the first European settlers arrived in North America is 60 million, probably the largest known aggregation of large animals. The bison (or "plains buffalo") provided the mainstay of the economy of Plains Indian tribes, but was slaughtered at an increasing rate during the nineteenth century as European agricultural settlements spread westward across the continent. By 1900 a low point had been reached, and the species was on the verge of extinction. Since then bison have been protected on government reserves and the number in managed herds now runs to several thousand animals. The dangerous decrease in range and number of the bison has parallels with the current situation for other large animals (notably the whale and the rhinoceros) in other environments today. [Data from J. A. Allen, in R. H. Brown, Ed., *Historical Geography of the United States* (Harcourt Brace Jovanovich, New York, 1948), p. 379, Fig. 9.]

some species have been severely reduced in number or wholly eliminated. Figure 8-1 shows the progressive destruction of the North American bison herds on the midcontinental plains during the nineteenth century. At the same time, cattle were being introduced into the same environment and today exist in far greater numbers than the species they replaced.

Second, there is the impact on plant populations. A similar process of selective destruction and expansion of plant populations has led to reorganizations in the balance of plant life. These reorganizations range from rather complete zonal changes, such as the replacement of the midlatitude mixed woodlands of Western Europe by an intensely cultivated mosaic of cropland, grassland, and woodland, to strictly local changes in species composition. These changes in land use are reviewed at length in Chapter 10.

Third, humans affect ecosystems by directly altering the inorganic environment. This has traditionally been achieved by interventions in the hydrologic cycle. Irrigation schemes ranging all the way from the primitive diversion of streams to massive basinwide projects are one aspect of such intervention (i.e., bringing water into dry areas to artificially boost plant production). The other side of the coin is the removal of surplus water from marshy or waterlogged areas by drainage and reclamation. Perhaps the most dramatic examples of reclamation are in coastal areas. In Holland, the history of land reclamation goes back to early dike-building projects in the eighth and ninth centuries. Reclamation of a large part of the Zuider Zee by the creation of extensive polders was begun in the early 1920s and is still continuing. In 1957 a plan was adopted for a 20-year program of reclamation in the Schelde and Rhine estuaries of south Holland which is now nearing completion. (See Figure 8-2.)

As human technological reach increases, the capacity to alter inorganic environments is accelerating. Projects for the major remodeling of terrain with massive earth-moving equipment and by the controlled use of explosives and attempts at small-scale climatic modifications are examples of this trend.

Figure 8-2.   Land reclamation.
One of the most spectacular examples of human intervention in the natural environment is provided by the reclamation of land from the sea in Holland. This extract from a seventeenth-century map of the Zeeland district in southwest Holland shows land reclaimed around the town of Terneuse on the southern side of the Schelde estuary. The method of reclamation was the building of protective banks (dikes) to give a series of polders from which water was drained by windmill-driven pumps. The reclamation, which began in the twelfth century and has continued at an accelerated rate in the twentieth century, has added over a half to the original land surface of the country. [Extract from a map by N. J. Visscher, 1655, at a scale of about 1:45,000.]

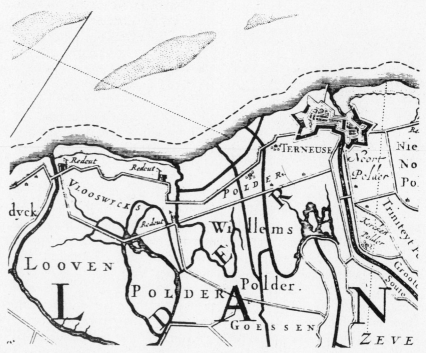

IMPORTANT TERMS USED IN THE STUDY OF POLLUTION

Biodegradable is the adjective applied to pollutants that can be decomposed by biological organisms.

Biological concentration is the process by which organisms concentrate certain chemical substances to levels above those found in their natural environment.

Biological magnification is the repeated concentration of chemical substances in food by each organism of a food chain.

DDT is the popular name for *dichlorodiphenyl-trichorethane*, a powerful insecticide developed in Switzerland in the 1930s.

Dioxin is a very powerful poison used in some weed-killers and found to cause certain deformities in fetuses.

Eutrophication is the excessive growth of algae in nutrient-rich waters leading to reduced oxygen levels and the death of many organisms.

Fallout is radioactive material that settles over the earth after an atomic explosion. Radiation from intense fallout can cause sickness and death and may affect inheritance in living organisms.

Greenhouse effect is the excessive accumulation of heat and water vapor in the earth's atmosphere due to the increased retention of solar energy by polluted air.

Methyl mercury is a highly toxic compound of mercury widely used as a pesticide.

Minimata disease is the name given to symptoms of mercury poisoning first recognized in Japan.

Particulates are tiny particles of solid or liquid matter released into the atmosphere through air pollution.

Photochemical smog is air pollution caused by the reactions between sunlight and particulates that produce toxic and irritating compounds.

Recycling is the reprocessing of waste products for reuse.

Teratogenic pollution is pollution causing birth defects.

Thermal pollution is the discharge of heat into waterways causing reduced oxygen levels and disrupting natural biological cycles.

Accidental Side Effects   To the changes that follow directed and purposeful human efforts at change we must add indirect impacts that have occured without people being aware of them. These accidents range from the spectacular, such as the much-publicized fouling of Lake Erie, to the insidious, such as the slow rise in DDT levels in some living species. (See the above discussion of pollution terms.) Books with titles like *Rape of the Earth, The Population Bomb,* and *Silent Spring* have attracted public attention to these growing byproducts of human activity. But it seems likely that many second- and third-order effects of humans on the environment remain undetected and surface only as links in complex ecological systems.

Any list of these indirect impacts would probably be incomplete. We can include the following: (1) accelerated erosion and sedimentation following changes in the vegetational cover of watersheds; (2) physical, chemical, and biochemical modifications in soils following cultivation or grazing; (3) changes in the quantity and quality of groundwater, surface water, and inland waters; (4) minor modifications of rural microclimates and major modifications of urban microclimates; (5) alterations in the composition of animal and plant populations, including both the elimination of species and the creation of new hybrids.

The extension and expansion of all five categories is possible. For example, the third category might be expanded to include not only proven effects on groundwater levels, such as the lowered level of water in the Texan aquifers (water-bearing rocks), but also the uncertain impact of toxic chemicals on lake waters. In a similar manner, human modifications of microclimates might conceivably be extended to higher regional levels if Soviet plans for large-scale river diversions in central Asia are ever carried out. As an extension of the fifth category, the long-term effects of fissionable materials on the genes of animals and plants can only be guessed.

Because a complete roster of impacts, both planned and accidental, is impractical, we shall examine some case studies to highlight the character of the interventions. We shall organize these cases around the idea of population densities; for, in general, the density of human numbers is an approximate indicator of the degree of environmental alteration. Other things being equal, the most crowded parts of the globe are those that have experienced the greatest environmental change.

8-2
Human Intervention at Low Densities

Fire and Shifting Cultivation   Typical of the permanent occupation of the earth at low densities are the tropical grasslands and forests. In both of these areas, one of the most important ways of modifying the environment has been by fire. Even before the advent of man, occasional fires were started by lightning bolts or volcanic action in most vegetated areas except tropical rain forests. The long-term impact of such periodic fires is difficult to judge, but there are some indications that species native to the chaparral of the summer-dry Mediterranean zones and the savannah of the winter-dry subtropical zones evolved in association with fire.

Primitive peoples undoubtedly caused fires themselves. Apart from accidental campsite fires, there were two basic types of purposeful burnings. The firing of native grassland in the dry season provided fresh growth for grazing herds; forest trees in the humid tropics were burned to permit cultivation. Vegetation in the humid tropics would not burn naturally, but, by girdling or felling, trees could be dried sufficiently in the dry season to burn. The newly opened areas of a forest allowed crops to grow for a few years before production dropped and the plot was abandoned. (See Figure 8-3.)

Shifting cultivation (sometimes termed *swidden*, or *slash-and-burn*) releases large quantities of soil nutrients which can be used by planted crops for a few years. However, these are quickly depleted and the soil and the vegetation can take 15 to 20 years to recover before cultivation can be repeated. (See Figure 8-4.) If the cultivation cycle becomes too short, as it is likely to do when the population rises and crowding increases, then there may not be enough time for the lands to regain their original levels of fertility.

Burning a 40-year-old tropical rain forest provides massive doses of nutrients that can be used by planted crops. The main elements, in proportion, are calcium (100 units), potassium (32 units), magnesium (13 units), and phosphate (5 units). Burning savannah woodland releases less than one-tenth as many nutrients, but a relatively greater share of potassium.

Fire continues to play an important part in crop strategies in many parts of the world. It eliminates unpalatable species and restricts the growth of woody and herbaceous plants. The exact role of fire in the creation and maintenance of grassland areas remains one of the puzzles of modern biogeography. The current trend in research has been toward concentration on the ecological role of humans in the creation of such areas, and we discuss this role in the next section.

The Ecology of Grazing Attempts to reconstruct the geographic distribution of vegetation at the end of the last Ice Age suggest about 30 percent of the land surface was then covered with open woodland and grassland. Early people cropped the wild animals of these areas for food by hunting, and for at least the last 5000 years have grazed their own domesticated animals over some of these same lands. Typically, the forage for animals varied with the

Figure 8-3. Shifting cultivation in the humid tropics.

This low-oblique photo shows secondary forest regrowth at various stages in New Guinea. Two existing patches of shifting cultivation are shown in the upper part of the photo, while the irregular outline of now-abondoned patches is shown below. [From J. W. B. Sisam, *Use of Aerial Survey in Forestry and Agriculture* (Imperial Forestry Bureau, Oxford, 1947), Fig. 59.]

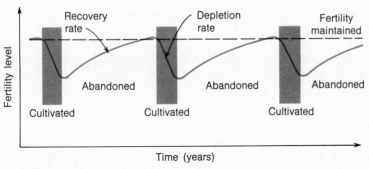

(a) Long cycle

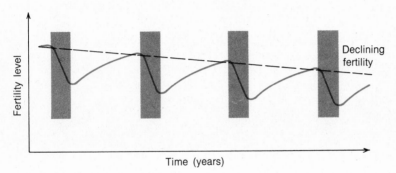

(b) Short cycle

Figure 8-4. Land rotation and population density.

The graphs show the relationship of soil fertility levels to cycles of slash-and-burn agriculture. In (a) fertility levels are maintained under the long cycles characteristic of low-density populations. In (b) fertility levels are declining under the shorter cycles characteristic of increasing population density. Notice that in both diagrams the curves of both depletion and recovery have the same slope.

seasons so that herds were moved from one area of good winter pasture to another of good summer pasture. These regular seasonal movements of herds form a type of rotational grazing termed *transhumance*. In biomes with a strong seasonal rhythym, these movements may be a very important part of the pastoral economy. For example, in the Mediterranean biome (see Table 4-2, on page 90), migration movements may span several hundred kilometers. In Alpine areas, differences in vegetation with elevation allow much shorter movements.

Even under modern intensive grassland conditions, the amount of energy which is stored by the grazing animal (and thus usable by humans) is very small indeed. Consider the situation shown in Figure 8-5 (a). The great bulk of organic matter in the grassland is stored not as grass but as organic matter in the soil. Grazing cattle represent only a very small fraction (less than 1/2500) of the total organic stock. If we try and increase the yield by pushing up the numbers of cattle on a grassland area, a point is soon reached where production crashes. [See Figure 8-5(b).] Overgrazing reduces the range of pasture grasses, and, in severe cases, may remove the grass cover altogether. Grazing is thought by some geographers to be a contributing factor in desertification—the expansion of the arid areas around the margins of the Sahara desert. Resting a grassland, grazing it in rotation, or supplementing the nutrient cycle with fertilizer are ways of coping with this problem.

The Creation of New Biotic Communities   There are two general effects of low-density human occupation on the composition of biotic communities, whether forest or grassland. First, humans tend to eliminate the more conservative, or aristocratic, elements in the biotic population—that is, those species with a low tolerance for fluctuations in moisture levels, high nutrient requirements, or little ability to withstand disturbance. Second, humans usually expand the numbers of the less conservative plants that have higher tolerances of drier, lighter, and more variable conditions. Where humans have been active for a long time, plant communities tend to be composed of a small number of extremely vigorous and highly specialized weeds; the secon-

Figure 8-5.   Grassland/cattle ecosystem.
(a) Flows within an intensively managed grassland/cattle ecosystem. Flows and storages are in calorie units standardized in terms of an original input of 1000 units of radiant energy. Note the very small share represented by grazing cattle. (b) Impact on the yield of livestock of changing the stocking level of cattle. Note the "crash" in production once a critical density level is passed. [Data from A. Macfadyen, in D. J. Crisp (ed.) *Grazing in Terrestrial and Marine Environments* (Blackwell, Oxford, 1964), Fig. 1, p. 5].

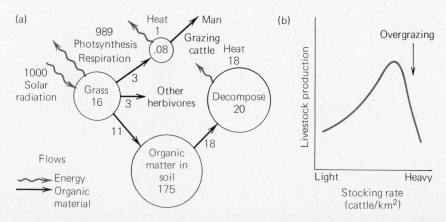

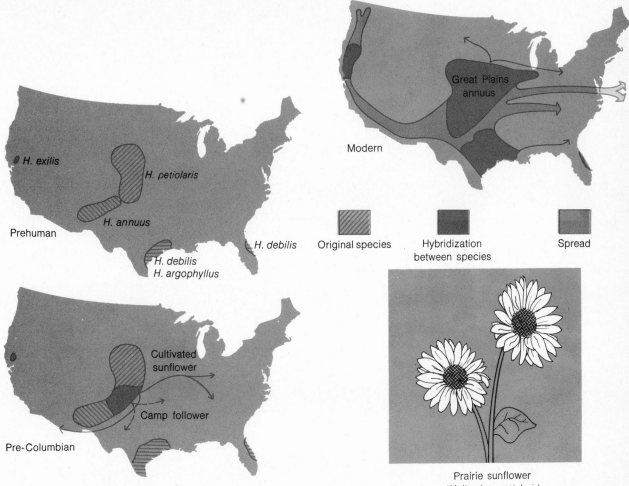

Prairie sunflower
(*Helianthus petiolaris*)

dary forest (jungle) typical of much of the tropics typifies this kind of biotic community. Many of the weeds are widely distributed and originated outside the areas where they now grow. Indeed, their distribution is itself a function of human intervention and the spread of *Homo sapiens*.

The creation of new types of plants, either domesticated or weeds, was probably a slow and continuing process in man's post-Pleistocene occupation of the earth's surface. Figure 8-6 shows the results of St. Louis botanist Edgar Anderson's reconstruction of stages in the spatial evolution and hybridization of one of these weeds, the various species of sunflower (*Helianthus*), in the United States. In this figure we see the mixing of two of the original five species in pre-Columbian times being followed by intercrossings between four of the five species in the present period. The evolution of the common weed sunflowers into what Anderson terms "superweeds" is continuing today. Mongrelization has increased their ability to colonize new areas like the Great Valley of California and the sandy lands of the Gulf Coast of Texas.

Figure 8-6. The human role in creating new biotic communities. The maps show the spatial extension and hybridization of the original distinct types of sunflowers in the United States. The longest arrow in the third map represents the overseas spread of sunflower hybrids, widely introduced into European gardens in the nineteenth century and now forming an important agricultural crop. [After Edgar Anderson, in W. L. Thomas, Jr., Ed., *Man's Role in Changing the Face of the Earth* (University of Chicago Press, Chicago, 1956), pp. 768–769, Figs. 150–152. Copyright © 1956 by the University of Chicago.]

Human intervention helped in forming the species by the creation of disturbed environments and by providing, either deliberately or accidentally, the possibility of hybridization between previously isolated species. Such intervention has important implications not only for the plant world but for the spread of microorganisms. New microorganisms, some of them disease carriers, may also evolve and hybridize in much the same way as Anderson's sunflowers.

The overall effects of human intervention at low densities appear to have been highly significant at the local level, but trivial globally. There certainly were changes; but, insofar as we can judge, they were largely beneficial and did not affect the long-term productivity of the areas occupied. Lest we should look back on this period as an ecological Eden, it is worth recalling that the technological achievements of these low-density populations also appear to have been somewhat sparse; civilization as we conventionally think of it was associated with a ruder disturbance of the natural environment.

8-3
Human Intervention at Medium Densities

Between the extremes of the empty lands and the crowded, heavily urbanized areas lie zones with a medium population density. Medium densities encompass all the ranges of population density that support permanent agriculture. These densities may, in fact, vary by a factor of 100. For example, the swidden agriculture of the northern Congo supports a density of about 8 people/km^2; by contrast, the intensive paddy lands of the Mekong Delta support about 800 people/km^2.

The Boserup Model: Agriculture and Rotation Cycles In the previous chapter (Section 7-2), we saw that Malthus thought food supply limited population size. A different view has been put forward by Esther Boserup. She suggests that a growing population stimulates changes in agricultural techniques so that more food can be produced. So, as population densities increase, the type of agriculture practiced tends to change accordingly.

Boserup has proposed a simple five-stage progression in which each step represents a significant increase in both the intensity of the cultivation system and the number of families it can support. Stage 1, *forest-fallow cultivation*, consists of 20-25 years of letting fields lie fallow after 1 or 2 years of cultivation. Stage 2, *bush-fallow cultivation*, involved cultivation for 2 to as many as 8 years followed by 6-10 years of letting lands lie fallow. In stage 3, *short-fallow cultivation*, there are 1-2 years when the land is fallow and only wild grasses invade the recently cultivated fields. In stage 4, *annual cropping*, the land is left fallow for several months between the harvesting of one crop and the planting of the next. This stage includes systems of annual rotation in which one or more of the successive crops sown is a grass or other fodder crop. Stage 5, *multi-cropping*, is the most intensive system of agriculture. Here the same plot bears several crops a year and there is little or no fallow period.

We can find examples of such systems by taking cross-sections through time or space. Thus, in Western Europe, we can trace the change from the forest-fallow system (stage 1) of the neolithic farmers to the short-fallow cultivation of the medieval three-field system (stage 3), in which one-third of the land area was left uncultivated each year. Present intensive cropping involving multicropping and supplemental irrigation represents a midpoint between stages 4 and 5. In the humid tropics, a cross-section through space reveals all five stages operating today.

The Hollow Frontier Not all agricultural systems fall neatly into simple rotational patterns. In some parts of the world, human intervention has been abrupt and episodic, with periods of intensive use being followed by periods of abandonment.

One example of such *episodic cycles* is provided by the history of some plantation crops in the humid tropics. For example, the growing of coffee (*Coffee arabica*) was introduced into southeast Brazil in the late eighteenth century. By the beginning of the nineteenth century, coffee-growing was still confined to the coastal lowland around Brazil's capital city, Rio de Janeiro. With escalating world demand, the area under cultivation increased rapidly; and by 1850 the coffee plantations had crossed the coastal mountain belt of the Sierra do Mar and had become well established in the foothills flanking the Paraíba River. (See Figure 8-7.) Geographers have mapped the spread of coffee plantations in the ensuing decades. Forests were felled and burned as the coffee frontier advanced for some 300 km (186 mi) along the Paraíba River to within a short distance of the city of São Paulo itself. Within another generation, the coffee frontier had moved northwest to Campinas and Ribeirão Preto, and with astonishing rapidity the plantation tract along the Paraíba collapsed. The coffee groves were abandoned to weeds and cattle, and the plantation houses and slave quarters to cattle ranches or to decay and the encroaching forest.

The environmental changes that followed the introduction and abandonment of coffee plantations are shown in Figure 8-7. The original forest cover which existed in the early 1800s was either cleared and planted with coffee or used as a source of charcoal and construction timber. Figure 8-8 identifies five ways in which the environment could be altered, which lead in turn to six main types of land use. Abandonment of areas once planted would, in the short run, bring about secondary forest areas dominated by rapidly growing species and, in the long run, some reestablishment of the slow-growing tropical rain forest appropriate to the area's prevailing climate and soil structure.

These land use cycles have led to a distinctive pattern of settlement. The term *hollow frontier* is a graphic description of a situation in which agricultural colonization proceeds as a wave leaving behind it a trough of worked-over land with a lower density of farm population. The century of movement in Figure 8-9 shows a simplified version of the Brazilian experience. There the

(a)

(b)

(c)

(d)

(e)

Figure 8-7. Changes in land use in the humid tropics.
European settlement of the humid tropics for plantation cropping
initiated an intricate cycle of changes in land use. (See the
diagram in Figure 8-6.) For the Paraiba Valley of southeast Brazil,
the main plantation crop was coffee (a). With the aging of the
coffee bushes (b) and the progressive abandonment of the hillside
plantations (c), the area has become largely grassland. The
grasslands are maintained by intensive grazing (d) and regular
burning (e). [Photos by the author.]

centers of colonization were the cities of Rio de Janeiro and Santos. Today the edge of coffee-growing settlement is some 700 km (450 mi) inland. In the longer term, the hollows may be filled with different forms of cultivation building up population levels.

Agriculture in the Ecological Food Chain   Agriculture can be directly incorporated into some of the ecological frameworks we met in Chapter 3. For an edible crop is also the beginning of a natural food chain, in which predatory and parasitic organisms compete for energy on every trophic level and divert calories away from human food supplies.

Some idea of the range and complexity of farming systems is given in Table 8-1. Four main types of farming are shown across the table, subdivided into those found in tropical and temperate latitudes. Different levels of land use intensity are given down the table. The numbers 1 through 4 indicate the type of ecosystem each farming system represents. These range from "wild" types such as reindeer herding or collecting fruit from wild trees at one end, to "permanent, man-directed" ecosystems such as rice cultivation at the other.

Despite the diversity of agriculture, a very few crops dominate the world's production totals. Grain yields dominate all other farm products in terms of tonnage with the "big three"—wheat, rice, and maize—making up three-quarters of grain output. Each of these crops shows great genetic diversity and can be fitted into very different natural environments and into various farming practices. Meat and animal products form a relatively small share in world food production overall, despite their great importance in specific regions.

Table 8-1 also describes each agricultural system in terms of a food chain in which humans are at the end of the chain. Four types of chains are set out in Figure 8-10. In the first, people consume food as herbivores and in the other three as carnivores. The efficiency of chain A, in which persons consume a crop directly is much greater than B, C, and D where they consume animal products. If you look again at Table 8-1, you will find examples of each of the four chains. Cattle ranching in the western part of the United States is a chain C system, while rice farming in Indonesia is a chain A system.

These examples give only a small indication of the rich variety of adjustments between agricultural systems and local environments. Ecologist

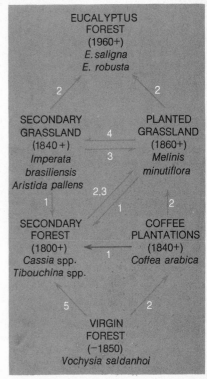

Figure 8-8.   Land-use cycles.
This flow diagram reconstructs the sequence of land use in the Paraiba Valley of southeast Brazil since 1800. Only representative plant species are shown. The numbers indicate how the environment was altered: (1) by abandonment, (2) by clearing and planting, (3) by burning, (4) by heavy grazing, or (5) by cutting trees for charcoal, timber, and so forth. [From P. Haggett, *Geographical Journal* **127** (1961), p. 52, Table 1.]

Figure 8-9.   The hollow frontier.
The land-use cycles described in the two preceding figures have left a characteristic settlement pattern in Brazil termed the hollow frontier. New waves of settlers have pushed further inland away from the centers of early agricultural colonization. The abandonment of some of the earlier coffee-growing areas has led to hollows in the density of farm population.

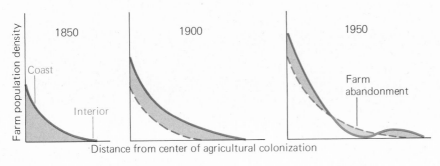

Table 8-1 Clarification of farming systems in terms of food chains

	Tree crops		Tillage with or without livestock	
	Temperate	Tropical	Temperate	Tropical
VERY EXTENSIVE EXAMPLES	Cork collection from Maquis in *southern France* **2**	Collection from wild trees, e.g. shea butter **1**	——	——
EXTENSIVE EXAMPLES	Self-sown or planted blueberries in the *north-east USA* **2**	Self-sown oil palms in *west Africa* **2**	Cereal growing in *Interior Plains of North America, pampas of South America*, in unirrigated areas, e.g. *Syria* **4**	Unirrigated cereals in *central Sudan* **4**
SEMI-INTENSIVE EXAMPLES	Cider-apple orchards in *UK;* some vineyards in *France* **4**	Cocoa in *west Africa;* coffee in *Brazil* **4**	Dry cereal farming in *Israel* or *Texas, USA* **4**	Continuous cropping in congested areas of *Africa;* rice in *south-east Asia* **4**
INTENSIVE EXAMPLES	Citrus in *California* or *Israel* **4**	Rubber in *south-east Asia;* tea in *India* and *Ceylon* **4**	Corn belt of *USA;* continuous barley-growing in *UK* **4**	Rice and vegetable-growing in *south China;* sugarcane plantations throughout *tropics* **4**

TYPICAL FOOD CHAINS (see Figure 8-10).

A	A	A, B	A

	Alternating tillage with grass, bush or forest		Grassland or grazing of land consistently in 'indigenous' or man-made pasture	
	Temperate	Tropical	Temperate	Tropical
VERY EXTENSIVE EXAMPLES	Shifting cultivation in *Negev Desert, Israel* **3**	Shifting cultivation in *Zambia* **3**	Reindeer herding in *Lapland;* nomadic pastoralism in *Afghanistan* **1**	Camel-herding in *Arabia* and *Somalia* **1**
EXTENSIVE EXAMPLES		Shifting cultivation in the more arid parts of *Africa* **3**	Wool-growing in *Australia;* hill sheep in *UK* (sheep in *Ireland*); cattle ranching in *USA* **2**	Nomadic cattle-herding in *east* and *west Africa;* llamas in *South America* **1**
SEMI-INTENSIVE EXAMPLES	Cotton or tobacco with livestock in the *south-east USA;* wheat with leys and sheep in *Australia* **4**	Shifting cultivation in much of *tropical Africa* **3**	Upland sheep country in *North Island, New Zealand* **2**	Cattle and buffaloes in mixed farming in *India* and *Africa* **4**
INTENSIVE EXAMPLES	Irrigated rice and grass beef farms in *Australia;* much of the *eastern* and *southern UK,* the *Netherlands, northern France, Denmark, southern Sweden* **4**	Experiment stations and scattered settlement schemes **4**	Parts of the *Netherlands, New Zealand* and *England* **4**	Dairying in *Kenya* and *Rhodesia* highlands **4**

TYPICAL FOOD CHAINS (see Figure 8-10).

A, B, C, D	A (C)	C (D)	C

Landscape type: **1** Wild, **2** Semi-natural, **3** Human-directed (temporary change), **4** Human-directed (permanent change).

Based on work by Duckham and Masefield as modified by I. G. Simmons, *The Ecology of Natural Resources* (Edward Arnold, London, 1974), Table 8.5, pp. 198–9.

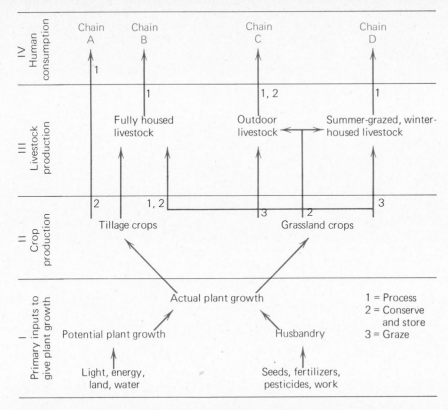

Figure 8-10. Food chains in agricultural production.
Note that whereas in A people directly consume tillage crops, in B, C, and D they consume animal products. These animal products are in turn formed from etiher tillage crops or by direct grazing. [From A. N. Duckham and G. B. Masefield, *Farming Systems of the World* (Chatto and Windus, London, 1970).]

Clifford Geertz, in his study of agricultural practices in Indonesia, has demonstrated clearly that the systems of swidden cultivation there actually simulate the exchanges of elements (among the atmosphere, vegetation, and soils) that occur in the tropical rain forest under natural conditions. Conversely, the terraced rice paddies in the same area represent an artificial system in which elaborate control of water, fertilizers, and weeds is necessary. Thus, although the swidden systems seem intuitively to be an unstable method of cultivation (since the land use changes every few years), and rice paddies a very stable one, the reverse may actually be the case. Replacement of the natural environment by a different ecology greatly increases agricultural production, but continuous work and the associated dense, rural population are required to maintain it.

Despite the very small percentage of the earth's surface occupied by cities, they have a profound effect on the environment. Within and around urban areas, the covering of rural land by city blocks proceeds faster as the cities grow larger. United States cities with populations of 10,000 have average densities of around 1000 people/km², those with populations of 100,000 have

8-4

Human Intervention at High Densities

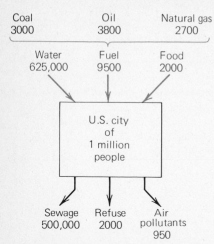

Coal	Oil	Natural gas
3000	3800	2700

Water	Fuel	Food
625,000	9500	2000

U.S. city
of
1 million
people

Sewage	Refuse	Air pollutants
500,000	2000	950

Figure 8-11.   The city as an ecosystem.
The maintenance of city life demands a
regular inflow of energy and nutrients and an
efficient waste disposal system. The figures
given are rough estimates in tons for an
American city of 1 million people. [Based on
data by J. McHale in I. G. Simmons, *The
Ecology of Natural Resources* (Arnold,
London, 1974), Table 13.7, p. 358.]

densities around 2500 people/km², and the larger cities of one million have
densities as high as 3500 people/km².

At these higher densities, the proportion of open space in the downtown
area dwindles; concrete and asphalt are almost everywhere. In the central
parts of large cities, up to 40 percent of the area may be covered with highways
alone. This artificial environment is ecologically highly unstable. As Figure
8-11 shows, a large city is dependent on a huge input of water, fuel, and food.
Equally, it needs to get rid of vast amounts of waste. If we cut off the water
supply, or turn off the power, or fail to collect the garbage, then city life is
soon disrupted. So, in looking at human impacts at the high-density level, we
shall concentrate on the environmental effects of the city.

Cities and Climatic Modification   The construction of large cities represents the most profound human influence on the climate of specific localities—an influence no less dramatic because it represents a a massive side effect
rather than an intentional change. For cities destroy the existing microclimates of an environment and create new ones. This is achieved by three
processes: the production of heat, the alteration of the land surface, and the
modification of the atmosphere.

The generation of heat within a city generally results directly from the
combustion of fuels and indirectly from the gradual release of heat stored
during the day in the city's material fabric (brickwork, concrete, etc.). Temperature studies reveal urban *heat islands,* caused by the fact that city temperatures are generally higher than those of surrounding rural areas. For
example, central London has a mean annual temperature of 11°C (58.8°F),
the surrounding suburbs have a mean annual temperature of 10.3°C (50.5°F),
and the rural areas have a mean annual temperature of 9.6°C (49.3°F). (See
Figure 8-12.)

These differences are at a maximum with calm conditions or low wind
speeds; wind speeds above 25 km or 15 mi per hour tend to blot out the heat
island effect. Contrasts for London are at their sharpest during the summer
and early autumn, thus indicating that thermal contrasts depend more on
heat loss from buildings by radiation rather than on combustion. There are,
however, marked variations between cities in different macroclimates and in
different topographic situations. In Japan the continued expansion of cities is
paralleled by increases in their mean annual temperature (e.g., Osaka's
temperature has risen by 2.5° [4.5°F] over the last century), but it is difficult to isolate the effect of suburbanization from other influences on
temperatures.

Cities also affect microclimates through their rugged artificial terrain of
alternating high and low buildings and streets. Even though we are conscious
of gusting winds being channeled along the canyon-like streets between high
buildings, the terrain of the city lowers average wind speeds compared with
those in surrounding rural areas. Average wind speeds for a site in central
London (7.5 km or 4.7 mi per hour) are substantially lower than for a

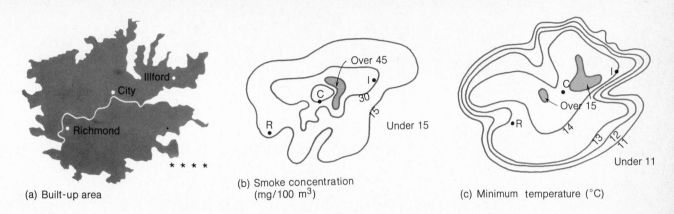

(a) Built-up area

(b) Smoke concentration (mg/100 m³)

(c) Minimum temperature (°C)

Figure 8-12.   Urban climates.
The map of Figure (a) shows the impact of London's built-up area on its atmospheric environment. Figure (b) shows the average smoke concentration from October 1957 to March 1958. Figure (c) shows the minimum temperature on June 4, 1959. [From T. J. Chandler, *Geographical Journal* **128** (1962), pp. 282, 295, Figs. 2, 13.]

suburban site (London Airport, 10.2 km or 6.3 mi per hour), although there are considerable variations according to the season and time of day. The effect of urbanization on rainfall is uncertain, but there are strong indications that, under certain conditions, cities in middle latitudes can cause sufficient local turbulence to trigger rainstorms. Significant variations among cities in different climatic zones are probable, and more research on comparative urban climatology is required. It would be dangerous to judge all the world's cities on the evidence provided just by London or Los Angeles.

City Wastes in the Atmosphere   The impact of cities on the atmosphere is particularly evident in the context of pollution. City atmospheres are polluted mainly by the emission of smoke, dust, and gases (notably sulfur dioxide). Pollution has three primary effects: It reduces the amount of sunlight that reaches the surface, it adds numerous small particles to the air that serve as nuclei for condensation and hence promote fogs, and it alters the thermal properties of the atmosphere. These three effects often combine to reinforce one another. Smogs, for example, reinforce the reduction of sunlight. The seriousness of this effect is indicated by the fact that British cities are estimated to lose between 25 and 55 percent of the incoming solar radiation from November to March. Although certain common features cause a high concentration of pollution (low wind speeds, temperature inversions, high relative humidities), there is considerable global variation in the severity and seasonal incidence of pollution conditions. For example, although Los Angeles smog is at its worst in summer and autumn, London smog is a winter phenomenon.

Although air pollution is most obvious when it soils buildings and reduces sunlight, its most important effects are on human health. The toxic effects of many pollutants are well known, although they reach dangerous concentrations only under unusual meteorological conditions. The most serious accumulation of pollutants over a large metropolis occurred during the London

SMOG FORMATION

When winds are strong there is seldom noticeable air pollution. Smoke, dust, and gases are rapidly mixed with a large volume of air and dispersed over a large region so that concentrations at any one point remain low. But the calm conditions and extremely light winds typical of high-pressure *(anticyclonic)* conditions favor the buildup of large and dangerous concentrations of pollutants. Normally, air temperature decreases with height (the average lapse rate is about 6.4°C or 43.5°F per km), so that the warm, polluted air over cities tends to rise and mix vertically [Figure (a)].

During anticyclonic conditions two types of *inversion layers* can interrupt this normal vertical dispersal of pollutants. First, a *high-level inversion* at 1000 or more meters (about 3300 ft) can be formed when calm, upper air falls to lower elevations where it is compressed and its temperature rises. Strong and persistent high-level inversions are typical of the eastern end of the Pacific anticyclone that extends over the Los Angeles Basin, particularly in summer. Second, a *low-level inversion* can be formed overnight by the rapid cooling of the ground. These shallow diurnal inversions affect only the lowest 100 m or so (about 330 ft) of the atmosphere.

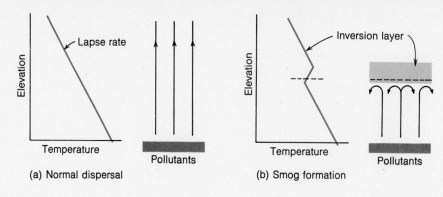

(a) Normal dispersal   (b) Smog formation

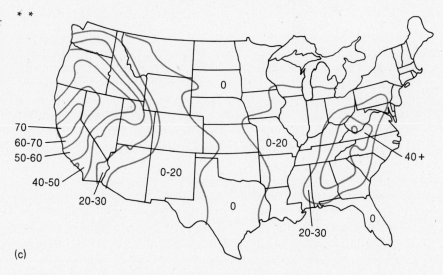

(c)

Inversion layers, from whatever cause, prevent the vertical dispersal of pollutants and raise the concentration levels [Figure (b)]. Topographic conditions such as narrow river valleys further constrain any dispersal and accentuate concentrations, as in the Donora Valley disaster in western Pennsylvania (October 26 to 31, 1948) and the Meuse Valley disaster in Belgium (December 1930). The map [Figure (c)] shows the average number of days per year when there are inversions and light winds (i.e., *potential* air pollution conditions) in various parts of the United States. [See R. A. Bryson and J. E. Kutzbach, *Air Pollution* (American Association of Geographers, Commission on College Geography, Resource Paper 2, Washington, D.C., 1968).]

smog of December 5 to 9, 1952. During a strong temperature inversion with little air movement, concentrations rose to six times their normal level, and visibility was reduced to a few yards over large areas of the city. (See the discussion of smog formation on page 184.) During the 5-day period, 4000 deaths were attributed to cardiac and bronchial ailments caused by the smog. The size of this environmental disaster led to legislation (the Clean Air Act of 1956) directly controlling the emission of pollutants through a series of "smokeless zones" where the burning of certain types of fuels was banned.

Cities and Accelerating Demands for Water   Outside the immediate limits of a city, the demands of its urban population lead to a series of impacts that are no less acute for being spatially distant. The insatiable needs of an urban economy for water, food, building materials, and minerals lead to long-range exploitation of the environment.

In the United States, south Florida and the southwestern states of California, Arizona, New Mexico, and Texas all face severe local water supply problems. As cities grow, water tables are being lowered (sometimes by more than a meter a year) as demand grows and cities are reaching out further into the mountains for fresh water. For example, water may be supplied from flooding a remote valley. The creation of artificial bodies of water by damming for a variety of purposes is rapidly increasing. Although the areas used for this purpose are small in relation to the total land area, this type of land use is important locally. For the United States as a whole, an area of not less than 40,000 km^2 (over 15,000 mi^2, an area larger than Belgium) is now covered with artificially impounded water.

The average per capita amount of water used in the world's Western cities is now around 600 liters a day. But the demands of the urban public for water are overshadowed by the needs of urban industry. Each ton of steel requires 100,000 liters, and each ton of synthetic rubber needs over 2,000,000 liters. Overall, the use of water is growing: It has trebled in the last 30 years and is expected to treble again in the next 30 years.

The most acute problems now being faced are not so much in the provision of water as in the disposal of polluted water. The average city of half a million people now produces over 1800 tons of solid waste each day and a further 190 million liters of sewage. The disposal of organic sewage poses less difficulty than the disposal of the inorganic metallic wastes of industry, and it is to this problem that we turn in Section 8-5.

8-5
Pollution and Ecosystems

In studying the effect of the very dense concentrations of people characteristic of our urban-industrial civilization on the environment, we are returning to the concern with pollution with which we began this chapter. Pollution *does* occur at lower population densities, but at such low rates and in forms so easily broken down that it fails to form the kind of ecological threat posed by the byproducts of urban dwellers.

The Pollution Syndrome   Any historian looking for a word to summarize the 1970s is likely to find the term "pollutant" (from the Latin *pollutus*, defiled) on the list. But despite the constancy with which it comes up, the term remains difficult to pin down. What do rising mercury levels in the ocean, rising noise around airports, higher temperatures in streams, and higher carbon dioxide levels in the atmosphere have in common? Perhaps the simplest answer is that each represents a substance which in terms of human environment, is in the wrong place, at the wrong time, in the wrong amounts, and in the wrong physical or chemical form.

One of the simplest illustrations of this definition is provided by thermal pollution in streams. Heat added to water is in no obvious sense a pollutant, and yet it changes the characteristics of that water as an environment just as surely as chemical contaminants or atomic radiation. Where does the heat come from, and just how does it affect an ecosystem?

Nearly all the waste heat entering our streams comes from industrial processes, and over three-quarters of this comes from electric-power generation. A single 1000-megawatt nuclear-generating plant may need around a million gallons of water each minute for cooling. (See the discussion of nuclear-power generation in Chapter 10.) The water emerging from the outlet pipes may be 11°C (almost 20°F) warmer than that entering the intake pipes.

The critical problem with warm water is what happens to its chemical structure. Warm water holds less dissolved oxygen than cold water but speeds up the metabolic rates of decay organisms in the water, which *increases* their demand for oxygen. These twin effects cause a marked reduction in the water's oxygen level and a fouling of the stream environment. The warmth of water is also critically related to the metabolic rates of fish populations. Spawning and egg development among salmon and most trout occur at around 13°C (55°F). Even small increases in heat may change the timing of a fish hatch and bring young fish prematurely into an environment in which their natural food sources have yet to arrive.

It would be misleading, however, to regard thermal pollution of streams as only harmful. Many fish species (e.g., catfish, shad, and bass) spawn and develop eggs at higher water temperatures (from 24°C [75°F] to 26°C [79°F] and grow very rapidly in temperatures up to 35°C (95°F). Broad changes in a stream's ecology following thermal pollution may actually increase its overall productivity if we count the increased growth of algae. Future research might well be directed toward making such increases in productivity available to human beings, thereby changing thermal pollution to thermal enrichment.

Chemicals in Food Chains   In the discussion above, we used thermal pollution to illustrate the sometimes equivocal nature of a "pollutant." There are other pollutants, however, whose damaging effect is direct and one-sidedly harmful. For example, of the 103 elements in the chemical table, about one in

Element	Symbol	Main link to pollution
Hydrogen	H	Constituent in pesticides
Carbon	C	Constituent in atmospheric pollution (carbon monoxide) and pesticides
Nitrogen	N	Constituent in photochemical smog
Oxygen	O	Constituent in atmospheric pollution (carbon monoxide and sulfur dioxide)
Phosphorus	P	Causes water pollution by excessive algal growth
Sulfur	S	Constituent in atmospheric pollution from coal-burning power plants
Chlorine	Cl	Constituent in persistent pesticides
Arsenic	As	Constituent in pesticides
Strontium	Sr	Radioactive isotope
Cadmium	Cd	Heavy metal; water pollutant from zinc-smelter wastes
Iodine	I	Radioactive isotope
Cesium	Cs	Radioactive isotope
Mercury	Hg	Heavy metal; toxic water pollutant from the manufacture of some plastics; pesticide
Lead	Pb	Heavy metal; toxic by-product of burning gasoline
Uranium	U	Radioactive element
Plutonium	Pu	Radioactive element

Table 8-2 The sixteen most common elements in pollution[a]

[a]Arranged in order of increasing weight.

eight plays a prominent part in environmental pollution. Table 8-2 lists the so-called "sordid sixteen."

We can divide these elements into three main groups. In the first group are elements like carbon, oxygen, phosphorus, and nitrogen, which are vital to all forms of biological life but which *can* form harmful compounds. In the second group come elements like strontium or uranium, which are important in radioactive pollution. In the third group are toxic chemicals like chlorine and arsenic used in insecticides and toxic heavy metals like mercury and lead.

It is elements in this last group that have proved the most dangerous to the structure of ecosystems in that their effects are insidious. We noted in opening this chapter the mystery that surrounded the Minimata disease. Although the chemical plant producing plastics on the edge of Minimata Bay was known to be discharging dangerous mercury compounds into the sea, the concentrations were so low—about 2 to 4 part per billion—as to be harmless if drunk in fresh water. Mercury levels in the affected fishermen were about four thousand times greater than this.

The process by which the concentration of mercury built up from harmless levels in seawater to crippling levels in human tissue is termed *biological concentration*. Each creature in a food chain collects in its tissues the mercury present in its own food, and it eats many times its own body weight in

Figure 8-13. Pollution in food chains. In this diagram of a food web in the Long Island estuary the colored arrows show the first links and the black arrows the subsequent links in chains of feeding. The DDT levels found in each organism are given in parts per million. Note the *biological magnification* of these DDT levels by the repeated concentration of this chemical substance by each organism in the food web. Compare the low concentration of DDT in the first column of marsh plants with birds in the last column. In some birds the levels are more than 1000 times higher than in the plankton and water plants.

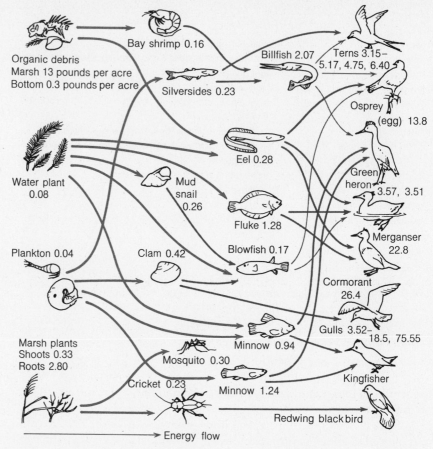

nutrients. Since the element is not excreted or broken down, it is passed up the chain in increasing concentrations. Predators like man, at the end of the food chain, are especially vulnerable to mercury poisoning since they consume large amounts of "mercury-enriched" food in which the levels of the metal are already high. In the case of Minimata, the situation was made worse by the local diet of fish, and particularly shellfish. Oysters have been shown to contain concentrations of insecticides as much as 70,000 times the concentration in seawater.

The widely publicized cases of mercury poisoning could be paralleled by the results of the buildup of persistent pesticides in the food chains. Figure 8-13 shows the buildup of DDT in part of the food chain in the Long Island estuary in the United States. Note the numbers indicating the DDT levels. These range from as low as 0.04 parts per million for plankton, at the bottom of the food web, to levels up to a thousand times greater in bird populations toward the top of the food web.

The effects of insecticides on ecosystems are often complex and indirect. Birds at the top of the predator tree (i.e., eagles and hawks) suffer the most. Yet the diminishing range of such birds is due less to direct poisoning than to the thinning of eggshells, which fail to hatch successfully. Both the bald eagle and the peregrine falcon have been eliminated from the northeastern United States, although changing habitats are likely to have had as much impact as food-chain pollution in this process.

Dimensions of Pollution in Time and Space   Studies of the lead content of the ice layers in the Greenland ice cap (Figure 8-14) show the sharp upturn in worldwide atmospheric lead associated with the beginning of the Industrial Revolution and with the widespread use of gasoline for automobile fuel. Lead has been one of the most useful heavy metals for some 5000 years. It has been widely used in pottery, household plumbing, paints, and insecticides. Unfortunately it is also extremely toxic, and the ingestion of excessive amounts can severely affect the human kidneys and liver, as well as the reproductive and nervous systems.

Although most lead is heavily concentrated near its source—so that downtown city dwellers, in areas where the concentration of automobiles is high, have levels of lead in their blood twice that of their suburban neighbors—it also gets caught up in the general atmospheric circulation of the earth. Fallout from the atmosphere affects all parts of the globe, and the fallout on Greenland shown in Figure 8-14 represents a gross underestimate of lead levels in the urban areas of the globe.

Pollution problems need to be evaluated in terms of four factors: (1) the nature and properties of the pollutant, (2) the space-time context of the

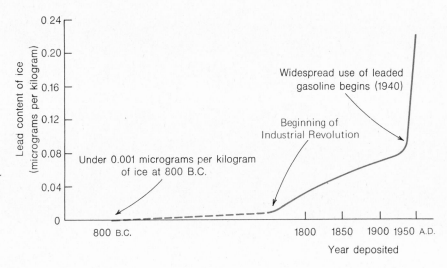

Figure 8-14.   Long-term lead pollution.
The chart shows the lead content of the Greenland ice cap due to atmospheric fallout of the mineral on the snow surface. A dramatic upturn in worldwide atmospheric levels of lead occurred at the beginning of the Industrial Revolution of the nineteenth century and again after the more recent spread of the automobile. [From M. Murozumi et al., *Geochim. Cosmochim. Acta* **33** (1969), p. 1247.]

Table 8-3   Multiple dimensions of a single pollution problem[a]

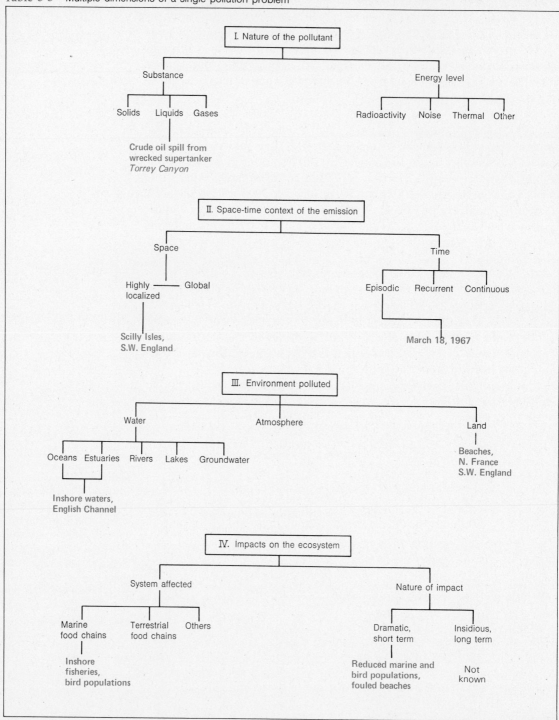

emission, (3) the specific environments affected by it, and (4) the impacts on ecosystems. It is important in studying the lead pollution problem to know that lead is a highly toxic heavy metal which has been continuously released into the environment for some thousands of years and that, with the advent of leaded automobile fuels, the amount of lead in the atmosphere has been increasing at an accelerating rate. It is also crucial to our understanding of the lead pollution problem to know that atmospheric lead is absorbed by animals through their lungs, and that its impact is long term and insidious rather than short term and dramatic.

Table 8-3 summarizes this multidimensional approach to pollution problems. It uses a single example, the oil pollution resulting from the wreck of the supertanker *Torrey Canyon* off the southwest shore of England, to illustrate the dimensional complexity of such problems. Other instances of pollution, such as noise pollution around a major airport or DDT buildups in marine food chains, can also be analyzed by using the approach in the table.

One difficulty we face in analyzing pollution problems is that the term "pollution" is now so overworked in the press and TV that it is difficult to establish a balanced view of the degree of environmental threat it represents. The amount of accurate information available on such subjects is less than the strong positions adopted by many people would lead us to suppose. More monitoring of the environment using various sensing techniques is clearly needed. Books like Rachel Carson's *Silent Spring* (published in 1962) certainly played a critical role in awakening both the scientific community and the political lobbyists to the actual and potential hazards that surround us. At the same time, we must note that much pollution is very short-lived and that many ecosystems have remarkable powers of recuperation. Pollution control may have to be carefully priced and the undoubted gains from a cleaner environment set against other desirable goals of the human population.

Summary

1. An understanding of the effect of human beings on the natural enviroment requires knowledge of our position within the structure of ecosystems. Most human-initiated modifications of natural ecosystems have been well intentioned, based on a desire to improve the short-run productivity of some part of the system. However, these modifications may involve impacts on other animal populations, on plant populations, and on the inorganic environment: such changes have had indirect side effects which are often not beneficial.

2. The degree of environmental change is directly related to population density.

3. Alterations in the environment in low-density population areas may come from (a) clearing and burning for swidden or slash-and-burn cultivation, (b) grazing and pastoral land use, and (c) the creation of new biotic communities where environments are disturbed. While on a global scale environmental change seems to be slight as a result of these kinds of human activities in areas of low population density, the effect is much stronger locally.

4. In areas with medium population densities, agriculture is more intensive. The different types of agriculture can be usefully described in terms of ecological food chains, and Boserup has defined five types of

agricultural cycles beginning with forest-fallow and ending with multicropping. Episodic cycles of intensive agricultural land use and later abandonment may lead to hollow frontier conditions.

5. Intensive environmental impacts are found in areas of high-population density, especially in cities where the natural environment is replaced by a built environment. These urban structures have a marked influence on the local climate, e.g., through the creation of heat islands. Atmospheric pollution from cities has an effect on sunlight reaching the surface, can cause fogs, and can change atmospheric thermal properties. Other environmental problems arise as a result of increasing demands for water by growing urban areas.

6. Pollution may be defined as occurrence of a substance which, in terms of human environment, is in the wrong place, at the wrong time, in the wrong amounts, and in the wrong physical or chemical form. The pollution problem is exemplified by thermal pollution of streams, which is not obvious, but because it changes the ecological properties of streams may have several indirect impacts. Chemical pollutants can be dangerous in food chains because of the possibility of biological concentration in the tissues of animals (like humans) high on the food chain. Accurate information about pollution is scarce, and more scientific monitoring is required before we can reach clear answers to the problem of balance between pollution and the welfare of the world's population.

Reflections

1. Human intervention in the ecosystem is sometimes considered benign, sometimes malignant. List examples of each type.

2. Is shifting cultivation a specifically *tropical* form of agriculture? If so, why? Do you consider it (a) wasteful and destructive or (b) a logical reaction to environmental conditions?

3. Take one example from the farming system listed in Table 8-1. Try and represent this as a food chain in the manner shown in Figure 8-10.

4. Los Angeles has a worldwide reputation for atmospheric pollution. List (a) the man-made causes of this pollution and (b) the natural environmental factors that add to the problem. Do you think that a solution can be found? Why or why not?

5. Debate the proposition that pollution is a luxury—a problem that exists only in the developed one-third of the world.

6. Using Table 8-4 as a guide, analyze the multiple dimensions of any single pollution problem you encounter in your own local area.

7. Review your understanding of the following concepts:

The Boserup model	atmospheric pollution
shifting cultivation	thermal pollution
land use cycles	biological concentration
hollow frontier	in food chains
heat islands	

One Step Further . . .

A number of the books we encountered on ecosystems in Chapter 3 also include some consideration of the human role. In particular, look at the introductions provided in
  Chute, R. M., Ed., *Environmental Insight* (Harper & Row, New York, 1971), esp. Part 3, and

Clapham, W. B., Jr., *Natural Ecosystems* (Macmillan, New York, 1973), Chap. 7.

An ecosystem approach to the major branches of human economic activity (farming, mining, industry, etc.) is given in

Simmons, I. G., *The Ecology of Natural Resources*, 2nd ed. (Arnold, London, 1980),

while agriculture is treated at length in
Grigg, D. B., *The Agricultural Systems of the World* (Cambridge University Press, Cambridge, 1974).

A good review of man's polluting impact in urban areas is given in
Berry B. J. L. and F. E. Horton, *Urban Environmental Management* (Prentice-Hall, Englewood Cliffs, N.J. 1974),

while more detailed chemical relationships for all the main environments are presented in readable but rigorous form in
Giddings, J. C., *Chemistry, Man and Environmental Change* (Canfield, San Francisco, 1973).

In addition to the regular geographical journals, keep a weather eye on Scientific American *(a monthly), which gives good coverage to ecological issues. Some of the major papers from past issues have been drawn together in*
Ehrlich, P. R., *et al.*, *Man and the Ecosphere: Readings from Scientific American* (Freeman, San Francisco, 1971).

Our planet has been aptly called "Spaceship Earth." It forms, overwhelmingly, a closed system as far as materials are concerned. Science fiction to the contrary, we have no present basis for believing that this essential isolation will be altered. . . . This earth is our habitat and probably will be as long as our species survives.

MARSTON BATES
The Human Ecosystem (1969)

In late 1859 the first production oil well was drilled at Oil Creek near Titusville in northwestern Pennsylvania. The rig struck oil at a depth of 20 m (about 66 ft). Twenty years earlier the oil had been merely a nuisance as a contaminant in salt wells or had been collected from surface seepages and sold in small bottles as "Rock Oil," a medicine with uncertain curative powers. Twenty years later 30 million barrels were being filled from wells around the world, 80 percent of it from Pennsylvania. A new era in world fuel resources was under way.

The story of oil is just one of the dramatic examples of the selective use humans have made of the earth's natural resources. We could supplement it with parallel studies—of copper, uranium, even sand—that depict a natural substance rising rapidly in people's estimation to become a valuable resource. Such resources raise various questions that we consider in this chapter. What are natural resources, and how do we measure them? What determines whether or not particular resources will be used? If we do use them, how long will they last? Such issues lead on to the question of conservation, and we look at the issues this in turn raises at the end of the chapter.

chapter 9

Resources and Conservation

9-1
The Nature of Natural
Resources

The language in which we talk about the earth's natural resources has become somewhat tangled. In particular, we tend to confuse the concept of potential resources—like the potential hydroelectric power of the Amazon River system—with resources that have actually been developed, such as the electric power produced from Niagara Falls. It is useful, therefore, to distinguish at the outset between three apparently similar terms: stocks, resources, and reserves.

Stocks, Resources, and Reserves The sum total of all the material components of the environment, including both mass and energy, both things biological and things inert, can be described as the *total stock*. In earlier chapters we saw that the prime source of energy on the earth is solar radiation. The fact that the earth receives 17×10^{13} kW per day of solar energy provides some theoretical upper limit to production systems dependent on that source of power. Then too, we argue that all material goods must be derived ultimately from the 6.6×10^{21} tons of matter that make up the planet Earth. (See Table 9-1).

In spite of its abundance, the vast proportion of the earth's total stock of matter and energy is of very little interest to us. Either it is wholly inaccessible with our existing technology (e.g., as is the iron and nickel core of the planet) or it is in the form of substances we have not learned to use. Resources are a cultural concept. A stock becomes a resource when it can be of some use to people in meeting their needs for food, shelter, warmth, transportation, and so on. The petroleum stocks of Texas were substantially the same in 1790 and 1890, but in the intervening period attitudes toward those stocks changed dramatically. Uranium ores provide a more recent example of the stock-to-resource transformation.

The transformation from a stock to a resource is reversible. Figure 9-1 is an aerial photograph of one of the most valuable resources of Neolithic Britain, the flint-axe mines near Brandon. As iron axes replaced flint, around 500 B.C., the resource lost its usefulness and rejoined the unvalued stockpile. Thus, we can define *resources* as that portion of the total stock which could be used under specified technical, economic, and social conditions. Resources as such are determined by human concepts of what is useful, and we can expect *resource estimates* to change with technological and socioeconomic conditions. In this context, *reserves* are the subset of resources available under

Table 9-1 Earth's most abundant materials

Elements in the earth's crust	Percentage by weight	Metals in sea water	Percentage by weight
Oxygen (O)	46.60	Sodium (Na)	10.60
Silicon (Si)	27.72	Magnesium (Mg)	1.27
Aluminum (Al)	8.13	Calcium (Ca)	0.40
Iron (Fe)	5.00	Potassium (K)	0.38
Calcium (Ca)	3.63	Strontium (Sr)	0.01

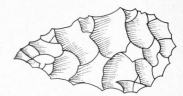
Flint axe

Figure 9-1. The changing status of natural resources.

Flint mines produced a key resource for Stone Age cultures but have been abandoned for over 2000 years. Grimes Graves, in eastern England, was one of the most important English sources of flint. Traces of mining activity are evident in the disturbed ground in the upper-middle section of the unforested area. [Royal Air Force photo. Crown copyright reserved.]

prevailing technological and socioeconomic conditions. They form the most specific but the smallest of the three categories and are relevant to one period of time only, the present.

Renewable and Nonrenewable Resources   Geographers classify natural resources in various ways, as Table 9-2 shows. The primary distinction made is between *nonrenewable resources*, which consist of finite masses of material like coal deposits, and *renewable resources*. Nonrenewable resources form so slowly that, from a human viewpoint, the limits of supply can be regarded as fixed. Some, like stocks of coal or metal ores, are unaffected by the passage of time, while others deteriorate. The planet's stocks of refined ore, for example, are reduced by oxidation. The stock of natural gas is reduced by seepages. Renewable, or flow, resources are resources that are recurrent but variable over time; an example would be water power. Flow resources are usually measured in terms of output over a certain time. For instance, the upper limit on world tidal power is about 1.1×10^9 kW per annum.

Renewable resources can be separated further into those whose levels of flow are generally unaffected by human action and those demonstrably affected. It is difficult to envision human beings ever being able to interfere with the world's potential tidal energy. In contrast, the yield of groundwater resources can be permanently reduced. Overpumping may irreversibly close fissures capable of storing water or, as in the coastal valleys of southern California, allow incursions of saline ocean water. Between these two extremes stand resources like forests where a reduced flow (e.g., due to overcutting) can be counterbalanced by subsequent remedial action.

Table 9-2 Types of natural resources

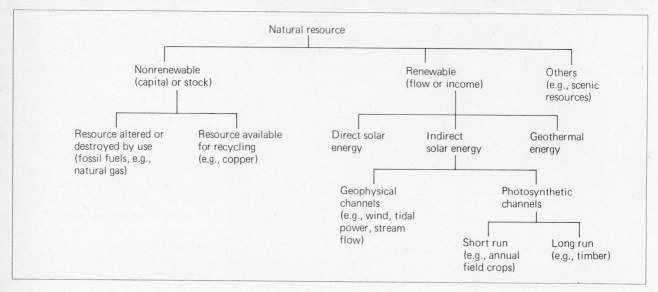

Estimating the Size of Reserves   How do we go about estimating the size of a particular reserve? We first need to know the distribution of the resource. Figure 9-2 indicates areas in which geologic conditions could produce petroleum. These areas are sedimentary basins where the organic products from which oil is derived were once deposited, compressed under other sediments, and preserved. The probable location of specific fields can be narrowed by geophysical surveys and confirmed by trial bores.

Whether a particular oil field will be used depends not just on geologic conditions. As Figure 9-3 indicates, we can consider the size of the reserves the field represents as a joint function of four factors:

1.   The quality of the oil, its chemical characteristics, and its freedom from impurities like sulfur;
2.   the size of the field and whether it is large enough to justify the capital investment needed to work it;
3.   the accessibility of the field, both in a spatial sense (i.e., its distance from refineries or consumers) and in a vertical, geologic sense (its depth); and
4.   the relative demand for oil as indicated by the prevailing price level.

Alteration of any of these four factors can change the size of the reserve. Note the effect of a low price in Figure 9-3 in reducing the estimated size of the reserve. We could go on to elaborate the relationship various ways. For example, we could define our third element, accessibility, as including strategic accessibility (which depends on who owns the field). Potential oil reserves in the North Sea are much more likely to be exploited because they fall within the ownership of countries (like Norway and the United Kingdom)

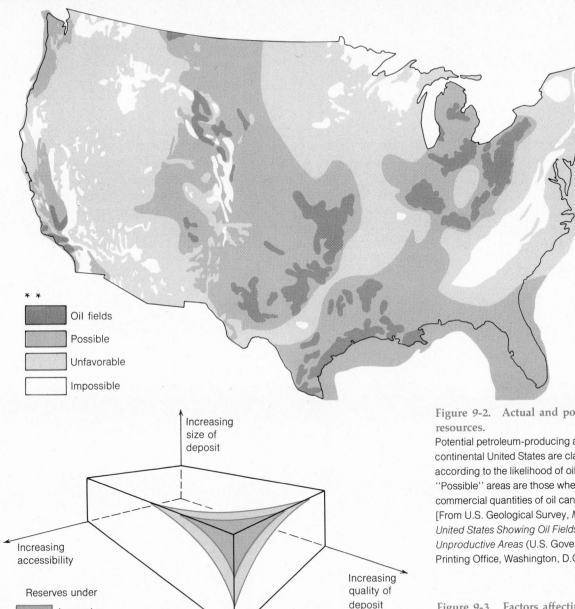

* *

Oil fields
Possible
Unfavorable
Impossible

Increasing
size of
deposit

Increasing
accessibility

Increasing
quality of
deposit

Reserves under

Low price
High price

Figure 9-2.   Actual and potential resources.
Potential petroleum-producing areas in the continental United States are classified according to the likelihood of oil finds. "Possible" areas are those where small commercial quantities of oil can be found. [From U.S. Geological Survey, *Map of the United States Showing Oil Fields and Unproductive Areas* (U.S. Government Printing Office, Washington, D.C., 1960).]

Figure 9-3.   Factors affecting size of reserves.
Hypothetical relationships between the size of a deposit, its quality, its accessibility, and the prevailing price level are indicated. The block indicates the total size of the world stock of a particular resource, and the colored "corner" the extent of reserves. High prices *increase* the area of reserves, low prices *decrease* the area of reserves.

that were wholly dependent on imported oil (see Figure 20-6). In cases like this the high cost of exploitation may be counterbalanced by strategic advantages to a country in being able to control its own sources of oil.

Criteria similar to those for estimating petroleum deposits can be used to gain a general idea of the size of current reserves of other resources. In the case of stock resources reserves are expressed as a finite total; reserves of flow resources are described in terms of the potential output in a particular time period. In both cases estimates of reserves are usually only approximate, refer to a specific time, and must be continuously updated as technical and market conditions change.

9-2
Limited Reserves? The Dilemma of Stock Resources

The extent of human exploitation of terrestrial resources is staggering, particularly in the most recent period of human history. We estimate that the world population doubled between 1800 and 1930 and doubled again between 1930 and 1975. Each human being whose existence is projected in the population models of Chapter 7 is going to require basic necessities such as food, water, shelter, and space (as well as a growing range of nonessential goods). As living standards rise, the pressure on resources created by an exponentially growing population is being increased still further by a rising per capita demand for resources. This demand is being met by a massive consumption of available natural resources.

The combined effect of increased population and resource consumption per capita has been a quintupling in the level of resource extraction between 1880 and 1980. The amount of most metals and ores used since 1930 is in excess of the combined amount used in all previous centuries. A projective study called *Resources in America's Future*, published in the mid-1960s by Resources for the Future, estimates that by A.D. 2000 the world will need a tripling of aggregate food output, a fivefold increase in energy, a fivefold increase in iron alloys, and a tripling of lumber output. Since more recent checks show these projections to be broadly on target in the late 1970s, we can expect a massive increase in the use of resources for the remainder of this century.

How long will reserves of the earth's nonrenewable resources last? There appear to be two kinds of answers. The first, which concerns the medium term (about 30 years), is primarily based on economics and is generally optimistic; the second involves a much longer historical period, is based on ecological arguments, and is less hopeful.

An Optimistic View   The classic economic test of increasing scarcity is a marked rise in the real cost of a product in comparision with the general price level. How do natural resources meet this kind of test? Changes in price levels for natural resource products since 1870 are shown in Figure 9-4(a) by fluctuations in the price of each commodity. Because the prices are plotted in ratio terms, the erratic movements represent considerable fluctuations. For example, the real prices of forest products are now over three times what they were in 1870. In comparison, mineral prices have declined and farm products have increased relatively little. Perhaps the most remarkable fact is that the prices for all resource products varied rather little over the past century. According to this conventional economic index of scarcity, natural resources do not appear to have become significantly scarcer since 1870.

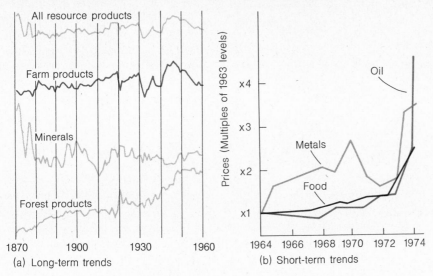

Prices (Multiples of 1963 levels)

(a) Long-term trends

(b) Short-term trends

Figure 9-4. Trends in resource prices. Figure (a) shows relative fluctuations in the deflated price of natural-resource products between 1870 and 1960. Figure (b) shows undeflated-price trends in natural-resource products over a ten-year period. Despite the continued energy crisis in the late 1970s, the relative price of oil in real terms is growing only slowly. [After H. H. Landsberg et al., *Resources in America's Future* (Johns Hopkins Press, Baltimore, 1963), p. 13, Fig. 4.]

It is important to bear this long-term trend in mind in looking at the leaps in prices so alarmingly reported in the press in the 1970s. Figure 9-4(b) shows price trends over the last ten years and emphasizes the surge in oil prices following the Arab-Israeli conflict in 1973. To see these changes in perspective, we need to recall that it is the relative price of resources that is at issue in determining scarcity. Figure 9-4(b) also uses a linear rather than a ratio scale on the vertical axis.

To understand the relative stability of natural-resource prices over the long term, we need to think about what happens when a rapid hike occurs in the price of a commodity—as happened in the case of tin in the 1960s or of oil in 1974.

As the price of a resource rises, a chain of compensating movements occurs. First, high prices bring greater care in the way resources are used. Supplies of products with high values per ton are carefully metered and their use carefully recorded; conversely, wastage results from low-value products. Water is a classic case of the effect of a change in attitudes toward a resource regarding the way in which it is used. Once free, its distribution is now metered; and the fact that people must pay for water discourages reckless use of it.

Another reason for stability in the price of resources generally is that various resources may be substituted for one another. An exponential increase in overall demand for one resource may be met by using other resources for the same purpose. The increased demand for textiles is balanced by a switch from natural fibers (such as cotton and wool) to synthetic fibers (such as Dacron, Orlon, and nylon) derived from coal, oil and even urine. Consequently, each resource has a convex consumption curve. Consumption does not fall because of a physical shortage of the old product but because it becomes more expensive in relation to the new substitutes. Ways in which substitution may occur are illustrated by the substitution which occurred in

Germany during World War II, when a variety of ersatz products, many based on chemical derivatives of coal, were produced. The changing pattern of world fuel requirements, another example of substitution, is summarized in Table 9-3.

Still another reason for stable resource prices is a switch in methods of extraction. Even when the same natural resources continue to have the same uses, considerable changes in methods of extraction can occur. In copper mining, since the last century, there has been a radical switch from the selective mining of high-grade deposits to the mass mining of low-grade deposits. In 1900 ore had to have a copper content of at least 3 percent to be worth mining; today, ore with a copper content of around 0.5 percent is being used. Oil drilling now involves less wasteful extraction methods than in the 1870s, and so on.

This switch to lower-grade resources significantly affects the total assessment of the reserves of a stock. The amount of the element in the earth may represent the total stock; in practice, however, the recoverable reserves will be a small fraction of the ultimate reserves. Figure 9-5 presents the theoretical distribution of an element as a frequency distribution diagram; generally, the element is so diffusely located that it is economically unrecoverable. Only when geophysical and geochemical processes concentrate the element in certain locations will recovery normally be possible. Other concentrations may be achieved biologically (e.g., as direct organic concentrations lead to coal, lignite, and petroleum deposits) or by mechanical means (e.g., fluvial sorting of gold into placer deposits).

Not all resources follow the simple arithmetic and geometric model in Figure 9-5. Two specific deposits that fail to show the regular (but inverse) relationship of ore quality to ore volume are the lead–zinc ores found in limestone, and mercury. Here there are abrupt changes in concentration ratios. Although the principle that ore resources increase at constant geometric rate as ore quality decreases can be a valuable guide to the abundance of some mineral deposits (notably iron, bauxite, magnesium, and copper) in the earth's crust, it would be dangerous to assume that it applies to all minerals, and more dangerous still to assume that it applies to nonmineral resources. Work remains to be done on developing precise models of the relationship

Table 9-3 Changing emphasis on world energy sources[a]

Energy source	Percentage contribution to total energy used					
	1875	1900	1925	1950	1975	2000 (est.)
Wood, vegetation	60	39	26	21	13	5
Coal	38	58	61	44	27	21
Oil	2	2	10	25	40	39
Natural gas	<1	1	2	8	15	15
Other sources (mainly hydroelectric and nuclear)	<1	<1	1	2	5	20

[a]Data compiled from U.N. and other sources.

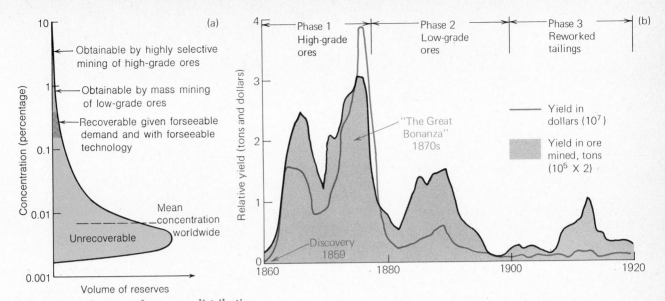

Figure 9-5. Patterns of resource distribution.
(a) The curve shows the general concentration and volume of mineral resources; the vertical axis is logarithmic. For the upper part the quality of ores (concentration) is inversely related to their quantity (volume of reserves). (b) Silver production from the Comstock Lode in Nevada, discovered in 1859. Mining has gone through three distinct phases, each using ore of lower quality. [Data from E. Cook, *Man, Energy and Society* (Freeman, San Francisco, 1976), Fig. 13.3, p. 394.]

between quantity and quality for the widest possible range of natural resources. Such models would be valuable guides to the probable future availability of some key resources.

A Pessimistic View If we take a much longer view of resource extraction, then our predictions may be more pessimistic. We noted in Chapter 7 that the present period of rapid population growth and resource exploitation, far from being part of the normal order of events and projectable into the future, is very abnormal. A continuation of the present rate of world population expansion would allow each person a share of only 1 m² of the earth's land surface—Antarctica and the Sahara included—in around 525 years.

Population growth is a useful starting point, for we have seen that the consumption of resources is partly a function of the rate of population increase. If we take the rates of energy consumption per capita as an example, the daily minimum needed by a primitive person to keep alive was equivalent to about 100 watts (for food). As other sources of energy, notably firewood, were added, the level rose to around 1000 watts per capita. Here the rate stayed until the continuous mining of coal (beginning about eight centuries ago) and the production of oil (beginning just over a century ago) brought an exponential increase in energy consumption of around 10,000 watts per capita. We can gain some idea of the recentness of the consumption of fossil fuel deposits by noting that half the cumulative total (or world) production of coal has occurred since the 1930s and half the oil consumption since the

PROJECTING FUTURE RESERVES

Assume that the volume of a mineral resource available in any area is finite and that discovery and production follow logistic curves over time. Then the rates of change of discovery, production, and reserves must follow a generally cyclic form. [See diagram (a) below.]

The peak of proved reserves comes where the curves for the rates of discovery and production intersect, with the production rate still rising but the discovery rate already on the decline. This intersection occurs roughly halfway between the two peaks. With similar curves for actual resources, we can project how far their cycle of exploitation has run. As an example, consider the curves for crude oil in the United States, exclusive of Alaska. [See diagram (b) below.]

This figure indicates that reserves probably reached their peak in the 1960s and are now on the decline. Problems in calibrating such curves, their applications, and the reservations needed in interpreting them are discussed in M. King Hubbert, *Energy Resources* (National Academy of Sciences–National Research Council, Washington, D.C., 1962).

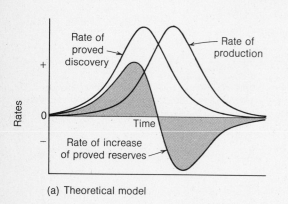

(a) Theoretical model

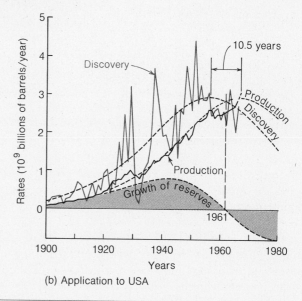

(b) Application to USA

1950s. Given the rates of consumption typical of the present decade, natural resources that took 100 million years to form by sedimentation will be consumed in about 100 years of industrialization.

The Problem of Dwindling Fossil Fuels   Energy resources are the key to the pace of resource extraction. The current form of world society is highly dependent on energy, and the future availability of other resources, organic and inorganic, is indirectly related to energy resources. Models for estimating the likely volume of fuels available have recently been developed. (See the above discussion of projecting future reserves.) These models allow us to project complete cycles of production for the major fossil fuels. Despite some variation, the results of the projections indicate that approximately 80 percent of the petroleum family's resources (crude oil, natural gas, oil from tarsands, and shale oil) will probably be exhausted in about a century. Similar calculations show that roughly 80 percent of the world's coal reserves will be depleted in about 300 to 400 years. From a historical perspective, the age of fossil fuels will be a limited one, lasting from about A.D. 1500 to A.D. 2800—a very short time even in terms of the brief human occupation of the planet.

Oil illustrates well the rapid geographic change in areas of supply which followed the working out of easily accessible fields. We noted earlier that in 1880 world production was 30 million barrels, 80 percent of it coming from Pennsylvania. The succeeding decades have seen a dynamic shift in both the quantity and the location of supplies. By 1910 the 1880 world total of 30 million barrels had increased tenfold, and by 1950 it had grown by over 100 times. New producing areas had sprung up all over the world: the Caucasus in southern Russia and Dutch Indonesia in the 1880s and 1890s, Texas and Oklahoma in the 1900s, Venezuela and the Middle East in the 1930s. Today the pattern continues to change. To the discoveries in Libya and Algeria in the early 1960s, we now must add those in Nigeria, the North Slope of Alaska, and Europe's North Sea. The dominant spatial fact which emerges from the changing historical geography of oil production is the increasingly important role of the Middle East in general and the Persian Gulf in particular. Gulf states now control about three-quarters of the world's petroleum reserves, a share which seems unlikely to change very significantly in the next decade or so.

Alternative Energy Sources? According to current projections, within the next two centuries there will be a need for a reliable source of energy to substitute for the fossil fuels. The possible sources are solar radiation, water power, tidal power, geothermal energy, atomic fission, and atomic fusion. The problem with the first four conventional sources is their scale. As Table 9-4 indicates, only hydroelectric power plants are currently capable of producing power on the kind of scale likely to be needed. To meet the needs of a large

	World annual capacity in 10^3 megawatts				
	Solar	Hydroelectric	Tidal	Geothermal	Nuclear
CURRENT (ca. 1970)	<0.1	152	<0.1	1.12	25
PROBABLE MAXIMUM	Small-scale special purpose	2860	64	60	Large, but highly dependent on technological developments

Table 9-4 The magnitude of new energy sources[a]

[a]Data from M. King Hubbert in *Resources and Man: A Study and Recommendations* (Committee on Resources and Man of the Division of Earth Sciences–National Research Council, with the cooperation of the Division of Biology and Agriculture. W. H. Freeman and Co. Copyright © 1969), Chap. 8.

city, a solar power plant of 10^{10} thermal watts would have to collect solar power over an area of the earth's surface equivalent to 6.5 km² (2.5 mi²).

The most important sources are therefore the two nuclear sources, atomic fusion and fission. Although the earth's resources of nuclear materials (uranium, thorium, and deuterium) are finite, they are large enough so that, given a reasonable rate of innovation in nuclear-reactor technology, the future scale of supply looks promising. The chief limitation to nuclear fuel may lie less in the adequacy of resources than in the safe disposal of radioactive wastes. (See Figure 9-6.)

The possible environmental threats from nuclear power production are immense and increase in proportion to the increasing number and size of the power plants. Radiation release, the amount of buried radioactive waste, and even the possibility of nuclear hijacking are likely to get more rather than less important in the next few decades. Yet the importance of success in nuclear power production is equally vital. The world's greatest potential source of energy is not fossil fuels, but the hydrogen locked up in the waters of the world's oceans. Hydrogen as a liquid or gas may well be the element powering our cars and heating our homes in the 21st century. The electrolysis of water to produce hydrogen is already being undertaken on a small scale where cheap power is available; if fossil fuel costs continue to rise and nuclear power costs continue to fall, the prospects for extracting hydrogen on a mammoth scale should improve.

Energy looks like the main resource bottleneck over the medium term. We could argue that since energy resources are only one of the sets of natural resources used by man, a pessimistic view in this area need not prejudice a more optimistic view of other resources. Unfortunately, this is not so. The main human achievements in these other areas of resource use have been

Figure 9-6. Nuclear power and environmental hazards.

(a) Main stages in the mining and production sequence for the use of fissionable fuel for energy production. The main environmental threats at each stage are indicated. (b) Sites of nuclear power plants (stage 3 in the production process) are adjacent to cooling-water supplies. Plants are commonly located in rural areas to minimize the population at risk from local leakage problems. The Indian Point nuclear complex at Buchanan, New York, is shown here. The number of such stations is expected to grow exponentially over the globe in the next few decades to meet surging power demands; the associated radiation hazards are likely to pose some major environmental dilemmas. [Con Edison photo.]

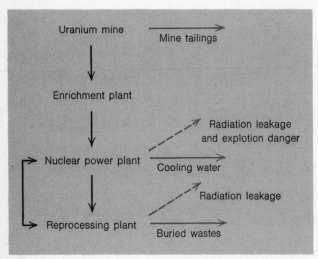

(a)

(b)

dependent on an expanding rate of energy consumption. For the last half-millennium, we have been drawing on rather readily available sources of fossil fuel energy stored up from geologic processes. The finite nature of these resources is causing some scientists to take a cautious, even gloomy, view of the future.

Renewable resources depend on the great energy cycles of the earth that we have already met in our consideration of global environments in Part One of this book. These are of two kinds: first, *physical* energy cycles related directly to solar energy, and second, *biological* energy cycles indirectly related to solar energy through photosynthesis.

**9-3
Sustained Yields: The Problem of Flow Resources**

Since we have already met some examples of people's use of earth's renewable resources related to physical cycles—solar power, water power, and tidal power—in our discussion of energy problems, we shall concentrate here on those related to biological cycles. In this category come plant and animal energy sources we tap through agriculture, forestry, and fishing. We shall also look briefly at the special problem of recreational resources, which lie on the border between renewable and nonrenewable resources. The key theme linking the examples chosen is that of how to maintain and increase the sustained yield from a particular resource.

The Green Revolution   Our first example of renewable resources is drawn from crop production in the agricultural sector. As we saw in Section 8-2, man's disturbing and mixing of plant species to produce new hybrids has been going on for many thousands of years. Plant breeding, albeit of an accidental kind, has been a camp follower of all human agriculture. Since Mendel (1822-1884) laid the foundation of plant genetics with his experiments in heredity, plant breeding has played an increasingly important role in increasing and maintaining agricultural yields; indeed, without it the Malthusian forecast of worldwide famines as a check on human population growth would surely have been fulfilled.

Perhaps the most striking example of the impact of new hybrids is in the recent *green revolution*. The green revolution is an evocative term used to describe the development of extremely high-yielding grain crops that allow major increases in food production, particularly in subtropical areas. In 1953 scientists in Mexico began to develop rust-resistant dwarf *wheats* which doubled Mexico's per-acre production in the next decade. After a major drought in India in 1965, Mexican dwarf wheat was widely planted in the northern part of that country with dramatic results in terms of wheat yield in areas like the Punjab. (See Figure 9-7.)

Rice development followed a course similar to but more laggardly than that of wheat. The Los Baños research institute in the Philippines was set up with Ford and Rockefeller Foundation backing in 1962 to develop varieties of improved rice (IR). The now famous IR-8 variety was spotted in 1965. Its first

Figure 9-7. Network of international agricultural research.
Major stations established since 1959, many under United Nations (FAO) sponsorship, with the leading research interests of each station. The oldest is that of the International Rice Research Institute (IRRI) at Los Baños in the Philippines, home of the new rice strains of the "green revolution." [Photo by Florith Botts, Nancy Palmer Photo Agency.]

harvest, from 60 trial tons of seeds, produced an astonishing sixfold increase of rice under field conditions. From these beginnings, other varieties of IR with important additional characteristics—better resistance to disease, better taste, and (very important in Asia) better appearance after cooking—were developed. Broadly speaking, the new varieties of rice have produced results wherever they were planted in the humid tropics. About 10 percent of India's paddy land is now planted with IR varieties, and the Philippines, once a major rice importer, is now near self-sufficiency in this area.

How far have the "miracle grains" of the green revolution proved an unmitigated blessing? Any seed which can (a) give two to four times the yield of indigenous grains, (b) has such a shortened growing season that two crops per year are often possible, and (c) has a wider tolerance of climatic variations must be welcome. At the same time, some severe side problems have appeared. The high yield is dependent on high applications of fertilizer and insecticides plus, in the case of rice, copious irrigation. Hence, innovation has been most rapid in the most prosperous areas and among the most prosperous farmers, and, in the short run, interregional and social gaps have widened. There is an urgent need for more widespread adoption of the new varieties in poorer sectors, but the fertilizer and water they need are still beyond the financial reach of many of the agricultural peasants of South Asia. At a quite different scale, traditional marketing patterns have been upset. Countries like Thailand and Burma—major exporters of rice—have found their traditional markets disappearing. Japan, normally a great rice importer, has bulging elevators and now looks for export areas.

The successes of the last twenty years have brought to the tropics and subtropics the kind of benefits which plant breeding has been bringing to the midlatitude farmlands for the last half-century. Given the much greater growth potential of the tropics (Figure 4-3) and the lower yields of indigenous crop varieties, the impact of the revolution has been dramatic. It remains to be seen whether demands for fertilizers and pesticides will take the edge off some of the more promising aspects of this important step in the development of the world's agricultural resources. It may well be that since increased agricultural production in this case depends on fertilizer supplies—much of which come from mineral sources—increased yields from a renewable resource may indirectly depend on a nonrenewable resource.

Sustained-Yield Forest Resources The view of forests as a renewable resource is a relatively recent one. For most of human history on the earth, people's activities have tended to reduce the world's forest cover. (See Chapter 10.) Heavy inroads were made to clear land, for fuel wood, and for construction materials. Destruction was encouraged by the belief that forest resources were inexhaustible, if not in the immediate neighborhood, then certainly in the world beyond.

Local devastation of forests and the increasing costs of bringing timber from far afield brought gradual changes in this attitude. As early as 450 B.C., official attempts were made to restrict the cutting of the cedars of Lebanon in the eastern Mediterranean. General signs of change did not appear, however, until the twelfth and thirteenth centuries, when drastic restrictions on cutting were imposed in central Europe. Then, gradually, the belief that forests must be protected as a finite and dwindling resource gave way to a belief in new planting and forest-management techniques. By the middle of the eighteenth century, timber was beginning to be regarded as a slow-growing but renewable crop rather than a nonrenewable source of fuel and timber.

Modern forest management follows two main ecological principles. The first is that of *sustained-yield*—the continous production of forest products from an area at some appropriate yield level. This is achieved through planned rotational systems (some with cycles of over 100 years), careful species selection, and protection of the timber crop from both fire and disease. There are presently available a wide range of ways of obtaining a continuous flow of forest products. Fast-growing species in the subhumid tropics (e.g., some members of the Eucalyptus family) may be cropped as frequently as every seven years. Douglas fir in the Pacific Northwest may be cropped by *patch cutting* (see Figure 9-8), after which timber from areas around each patch naturally re-covers the cleared areas.

The second ecological principle followed in managing forests as a renewable resource is that of *multiple use*. A forest's yield may be measured in more than wood products, and the objective of multiple-use forestry is to maximize the total flow. Thus, timber production may have to be balanced against the role of the forest as a protection against erosion and pollution, as a wilderness or wildlife refuge, or as a recreational area. Not all such uses may be mutually compatible, and we shall look later in the book (in Chapter 22) at the planning problems that the use of land for multiple purposes entails. Indeed, the role of a forest area in reaction may be so important that commercial logging may have to be stopped. The precedent set by the United States in 1872 when it established Yellowstone National Park has been followed frequently in the last century. The United Nations *List of National Parks* contains the names of some hundreds of parks and refuges in scores of countries around the world.

Sustained-Yield Recreational Resources   As the demand for leisure increases, the role of forests as primarily recreational areas is sure to increase. Sustaining the yield of a forest given an increasing number of picnickers,

Figure 9-8.   Sustained yield forestry. The early history of lumbering in America shows that forests were treated as a stock resource to be clean-felled. Today conservation programs insure that forests are husbanded as a flow resource. Patch cutting is a technique used for Douglas fir lumbering in the Pacific Northwest; it allows natural regrowth of the felled areas from the surrounding untouched ring of forest. [Photo by James H. Karales, DPI.]

ramblers, and the like may prove even more difficult than maintaining a sustained timber flow. Most countries have now set aside substantial tracts of country for recreational use (see Figure 9-9) but their management poses increasingly severe problems. A shorter working week, increased leisure time,

Figure 9-9. Conservation of recreational resources.

Most countries have now introduced legislation conserving areas of land for public recreational uses. These maps show the situation in England and Wales in 1970. Not all the proposed "green belt" areas have been formally approved. [From J. A. Patmore, *Land and Leisure* (Penguin, Harmondsworth, 1970), p. 179, Fig. 62.]

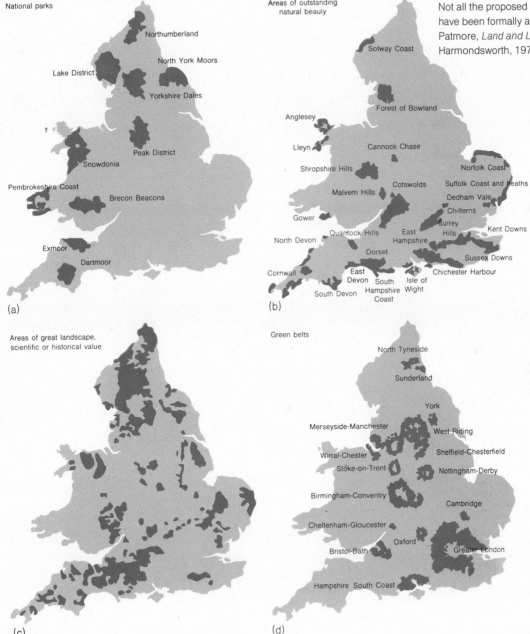

(a) National parks

Northumberland
North York Moors
Lake District
Yorkshire Dales
Peak District
Snowdonia
Pembrokeshire Coast
Brecon Beacons
Exmoor
Dartmoor

(b) Areas of outstanding natural beauty

Solway Coast
Forest of Bowland
Anglesey
Lleyn
Cannock Chase
Shropshire Hills
Norfolk Coast
Cotswolds
Suffolk Coast and Heaths
Malvern Hills
Dedham Vale
Gower
Chilterns
Quantock Hills
Surrey Hills
Kent Downs
North Devon
East Hampshire
Dorset
Sussex Downs
Cornwall
East Devon
South Hampshire Coast
Isle of Wight
Chichester Harbour
South Devon

(c) Areas of great landscape, scientific or historical value

(d) Green belts

North Tyneside
Sunderland
York
Merseyside-Manchester
West Riding
Wirral-Chester
Sheffield-Chesterfield
Stoke-on-Trent
Nottingham-Derby
Birmingham-Conventry
Cambridge
Cheltenham-Gloucester
Oxford
Greater London
Bristol-Bath
Hampshire South Coast

and falling travel costs have created a surging demand for increased recreational facilities.

Questions of Resource Demand   Table 9-5 summarizes some of the current trends in the demand for outdoor activities in the United Kingdom. The figures for annual growth rates are based on varying periods of time in the 1960s and should be treated with caution. They indicate, however, the nature of the pressures and the special demands for use of water areas.

We can illustrate the impact of demand on a recreational resource by going back to the lake we used to illustrate ecosystem concepts in Chapter 3. Let us suppose the demand in this case is for sailing. On a weekend day with good sailing conditions, a number of folk will bring their sailing dinghys to the lake. Figure 9-10(a) shows this influx of people building up over the day simply as a line that rises until a saturation point is reached. How much sailing enjoyment does each family receive? Clearly, at the beginning [point A in Figure 9-10 (b)], the early arrivals have the lake to themselves. But as numbers build up a point is reached (point B) at which boats begin to impinge on each other's territory and the amount of sailing each family can do is reduced. If numbers were to continue to build up, the lake would finally be so jammed with small boats (point C) that no sailing at all would be possible. At this point the enjoyment of each family would presumably have dropped to zero! (Note that we could apply exactly the same argument to the crowded beach we described in Chapter 1).

How does this crowding affect the "recreational yield" of the lake? We can define the *yield* as the "the number of boats $\times$ the amount of sailing activity" and plot its curve in Figure 9-10(c). The curve begins at zero (no boats on the

Table 9-5 The growth in the demand for recreational resources [a]

Growth rate	Urban resources	Countryside resources	Water-area resources
More rapid than population growth	Athletics Golf (8)	Motoring Mountaineering Skiing Camping Horse riding Nature study Gliding (10)	Diving (24) Canoeing (18) Sailing (7) Angling (7)
Similar to population growth	Gardening Major team games Swimming	Walking (3) Hunting	
Slower than population growth	Major spectator sports	Cycling(−2) Youth-hosteling	

[a]General patterns of growth in the United Kingdom. Where figures in parentheses are given they indicate the average annual increase as a percentage and are based on different periods.

SOURCE: J. A. Patmore, *Land and Leisure* (Penguin, Harmondsworth, 1972), p. 49.

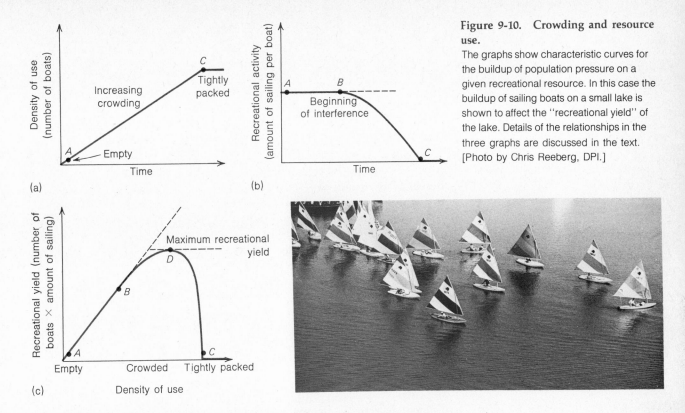

Figure 9-10. Crowding and resource use.
The graphs show characteristic curves for the buildup of population pressure on a given recreational resource. In this case the buildup of sailing boats on a small lake is shown to affect the "recreational yield" of the lake. Details of the relationships in the three graphs are discussed in the text. [Photo by Chris Reeberg, DPI.]

lake), rises to a maximum, and then falls back to zero (when the lake is packed with boats but no sailing is possible).

To obtain the maximum yield from the lake, we would like to keep the number of people using it around point D at the top of the yield curve. But how do we do this? A rule of "first-come, first-served" is one way; charging a fee, rationing, or exclusive club memberships are others; but each has undesirable social implications. Of course, no sensible family would launch a dinghy onto an already crowded lake. Thus, we might expect feedbacks to operate which would keep the maximum population at around point E. Latecomers would be likely to give up the idea of going sailing or go on to a more distant lake. But if the lake has other uses, if it is used for fishing or as a water supply, then even point D may be too high if it causes too much disturbance for the lake to be used for these other purposes. (Compare this situation to the Chew Valley Lake problem discussed in Chapter 22.)

Once again, the problems of the lake are representative of other and wider problems. Pressures on beaches, on wilderness areas, and on historical sites all raise broadly similar problems of trying to preserve resources—and yet obtaining the maximum yield from them.

Questions of Supply   One difficulty in making assessments of the available supply of recreational resources is that they are cultural assessments. The same qualities may be judged quite differently by different cultures. One of the simplest ways of illustrating this is to show the changing assessment of the same area over time.

Consider Figure 9-11(a). William Brockedon's engraving of the Val d'Isère in the French Alps was made in 1829. It shows a lyrical summer scene with the emphasis on pastoral tranquility in the foreground and the majesty of Mont Blanc beyond. These were the Alps which the painter John Ruskin would write of later as "alike beautiful in their snow, and their humanity." This is a view which has persisted until today. It is not, however, the only view of the Alps. Travelers in the eighteenth century, anxious only to cross the Little St. Bernard Pass, would curse "this awful place" in their journals, and hurry on to the welcoming towns of the Italian plains beyond.

So how do geographers pin down what they mean by a recreational recourse? Let us begin with the simple case of "wilderness areas." Figure 9-13(b) shows an empty forest and lake area on the borders between Canada and the United States made up of the Quetico Provincial Park in Ontario and the Boundary Waters Area in Minnesota. This area has been officially designated as a wilderness by the two state governments, so its boundaries represent the official image of a wilderness area. But how do these boundaries match the view of the people who actually use the area? What areas do the campers, canoeists, and fishermen consider to be wilderness, and what essential qualities should a wilderness environment have?

The main research problem is one of converting anecdotal observations and subjective hearsay into some form of quantitative yardstick that allows different assessments of environments to be metered and mapped. The solution adopted was to interview a representative sample of the groups using the area. Respondents were invited to say whether they thought they were in a wilderness area and to indicate on a map where the boundaries of the area lay. The term "wilderness" was always left to the respondents to define, implicitly or explicitly.

By superimposing the maps of the respondents on one another, we can construct contour maps of the area showing the percentage of parties visiting the area that described it as being "in the wilderness." The canoeists' contour maps have an intricate pattern [Figure 9-11(b)] closely related to the waterways and sensitive to distance from the canoe trails. The boundaries of the wilderness for other users (campers, motorists, resort guests) were less sharply defined and were related more to roads, crowding, and noise levels in the less remote areas.

From a planning viewpoint, it is worth recording that all the boundaries drawn by users of the area differed from those drawn by the resource managers of the wilderness area. Such users' responses, when translated into isarithmic maps, are clearly of considerable potential value in the future zoning of areas for recreational use. Also, these responses imply that some measurement of natural resource potential is possible even for such elusive things as the quality of a landscape.

(a)

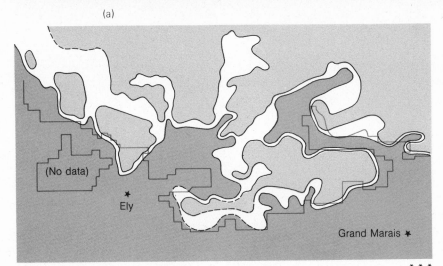

(No data)

Ely ✶

Grand Marais ✶

✶ ✶ ✶

Proportion of canoeists regarding area as wilderness

(b) ⬜ Over 90%   ▨ Under 50%   ⌐‾⌐ Park boundary

Figure 9-11. Perception of wilderness areas.

(a) Once considered a natural hazard, the European Alps are now regarded as a recreational resource. This early nineteenth-century engraving is symbolic of this change in viewpoint. Its sweeping views comprise a number of well-harmonized elements—the sheep and distant church giving an air of pastoral peace and tranquillity, the two small human figures underlining the smallness of humanity in relation to the grandeur of nature. [William Brockedon, 1829.] (b) Map shows canoeists' perceptions of "wilderness" areas within the Quetico-Superior Park on the boundaries of Minnesota and Ontario. Shading indicates the percentage of canoeists who recognized the area as a "wilderness." This map differs from that constructed by other campers using automobiles or boats. [From R. C. Lucas, *Natural Resources Journal* **3** (1962), p. 394, Fig. 3. Reprinted with permission.]

9-4

Conservation of Natural Resources

(a)   Pacific Northwest sardine
       (Sardinops caerulea)

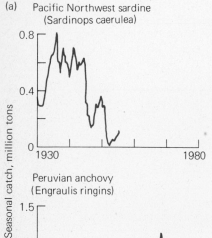

Peruvian anchovy
(Engraulis ringins)

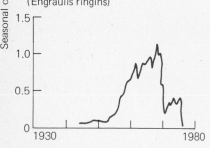

Antarctic/Pacific blue whale
(Balaenoptera musculus)

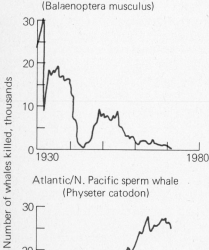

Atlantic/N. Pacific sperm whale
(Physeter catodon)

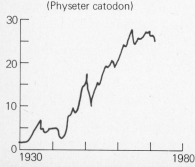

The notion of conserving the earth's natural resources is immediately appealing. No reader would willingly see Great Plains topsoil eroded, Tennessee hillsides filled with gullies, or redwood groves needlessly felled. Indeed, the destruction of animal population, like certain species of whales, has become an emotional public and political issue. (See Figure 9-12.) But what about a natural gas field in Oklahoma? What are the special merits of leaving the gas in the ground? Who benefits?

Some Definitions   Let us begin by trying to define resource conservation. One oft-quoted definition runs as follows: "*Resource conservation* is the scheduling of resource use so as to provide the greatest yield for the greatest number over the longest time period." This definition fits renewable resources nicely and encompasses the targets we met in our discussion of sustained-yield forestry, though it is less easy to apply to nonrenewable

Figure 9-12.   Collapse of overused biological resources.
(a) Fishing and whaling have been a traditional source of food and raw materials for man. The last half-century has seen a striking pattern of rise and fall in some marine species. Note that information is not available for exactly comparable time periods for the four species shown. (b) Whales were common off the coast of Western Europe up to the middle seventeenth century. A Dutch engraving of a stranded sperm whale (1958). [From I. G. Simmons. *The Ecology of Natural Resources* (Arnold, London, 1974). Figs. 9-4, 9-5, pp. 237–8. Engraving by courtesy of the American Museum of Natural History.]

(b)

resources (e.g., our gas field). Restricting the use of a finite resource means that we are saving for future generations some portion of a resource that would otherwise have been used in in the present generation.

Over the short run, this idea is appealing, since it encourages the use of other resources and stimulates substitution. However, we have already noted that few natural resources of the past were ever fully worked out (though some natural gas fields may stand as an exception); instead, they were priced out of production. Such is the rapidity of technical substitutions that resources saved by an overcautious conservation policy may never be used at all. Indeed, by limiting the investment of the present generation in them, we may actually be making future generations poorer. Clearly, this is a hard view to accept, but if we really wish to make the best use of nonrenewable resources, it makes sense to use up the cheapest and best ones first and then go on to the more inaccessible ones. Timing is of the essence; and a more useful definition of resource conservation than that above is simply that it is "the optimal timing of the use of natural resources." And along with acceptance of this revised definition goes acceptance of the possibility that the optimal time for the use of some nonrenewable resources could be right now.

The Conservation Movement   Even if preservation for preservation's sake is untenable for nonrenewable resources, there remains a wide range of resource areas where preservation is both sound ethics and sound economics. The richness of natural recreational areas in the United States owes much to the ethical views of Gifford Pinchot. Head of the U. S. Forest Service from 1898 to 1910, Pinchot caught President Theodore Roosevelt's interest and aroused his enthusiasm. Roosevelt added more acres of the United States to its protected forest lands than all the presidents before and after him.

Presidential concern with conservation has, however, a long history. George Washington was acutely anxious at the loss of soil from his Mount Vernon estate and had his servants bring up mud from the Potomac River to fill in gullies. Nearly two centuries later, in 1970, President Richard Nixon set up a National Environmental Protection Agency. In the intervening period, and particularly since Pinchot, legislation governing the use of all kinds of natural resources has been passed by the Congress. The National Park Service was established in 1916 (Figure 9-13), and the first international treaty relating to resource protection (with Canada, on protecting migrating birds) was signed in the same year. The 1930s were a decade of special concern with soil erosion and water control and saw the passing of the Taylor Grazing Act (1934), the setting up of the Soil Conservation Service (1935), and the passage of the Flood Control Act (1936). In the last thirty years, the emphasis has swung to a concern with conserving waste from the use of mineral resources, to protection from pollution (via the 1970 Clean Air Act), and to general improvements in recreational facilities and improved environmental standards.

Federal legislation on conservation in the United States has been important on a global scale. It not only initiated complementary legislation at the state level *within* that country, but was widely copied in other areas. The

Figure 9-13.   Conservation as a popular movement.
The National Parks set of ten stamps issued in 1934 was the first of what is now a commonplace theme in postal designs in countries around the world. Naturalist John Muir described the Yosemite Valley in the 1860s and the area was given by Congress to California as a public park and recreation area. It was created a National Park in 1890, and has since been extended to cover an area of 3000 km^2 (1200 mi^2). The National Parks movement, together with other ecological societies (such as the Audubon Society, the Sierra Club, and the World Wildlife Fund), is now a major force in the conservation of the world's flora and fauna. [Stamp reproduced courtesy of the U.S. Postal Service.]

Tennessee Valley Authority (TVA) set up by New Deal legislation in 1933 served as a basinwide "demonstration farm" for watershed-control programs elsewhere. The São Francisco Valley scheme in Brazil, the Gal Oya scheme in Sri Lanka, and the Snowy Mountain scheme in Australia are examples of the scores of developments that followed, in some measure, the TVA lead.

Currently, interest in conservation is expanding spatially in two ways. First, at the global level, there is increasing concern with international cooperative action on the wise use of resources and on environmental protection. United Nations agencies are now playing a key role in setting the world on the long road toward acceptance of international standards of air pollution control and joint action on the use and abuse of ocean resources. Second, there is unprecedented conservation activity at the local community level. Locally active groups involved in conservation-minded activities range from the Sierra Club and Friends of the Earth, right through to local Women's Institutes in English villages valiantly protecting their oaks and elms from the chain-saws. This significant increase in public support for conservation measures at both international and local levels represents a most encouraging shift in people's view of their place in the ecosystem. Whether it is enough of a shift in view of the unprecedented inroads now being made on all our natural resources—both stocks and flows—is still in doubt.

Summary

1. Natural resources are that part of the material environment that has utility in human terms; while all of the material environment comprises the world's total stock, most is of little interest to human beings.

2. Resource estimates may change as technology and socioeconomic conditions change. Reserves are the part of resources that people can presently use. Some resources are fixed in amount and are termed nonrenewable. Others are renewable, and may vary in amount over time; these are recurrent. Reserve size estimates for any resource are based on a number of factors such as distributions, quality, accessibility, and demand.

3. The question of longevity of nonrenewable resources is answered, first on economics over a single generation, and second, on ecology over a longer period of time: the former view is broadly optimistic, the latter somewhat pessimistic. From the short range point of view, the earth's nonrenewable resources have been relatively stable in terms of cost over the past century, and may be further stabilized on the basis of the existence of alternative resources.

4. From the ecological point of view, resources such as fossil fuels are expected to last for about 400 more years. A need for substitues for these fossil fuels will develop over the next 200 years. Solar energy, water power, tidal power, geothermal energy, atomic fission, and atomic fusion are potential substitutes.

5. The major problem for renewable resources is the increase and maintenance of a sustained yield from any one resource. In agriculture, people have been able to use hybrid plants to create the green revolution. Large increases in food production are made possible through the development of extremely high-yielding grain crops (e.g., wheat and rice). New ideas in forestry management have brought about the ecological principles of sustained yield and multiple use. Increasing use of forest for recreation causes some difficulties in maintaining a sustained yield in some areas. The pressure placed on recreation resources by outdoor activities may be defined on the basis of recreational yield.

6. The common definition of resource conservation states that resource use should be organized so that

the largest number of people receive the greatest yield through the longest period of time. Renewable resources fit well within the definition; however, a more appropriate definition for nonrenewable resources states that they should be conserved through optimal timing of use.

Reflections

1. List the main characteristics that separate *renewable* from *nonrenewable* resources. Can you think of any examples of resources that (a) fall on the borderline between the two types or (b) have changed from one type to the other as human attitudes toward them have altered?

2. Debate the contention that "resource shortages are simply the result of prices that are too low." What happens when the price of a natural resource rises? Does this have geographic implications for the spatial pattern of its production?

3. Review the natural nonrenewable resources of your own state or local area. Look for examples of resources which were once important but are no longer so. What factors account for this?

4. Follow up on the problems of nuclear-energy production by reading some of the material suggested in "One Step Further...." What are your own views on the balance between the advantages of this fuel source and its possible environmental hazards? Find out if nuclear power plants are planned for your own area and, if they are, where they will be located.

5. Evaluate the model of the recreational yield of the lake in Figure 9-10. How would you go about restricting the use of a wilderness area to ensure that it was not degraded by overuse? Who benefits from such restrictions? Who pays for them?

6. Think critically about the conservation movement. Would you favor a $10 or a $100,000 fine for shooting members of an endangered bird species (e.g., bald eagles)? Is there a price limit on environmental protection in terms of other social goods? How much conservation can we afford?

7. Review your understanding of the following concepts:

the planet's total stock resources
reserves
resource substitution
fossil fuels

the green revolution
sustained-yield forestry
recreational yield
resource conservation

One Step Further . . .

A useful way to begin is to browse through readings that give a wide range of approaches to the resource–population relationship, such as
Burton, I., and R. W. Kates, Eds., *Readings in Resources Management and Conservation* (University of Chicago Press, Chicago, 1965).

The notion of resources, how they are defined, and how we use them is presented in a major work by a geographer
Simmons, I. G., *The Ecology of Natural Resources*, 2nd ed., (Arnold, London, 1980).

Future patterns of natural-resource availability are discussed at length in
Fisher, S. C., *Energy Crisis in Perspective* (Wiley, New York, 1974).
Landsberg, H. H., *et al., Resources in America's Future: Patterns of Requirements and Availabilities 1960–2000* (Johns Hopkins Press, Baltimore, Md., 1964).

The special problems of mineral resources and recreational resources, respectively, are attractively reviewed in two recent paperbacks by geographers:
Warren, K., *Mineral Resources* (Penguin, Harmondsworth, 1973), esp. Chaps. 10 and 11; and
Patmore, J. A., *Land and Leisure* (Penguin, Harmondsworth, 1972), esp. Chaps. 1, 2, and 6.

Important papers on the fuel and power problem from Scientific American *are now available in book form in*
Energy and Power *(Freeman, San Francisco, 1971).*

Geographers have traditionally played a leading role in resource evaluation and in the conservation movement. In addition to the regular geographic serials, you should also browse through one of the more popular resource-oriented journals such as The Ecologist *(published quarterly).*

Not here for centuries the winds shall sweep
Freely again, for here my tree shall rise
To print leaf-patterns on the empty skies
And fret the sunlight. Here where grasses creep

Great roots shall thrust and life run slow and deep:
Perhaps strange children, with my children's eyes
Shall love it, listening as the daylight dies
To hear its branches singing them to sleep.

MARGARET ANDERSON *(1950)*

chapter 10
Our Role in Changing the Face of the Earth

Our opening quotation and our opening photograph in this chapter stand in stark contrast. Biogeographer Margaret Anderson catches gentle moments of reflection on the planting of a young tree and picks up the theme of people coaxing environmental change little by little. Our photograph peers down at an altogether cruder scene. Here a copper-rich hillock near Butte, Montana, is being gouged and blasted away in successive strips until it now forms one of the greatest man-made hollows on the planet. Thus the threads of the last chapters—increasing human numbers, human interference in the structure of ecosystem, and the accelerating search for resources—begin to intertwine here, as we see the landscape change.

Landscape change and the human role in changing the face of the earth have been a constant theme in geographic writings across the centuries. In the United States one of the earliest volumes on these topics came from a Vermonter, George P. Marsh. In his *The Earth as Modified by Human Action* (1874) Marsh drew attention to the unsuspected importance of human intervention in shaping what had hitherto been regarded as natural America. In the present century Marsh's ideas have been developed in two ways: First, by detailed historical reconstruction, geographers have estimated the magnitude of the effects of human intervention. This reconstruction has been facilitated by modern techniques of measurement and monitoring. Second, academic (and, increasingly, public) concern over the harmful effects of human intervention on environmental quality has created a convergence of disciplines. Contemporary geographic work in this area draws not only on its own long traditions, but on parallel work in such fields as biology and civil engineering.

In this chapter we look at the net effects of all the ways we have influenced our environment in terms of how we have changed the landscape. For it is in the changing face of the land that we can see most clearly the spatial effects of the several processes disen-

221

tangled in Chapter 8. However, measuring changes in land use turns out to
be a more perplexing problem than it
appears at first sight. As we noted in
Chapter 5, environmental change also
can have natural causes, and in any
situation these two sources of change
—human and natural—must be singled
out and identified. We shall also, in this
chapter, assess the extent of human intervention on different geographic
scales, from the global to the local-

township level; and we shall pay special
attention to the present and future roles
of human beings like ourselves in
changing the face of the United States.

Finally we look back on the last few
chapters and in Section 10-4 consider
the "human-environment controversy."
Just how do geographers see people:
dominating the environment, or being
dominated by it, or neither?

10-1
Difficulties in Interpretation

We have already encountered, in the last two chapters, some dramatic examples of landscape change. Figure 10-1 shows an aerial view of Brazilian area
discussed in Chapter 8. Just over a century ago, this rolling country in the
Paraíba Valley (midway between the cities of São Paulo and Rio de Janeiro)
was covered with a high tropical rain forest. In the 1850s it was caught up in
the great swath of forest felling and burning that preceded the coffee boom in
this part of Brazil. The boom lasted hardly a generation, and by the 1890s the
plantation houses and the slave quarters were being turned into cattle
ranches. In the meantime, the coffee frontier was moving into virgin country
hundreds of kilometers to the west.

The environment that we see in the photograph has experienced faster and
more radical changes in the last century than in the tens of thousands of years
humans have previously lived there. Instead of the original forest cover, we
now have a mixture of fire-controlled grassland and scrub forest. Replacing
the original soils on the steep hillsides and debris-choked valleys is a thinned
and denuded skeletal soil. Molasses grass from South Africa and eucalyptus
from Australia are helping to stabilize the vegetational community and to
make it of some continuing use to people.

The story of the Paraíba Valley could be retold with suitable changes
throughout the rest of eastern Brazil. (See the maps in Figure 10-1.) Indeed,
Brazil's story of widespread deforestation and consequent environmental
changes is typical throughout the heavily populated areas of both the tropics
and the midlatitudes. Over much of these areas, the environment is increasingly a human artifact. Some of the effects of our impact on the changing
landscape are immediate and apparent. The suburban sprawl of Los Angeles
into the San Fernando Valley, the flooding of Lake Nasser above the Aswan
dam on the Egyptian Nile, and the reforestation of Scottish moorlands are
plainly visible evidence of ongoing processes. It is when we extend our investigation to eras before our own that we run into more difficult problems in
interpreting change.

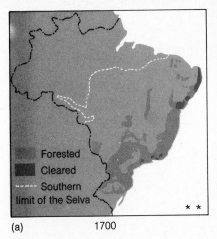

(a)   1700

Forested
Cleared
Southern limit of the Selva

(b)   1800

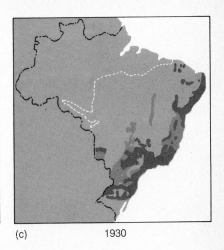

(c)   1930

(e)

(d)   1950

Figure 10-1. Man's impact on vegetational cover.
The maps show the clearing of the Brazilian forests over 250 years of European settlement from (a) 1700 to (d) 1950. Note that the clearing has been confined to the eastern area; the tropical rain forest (selva) of the Amazon basin is largely untouched. The photograph (lower right) shows a sample of heavily eroded terrain in the Rio de Janeiro-São Paulo area. There are still a few remaining woodlots, much degraded remnants of the original forest cover. [Photo by Services Aerofotogrammetricos, Cruzeiro do Sul, S. A. After P. E. James, *Geographical Review* **43** (1953), p. 309. Fig. 3-6. Reprinted with permission.]

Here we look at two basic questions. First, how do we establish the facts of past change? Second, once we have established the facts of change, how do we know what parts of the environmental change we have measured to credit (or debit) to human activities?

Historical Evidence   As we saw in Section 5-1 various methods have been developed to reconstruct the nature of past environments. Table 10-1 summarizes the methods available for reconstructing changes in land use. Note that as we approach the modern period, the variety and accuracy of the evidence tends to increase. How have geographers used these tools, and what results have they found? Let us illustrate the answer by taking a representative case and looking at it in some depth.

European historical geographers, led by H. C. Darby at Cambridge University, have diligently explored manuscript archives from the medieval period in reconstructing the massive changes in the forest cover of Western Europe. The changes can be inferred from an interpretation of such sources as the great survey carried out in Norman England nearly nine centuries ago, during the Domesday Inquest of 1086. For example, one of the questions asked by the Royal Commissioners was "How much wood in this place?" The form of the answer varied among the thousands of replies received from small hamlets, villages, and towns across the country. Some respondents specified the amount of woodland in quantitative terms in one of two ways: by its area or linear dimensions, or by the number of hogs it would support (by feeding on acorns or beech mast). Still other respondents simply stated that there was enough wood for fuel, mending fences, or repairing houses. Darby's patient scrutiny and assembly of detailed and dissimilar information built up a picture of the area of woodlands in early medieval England that shows marked local and regional variations in the pattern of heavily and lightly wooded tracts. (See Figure 10-2.)

Table 10-1   Methods of reconstructing land use

Contemporary changes (past 10 years)	Recent historic changes (100 years ago)	Remote historic changes (1,000 years ago)	Prehistoric changes (10,000 years ago)	Postglacial changes (100,000 years ago)
Direct observation			←——————— Radiocarbon dating ———————→	
Aerial photographs				
←——— Regular censuses and surveys ———→		Occasional cross sections (e.g., 1086 Domesday survey)		
	Comparison of maps			
	←————— Written accounts —————→			
		←——— Pollen analysis, study of macroscopic remains, ———→ lake and bog deposits, tree rings		

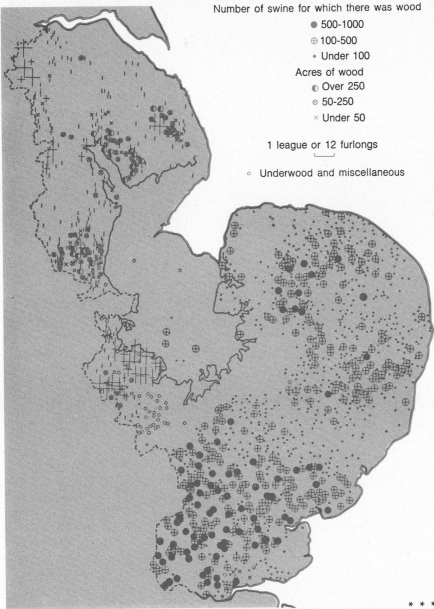

Number of swine for which there was wood
- ● 500-1000
- ⊕ 100-500
- · Under 100

Acres of wood
- ◐ Over 250
- ⊗ 50-250
- × Under 50

1 league or 12 furlongs

○ Underwood and miscellaneous

Figure 10-2.   Documentary evidence of change.

This reconstruction of the eleventh-century forest cover of eastern England is based on the Domesday survey of 1086. The different symbols reflect the fact that the extent of woodland was recorded in different terms in different parts of the area. For example, over much of the east, woodland was recorded in terms of the number of hogs (swine) it would support, while in the west it was more common to describe its size in terms of now-obsolete measures of length (leagues and furlongs). The exact meaning of certain woodland terms such as "underwood" is uncertain. Despite all these difficulties the main regional contrasts in timber cover stand out. Note that the empty area in the center of the map, now the English fenland, was largely undrained marshes with some woodland on the few "islands" within the marsh. Some of these islands had settlements and are shown toward the southern end of the marshland. [From H. C. Darby, *The Domesday Geography of Eastern England* (Cambridge University Press, Cambridge, 1952), p. 363, Fig. 106.]

* * *

Detailed statistical evidence for such an early period in landscape evolution is highly unusual. Even when documentary evidence is missing, however, some land uses can be determined from place names. Figure 10-3 presents the distribution of names of hamlets and villages in an English county with characteristic woodland place names: *leah, feld, wudu,* and *holt.* The location of these names is well to the north of a band of names (*tun, ingham,* and *ham*)

Figure 10-3.   Cross checking of evidence.

Geologic factors (a), as well as place-name and Domesday evidence (b-d) for Middlesex County in southeast England, suggest that there was a heavy forest in the northern half of the country but a more broken cover in the south during the eleventh century. [After H. C. Darby, in W. L. Thomas, Jr., Ed., *Man's Role in Changing the Face of the Earth* (University of Chicago Press, Chicago, 1956), p. 192, Fig. 55. Copyright © 1956 by The University of Chicago.]

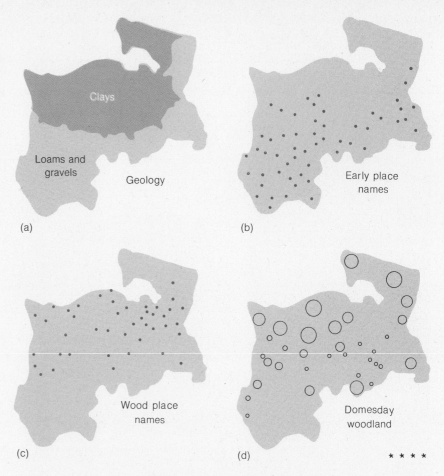

indicating early settlement. There is a contrast between early settlement of areas of light gravel and loamy soils in the south and later settlement of areas of intractable heavy London clay to the north. The distribution of this place-name evidence acts as a cross check on the supports the evidence of the Domesday records.

The reconstruction of successive patterns of woodland distribution in medieval Europe indicates that the clearing process was not one of continuous advance before the axe and the plough. Working at the Würzburg Geographical Institute in Germany, Helmut Jager has found an oscillating pattern in the upper Weser Basin of Germany. As over much of Western Europe, the destruction of war, great plagues, and population declines were reflected in changes in land use; land returned to woodland at times of reduced economic activity and was cleared during prosperous periods when population increased.

The possibilities and limitations of archives as sources of information exemplify the kinds of characteristics peculiar to a single source of evidence

Figure 10-4.   Landscape change shown by aerial photographs.
Aerial views may sometimes show features of geographical interest not apparent from the ground. An old pattern of medieval fields (cultivated in long, parallel strips) shows up beneath the modern field boundaries in an English midland village (Husbands Bosworth, Leicestershire). Note how the low early morning sun casts shadows which emphasize small, otherwise hard-to-see changes in surface relief. [From Cambridge University Collection. Copyright reserved.]

about a past environment. Some of the possibilities and limitations are summarized in Table 10-1 and illustrated in Figure 10-4. In practice, the reconstruction process involves using various types of evidence to verify estimates derived from a single source. It is this cross checking of evidence from various sources that allows a final sequence of landscape changes to be confirmed.

Human or Natural Changes?   Changes brought about by human intervention and those arising from natural causes may be difficult to separate. Consider, for example, the effect on a small watershed of changes in the rhythm and amount of precipitation (Figure 10-5). Suppose there is a poleward displacement of the present climatic zones that brings semiarid conditions into forested areas. Instead of a year-round pattern of rather light rainfall, we now have a highly irregular pattern of precipitation (occasional large storms separated by droughts). What other changes are likely to follow?

We would expect the droughts and lower precipitation to reduce the moisture available for plant growth and thereby alter the character of the

Figure 10-5. Natural changes in the landscape.

Environmental changes causing a diminished vegetational cover lead to severe gullying on the slopes and aggradation in the main valley of a small watershed. *Alluvial fans* describe the fan-shaped spreads of eroded materials that form at the lower end of the new channels or gulleys. [From A. N. Strahler, in W. L. Thomas, Jr., Ed., *Man's Role in Changing the Face of the Earth* (University of Chicago Press, Chicago, 1956), p. 635, Fig. 124. Copyright © 1956 by the University of Chicago.]

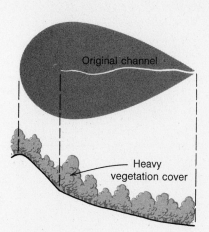

Original channel

Heavy vegetation cover

(a) Original morphology

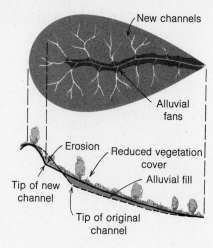

New channels

Alluvial fans

Erosion   Reduced vegetation cover

Tip of new channel

Alluvial fill

Tip of original channel

(b) Accelerated erosion

vegetation in the drainage basin. As a result, forest species might be replaced by drought-resistant shrubs. This change would decrease the ability of vegetation to absorb precipitation falling on the soil surface during storms, and increase the amount of material washed downslope. Figure 10-5 shows a sequence of morphological changes within the drainage basin. Increased erosion is followed by severe gullying on the slopes, which in turn is accompanied by the dumping of eroded material in the main valleys (aggradation). The number of streams per square kilometer is increased, and the slopes of both the debris-choked channels and the eroded valleysides are steepened.

Such cycles of aggradation and erosion are well documented for many parts of the world. In the semiarid areas of the southwestern United States, geomorphologists have pieced together a detailed sequence of cycles of gullying (arroyo cycles), although there is considerable dispute over whether these are due to natural variations in rainfall or to alterations in grazing pressures during the post-Columbian period of settlement. More rapid adjustments of the characteristics of channels can be observed over much shorter time periods. Near Ducktown, Tennessee, the denuding of vegetation caused by noxious smelter fumes has transformed a region of few channels and low slopes to one of numerous channels and high slopes. At the same time these gross changes in vegetation and morphology are taking place, more detailed alterations in the structure of the soil and the characteristic pattern of stream flows also occur.

Interpreting environmental change demands, therefore, that we establish *geologic norms*—that is, expected long-term rates of environmental change under natural conditions (discussed in Chapter 5). Only then can we begin to assess the ways in which human activities have accelerated or distorted these norms.

How much change has there been in the landscape in the past? To give exact figures on the amount of change induced by the kinds of processes we have described is a much more difficult task than simply itemizing examples of change. Many changes have only recently begun to be viewed as important, and so the desire to measure them is a relatively new phenomenon. However, we can make gross estimates by examining the changes in boundaries between major ecological zones. Within these zones we can get a rough idea of the minor variations that must have accompanied shifts in boundaries by looking at case studies of smaller areas.

On the Global Level   On the global level, the evidence is particularly difficult to piece together. Table 10-2 lists the assumed distribution of natural vegetation before it was disturbed to any great extent by human beings. Roughly one-third of the earth's surface was covered by forest; another third was split into polar, mountain, and desert zones; and the remainder was largely open park and grassland. The exact estimates given by different authorities vary, but the general proportions given in the table appear to be of the right order of magnitude.

How have these proportions changed under human impact? Maps of world population density (turn to the atlas section) indicate that well over one-third of the earth—the polar areas, the mountains and the dry zones—is unpopulated or very lightly populated. The massive concentrations of population are in environments that orginally supported forests or grasslands. The United Nations Food and Agriculture Organization (FAO) estimates that about 10 percent of the world's land is planted with crops, about 25 percent is forest, and a further 20 percent is grassland (covered by grasses, legumes, herbs, and shrubs). It is difficult to compare the estimates of natural vegetation in Table 10-2 with the FAO figures for current land use because slightly different classifications of land were used. What a comparision suggests, however, is that the natural vegetation has been changed mainly in forest areas. Within these zones, the changes have been highly selective and confined mainly to midlatitude woodlands in eastern North America, Eu-

Original cover (percentage) (about 10,000 B.C.)		Disturbed cover (percentage) (land use, in 1970s)	
Forest	33	Forest, woodland, and natural rangeland	36
Open woodland and grassland	31		
Desert	20	Desert	19
Arctic and alpine	16	Arctic and alpine	16
		Cropland	10
		Pasture and meadow	19

Table 10-2   Global changes in land use[a]

[a]Figures are averages based on a series of alternative estimates.

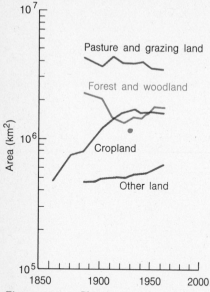

Figure 10-6.   Changes in land use in the United States since 1850.

Note that the vertical scale is logarithmic. [Data from M. Clawson et al., *Land for the Future* (Johns Hopkins Press, Baltimore, 1960), p. 39, Table 4.]

Atlas cross-check.

At this point in your reading, you may find it useful to look at the United States and Canada Environments map in the atlas section in the middle of this book.

rope, and eastern Asia, and to the monsoon woodlands of South Asia. Large areas of the tropical rain forest and the circumpolar boreal forest remained lightly touched by human intervention until the last few decades.

The reduction in the world's grassland zones has likewise been highly concentrated. The many dramatic changes in midlatitude grasslands, such as the North American prairies or New Zealand's Canterbury Plains, stand in contrast to the less dramatic alterations of the tropical grasslands and savannah zones. Of course, the estimates here refer only to the changes in cover for very general types of vegetation. As we saw in Section 8-2, significant alterations in species composition can occur in areas where the general appearance of the vegetation apparently has been unchanged.

The Subcontinental Level   At the subcontinental level the pattern of changes in land use becomes clearer. Figure 10-6 graphs more than a century of figures for categories of land use within the United States. The statistics vary in accuracy from one category to another and over the survey period, but they generally improve in reliability during the last half-century. We can see in these figures three types of environmental changes.

First, there was a phase of cropland expansion which was already under way by 1850 and continued to 1920. The period of this expansion corresponds to the frontier era, when the limits of settlement were extended from the Atlantic seaboard and trans-Appalachian areas to the Midwest, the Great Plains, and the Pacific and Mountain areas. During this period, the total amount of cropland increased fourfold, despite the fact that land was going out of cultivation in the East, particularly in New England.

Second, since 1920 the amount of cropland has remained relatively stable. Additions to existing cropland, such as western irrigated areas, have been counterbalanced by the abandonment of farms and the encroachment of housing on rural areas in the East. The expansion in farmland during this period has been due primarily to changes in the ownership of grazing land; that is, land previously in the public domain (mainly Federal land) or owned by the railroad companies has been incorporated into farms. Much of this change has been concentrated in the Great Plains area. The total area of grazing land as an environmental category has not changed much over this period if we consider both farm pasture and nonfarm grazing land.

Third, the composition of forests has changed drastically. Although the amount of forest land outside the farms has diminished by about one-third, the total area of woodland and forest (despite fluctuations) is much the same now as it was in 1850. Since that time the area of virgin timber has shrunk considerably. But the area devoted to forestry in the sense of managed timber resources has grown, and second-growth timber on abandoned farmland is on the increase. (See Figure 10-7.) The pattern of change, both in the total area of forest land and in the fraction of the original forest land remaining, has been most noticeable in the eastern half of the country. Here forests were cleared both for lumbering and for cropland. Commercial lumbering showed

(a) (b)

a distinct east-to-west shift across the country, with New England and the northern Appalachians forming the original center. The first westward movement was into the Great Lakes regions, where white-pine cutting was at its maximum from 1870 to 1890. By 1900 the southern pine region was the chief source of lumber for the national market. The extensive logging of Douglas fir and the shift of emphasis to the Pacific Northwest happened largely after World War I.

The key shifts in land use which began with the general European settlement in the 1700s largely came to an end by 1910 or 1920. Since then, the most significant changes have centered on the small area termed "other" land in Figure 10-6. This land includes urban, industrial, and residential areas (outside farms), parks and wildlife areas, military land, land for roads and transportation, and so on. The rapid growth of urban areas on the one hand and reserved wildlife areas on the other represent the extreme (and compensating) poles of environmental intervention.

On the Local Level   A small-scale example of changes in land use within the United States is provided by Figure 10-8. This shows the progressive deforestation of a sample area of approximately 10 km² (about 4 mi²) in southwest Wisconsin, Cadiz township, over a 120-year period of European colonization. The map for 1831, before agricultural occupation began, was compiled from the original government land survey. Apart from some prairie and oak-savannah in the southwest corner, the area was covered by upland, deciduous, hardwood forest. By 1882 about 70 percent of the forest area had been cleared for cultivation, and the boundaries of the cleared area reflected the township and range system of land division. (See Section 2-2, esp. Figure 2-7.) By 1902 the forested area had been reduced to less than 10 percent of the total area; it comprised about 60 small woodlots averaging less than 40 acres each. Continued felling of trees for firewood and occasional saw timber, plus heavy grazing by cattle, caused a decrease to less than 4 percent by the 1950s. This

Figure 10-7.   The extension and retreat of settlement and its impact on woodland.
(a) In the Matanuska Valley, Alaska, modern farm settlement is systematically replacing forest by cropland. (b) In southern Illinois, oak and hickory encroach on abandoned fields. Note the relatively uniform texture on the photos shown by the virgin timber stands in Alaska. In contrast the varied texture of the Illinois woodlands shows trees at various stages of regrowth linked to the different times at which fields were abandoned. In both photos the regular checkerboard pattern of fields and roads indicates that land divisions were laid out on the "township and range" system discussed in Chapter 2, Section 2. [U.S. Department of Agriculture photographs.]

Figure 10-8.   Changes in land use on the local scale.

The maps show changes in the wooded area of Cadiz Township, Wisconsin, since the beginning of European settlement. The colored area represents land remaining forested or reverting to forest in each year. [From J. T. Curtis, in W. L. Thomas, Jr., Ed., *Man's Role in Changing the Face of the Earth* (University of Chicago Press, Chicago, 1956), p. 726, Fig. 147. Copyright © 1956 by The University of Chicago.]

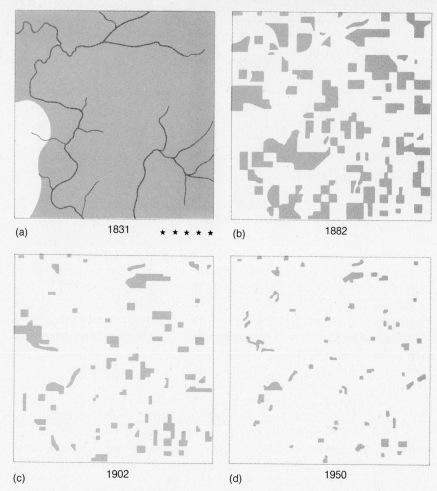

(a)   1831   ★ ★ ★ ★ ★   (b)   1882

(c)   1902   (d)   1950

diminution in forest land was due largely to a reduction in the size of individual woodlots rather than to the elimination of entire woodlots. The number of these has remained stable in this century.

Detailed investigations by ecologists at the University of Wisconsin have highlighted the significant effects of this environmental transformation. Decreases in the amount of water stored in the subsoil as agricultural fields replaced timber caused springs to dry up; the number of permanently flowing streams had decreased to around a third by 1935. The separation of the forest into isolated blocks reduced widespread burning, while fencing diminished the level of grazing within the woods. Both influences led not only to changes in species composition but to a denser forest cover with more mature trees per acre than under the original, pre-1831 conditions. Animal populations underwent similar changes. Species that were adapted to *edge conditions* (i.e., conditions at the boundary between woodland and open land) benefited from the increased length of the perimeter of the forest caused by its subdivision.

To sum up: It is clear that gross changes on the global level conceal subtle environmental adjustments on lower spatial levels. This local case study is one of scores of similar investigations of different ecological zones that illustrate the fine level of adjustment within ecological systems, and the key role that human intervention plays in their shaping. Understanding the interplay of human intervention and environmental adjustments gives us a comprehensive view of the nature and magnitude of changes in land use.

In the last few chapters we have looked at the ways in which human populations have responded to the challenge set by the global environment. As in Chapter 1, we have found that relation to be a two-way one: people being influenced by environment and changing that environment in turn.

10-3 The Human-Environment Controversy

The Historical Debate   Historically, the "chicken-and-egg" relationship between people and environment has always puzzled geographers. Did the environment control us? Or was environment an opportunity for conquest? Let us go back to the mid-nineteenth century, when English biologist Charles Darwin published his *Origin of Species* (1859). The book, based on worldwide evidence from Darwin's travels, made a strong impact on geographers. Its theme of competition between species for limited resources and the selective survival of the better-adapted species held special attraction for a German geographer, Friedrich Ratzel. In his *Anthropogeographie,* published a quarter of a century after Darwin's book, Ratzel argued that the distribution and grouping of human population on the earth's surface can only be understood in the context of the physical environment. He was particularly concerned with (a) those push-and-pull factors which had led to major migrations and (b) the physical conditions under which major civilizations had been able to develop.

Ratzel's ideas were enthusiastically taken up in the United States by Ellen Churchill Semple. She attended his lectures at Leipzig during the 1890s, and on her return home set out to introduce the ideas in American geography. Her *Influences of Geographic Environment* (1911) represents the fullest, best-documented, but perhaps most extreme arguments for what came to be termed *environmental determinism.* (See discussion on next page.) On the opening page of her book she set out her position in no uncertain way: "Man is a product of the earth's surface. This means not merely that he is a child of the earth . . . ; but that the earth has mothered him, fed him, set him tasks, confronted him with difficulties . . . and at the same time whispered hints for their solution." From this the book goes on to illustrate how the different major environments—oceans and continents, mountains and plains, warm climates and cold—have shaped the history of the human groups that had occupied them.

ENVIRONMENTALISM: THE MAIN "SCHOOLS OF THOUGHT"

Cognitive behavioralism holds that the impact of environment on people is partly dependent on their perception (cognition) of the resources and barriers it poses. See Figure 10-10. This view is widely held today.

Human ecology envisages reciprocal reactions between human and environment, like those of other plant and animal species. This view is associated with the Chicago geographer Harlan Barrows (1877–1960).

Physical determinism holds that the environment largely controls human development. It is associated with the German geographer Friedrich Ratzel (1844–1904) and his American disciple Ellen Churchill Semple (1863–1932).

Possibilism argues that the environment offers sets of possibilities but that the choice between them is determined by human beings. The French historian Lucien Febvre (1878–1956) was one of the strongest proponents of this view.

Scientific determinism is a variant of physical determinism in which the argument proceeds from the statistical analysis of sets of data rather

than from individual case studies. Yale geographer Ellsworth Huntington (1876–1947) was the leader of this school of thought.

Stop-and-go determinism holds that people determine the rate but not the direction of an area's development. The term was coined by Australian geographer Griffith Taylor (1880–1963).

The ideas of the various schools are discussed further by R. E. Dickinson in *The Makers of Modern Geography* (Routledge, London, 1969).

It was inevitable that there would be a pendulum swing away from the more extreme views of the Ratzel-Semple school. This came in 1924 with Lucien Febvre's *Geographical Introduction to History*. A French historian deeply interested in geographical problems, Febvre argued for an alternative view. He saw the earth's environment as presenting not necessities, but possibilities. By citing examples of quite different human developments in the *same* types of environment, Febvre was able to develop counter arguments to the earlier views.

Environmental determinism was not, however, a dead issue, and the interwar period saw some further major statements. Conscious that Ratzel had been writing at a time of very sparse global information, the Yale geographer Ellsworth Huntington set out to retest some of his hypotheses using better statistical data. In particular he set out to measure more accurately the way in which climate affected the ability of humans to perform work—both physical and mental. He backed up these studies with extensive research, particularly in Central Asia, to try and check out the importance of climatic change in determining major migrations out of that area.

Another challenging variation on the environmentalist theme was introduced by the Australian geographer Griffith Taylor (Figure 10-9). His studies in the Antarctic and the Australian outback suggested to him that the environmental conditons indicated certain directions along which a country's development could go. Humans were able to accelerate, slow, or stop the progress of development along a particular path, but not change the path. Taylor suggested we were like a traffic controller in a large city, altering the *rate* but not the *direction* of progress. In consequence of this analogy, this

Figure 10-9. Griffith Taylor.
The commemoration of Griffith Taylor on an Australian stamp issued in 1976 marks a tribute to one of geography's most active and outspoken exponents of the importance of the environmental view. English-born (1880) but a Sydney University graduate, he was a member of Scott's last expedition to the Antarctic. After World War I, as professor of geography at Sydney, he became involved in public debates on the potential habitability of the tropical and inland region of Australia for European settlement. Government policy at that time was to attract migrants, but Taylor was critical of the exaggerated claims then being made about the settlement possibilities of these areas. His assessments later turned out to be considerably more accurate than official views, but the controversy played some part in his decision to leave for a Chicago University chair in 1928. He later moved to Canada to create that country's first geography department at Toronto University. His final years were spent back in Australia, where he died in 1963. He describes his life vividly in an autobiography, *Journeyman Taylor*. [Stamp reproduced courtesy of the Australian Post Office.]

view of human-environment relations is often called *stop-and-go determinism!*

Current Views The historical debate which lasted from the middle nineteenth century through to World War II appears to be based on a fallacy. Like the argument over human intelligence (i.e., whether intelligence is a product of our genes or our early upbringing) it assumes a duality. Environment is in many senses inseparable from people. We can illustrate this idea with an example. One of the classic areas frequently quoted by the environmental determinists was the Australian desert. An area in which most readers of this book (and certainly the writer!) would starve in a few days might be rich in resources to Aborigine tribes with youngsters taught from childhood to recognize potential waterholes and buried food sources such as tree roots or insect grubs. Conversely, the tribal group might be wholly unaware that their hunting grounds are underlain by uranium deposits of great value to the western man.

The modern view of environmentalism is that summarized in Figure 10-10, and discussed in an elementary way in Chapter 1. It shows the environment composed of two sections: (a) *natural phenomena* in the sense of the totality of the world, and (b) the *perceived environment* and *hidden environment* as those parts of the natural phenomena known and not known to us. (You might like to think of this as being like an iceberg, with only a fragment of ice

Figure 10-10. Human-environment relations.

(a) Environments may be thought of in two segments, the perceived and the hidden. (b) Perceptual boundaries may also alter from one cultural group to another.

Figure 10-11. Human perception of future flood hazards.

(a) The curve shows the frequency of floods in 496 urban places in the United States. Most places for which flood-frequency data were available have two or three floods each year. (b) The degree of adjustment to the hazard in three places with three different experiences of flooding is illustrated. The height of each column reflects the number of respondents in each place who fail to perceive a threat, who perceive a weak or strong threat (two levels of "perceived"), or who adjust to the hazard. The letters and shading identify the three locations named in (a). [From R. W. Kates, *University of Chicago Department of Geography Research Paper* **78** (1962), Fig. 9.]

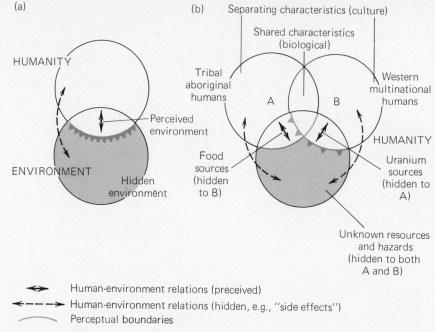

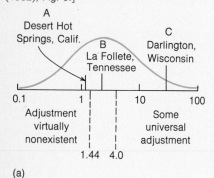
(a)

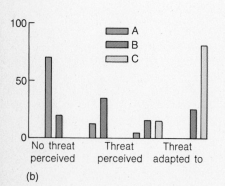
(b)

visible above the surface.) The two-way relations that go on between the two parts are clearly not restricted to the perceived environment. Minimata disease (discussed in Chapter 8) illustrates a case where there were two-way side effects which only later became a part of the perceived environment.

We show in Figure 10-10(b) a two-group situation. The "western multinational" person and "Aborigine tribal" person share certain common biological characteristics (shaded) and see part of the harsh environment in the same way. But the buried eggs and the buried uranium lie in different sets. The critical ingredient separating them is the bundle of attitudes which form part of their culture.

Even within western society there may be major differences in reactions to a given environment. For example, many human groups occupy environments that are dangerously unpredicable. River valleys may be subject to flood, coastal settlements are at risk of high tides, and semiarid areas may be hit by drought. Geographers Gilbert White and Ian Burton have investigated the ways in which such dangers are viewed. They began by looking at the likelihood of flooding in different areas and constructed a *risk curve* [Figure 10-11(a)] for the 498 urban communities in the United States that are built on river floodplains (and have flood-frequency data). Some of these urban places were reliable to have floods only once in ten years, while in others waters rose to danger levels dozens of times in a single year. For most cities in the sample, the likely number of floods each year was two or three.

As Figure 10-11 indicates, human responses and adjustment to the known danger of flooding do not increase consistently as the risk becomes greater. Until the environmental stress builds up to the point where the likelihood of damage is regular and recurrent, little or no adjustment to the possibility of flooding takes place. Figure 10-11(b) shows the degree to which inhabitants of three United States communities where the likelihood of flooding is different "perceive" a flood threat. Darlington, Wisconsin, can expect 20 floods in any 10-year period, whereas Desert Springs, California, can expect only one flood in the same period. When the probability of the recurrence of a hazard is high, as it is in Darlington, the danger is widely perceived but evaluated in different ways. Table 10-3 lists four of the ways in which individuals respond to the possibility of recurring natural disasters. Each response represents an optimistic rationalization for continuing to live in a hazard area. It is interesting that the range of responses is great when the probabilities of a recurrence are moderate; responses tend to be more uniform in high-risk and low-risk areas.

Flooding further stresses the essential interdependence of human being and land. By changing the face of the earth—building levees and dams—we have been able to occupy many of the flood plains of the world's great rivers. In doing so, the potential threat of flooding is *increased* as mud which would have been spread valleywide by the seasonal floods is forced to accumulate in the stream channel itself. To build the levees higher pushes the risk of an occasional big flood still higher, and so on.

Table 10-3   Human reactions to irregular natural hazards

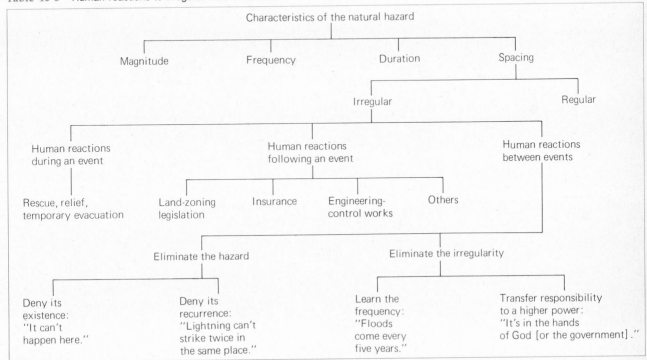

SOURCE: Adapted from I. Burton and R. W. Kates, *Natural Resources Journal* **3** (1964), p. 435.

Coping with environments, risky and otherwise, shows spatial variations around the world linked to the attitudes adopted by the groups living in those environments. It is misleading therefore to think of overall human-environment relations. Rather, as Figure 10-10 suggests, we have to look at the relationship between particular groups and an environment. It is to these group differences that we now turn in the third part of this book.

Summary

1. Over much of the earth's land surface, the landscape we now see is not a natural one, but rather, a man-made artifact. Only the polar ice caps, tundra, deserts, and high mountain zones remain as genuinely wild landscapes.

2. Although evidence of our impact on the natural environment is a matter of everyday observation, the cumulative impact of the human population over longer periods of time requires careful research and interpretation. Historical evidence of landscape change can be found in archival material such as the English Domesday Survey, through the study of place names, the reconstruction of patterns of succession in woodland distribution, and the use of aerial photographs.

3. The interpretation of landscape change is made more difficult by the similarity between some of the effects of both man-made and natural activities (discussed in Chapter 5). Geologic norms serve as a basis for separating changes caused by human intervention from those due to natural forces.

4. For the earth as a whole estimates of the amount of natural landscape change that has taken place in the past can only be in broad-brush terms. Research seems to indicate that the single most important change has taken place through the reduction in forest areas.

5. For a subcontinental area such as the United States, research has shown environmental changes from 1850 to 1900 to reflect cropland expansion, and after 1920 to be based on a stabilized cropland area. In addition there has been a great change in forest land composition. While no major shifts in land use in the United States are seen as occurring before the end of the century, minor changes should occur as follows: a continuing increase in urban areas: modest increases in managed forestry and recreation areas; and modest declines in agricultural areas.

6. There has been a longstanding controversy within geography over the ways in which the human population and the natural environment have been related. Distinct schools of environmentalism have grown up, each arguing for a different balance of influence between one component and the other.

Reflections

1. What have been the main changes in rural land use in your own state or district over the last hundred years? Are (a) similar or (b) different trends likely in the next decade?

2. How do geographers know what changes have occurred in land use? Compare the evidence available on changes over (a) a 10-year period, (b) a 100-year period, and (c) a 1000-year period.

3. Obtain a map which shows the distribution of forested land in your own country. Why are the forested areas located where they are? Is their distribution the result of (a) natural environmental conditions or (b) human intervention? How is the pattern changing?

4. Look at the growth of your own city or community. How large is its population expected to be in 1980, 1990, and

2000 A.D.? Calculate, roughly, the number of new homes that will be required to house these extra people and the additional building land that will be needed. What land will be used for this purpose? What alternatives are there?

5. Review your understanding of the following terms:

land use
Domesday survey
geologic norms
place-name evidence

edge conditions
environmental deter-
  minism
stop-and-go determinsim

One Step Further . . .

The standard geographic account of the topics discussed in this chapter is provided by
    Thomas, W. L., Ed., *Man's Role in Changing the Face of the Earth* (University of Chicago Press, Chicago, 1956).

This is a splendidly produced book with a wealth of talented authors and is strongly recommended for browsing. Another thoughtful book on the same theme is
    Wagner, P. L., *The Human Use of the Earth* (Free Press, New York, 1960).

For a massive survey of Western philosophers' views of human beings as agents of terrestrial change, look through
    Glacken, C. J., *Traces on the Rhodian Shore* (University of California Press, Berkeley, 1967).

Detailed accounts of changing land use in specific areas of the United States are provided in
    Marschner, F. J., *Land Use and Its Patterns in the United States* (U.S. Department of Agriculture, Washington, D.C., Handbook 153, 1959) and

Clawson, M. et al., *Land for the Future* (Johns Hopkins University Press, Baltimore, Md., 1960).

An engaging book which explores the role of humans and environment is
    Tuan, Yi-Fu, *Topophilia: A Study of Environmental Perception, Attitudes, and Values* (Prentice-Hall, Englewood Cliffs, N.J., 1974).

Research reports on human impact on environment occur regularly in all the main geographic serials. You should also browse through biological journals like Ecology *(a quarterly), to see something of the research in neighboring scientific fields, and the* Journal of Historical Geography *(also a quarterly) to see work on past changes in land use.*

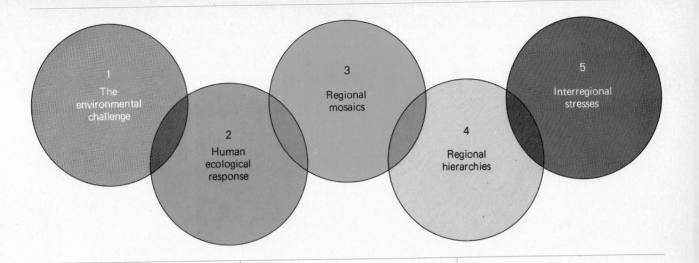

In Part Three we turn from an ecological view of people in relation to the environment to a cultural view of human organization of the earth's surface. In *Cultural Fission* (Chapter 11) we look at the geographic implications of our nonbiological diversity in terms of such characteristics as religion and language. Just how these affect people's spatial behavior and the way in which they give rise to a highly differentiated system of cultural regions forms the main substance of this chapter. *World Cultural Regions* (Chapter 12) takes the study of cultural regions further by looking back at the ways in which these early spatial forms began to crystallize. Special attention is paid to the explosive growth of Europe since

the 1500s, and the dramatic way in which it has shaped the world geography of the present century. To illustrate this theme, the ways in which the main cultural regions of the United States—areas like New England, the Midwest, and California—emerged are studied in more detail. *Spatial Diffusion* (Chapter 13) looks at how information and innovations pass from one cultural region to another, bringing changes both trivial and lethal. We examine the geographic theory and models developed to predict waves of innovation and suggest how the models can be used. Together, these three chapters describe a regional mosaic forged by human cultural differences but becoming increasingly mixed and muddied by forces overriding older cultural barriers.

part three

Regional Mosaics

Although I am unborn, everlasting, and I am the Lord of all, I come to my realm of nature and through my wondrous power I am born.

BHAGAVADGITA

When the author engaged, usually unsuccessfully, in the schoolyard scuffles beloved of all small boys, the key word to know was "faines." Children, happily, are one of the few remaining uncivilized tribes, and—at least in English village schools in the 1930s—they still retained the local dialect truce words. Shouting "faines" allowed a small boy at the bottom of a pile of bodies a respite or truce. It was important, however, to know your location when you called for time out. In the next county to the west the truce word was "barsy," and in the next county to the east it was "cree"!

Such local differences in schoolyard language over a few tens of miles in as relatively homogenous a country as England are a microcosmic example of the vast cultural differences that shatter the earth's population into a regional mosaic of immense intricacy and complexity. Jew and Gentile, Moslem and Hindu, WASP and Afro, Maoist and Mennonite, all serve to recall the countless ways in which the single biological species of humans seek to separate and divide itself into different stereotypes.

So in this chapter we move away from the biological and ecological view of the human population which dominated Part Two of the book and look at the geographic implications of their nonbiological diversity. In this first chapter of Part Three we try to clear the ground by defining some of the ways in which humanity differs. We begin by asking, "What do we mean by cultural differences?" and "How can we define them?" Once these points have been settled, we go on to look at the spatial pattern of some important cultural differences. Here the critical questions are, "How does culture vary spatially?" and "Why is it critical for geographers to know about cultural variations?" Finally, we try to use our concepts of culture to build a code by which geographers can separate one part of the earth from another by identifying a system of cultural regions.

chapter 11
Cultural Fission
Toward Regional Divergence

243

In this chapter more than in any other, the cultural biases of the author must show. No one can shake free from deep-rooted and unconscious prejudices that come from being reared in a particular social and cultural setting. Readers are invited to challenge the judgments made here, and to submit for discussion alternate views of the human cultural mosaic based on their own experience.

11-1
Culture As a Regional Indicator

In our chapter title and throughout this third part of the book, we use the term "culture." Just what do geographers mean by this term? Since culture turns out to be a complex word, we shall begin with some negatives before going on to a positive definition.

Culture: Some Negative Conclusions   Let us begin by defining what culture is not. First, culture does not mean simply an interest in artistic pursuits. Though the phrase "cultural activities" is commonly used to describe musical, literary, and artistic efforts, this is a special use of the term culture in a very limited sense. Although such activities may have a distinct spatial pattern and may define certain cultural boundaries (e.g., the geographic extent of interest in a virulent variety of bagpipe music may provide a clue to the distribution of folk of Scottish ancestry), this view of "culture" is too narrow for our purposes.

Second, culture does not mean race. *Race* is a biological term used to classify members of the same species who differ in certain secondary characteristics. Humans are a single species (*Homo sapiens*) with a common chromosome number (46), and fertile interbreeding among all its billions of members is possible. Nonetheless, specific biological differences do separate people into recognizably distinct subgroups. (See Chapter 12, Section 12-1.) Such differences range from (a) variations in external features (e.g., in skin pigmentation, in the shape of the skull, in hair type, and in eyefolds) to (b) differences in internal features (e.g., blood types).

Blood types may be an important indicator of group differences that have been maintained for a long time. Thus the Basque-speaking population of southwest France and northern Spain has a very high proportion of Rhesus-negative blood types compared to the population throughout most of Europe. Moreover, Rhesus negativism in blood groups is a peculiarly European trait; it is extremely rare among Asians, Africans, and American Indians.

Racial differences of this kind are almost certainly due to long periods of isolation in which genetic variations are accentuated by generations of interbreeding. This hypothesis seems to be supported by the spatial distribution of certain genetically determined diseases. Some of these diseases are highly localized: Kuru, a progressive disorder of the nervous system, is so far known only among the Fore peoples of eastern New Guinea. Other diseases have distinct but intercontinental distributions. Sickle-cell anemia is a disorder of the blood cells found in much of Africa south of the Sahara (and by transfer, in the black population of the Americas) and in another broad belt running from India to Indonesia.

Whether genetic drift (random changes of gene frequencies) permits long-term adaptation of groups to particular environments is not clear. Some differences have been medically established—for example, differences in the sweating capacity of blacks and whites, the fact that Eskimos' skin temperatures remain higher than normal in cold weather, and the lower metabolic rates of Australian aboriginals at night, which allow them to withstand low night temperatures. The cases of genetic adaptability stand in contrast to the extraordinarily widespread ability of people of all kinds to adapt *technically* to extreme environments by devising life-support systems that range from parasols and the fur wrap to spacesuits for moon-walkers.

Culture: Some Positive Conclusions Considerable efforts have been made by cultural geographers to define culture in a precise and positive way. We can perhaps summarize their view by saying that *culture* describes patterns of learned human behavior that form a durable template by which ideas and images can be transferred from one generation to another, or from one group to another. Three aspects of this definition need further accenting. First, the transfer is not through biological means. The *same* newborn child will grow up with quite *different* sets of cultural characteristics if it is reared in different cultural groups. Second, the main imprinting forces in cultural transfers are symbolic, with language playing a particularly important role. (By "imprinting" is meant the spontaneous acquisition of information, particularly those habits of speech and behavior acquired in the early years of life.) Third, culture has a complexity and durability which make it of an entirely different order from the learned behavior of other, nonhuman animals.

The diversity of human cultures and their innate complexity is staggering. Anthropological research by such scholars as Claude Lévi-Strauss has long since done away with the notion that there are any "simple" cultures. Even in the smallest and most "primitive" cultural groups (e.g., a small Amazonian hunting tribe or the population of a Micronesian village), there is a massive amount of cultural information to acquire. The child growing up in the simplest society slowly acquires millions of pieces of a cultural pattern, which will be duly passed on (albeit in some modified form) to the next generation. We should note (a) that such transfers are independent of formal education (in the western sense of going to school) and (b) that the transfer process is always incomplete. Thus the culture of the group is always several times larger than the culture of the individual. The most distinguished Harvard professor or the oldest village elder can never hope to acquire, in a lifetime of study, more than some fractional part of the "genetic code" of the culture of which each is a part.

Huxley's Model   If culture is so complex and all-embracing a thing, is there any hope of disentangling it into simpler, molecular units which we can study and comprehend? Let us look very briefly at the solution of one man. One of the simplest ways of categorizing culture was proposed by English biologist Julian Huxley in a comparison of cultural and biological evolution. Huxley's model has three components: mentifacts, sociofacts, and artifacts. (See Figure 11-1.)

Mentifacts are the most central and durable elements of a culture. They include religion, language, magic and folklore, artistic traditions, and the like. They are basically abstract and mental. They relate to the human ability to think and to forge ideas, and they form the ideals and images against which other aspects of culture are measured.

Sociofacts are those aspects of a culture relating to links between individuals and groups. At the individual level they include family structures, reproductive and sexual behavior, and child rearing. At the group level they include political and educational systems.

Artifacts are material manifestations of culture. Sometimes termed "cultural freight," they include those aspects of a group's material technology which allow basic needs for food, shelter, transport, and the like to be filled. Systems of land use and agricultural production are cultural artifacts, as are tools and clothing of a particular design.

Like all such schemes, Huxley's model is only an approximation of reality. In practice, we find aspects of culture in which the three components seem tangled into an intractable knot and others where a strand can be pulled free and studied individually.

Figure 11-1.   The components of culture.
Wilbur Zelinsky has created a model of culture using a three-dimensional cube which can be analyzed in terms of (a) its intrinsic components, or (b) the cultural characteristics of a given region, or (c) the cultural characteristics of a distinctive group or subculture. Note how the three approaches interlock, so that we can study small cubes within the master cube, that is, the cube of the spatial distribution of terracing in areas of Chinese culture within the cultural "realm" of Southeast Asia. [From W. Zelinsky, *The Cultural Geography of the United States* (Prentice-Hall, Englewood Cliffs, N.J., 1973), p. 73, Fig. 3-1.]

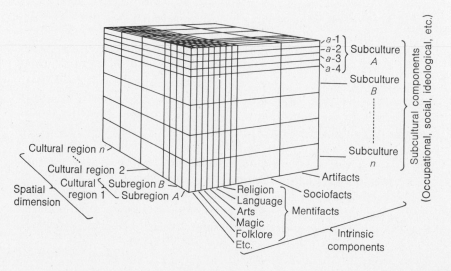

Does culture have a specifically geographic pattern? To answer this basic question, we shall take one of the most important of cultural differences, language. In studying language remember that we are using this simply as one example from the multitude of other mental, social, and material examples that might suggest themselves. But the questions we shall ask will be general ones about spatial stability, order, and change. These may provide a basis for your own analysis of other cultural elements that may interest you more.

The Scale of Language Variations Language is the essential linking device in human cultures, enabling members of a group to communicate freely with each other. It is also a barrier in that members of one language group cannot communicate with members of another language group.

Just how many different languages are spoken today depends partly on our definition of language. Table 11-1 gives the proportions for the world's major languages. If we leave out minor dialects, there are still around 3000 different languages in current use. At least another 4000 were once spoken but have now gone out of use. If we define language still more widely to include the various scientific "jargons"—for instance, those used by geographers—then the number becomes incalculable.

One of the simplest ways of classifying languages is according to the number of people who speak them. *Global languages* are spoken by very many people indeed; *local languages* are spoken by very few. English, a global language, is the primary tongue of around 400 million people (almost 1 out of every 10 persons in the world) and serves as a second language for many more. However, the most widely spoken language (though those who speak it are more concentrated spatially) is Mandarin Chinese, which, with its many dialects, is the language of about 825 million people in East Asia. Figure 11-2 shows the principal languages spoken on the Indian subcontinent. Were we to rank the world's languages, putting the global ones like English and

11-2
A Geography of Language

Atlas cross-check.
At this point in your reading, you may find it useful to look at the World Languages map in the atlas section in the middle of this book.

Table 11-1   Global language groups

Language family	Estimated share of world's language	Typical languages	Area
Indo-European	47	German, English, Hindustani	India, Europe, America, Australia, S. Africa
Sino-Tibetan	25	Chinese, Thai, Tibetan	East Asia
Afro-Asiatic	5	Arabic, Hebrew, Berber	N. Africa, Middle East, Ethiopia
Dravidian	5	Tamil, Telugu	S. India, Ceylon
African	5	Swahili, Hausa	Africa S. of Sahara, but excluding Ethiopia
Malayo-Polynesian	5	Indonesian	Malaysia, Indian Ocean and Pacific Ocean islands
Japanese-Korean	4	Japanese, Korean	E. Asia
Uralic-Altaic	3	Mongol, Finnish, Turkish	E. Europe, SW. and Central Asia
Mon-Khmer	1	Cambodian	SE. Asia
American Indian	—	Over 1200 different languages	North and South America

Figure 11-2. Linguistic differentiation of cultural areas.
The map shows the principal languages used in the Indian subcontinent. Darker shading is used for areas in which two or more languages are spoken. Note that only the major languages are shown and that English is widely used for government and business. Tamil speakers make up a significant minority in the dominantly Singhalese-speaking island of Sri Lanka (Ceylon).

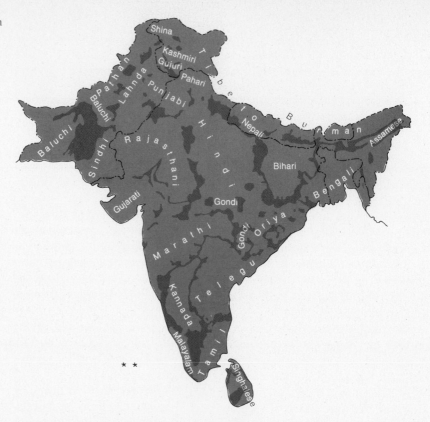

Chinese at the top, these would come about halfway down the list. At the bottom of the list would be the really local languages. Research in New Guinea has revealed some wholly distinct languages (uncomprehended by neighboring groups) confined to single valleys spoken and understood by only a few hundred people, and having a spatial extent of less than 65 km² (25 mi²). Actually, a few languages are spoken by a disproportionately large percentage of the human population. The top 14 languages are spoken by 60 percent of the world's people. At the other extreme, the bottom 500 are divided among no more than 1 million people in the remoter parts of Asia, Africa, South America, and Australasia.

The Origin and Dispersal of Languages Language illustrates clearly a second theme in cultural geography—the origin and dispersal of cultural elements. The questions we must ask to understand this process are about the relation of one language to another.

Extensive linguistic research has revealed that many different languages seem to have emerged from a common stock. For example, it is possible to trace back Indian languages in the northeastern United States like Cayuga, Seneca, and Tuscarora to a common Iroquoian stock which has some linguistic connections with Sioux language groups further west. However, the language group whose evolution we know most about is the Indo-European family of languages. (See Figure 11-3.) A wealth of written records in these languages has allowed us to unravel slow linguistic drifts over the centuries. Despite the fact that these languages are spoken by half the world's people they still have many simple, basic words in common. The English word "mother," for example, is "Mutter" in closely related German. It is also recognizable in quite different subgroups—in the Romance group, in the Spanish "madre"; in the Balto-Slavonic group, in the Russian "mat," and in the Indo-Iranian group, in the Sanskrit "mata." Even in Greek, whose position in the family language tree is still hotly debated, mother is "meter."

The distilling out of the main language groups shown in Figure 11-3 was a slow process, taking over tens of thousands of years. Changes over a much shorter period are noticeable in dialects within a language, but languages themselves are stable enough to provide useful spatial signals of the migrational history of various groups. Thus the language differences between the Gilbert and Ellice Islanders in the Pacific provide evidence of long separation. (See the discussion of Kon-Tiki voyages in the Pacific in Section 13-4).

Forces of Spatial Change   Cultural patterns are clearly not static in either time or space. The proportions of the world's population speaking each of the

Figure 11-3. Linguistic origins and differentiation.

About one-half of the world's 4.3 billion people speak languages in the Indo-European family. The chart shows the main languages in this family and the links between them. Note that some European populations speak languages outside the Indo-European group (e.g., Finnish and Basque).

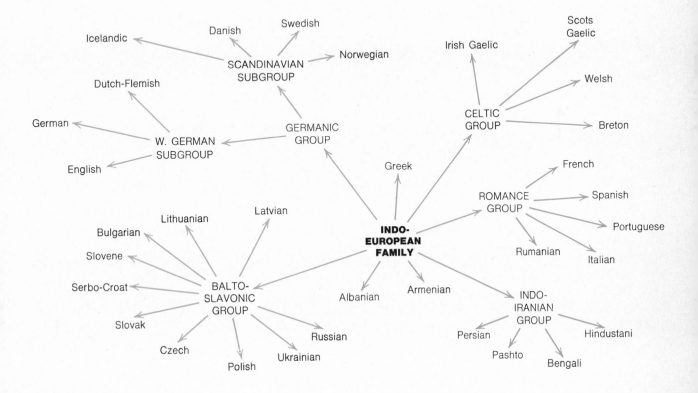

major languages is changing, and their spatial distributions are waxing or waning. Patterns of language are changed not only by the demographic forces of birth and death—which affect mainly our "first," or "native," language—but by the agressive spread of second languages. Currently, the proportion of English-speaking persons is rapidly increasing in the urbanized and "western-ized" world. Figure 11-4 summarizes the main forces at work hammering out these changing linguistic patterns. Try to work your way through the different branches of this tree and think of examples that would illustrate each type of spatial change in language.

At the same time as global languages are spreading, some small languages are slowly dying out. The Celtic languages of Western Europe, for example, have been losing ground for centuries to the more aggressive English and French tongues. Celtic languages were once spoken in the western parts of the British Isles, the Brittany peninsula in France, and northwest Spain. One of the Celtic languages—Cornish—was confined to the extreme southwest corner of England. Until the fifteenth century, Cornish was spoken over most of the county of Cornwall; but by 1600 it was heard only in the extreme west. The mining industry brought an increasing number of English-speaking outsiders into the area, and by 1800 the language was virtually dead. The last Cornish-speaking person died in the 1930s. Even much stronger Celtic languages like Welsh are now confined to a part of their original area. (See Figure 11-5.)

Figure 11-4.   Spatial changes in languages.

The chart gives a schematic view of some of the main forces causing linguistic changes. Part of eastern Canada is bilingual (speaking French and English). Creole languages (e.g., the French creole in Haiti) have emerged from the mixing of French or Spanish with Caribbean languages. "Pidgin languages" are very basic forms of English spoken as a trading language in many Pacific island communities. "Loanwords" are borrowings due to human communication. (The words *jazz* and *taxi*, for example, are loanwords common to very many languages.)

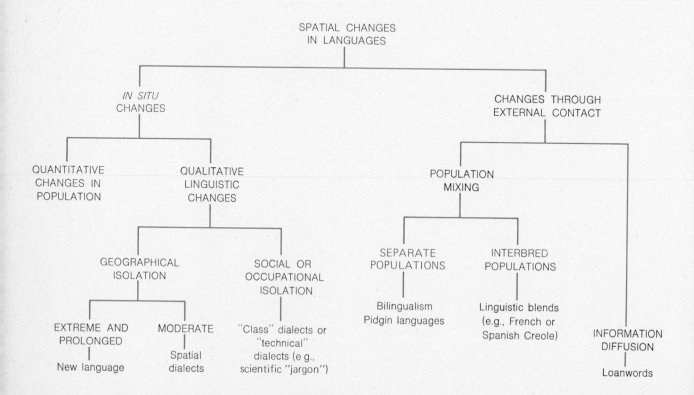

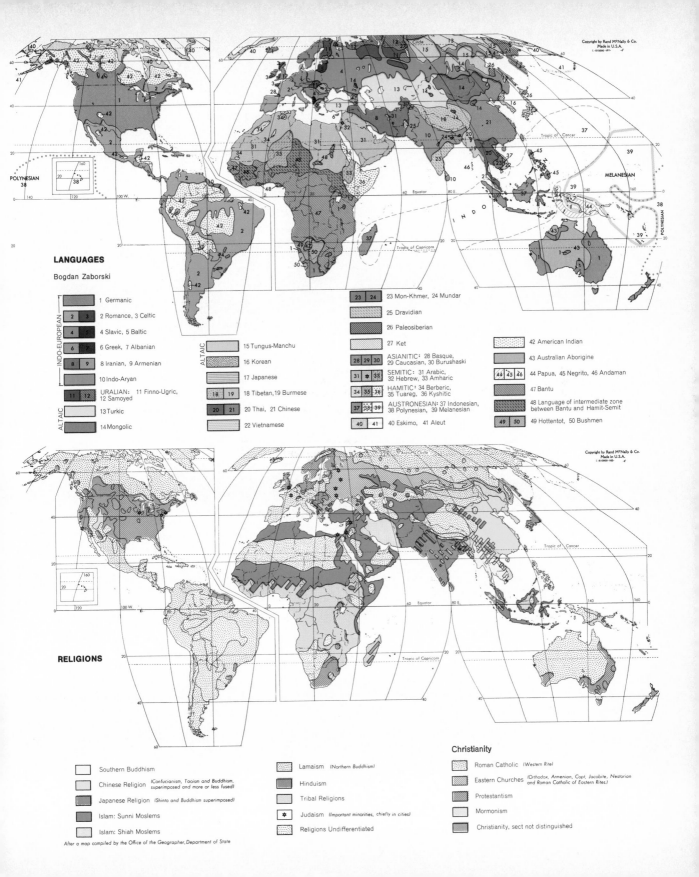

Copyright by Rand McNally & Co.
Made in U.S.A.

LANGUAGES

Bogdan Zaborski

INDO-EUROPEAN
1 Germanic
2 Romance, 3 Celtic
4 Slavic, 5 Baltic
6 Greek, 7 Albanian
8 Iranian, 9 Armenian
10 Indo-Aryan

URALIAN: 11 Finno-Ugric, 12 Samoyed

ALTAIC
13 Turkic
14 Mongolic
15 Tungus-Manchu
16 Korean
17 Japanese
18 Tibetan, 19 Burmese
20 Thai, 21 Chinese
22 Vietnamese

23 Mon-Khmer, 24 Mundar
25 Dravidian
26 Paleosiberian
27 Ket

ASIANITIC: 28 Basque, 29 Caucasian, 30 Burushaski
SEMITIC: 31 Arabic, 32 Hebrew, 33 Amharic
HAMITIC: 34 Berberic, 35 Tuareg, 36 Kyshitic
AUSTRONESIAN: 37 Indonesian, 38 Polynesian, 39 Melanesian
40 Eskimo, 41 Aleut

42 American Indian
43 Australian Aborigine
44 Papua, 45 Negrito, 46 Andaman
47 Bantu
48 Language of intermediate zone between Bantu and Hamit-Semit
49 Hottentot, 50 Bushmen

POLYNESIAN 38

MELANESIAN

POLYNESIAN 38

Copyright by Rand McNally & Co.
Made in U.S.A.

RELIGIONS

Southern Buddhism
Chinese Religion *(Confucianism, Taoism and Buddhism, superimposed and more or less fused)*
Japanese Religion *(Shinto and Buddhism superimposed)*
Islam: Sunni Moslems
Islam: Shiah Moslems

Lamaism *(Northern Buddhism)*
Hinduism
Tribal Religions
Judaism *(Important minorities, chiefly in cities)*
Religions Undifferentiated

Christianity

Roman Catholic *(Western Rite)*
Eastern Churches *(Orthodox, Armenian, Copt, Jacobite, Nestorian and Roman Catholic of Eastern Rites)*
Protestantism
Mormonism
Christianity, sect not distinguished

After a map compiled by the Office of the Geographer, Department of State

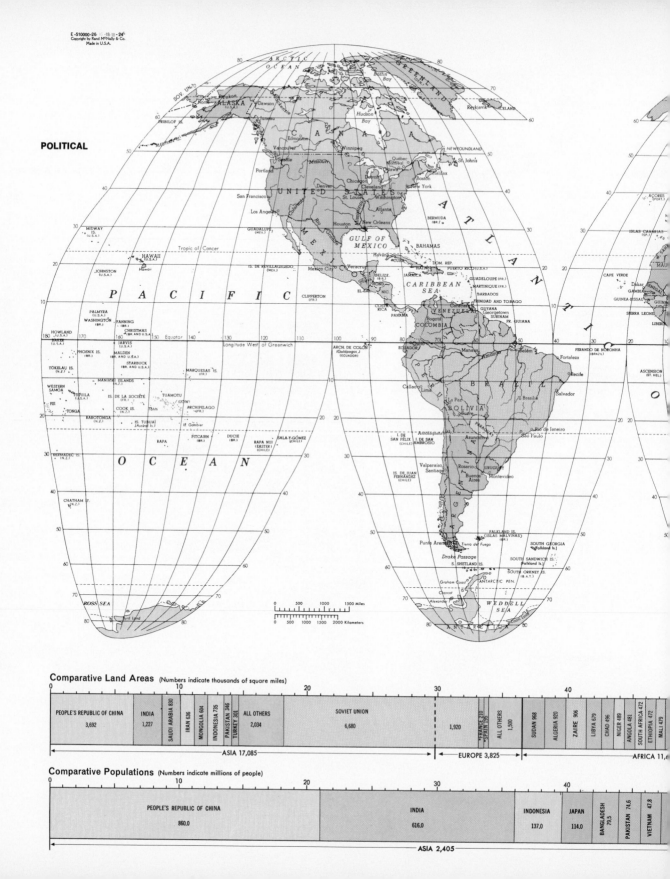

POLITICAL

E -510000-26 ... -18-18-24
Copyright by Rand McNally & Co.
Made in U.S.A.

Scale:
0   500   1000   1500 Miles
0   500   1000   1500   2000 Kilometers

Comparative Land Areas (Numbers indicate thousands of square miles)

0	10	20	30	40

| PEOPLE'S REPUBLIC OF CHINA 3,692 | INDIA 1,227 | SAUDI ARABIA 830 | IRAN 636 | MONGOLIA 604 | INDONESIA 735 | PAKISTAN 346 | TURKEY 301 | ALL OTHERS 2,034 | SOVIET UNION 6,680 | 1,920 | FRANCE 210 / SPAIN 195 | ALL OTHERS 1,500 | SUDAN 968 | ALGERIA 920 | ZAIRE 906 | LIBYA 679 | CHAD 496 | NIGER 489 | ANGOLA 481 | SOUTH AFRICA 472 | ETHIOPIA 472 | MALI 479 |

◄—————— ASIA 17,085 ——————► ◄—— EUROPE 3,825 ——► ◄—— AFRICA 11,6...

Comparative Populations (Numbers indicate millions of people)

0	10	20	30	40

| PEOPLE'S REPUBLIC OF CHINA 860.0 | INDIA 616.0 | INDONESIA 137.0 | JAPAN 114.0 | BANGLADESH 79.5 | PAKISTAN 74.6 | VIETNAM 47.8 |

◄—————————— ASIA 2,405 ——————————►

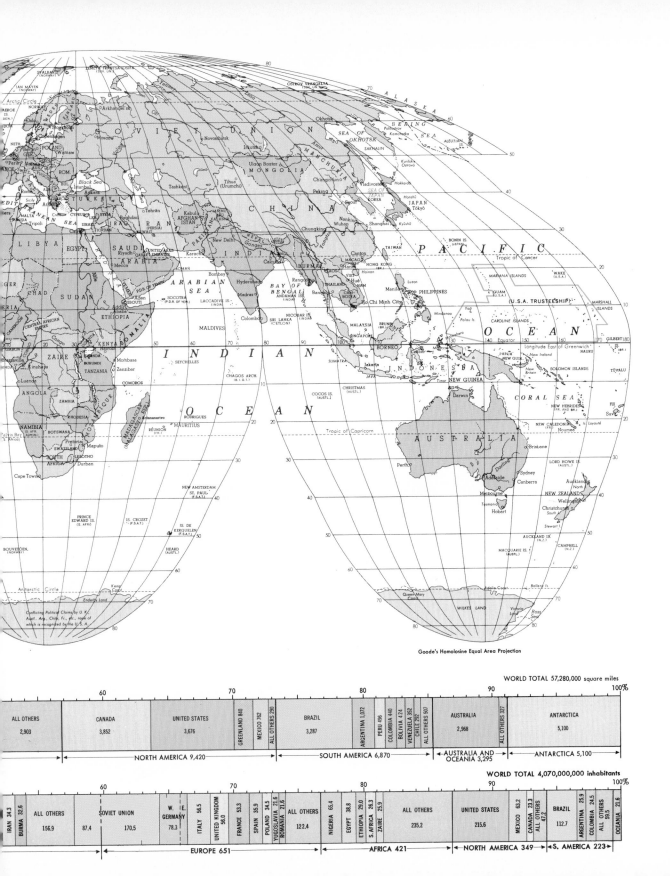

Goode's Homolosine Equal Area Projection

WORLD TOTAL 57,280,000 square miles

| ALL OTHERS 2,903 | CANADA 3,852 | UNITED STATES 3,676 | GREENLAND 840 | MEXICO 762 | ALL OTHERS 290 | BRAZIL 3,287 | ARGENTINA 1,072 | PERU 496 | COLOMBIA 440 | BOLIVIA 424 | VENEZUELA 352 | CHILE 292 | ALL OTHERS 507 | AUSTRALIA 2,968 | ALL OTHERS 327 | ANTARCTICA 5,100 |

— NORTH AMERICA 9,420 — SOUTH AMERICA 6,870 — AUSTRALIA AND OCEANIA 3,295 — ANTARCTICA 5,100 —

WORLD TOTAL 4,070,000,000 inhabitants

| IRAN 34.3 | BURMA 32.6 | ALL OTHERS 156.9 | 87.4 | SOVIET UNION 170.5 | 78.3 W. & E. GERMANY | ITALY 56.5 | UNITED KINGDOM 56.0 | FRANCE 53.3 | SPAIN 35.9 | POLAND 34.5 | YUGOSLAVIA 21.6 | ROMANIA 21.6 | ALL OTHERS 122.4 | NIGERIA 65.4 | EGYPT 38.8 | ETHIOPIA 29.0 | S. AFRICA 26.3 | ZAIRE 25.9 | ALL OTHERS 235.2 | UNITED STATES 215.6 | MEXICO 63.2 | CANADA 23.3 | ALL OTHERS 47.2 | BRAZIL 112.7 | ARGENTINA 25.9 | COLOMBIA 24.5 | ALL OTHERS 59.5 | OCEANIA 21.6 |

— EUROPE 651 — AFRICA 421 — NORTH AMERICA 349 — S. AMERICA 223 —

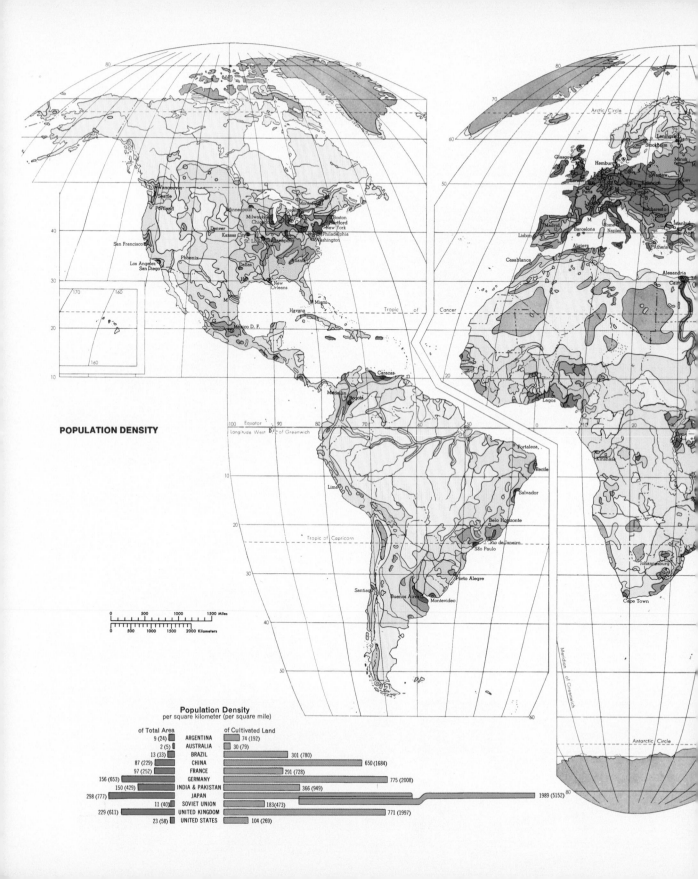

POPULATION DENSITY

Population Density
per square kilometer (per square mile)

of Total Area		of Cultivated Land
9 (24)	ARGENTINA	74 (192)
2 (5)	AUSTRALIA	30 (79)
13 (33)	BRAZIL	301 (780)
87 (229)	CHINA	650 (1684)
97 (252)	FRANCE	291 (728)
156 (653)	GERMANY	775 (2008)
150 (429)	INDIA & PAKISTAN	366 (949)
298 (777)	JAPAN	1989 (5152)
11 (40)	SOVIET UNION	183 (473)
229 (611)	UNITED KINGDOM	771 (1997)
23 (58)	UNITED STATES	104 (269)

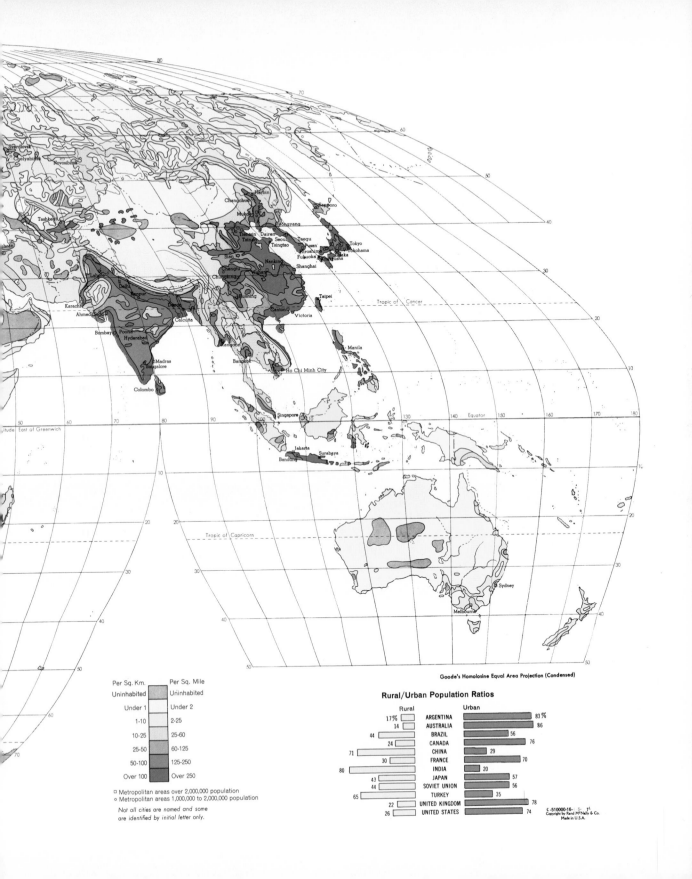

Spverdlovsk
Chelyabinsk
Novosibirsk
Tashkent
Tehran

Lahore
Delhi
Jaipur
Karachi
Ahmedabad
Bombay
Poona
Hyderabad
Madras
Bangalore
Colombo

Harbin
Changchun
Mukden
Pyongyang
Tientsin Dairen
Tsingtao
Sian Seoul Taegu
Chengtu Nanking Pusan
Chungking Wuhan
Kunming
Canton
Victoria
Rangoon
Bangkok
Ho Chi Minh City

Sapporo

Tokyo
Yokohama
Hiroshima Osaka
Fukuoka Nagoya
Shanghai
Taipei

Tropic of Cancer

Manila

Singapore
Jakarta
Bandung
Surabaya

Equator

Tropic of Capricorn

Sydney

Melbourne

Goode's Homolosine Equal Area Projection (Condensed)

Latitude East of Greenwich

Per Sq. Km.	Per Sq. Mile
Uninhabited	Uninhabited
Under 1	Under 2
1-10	2-25
10-25	25-60
25-50	60-125
50-100	125-250
Over 100	Over 250

□ Metropolitan areas over 2,000,000 population
○ Metropolitan areas 1,000,000 to 2,000,000 population

*Not all cities are named and some
are identified by initial letter only.*

Rural/Urban Population Ratios

	Rural		Urban	
ARGENTINA	17%			83%
AUSTRALIA	14			86
BRAZIL	44		56	
CANADA	24			76
CHINA	71		29	
FRANCE	30		70	
INDIA	80		20	
JAPAN	43		57	
SOVIET UNION	44		56	
TURKEY	65		35	
UNITED KINGDOM	22			78
UNITED STATES	26		74	

C-510000-16-
Copyright by Rand M°Nally & Co.
Made in U.S.A.

	Urban
	Cropland
	Cropland & Woodland
	Cropland & Grazing Land
	Grassland, Grazing Land
	Forest, Woodland
	Swamp, Marshland
	Shrub, Sparse Grass, Wasteland (pattern)
	Barren Land

Polyconic Projection

| 0 | 50 | 100 | 200 | 300 | 400 Miles |
| 0 | 75 | 150 | 300 | 450 | 600 Kilometers |

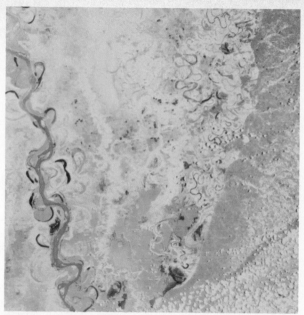

Western United States. The distinctive shape of the Salton Sea (center) shows black in California's Imperial Valley, while the checkerboard of irrigated fields shows green crops as red. Note the abrupt change in cultivation along the line of the Mexican border. Mountain ranges surrounding the valley show blue and white where bare, red where tree- or scrub-covered. (NASA photo.)

Central Canada. Three types of land use show up clearly on the high plains of southern Alberta: irrigated fields with corn, sugar beets, and potatoes (red), dryland wheat farming (dark tan to brown), and the pasture and rangeland used for stock raising (light tan). Rivers dammed to build the reservoirs to provide irrigation water show black. (NASA photo.)

Eastern United States. Old meander loops show the unstable history of the Mississippi River (left) and the other streams crossing the Yazoo basin, Mississippi. The many bare fields in the basin show as tan, while the oak-hickory and pine forests on the bluffs and upland (right) show red. Fine-weather cumulus clouds (white) dot the landscape. (NASA photo.)

North Africa. Long, parallel sand ridges (upper) mark a major Saharan dune field in southern Algeria. Although now a desert, the record of wetter climates in recent geologic history of this area is shown in the intricate branching pattern of old stream channels dissecting a plateau (lower left). Sand streaks in the lower right corner suggest dominant winds blow from the southwest. (NASA photo.)

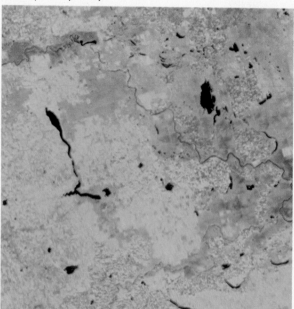

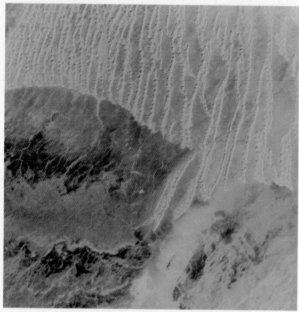

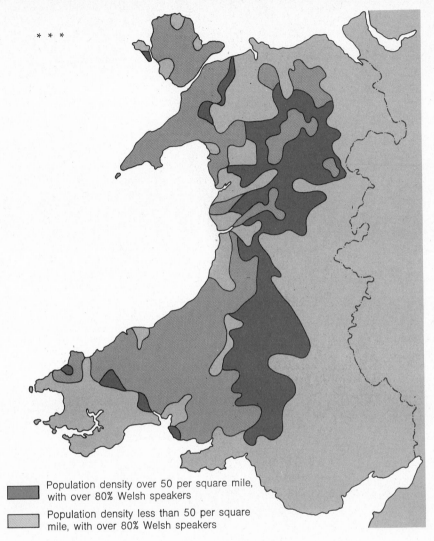

Figure 11-5. Contraction of culture areas.

The core of the Welsh language and characteristic Celtic culture has retreated westward and now lies well inside its traditional political boundary with England. Vigorous steps are being taken to halt the decline of the language as part of a resurgent Welsh nationalist movement, but the cultural forces bringing about the decline of local languages are rather pervasive. "Welsh speakers" are defined as those persons with an ability to speak the language; the use of the phrase does not necessarily imply that Welsh is the primary language in an essentially bilingual population. Although the main area of dominant English speakers lies along the England-Wales border and in the industrial south, there is one important exception: Southwest Wales (Pembrokeshire) has had a high proportion of English speakers since it was colonized in the eleventh century. [From E. G. Bowen, *Institute of British Geographers, Publications* **26** (1959), p. 4, Fig. 2.]

Population density over 50 per square mile, with over 80% Welsh speakers

Population density less than 50 per square mile, with over 80% Welsh speakers

Only in Ireland, where Irish (also called Gaelic, or Erse), has been revived and taught in the schools as part of a program to stimulate the national sense of identity, has the language held its ground.

Of course, we cannot be sure how persistent such trends will be. Once a language becomes threatened with extinction, then there is likely to be a strong movement to preserve it by those who see language as an essential indicator of cultural distinctiveness.

Language Differences as Spatial Barriers One important divisive effect of language is to act as a barrier or filter to spatial interaction between regions. This effect can be clearly seen in Canada where there is a strong division between French-speaking Quebec and English-speaking Ontario. In this case, the duality is almost complete and the only major zone of language overlap is in the Ottowa valley and in a few areas of "loyalist" settlement in southwest Quebec.

Some indication of the distorting effect of the Quebec-Ontario border has been uncovered by comparing the observed and expected contact between cities there. ("Expected" contact assumes that city-size and distance control the flow of information between cities: See Section 18-1 for a detailed account of how this is calculated.) In one study, contacts between Montreal and surrounding cities were measured in terms of long-distance telephone traffic. The expected interactions given by a population-distance formula are presented in Figure 11-6, along with the actual interactions. The traffic between Montreal and other cities in Quebec was from five to ten times greater than the traffic between Montreal and cities with comparable population-distance values in the neighboring province of Ontario.

If is difficult to disentangle how much of the Quebec-Ontario difference was due to language differences and how much to the political separation into two different provinces. The strength of the provincial Quebec-Ontario barrier in blocking interaction was overshadowed by the blocking effect of the international boundary to the south. Montreal's traffic with comparable cities in the United States was down to only one-fiftieth that of its traffic with places in Quebec.

Language barriers may have a one-way flap. Many Japanese businessmen read and speak English fluently; very few American businessmen have a comparable facility in Japanese. The one-sidedness of certain interregional flows in trade may reflect, in some small part, the way in which language barriers may be crossed more easily in one direction than in another.

To sum up, language is one of the most important but paradoxical of cultural elements. It binds together certain parts of the human population but separates others. It shows rapid spatial evolution and change and yet is one of the most persistent of cultural features. Certainly, any geographer trying to understand the mosaic of world cultural regions must include language in the analysis.

Figure 11-6.   Boundaries as filters. The impact of the Quebec-Ontario boundary on spatial interaction is measured by the number of telephone calls between the city of Montreal, Canada, and other centers in the two provinces. Actual calls are plotted on the vertical axis against expected calls on the horizontal axis. (Expected interaction is given by a gravity model using the size of the other centers and the distance from Montreal.) Note how the calls from Montreal to its neighboring cities *within* the province of Quebec (a) are much higher than those to neighboring cities in the adjoining province of Ontario (b). [From J. R. Mackay, *Canadian Geographer* **11** (1958), p. 5.]

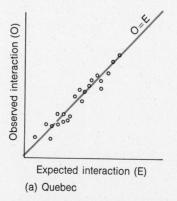

(a) Quebec

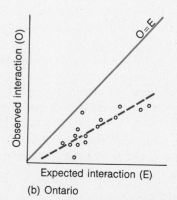

(b) Ontario

If cultural variations were simply a matter of language differences, the geographer's job would be an easier one. In fact, systems of belief cut clear across language barriers. Albanians and Chinese may hold similar shades of Marxist political belief, congregations in Zaire and Connecticut may share the same Catholic creed.

Spatial Variations in the Major Religions Each of the world's main religions has a distinctive geography. Christianity's more than one billion adherents are located largely in Europe and the Near East, the Americas, and Australasia. Islam has diffused from its birthplace in western Arabia through the northern half of Africa, central Asia, and India, and into Indonesia. Hinduism and Buddhism are highly localized, the former being largely confined to the Indian peninsula and the latter to East Asia. (See Figure 11-7.)

We can, of course, divide each major religion into its various subgroups. If we examine the Christian subgroups within the United States, we find a strong zonal pattern. The Roman Catholics are strongly represented in New England and the industrial northeast; the Baptists in the southern states and Texas; the Lutherans in Wisconsin, Minnesota, and the Dakotas; the Mormons in Utah. Even within a metropolitan area geographic differences in religion may occur, with Christian Protestant churches most frequent in the wealthy suburbs.

11-3
A Geography of Belief

Figure 11-7. Cultural core areas.
The map shows Old World core areas for the world's main religions. The figures indicate the approximate number of adherents of each religion, in millions, in the 1960s. The lines showing the expansion of Buddhism (with dates) are approximations only.

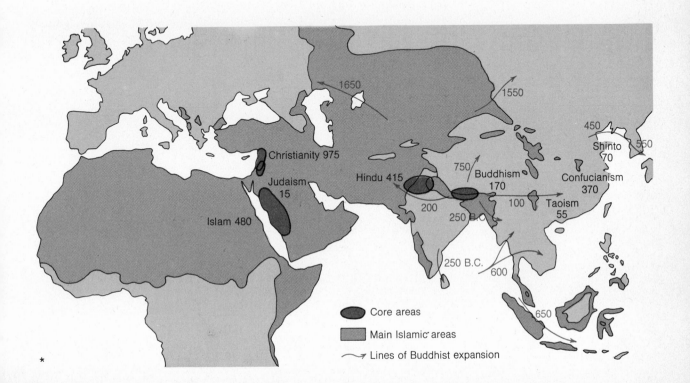

Core areas

Main Islamic areas

Lines of Buddhist expansion

Atlas cross-check.
At this point in your reading, you may find
it useful to look at the World Religions map
in the atlas section in the middle of
this book.

Why are these variations important in determining the cultural mosaic of the world? Religion's role in group organization, its close relationship to politics and the state, and the attitude of churches toward change and development are part of the answer. Many of the great political conflicts of world history have had a religious basis, and lines of conflict (like those between Israel and Egypt or within Northern Ireland) may still be drawn up along religious divides. Some aspects of religion's role in determining geographic divisions will be considered in our treatment of boundary conflicts in Chapter 19.

Geographical Impact of Religious Beliefs   Although beliefs may take a bewildering variety of forms, it is religious belief that has played the central part in our cultural differentiation. Like the political beliefs discussed in Part Five of this book, religion gives a system of interlinked values that enter into several aspects of human geography.

Let us take two examples to illustrate the connection.

Belief and the Use of Agricultural Resources   Religious beliefs affect agricultural development indirectly through constraints on diet and through the symbolic significance given to animal life. For although humans have been biologically designed as omnivorous animals (to judge from our teeth), a major portion of the world's population restricts its diet in some degree. The world's 270 million Buddhists are generally vegetarians, its 620 million Hindus may not eat beef, and its 17 million Jews may not eat pork. Smaller groups may have still more precise rules; India's Jain communities (with about 3 million people) are forbidden to kill or injure *any* kind of living creature.

As a result of these views, cattle throughout most of India are used only as draft animals and to some extent for milk production. Under the Hindu doctrine of *ahimsa*, the slaughter of cows is prohibited in many of the Indian states. As a result, aging and unproductive cattle add to the pressure on grazing resources and estimates of the number of "surplus" cows run to between one-third and one-half of the total.

An extreme view of the importance of cattle is taken by the herding tribes of eastern and southern Africa. Among the Pakot of Kenya, the number of cattle a man has is directly related to his prestige and wealth, and cattle serve as the means of exchange, most notably in the purchase of brides. Numbers rather than quality appear to be important, and this has a bad effect on the standards of livestock and the amount of grazing per animal. Although the attitude toward cattle among the herding tribes has a religious component (cattle having been said to be "the gods with the wet nose"), the emphasis on numbers appears to have more to do with their convenience as a means of exchange.

More specifically, religious significance attaches to the Moslem view of the pig as an unclean animal. Thus in Malaya pigs are reared for food only in the Chinese enclaves; the native Malay population follows the Islamic code and deprives itself of an important source of food.

Belief and modernization   Religions of most kinds lay heavy emphasis on continuity, tradition, and strict adherence to long-established patterns of behavior. They have acted, and continue to act, as a vital stabilizing influence, or—depending on one's point of view—an inhibiting drag on change.

Religion is often held to be a major factor inhibiting the spread of family-planning practices. The moral values attached to the human fetus in Roman Catholic doctrine serve as a major barrier to the spread of abortion and certain contraceptive methods. This barrier may operate at the individual and family level for members of the Catholic faith or may become a matter of national policy in countries where there is a strong link between the Catholic Church and the state. Thus contraceptive devices are banned in Ireland, and different attitudes are taken on abortion laws in the various states of the United States.

It is difficult to determine the importance of these attitudes from a strictly demographic viewpoint. Population-control practices are clearly described in the Old Testament and in Egyptian wall paintings dating from 5000 B.C. There is ample evidence that human groups throughout history have been able to control family size when this was considered desirable. Attitudes toward what is the most desirable family size are demonstrably more important than which birth-control method is followed. Thus in Europe, a continent with lower birth rates than any area of comparable size in the world (generally about 8 per thousand), there is no major difference between Catholic and non-Catholic populations at the national level. Countries in which contraceptives and birth-control information are banned or restricted have birth rates just as low as those where both are freely available.

In other aspects of human behavior, the influence of religious beliefs on the acceptance of innovations is clearer. Let us take a specific regional example. About 20,000 of the world's 370,000 Mennonites follow the Amish religious code. The Mennonites emerged in the early sixteenth century in Switzerland as a nonconformist branch of the Protestant church, with half their numbers in North America today. The Amish represent an extremist breakaway from that Mennonite movement and are today concentrated in farm communities in certain counties of Pennsylvania and Indiana. Located in the midst of the most highly modernized and swiftly changing regions of the world, Amish communities stand out as islands of tradition. Services are conducted in Pennsylvania Dutch (Palatinate German with some English mixed in), traditional plain clothes continue to be worn, telephones and electric lights are shunned, and the horse and buggy continue to serve as a means of transport instead of the all-pervasive automobile. Here, religious beliefs serve

(a)
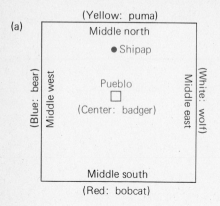

(b)
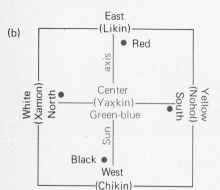

(c)
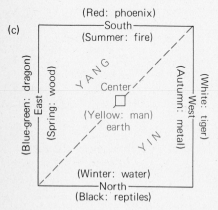

Figure 11-8. Cultural variations in the world view.
World views of (a) Pueblo Indians of North America, (b) Maya of Central America [A.D. 600–900], and (c) traditional Chinese. Note the similarities in that spatial structure is oriented to the four cardinal directions—north, south, east, and west. The associations with each direction vary. [From Yi-Fu Tuan, *Space and Place* (University of Minnesota Press, Minneapolis, 1977), Fig. 9, p. 94 and Joyce Marcus, *Science* **180** (1973), Fig. 2. Reprinted with permission of the American Association for the Advancement of Science. Copyright © 1973 by the American Association for the Advancement of Science.]

as a cement which continues to hold together a human group with a behavior pattern more reminiscent of seventeenth-century rural Europe than America in the mid-1970s.

Belief and World Views At the most generalized level, systems of belief condition our view of the world—and our place in it. We want to order our experiences of the world, and build a picture of the world around ourselves.

Consider Figure 11-8. This shows the world views of two American Indian cultures and of the Chinese. In each case the center of the world is the human being. Around this is a four-sided framework linked to the four cardinal directions—north, south, east, and west. Note that in all three cases direction is associated with a particular color. The Pueblo Indians add an animal and the Chinese have a most complex set of associated ideas. Thus the west is associated not only with a color (white) and an animal (the tiger) but is the direction of metallic autumn, symbolic of both harvest and war. This is also the direction of twilight, of memory and regret, of past mistakes which cannot now be altered.

Ideas of the world have played a significant part in the design of individual houses, settlements, and cities. Many were designed as small-scale replicas of the world picture. Similar ideas permeated Greek civilization while the Christians were later to build their churches with distinct east–west orientation. In China today the world view remains important in building construction. For example, when the recently founded Chinese University of Hong Kong came to lay out its new buildings—including a geography department —at Sha Tim, an unusual locational factor had to be considered. Sites were chosen in which *feng-shui* ("local currents of the cosmic breath") were harmonious. Chinese landscapes continue to have paths, structures, and woodlots designed to blend with the natural landscape rather than dominate it. This response to the environment is not uncommon. For most of the period of human occupation of the earth, most of its peoples believed that individual natural elements—trees, springs, and hills—had guardian spirits. Before such objects could be used, these spirits had to be raised and mollified.

When, in Lewis Carroll's *Through the Looking Glass,* Alice objects that "glory" doesn't mean the same as "a nice knockdown argument," she elicits Humpty Dumpty's evasive reply: "When I use a word, it means just what I choose it to mean—neither more nor less." The word *region* has caused centuries of "nice knockdown arguments" among geographers.

11-4
The Nature of Cultural Regions

Arguments Over Regional Boundaries We can illustrate the nature of regional arguments by Figure 11-9, which shows how the Great Plains region of the United States has been defined at various times by various scholars. The discrepancies in the three maps have arisen in two ways. First, there is a difference in the criteria used to define the region. The first map describes the limits of the characteristic vegetation of the region and the second the limits of the typical landforms. (You may like to refresh your memory of the character of this part of the United States by turning back to Figure 4-15, which shows a cross section of that region from the Rocky Mountains east to the Mississippi River.) The boundaries of the region in the last map are based on the spatial extent of a particular American Indian culture, that of the Great Plains tribes.

Second, seven different definitions of boundaries, by seven leading scholars, were used for each map. Although the maps may seem to support the theory that "the number of different boundaries for any region is equal to the square of the number of geographers consulted," the area of agreement is considerable. Areas within the shaded zone represent a consensus on what to *include;* areas outside the solid line represent agreement on what to *exclude.* Between lie the areas of disagreement. If we were to trace the boundaries of the region in each of the three maps and superimpose them on one another, we would even find a small area that all 21 scholars would be happy to call part of the Great Plains. We may regard such areas (like western Nebraska and eastern

(a) Region of distinctive vegetation

(b) Region of distinctive landforms

(c) Region of distinctive cultural elements

Figure 11-9. Regional arguments. Alternative definitions are possible even for major geographic regions. The American Great Plains region is defined in terms of three of its characteristics: its distinctive vegetation, landforms, and culture. Each map is based on definitions given by seven leading authorities. [After G. M. Lewis, *Transactions of the Institute for British Geographers,* No. 38 (1966), pp. 142–143, Fig. 11-13.]

Area included within all definitions

 Boundary of area excluded by all definitions

Montana) as particularly representative of the special geographic flavor of the Great Plains. Geographers use the term *core area* to define such regional heartlands, which have a special significance in our study of the regional mosaic of the earth.

Types of Regions   We said earlier in this book that a *region* is any tract of the earth's surface with characteristics, either natural or of human origin, which make it different from the areas that surround it. In Chapter 3 we saw examples of ecological regions and in this chapter examples of cultural regions.

Geographers also draw a distinction between regions on two other grounds. The features which distinguish regions may be singular or plural. Kniffen's region of covered bridges (which we shall meet later in this chapter in Figure 11-11) is a *single-feature region*, while the Great Plains core area (Figure 11-9) defined by the overlap of three sets of features, is a *multiple-feature region*. Since culture is such a multifaceted concept, most cultural regions tend to have multiple rather than single distinguishing features.

A distinction also is drawn between the degree of spatial organization within regions. *Uniform regions* are defined by the presence or absence of a particular distinguishing feature. A "covered bridge" region is an area in which this element is part of the rural landscape. The boundary of such regions is rather clear-cut compared with that of, say, the Great Plains region, in which we have a clearly defined core and a gradual weakening of "Great Plains" features as we move outward from that core. Regions of this type are termed *nodal* regions. (Some geographers prefer the term "focal" or "functional" region.) The center is well defined but the regional characteristics die out toward a periphery in a way which makes it very difficult to plot an outer boundary. Perhaps the best examples of nodal regions are the urban regions we shall study in Part Four.

You should not think of the worldwide regional mosaic as a collection of separate, nonoverlapping units with sharp boundaries. Rather, it is a mixture of uniform and nodal regions. An appropriate analogy is not the stained-glass window of your church, but rather the litter of overlapping papers on your desk! The earth's surface is so large and its cultural fissioning so complete that the temptation is always there to break it up into a very fine mesh of regions. But the parts must not be too small. A world system of 999 cultural regions may be an athletic geographic feat, but it is self-defeating if our purpose is to provide a shorthand guide to the world's diversity. Too much information (too many regions) can be as much of a problem as too little, and geographers are always searching for the right balance. The ideal set of regions is one which has just enough differentiation to serve our purposes—but not a boundary more.

The Problem of Cultural Regions   If we take the individual cultural elements such as language and religious belief and add to them ethnic differ-

(a)

(b)

(c)

Figure 11-10. Cultural landscapes.

Patterns of fields and farms reflect the manner and timing of agricultural settlement. (a) Irregular shapes and sizes of fields in a long-settled area of southwest England. (b) Regular 40-acre fields in an area on the Illinois border settled when the township and range system was used for land surveying. (See Figure 2-7.) (c) A pioneer Canadian settlement in the Lake St. John lowland with farm boundaries perpendicular to a road. [Photos courtesy of the United Kingdom Department of the Environment (a), U.S. Department of Agriculture (b) and Canadian Department of Energy, Mines, and Resources (c).]

ences, then we have some of the ingredients for a system of cultural regions. Various proposals for systems of cultural realms have been put forward.

Landscape and "Pays"   Geographers show special interest in the visible impact of mentifacts, sociofacts, and artifacts on the earth's surface. As we saw in Chapter 10, large areas of the world's landscape have been shaped and patterned by human activity. In rural areas, special attention has been paid to the patterns of fields and farms, roads and boundaries. Different cultural groups have had different ways of settling the land and different ways of fixing boundaries, so that strong contrasts in the *cultural landscape* are observable from the air. Figure 11-10 gives typical examples of such contrasts.

The French word *pays* describes one sort of culture region, with names like Brie and Medoc familiar to lovers of cheese and wine. It was first used by French geographers in the nineteenth century, notably Paul Vidal de la Blache (Figure 25-3) the Sorbonne professor, to describe small distinctive regions within that country. The distinction was due not just to some striking physical difference or cultural difference, but to a blending of the two in terms of the landscapes of the area.

Other geographers have taken individual elements in the cultural landscape and traced their origin and spread. Evidence on the spread of covered bridges in America has been carefully assembled by Louisiana geographer Fred Kniffen. Originally developed in Switzerland, Scandinavia, and northern Italy, this type of bridge was first used in the United States in an area running from southern New England to eastern Pennsylvania. As the datelines in Figure 11-11 show, it spread rapidly through the Midwest and into the southern Piedmont, but only very slowly into northern New England. About 1850, when the number of these bridges was expanding most rapidly in the East, new centers of growth emerged in the western United States—notably in Oregon's Willamette Valley. Thus do artifacts, like bridges and barns, fields and fences, house types and street patterns, provide clues to the limits of cultural regions.

Regional Images   Another approach to the cultural region is through its image. How do we view the different parts of our own countries? What is our mental map of such regions as northern New England or the bush country of Australia?

Geographers Peter Gould and Rodney White already have attempted to establish the mental maps of the different parts of Britain. Cross sections of young people (aged 16 to 18 years) from 20 schools in Britain were used to determine views of the country. All the individuals were asked to rank the counties of England, Wales, and Scotland in terms of their desirability as places to live and work. Statistical analysis of the results for each school yielded surface maps of the areas' perceived desirability (Figure 11-12). Although the maps for each school differed, they had several common elements. First, each map gave a high rating to areas immediately adjacent to the

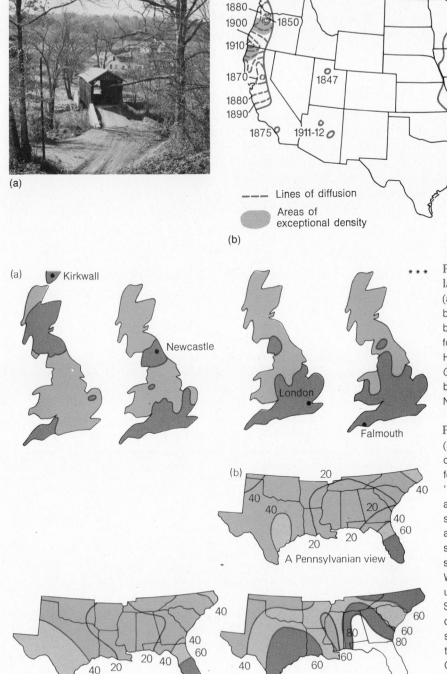

(a)

***

Lines of diffusion

Areas of exceptional density

(b)

(a) Kirkwall

Newcastle

London

Falmouth

(b)

A Pennsylvanian view

A Minnesota view

An Alabama view

Figure 11-11.   Elements in the cultural landscape.
(a) A typical example of a covered road bridge. (b) Localities with covered road bridges in the United States, with timelines for the bridge's spatial diffusion. [Photo by Hugh Mackey, DPI. Map from F. Kniffen, *Geographical Review* **41** (1951). Copyright by the American Geographical Society of New York.]

Figure 11-12.   Regional images.
(a) The maps show contrasts in the images of British regions held by senior pupils at four schools in different locations. The "most desirable" areas of the country according to pupils at each school are shaded. The shading reflects both an affection for the local area surrounding each school and a generally higher ranking of the southern part of the island. [After R. R. White, Pennsylvania State University, unpublished master's thesis, 1967.] (b) Similar maps in which university students on different campuses were asked to mark the states within the United States. Only views of the South and Texas are shown. [After P. Gould and R. R. White, *Mental Maps* (Penguin, Hamrondsworth, 1974), Fig. 4.2-3, pp. 98–99).]

school. This attachment to the home country was present in all the groups studied. Second, the south of Britain was perceived as being generally more desirable than Scotland and the north, expecially by students at southern schools like Falmouth. Third, London was seen as either very attractive or very unattractive. Other maps indicated a strong preference for well-known vacation areas such as the southwest peninsula.

The notion of regional perception has also been extended to the United States. Let us take Kevin Cox's measure of a group of Ohio students' perceptions of regional differences within the United States. A list of the 48 states of the conterminous United States was given to each student in a class. Each state was taken in turn as an anchor state, and class members were asked to underline the three states considered to be the "most similar" to it. No maps were available and definitions of similarity were left to each individual to resolve.

From this rapid and impressionistic clustering of states, the major regional differences within the country could be separated out. In order of priority these were: first, a split along the Mississippi separating the "West" typified by the state of Colorado from the "East" typified by Connecticut. Second in importance was a distinction between the "South" (typified by Georgia) and "New England" (typified by Maine). The distinction was less strong than the east–west split and values merged from north to south rather than breaking sharply at the Mason–Dixon line. Florida was *not* included as part of the South. While the first two distinctions were bipolar, separating two regions of approximately equal size, the third distinction is unipolar. This separates out a Great Lakes division (typified by Michigan) from the rest of the United States. A fourth but less significant regional boundary separates a midcontinental area from the rest of the United States.

By repeating the tests of Ohio students on other groups we can see how regional concepts are related to an individual's view of the world. At what age do regional concepts clarify? How stable are they over time? Is there any difference between the male and the female view of the world? We may also compare regional concepts as seen from different locations to test if we each make our distinctions and splits along the same lines. Images and stereotypes, like those of the Old South or the Australian outback, may persist long after the reality has changed. These ideas affect our decisions on migration ("New Zealand is a good country in which to raise children") or investment ("Paraguay is a risky place in which to invest capital"), and so shape the world around us in ways that are persistent and persuasive.

Cultural Regions: The Meinig Model   Although the definition of a culture region is not an easy task, there are cases where the structure is relatively clear-cut. American cultural geographer Donald Meinig has analyzed a number of cases where the area occupied by a culture has grown sequentially from the locality of its origin or hearth. There the culture region can be separated into an irregular series of concentric regional shells.

We can illustrate the Meinig model by taking his simplest case, the Mormon culture region. We choose this (a) since its origin in both time and is so precise and (b) since its culture is so distinct. On July 24, 1847, Brigham Young gained his first view of the Valley of the Great Salt Lake and said, "This is the place." So the Mormon region began to take shape around a clearly defined hearth. The Church of Jesus Christ of the Latter-Day Saints is a cohesive and highly self-conscious religion with distinctive secular impact in terms of family size, social order, and agricultural practice.

The spatial structure of the Mormon culture region is fourfold. (See Figure 11-13.)

(a) Core

(b) Domain

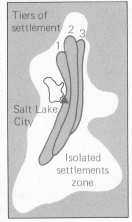

Figure 11-13. Structure of a culture region.

Donald Meinig's study of the Mormon culture region in the western United States forms a fourfold framework. This extends from the *core* area based on Salt Lake City (a) through the *domain* (b) largely confined to the present state of Utah. Beyond the domain lies the *sphere* (c) of outer colonies, and beyond that the highly dispersed *outliers* (d) of the Mormon church in the rest of the United States and Western Europe. [From D. W. Meinig. *Annals of the Association of American Geographers* **55** (1965), Fig. 7, p. 214. Photo from De Wys Inc.]

(c) Sphere

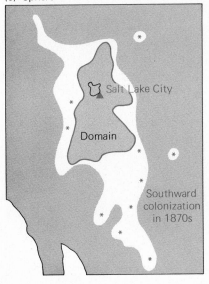

(d) Outliers

Core   The central area of the Mormon culture region is where the Mormon population is densest, the religion is most dominant, and the history of occupation by the group the longest. This core is formed by the Wasatch Oasis, a 210-km (130-mi) strip along the base of the Wasatch Mountains east of Salt Lake City. The area is growing rapidly in population, and for the last half-century has contained about 40 percent of the total Mormon population of the United States. It is a principal center of gentile immigration into the essentially Mormon region, but the Mormon–gentile ratio nevertheless has tended to remain rather constant. The links between the core area and the rest of the Mormon region are principally those of information, finance, and organization. The cultural links are symbolized by twice-yearly meetings when, every April and October, representatives from all Mormon communities meet in Salt Lake City to decide church policy.

Domain   Areas where the Mormon culture is dominant but with less intensity than in the core, and where local differences in social organization are evident, Meinig terms the Mormon domain. The domain extends outside Utah, notably into the river country of southeast Idaho and contains an area more than 20 times the size of the core. It contains a little over a quarter of the total Mormon population, largely in rural settlements, and has a small proportion of gentiles. As Figure 11-13(b) shows, the spatial pattern of colonization was affected by the north–south trend of the mountains and proceeded in a series of "tiers" from west to east. For the first 20 years after the foundation, settlements were confined to the first tier, but by 1870 (only 12 years later) the second had been occupied and the third was being entered.

Sphere   Outer zones of influence and peripheral contact, where the Mormons form significant local minorities, are termed the sphere. The Mormon sphere forms a fringe all the way from eastern Oregon to northern Mexico, greatly extended in the south and representing the last wave of rural Mormon expansion in the late nineteenth century. About 13 percent of the Mormon population lives in the sphere, where they form a varying minority proportion of the local population. As Figure 11-13(c) shows, the sphere is made up of discontinuous pockets of settlement, rather than a continuous low-density fringe. Each pocket has its distinctive history. For example, the San Juan colony in southeast Utah was established in 1879 after a famous trek across very inhospitable canyon-intersected country. The "girdle of wastelands" that surrounded the domain ensured that these pockets would be separate from the major area of colonization.

Outliers   Outside the sphere are small outliers (outlying areas) containing the remaining fraction of the Mormon population. The key outliers are in

Pacific coast cities, notably Los Angeles. In the last two decades the outliers have been extended outside North America to England, Switzerland, and New Zealand. The outliers are different in several ways from the other three zones. They are urban rather than rural, and the links to the core are long-distance and intercity rather than short-distance and rural-to-city. Third, the Mormons form a tiny minority of the population of each city. As Meinig puts it, the outlier is not a further expansion of Zion but a dispersion into Babylon.

The distinction between the first three zones (core, domain, and sphere) and the fourth (outliers) is of general significance to cultural geographers. In terms of spread, the contiguous spread of the nineteenth century has given way to the hierarchic spread of the twentieth. This "bursting" of the actively growing culture may be seen as something like the process of *metastasis* in biology. A well-known example is that of cancerous cell growth. In metastasis, the cancerous cells begin by local growth but then break away from this original lesion and are carried in the blood or lymph systems to distant parts of the body, where they set up new lesions. Certain culture elements can also get caught up in the powerful circulation movements that go on between large cities and be rapidly transferred to distant parts of the globe. We shall return to this topic in Chapter 13.

Meinig's analysis clarifies the spatial structure of one of the most distinctive regions to emerge on the subnational level within the United States over the last century and a quarter. In the Mormon cultural area the theological base affects significant aspects of the demography, economic organization, and political viewpoints of the population in the southwest. Although Meinig has used his method only to analyze other areas of Western culture—notably Texas—it has obvious revelance to other nonwestern cultures as well.

Summary

1. The human population, although biologically uniform, is broken up by culture into distinct groups at several spatial levels.

2. The concept of cultures as used by geographers describes patterns of learned human behavior that form a durable template by which ideas and images can be transferred from one generation to another, or one group to another. Understanding the complexity of culture is difficult without some simple categorization: Huxley's mentifacts, sociofacts, and artifacts are one such set of components. The link between culture and ethnicity is difficult to unravel in that minor biological differences may be visual signals identifying a number of associated cultural characteristics.

Together these give rise to distinctive plural societies in many geographical areas.

3. Languages vary in spatial extent from global languages like English down to small languages understood over only a few square miles. However, the spatial structure is continually changing with some languages aggressively spreading while others are in retreat. Language differences act as barriers or filters to spatial interaction.

4. Geographical variations in systems of belief have immense importance on attitudes toward resource use and attitudes toward innovations. For example, religious-based dietary restrictions are shown to have an indirect effect on agricultural development,

especially in regard to the use of livestock. Some cultures accumulate resources for the use of following generations, while others use resources for spiritual benefits in other worlds. Each culture has a distinctive world model.

5. Despite apparent disagreement among geographers as to the definition of the borders of a region, there is considerable agreement regarding core areas. Regions are distinguished by geographers as being (a) either single-featured or multiple-featured, and (b) either uniform or nodal (functional).

6. The Meinig model has proved particularly useful in the analysis of an *evolving* cultural area. It analyzes growth in terms of three minor contiguous zones (the core, domain, and sphere) and a discontinuous outer zone (outliers).

Reflections

1. Consider whether your own county, state, or province can be divided into distinct cultural regions. What bases are there for such divisions? Sketch in some tentative boundaries and compare them with those proposed by other members of the class.

2. Identify the main types of housing in the community around your campus. To what extent are housing characteristics and designs a useful indicator of culture?

3. List (a) the languages you yourself speak and read, (b) those your parents speak and read, and (c) those your grandparents speak and read. How far does your list reflect birthplaces and family migration? Is the range of languages known decreasing, fairly constant, or increasing over the three generations?

4. Consult the Yellow Pages of your local telephone directory and plot the distribution of churches of different denominations in your city or country. Do the different denominations have distinctive spatial patterns? Can you think of any likely causes for these patterns? Check your local findings by looking at maps of church membership for the whole of your country in its National Atlas or a similar source.

5. Use either a place-name dictionary for your own country *or* a local county history to check on the place names of settlements in your own locality. How far do the names of places reflect (a) the origins and language of their founders and (b) the period when they were founded?

6. Review your understanding of the following concepts:
    race   plural societies
    the Huxley model   cultural traits
    culture   uniform regions
    cultural regions   nodal regions
    cultural landscapes   core, domain, and sphere

One Step Further . . .

Excellent general introductions to the spatial diversity of human cultural groups are given in two texts:
>   Broek, J. O. M., and J. W. Webb, *A Geography of Mankind* (McGraw-Hill, New York, 2nd Ed., 1973) and
>   Spencer, J. E., and W. L. Thomas, *Introducing Cultural Geography* (Wiley, New York, 1973).

Follow these up with a look through some of the papers in a very useful set of readings,
>   Wagner, P. L., and M. W. Mikesell, Eds., *Readings in Cultural Geography* (University of Chicago Press, Chicago, 1962).

For a further discussion of some of the more specialized topics touched on in this chapter, see
>   Sopher, D. E., *Geography of Religions* (Prentice-Hall, Englewood Cliffs, N.J., 1967) and
>   Stewart, G. R., *Names on the Land: A Historical Account of Place-Naming in the United States* (Houghton Mifflin, Boston, Rev. Ed., 1958).

Try especially hard to look at atlases showing the distribution of major cultural elements—patterns of settlement, languages and dialects, religions, voting behavior, ethnic groups, etc—for your own country. Few national cultural geographies exist, but for United States students, one is

Zelinsky, W., *The Cultural Geography of the United States* (Prentice-Hall, Englewood Cliffs, N. J., 1973).

This is a most exciting book that leads you into both the highways and the fascinating byways of your country. Compare Zelinsky's regions with those given in

Castil, J., *Cultural Regions of the United States* (University of Washington Press, Seattle, 1975).

Regular research tends to be published in the main geographic journals, but look also at Landscape *(a quarterly) for articles on the cultural landscape.*

> "I will tell the story as I go along of small cities no less than of great. Most of those which were great once are small today; and those which in my own lifetime have grown to greatness, were small enough in the old days."

HERODOTUS
(*ca.* 440 B.C.)

In an age of transistor radios, packaged holidays, and earth-circling satellites, the world seems sadly shrunken and homogenized. The same Coca-Cola cans litter the high tide lines from Coney Island to the Congo, and the same advertising jingles are crooned on the television sets. The urban-industrial forces that appear to be shrinking and standardizing the world's culture are very powerful ones, and we shall be looking closely at their increasing impact on world geography in Part Four.

At the same time, the resilience and continuity of the variety of human cultures remains impressive. There are few signs that the American melting pot—still less the global melting pot—has reached a temperature in which the kinds of differences we saw in the last chapter will be dissolved. On the contrary, the 1970 Census of the United States recorded a small swing toward ethnicity since 1960, with more individuals wishing to stress their American Indian, Finnish, or Hawaiian fore-

bears. In an age where there are strong trends to "sameness," the scarcity value of some aspects of our differences appears to be rising.

Since it seems that cultural differences and cultural regions are here to stay, it is natural for geographers to want to know how they originated and how they have changed over time. In this chapter we take a historical perspective and look at three basic questions. First, how and where did the great cultural differences we now see originate? Second, how did change occur and some cultural regions spread at the expense of others? Third, how stable is the present pattern of cultures, and how likely is it to persist? The present mosaic of regions reflects not only the pressures and balances between people and the environment today; it also reflects the tens of thousands of years of past population growth, cultural differentiation, and migration.

Of course, it would be impossible, in the span of a single chapter, to cover all these topics. The course we shall follow is a compromise. We shall tackle questions of origin from a global view-

chapter 12
World Cultural Regions
The Emerging Mosaic

point and choose one cultural region (Western Europe) for our study of the diffusion of cultural elements. Many areas have been affected by European culture, and we shall choose the situation in the United States as an example of the way various factors affect the persistence of cultural regions. Thus we shall move from a global to a subcontinental scale in our discussion, and finally to a national and regional scale. To use a photographic analogy, we shall change our focus and lens as we examine each of the three major questions raised in this chapter.

12-1
The Question of Origins

We can split the question of origins into three parts. Where did the first clusters of the human population form on the earth? Where were the first centers of agricultural innovation? And, finally, where, and at what stage, did urban cultures come into the picture? The order of these questions is a historical one.

The Origins of the Human Population   Let us begin our analysis of this question by recapping some facts we met in Chapter 7. There we noted that the mammal that we call *Homo sapiens* is a very recent arrival on the earth. According to most current estimates, the earth is about 4.5 billion years old. The first living forms, algae and bacteria, originated about 2.2 billion years ago. Just which date is assigned to the origin of the human species depends on which skeletal remains archeologists are prepared to call human. Several manlike species emerged during the last 3.5 million years of the most recent geologic period (the Pleistocene). *Homo sapiens* can be certainly traced back to one of the interglacial periods about 1.5 million years ago (although most recent discoveries are pushing this date back). If we consider earth history as a clock measuring out 24 hours, with the great geological periods represented by portions of the various hours and present time represented by midnight, we must place the advent of humanity at only a few moments before midnight.

The notion that human beings originated in a single area, let alone the possible location of such an area, has been a matter of acute archeological debate. From current archeological evidence it looks as if we originated in the Old World rather than the New. (See Figure 12-1.) Recent research is tending to narrow down the location of the source area to tropical Africa in general and to East Africa in particular. Asia is now regarded as a secondary rather than a primary source, and Europe is no longer seriously in the running. We shall use the term *hearth* to describe a center of evolution, using it to describe not only centers of biological or genetic evolution (for plant and animal species) but also centers of cultural evolution (e.g., for agricultural methods or city living).

The differentiation of humans into the three major races—Caucasoid, Mongoloid, and Negroid—may well have occurred at the same time that people evolved from the hearth areas. Toward the end of the Pleistocene era

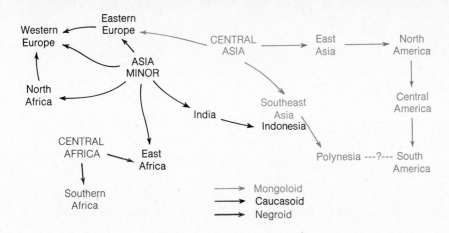

Figure 12-1. Early migration of human groups.
The probable migration path of the three main racial elements in the human population is shown here schematically. Remember in looking at the diagram that the archeological evidence over most of the world is very thin, and that the links may be altered by future research: East Africa may be a center rather than a dead end. However, the North American Indian is clearly of Mongoloid provenience.

(around 25,000 B.C.), human groups had spread to most of the land masses, except Antarctica. Figure 12-1 charts the likely sequence of migration from the Old World. Migration across very wide sea areas posed problems for early peoples and the spread of population is thought to have followed island chains (using them as stepping stones). As we saw in Section 5-1, sea levels fluctuated considerably during the Pleistocene epoch and so-called "land" corridors were opened up during the major ice ages when more of the earth's water was locked up in the ice sheets and ocean levels fell. Certainly the Bering Strait between Siberia and Alaska was replaced by a land corridor at these times. We must stress, however, that the routes shown on Figure 12-1 are only tentative, and will, in all probability, have to be substantially revised as more genetic evidence is gradually accumulated. It seems probable that Western Europe, southern Africa, and Australasia were all peripheral areas (i.e., at the end of migration routes) and that the Americas were colonized from East Asia, probably at a late stage.

Many of the smaller and more remote islands in the world's oceans were reached only in the recent past. Radiocarbon dating reveals that the first settlements on the Hawaiian Islands may have been as late as A.D. 1200. Charles Darwin saw the possibilities that island chains offer for research during his visit to the Galapagos in 1835. Since then, an increasing amount of research is being done on the spread of human populations through island chains. (See Section 13-4, which reports research in the Pacific Ocean.) Figure 12-2 throws some light on the question of island colonization by using a model first developed to explain the varying number of plant and animal species found on different islands. Generally, large islands which are close to the continental land masses have a much richer range of fauna and flora than islands which are small and remote. Why should this be? If you follow the figures and caption in Figure 12-2, you'll see one possible explanation. So far as human beings are concerned, the same kind of rules appear to apply. If we measure human variety by means of genetic terms, it is the small, remote

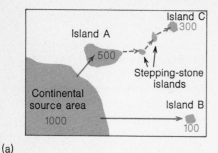

(a)

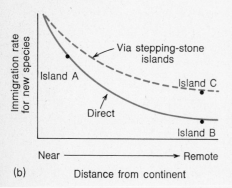

(b)

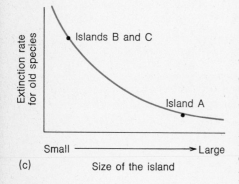

(c)

Figure 12-2.   Island colonization models.

Since Charles Darwin's *Origin of Species* (1859), islands have held a special fascination for biogeographers. Typical relations between a continental land mass and some offshore islands are shown in (a), where the figures indicate the number of different species of plants or animals found on each. The number of species on each island is determined by the *immigration rate* (the number of new species which arrive from the continent over a given time) and the *extinction rate* (the number of existing species on each island which fail to establish themselves and die out in a given time). Immigration rate is inversely related to the distance of the island from the continent (b), so that nearby islands have more species arriving (whether by wind, wave, or animal dispersal) than do remote islands. The upper curve shows that this simple relationship may be modified by "stepping-stone" islands which allow species to migrate by a series of shorter steps. Extinction rate is indirectly related to the size of the island (c) so that more species die out on small than on large islands. This fact is largely due to the narrower range of ecological conditions normal on small islands. If we now combine these two factors—island accessibility and island size—we have a reasonable explanation for the kind of variation in species number so generally observed. Island A is both nearby the continental source and large in area and so has many species, while island B is both small and remote and thus has a low number of species. Island C is also small and remote but is better linked to the continental source via the intermediate chain of islands, so it has a higher number of species than B. [From R. H. MacArthur and E. O. Wilson, *The Theory of Island Biogeography* (Princeton University Press, Princeton, N.J., 1967), p. 22, Fig. 8.]

islands which display the smallest range in blood types or biochemistry. Whether the model would also help us to understand cultural variability is more debatable.

Estimates of the total world population in this early period are necessarily vague. If people lived in small food-gathering and hunting communities, we can assume that the average population densities might have been similar to those of surviving preagricultural groups like the Australian aborigines, who have an average population density of about 50 km² per person (2 mi² per person), or the Haida Indians on the northwest coast of North America, who have a population density of 1.7 km² per person (0.6 mi² per person). These densities suggest a total population of only around 5 million by the end of the preagricultural period.

The Origin of Agricultural Hearths   Geographers commonly divide human culture into four distinct technical stages. These are (1) food-gathering and hunting cultures, (2) herding cultures, (3) agricultural cultures, and (4) urban cultures. Each stage is matched by an increasing complexity of material goods and social organization, by increasing ability to support high population densities, and by ever greater interference with the natural environment (see Chapter 8). Not all cultures need to go through all stages.

The origin of the food-gathering and hunting cultures is the same as that of the human population itself; the first human groups in East Africa supported themselves in this way. Little is known about the earliest domestication of animals and the origin of herding cultures; some authorities regard this as a late, rather than early, stage in human cultural development. Most debate has centered on the origins of the third cultural stage—agriculture—and it is to this controversy that we first turn.

The origin and location of the world's agricultural hearths have been the subject of intense academic debate. Archeological evidence indicates the domestication of plants and animals by 8000 B.C. in the hills of what are now Iraq and Iran. Other finds reveal some similar activity in scattered spots in India, northern China, and central Mexico. It seems likely that wheat and barley were cultivated in the Middle East at an early date, and that the cultivation of corn by the Indians of Central America came later. Little is known of the early beginnings of rice cultivation in Asia, but new archeological finds and new ways of dating finds may yet enable us to revise and rewrite the fragmentary story of the development of agriculture.

The Sauer Hypothesis   Despite the scarcity of firm evidence, there has been plenty of conjecture on the location of the first agricultural communities. In the sweeping survey *Agricultural Origins and Dispersals*, Berkeley geographer Carl Sauer (Figure 12-3) argued for separate hearths of domestication in both the Old and New Worlds, outside the conventional hearth areas. As Figure 12-4 illustrates, he places the Old World hearth in South Asia and the later New World hearth in the valleys and lowlands of the northern Andes. Sauer chose these areas on the basis of five criteria. First, the domestication of plants could not occur in areas of chronic food shortages; the domestication of crops and animals implies experimentation, and a sufficient abundance of food so that the experimenters can wait awhile for results. Second, hearths must be in areas where there is a great variety of plants and animals and thus a large enough gene pool for experiment and hybridization to occur. Third, large river valleys are unlikely hearth areas because their settlement and cultivation require rather advanced techniques of water control. Fourth, hearths must be restricted to woodland areas where spaces can readily be cleared by killing and burning trees; grassland sod was probably too tough for primitive cultivators. Finally, the original group of cultivators had to be sedentary to stop crops from being consumed by animals. The main nomadic groups probably did not meet this requirement; nor, probably, did the areas they inhabited (see Figure 12-4) meet the first four requirements.

By combining these criteria, Sauer chose as his hearth areas the most probable environments for agricultural innovation. They had the climatic range to induce diversity and the rivers to provide a regular supply of fish for a sedentary settlement. Here wild plants were developed by centuries of selecting, propagating, and dividing. In Sauer's view the seed agriculture of the Middle East, China, and Central America is a much later, and more sophisticated, outgrowth of the activity at the two earlier centers of this type of agriculture in central Mexico and Asia Minor. The lively and sometimes hostile response of archaeologists to this view suggests that the debate is still

Figure 12-3.   Carl Ortwin Sauer. For more than half a century, Sauer's research on the early relations of man and plants has placed him at the forefront of cultural geographers. Many of his more speculative and controversial views on the location of agricultural hearths and on the antiquity of human colonization of the New World have been supported by recent archeological evidence. Sauer's very extensive writings, published from 1911 until after his death in 1975, show a sensitivity to the diversity of human cultures and the way in which they have adapted to and changed their landscapes over time. The regional focus for his writings was originally the Midwest, but later shifted to Mexico, the Caribbean, and South America. As professor of geography at the Universtiy of California at Berkeley from 1923 to 1957, he built one of the most distinctive and distinguished graduate schools in the United States. (See Figure 25-6.) A review of Sauer's work is given by John Leighly in the *Annals of the Association of American Geographers* **66** pp. 337–348 (1976). [Photograph courtesy of the late Mrs. C. O. Sauer.]

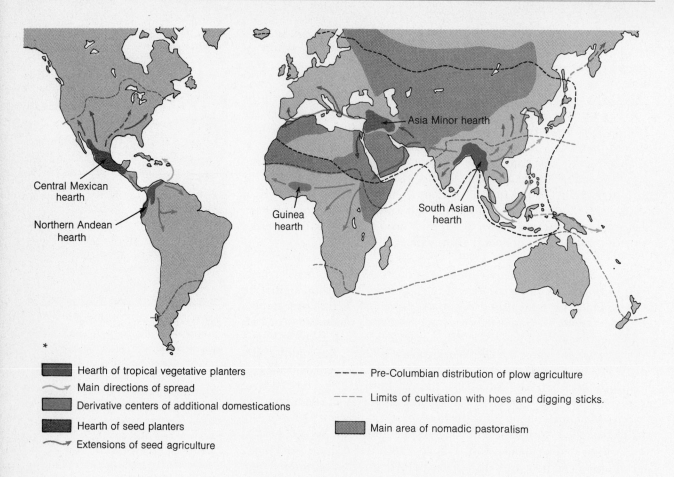

Hearth of tropical vegetative planters

Main directions of spread

Derivative centers of additional domestications

Hearth of seed planters

Extensions of seed agriculture

- - - - Pre-Columbian distribution of plow agriculture

- - - - Limits of cultivation with hoes and digging sticks.

Main area of nomadic pastoralism

Wild Einkorn
(Asia Minor 7000
B.C.) One of the
ancestors of
modern wheat

Figure 12-4. Agricultural origin and dispersal.

This world map shows in a highly generalized form the supposed main features of agricultural diffusion in the pre-Columbian world. It is based on the views of geographers Carl Sauer and Eduard Hahn and geneticists such as N. I. Vavilov. Note that there are two main hearths shown in the Old World: (1) tropical southern Asia as a center for agricultural systems based on reproduction by vegetative planting (e.g., subdividing an existing plant into several parts, each of which grows into a new plant); and (2) subtropical Asia Minor as a center for agricultural systems based on reproduction by planting seeds. This fundamental distinction between the two types of plant propagation is repeated in the two centers (again, tropical and subtropical) suggested for the New World. The probable main lines of spread around each hearth are shown together with a secondary hearth area in West Africa. The map also shows the important technological distinction between methods of cultivation based on hoes and digging sticks (found in both Old and New Worlds) and that based on the plow (found only in the Old World in pre-Columbian times). Areas of nomadic pastoralism are those where herding of animals by migratory peoples had been established in the Old World. Compare this map with the detailed table of plant origins given later in this chapter (Table 12-2). [After E. Isaac, *The Geography of Domestication* (Prentice-Hall, Englewood Cliffs, N.J., 1970), p. 41, Fig. 3.]

wide open. Sauer's special contribution was to argue that, since evidence is so thin from this very distant period, we need to think hard about the geographical conditions under which domestication might occur.

The Spatial Impact of the Agricultural Revolution   Whatever the precise location of the first agricultural communities, the impact of permanent agriculture on the spatial organization and density of the human population is clear. It increased the reliability of the food supply as well as the volume of food, so that more people could be supported by a given area. They no longer needed to concentrate totally on food production and could branch out to nonagricultural crafts. As food surpluses became available, goods began to be exchanged. Pottery, weaving, jewelry, and weapons were bartered and traded over long distances.

The impact of these changes on human spatial organization was two-fold. First, the centrifugal forces that scattered small numbers of people over large areas were weakened; isolation gave way to contact, and some degree of agglomeration into settled agricultural villages became possible. Second, the population densities rose in some areas to levels several hundred times greater than those of the preagricultural communities. For example, we know from archeological evidence that the hill-farm communities of northern Mesopotamia had densities of approximately 70 people/km^2 (180 people/mi^2) around 8000 B.C.

By 4000 B.C. the total population of the globe had probably reached around 87 million. The greater part of this population was probably concentrated in areas where village agriculture was intermingling with and slowly replacing food-gathering and hunting. Such areas certainly included a belt stretching from Western Europe and the Mediterranean through the Middle East to western India, northern China, Indonesia, and Central America. Outside this area population changed little from its preagricultural pattern. The extreme zones of the Arctic and Antarctic, together with the more remote ocean islands, remained wholly unoccupied.

The Origin of Urban Hearths   Although the evidence on the origin and early growth of cities is more plentiful, its interpretation has led to academic controversy hardly less acute than that over agricultural hearths. Specific evidence of urban forms is available for several sites in the Tigris-Euphrates Valley for the period 3000 to 2500 B.C. Calculations based on the size of these built-up areas yield probable populations of around 50,000 for Uruk and 80,000 for Baghdad in that period. Less controversy surrounds the time of early urban centers (although new excavations in Asia Minor indicate that they may be older than they were first thought to be) than how they fit into a developmental sequence.

The Developmental Sequence   Figure 12-5(a) shows a highly generalized version of the traditional view of human developing resource-organizing

Figure 12-5. The position of cities in the developmental sequence.
Figure (a) shows the traditional main stages and processes in the developmental sequence. Figures (b), (c), and (d) provide alternative models of the place of the cities in human development. Of course, the models shown here are highly simplified. Herding, for example, has a very complex origin and probably developed in different ways in various areas.

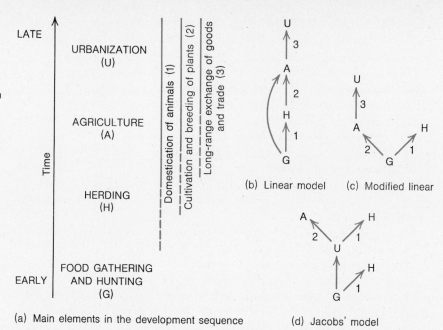

(a) Main elements in the development sequence

(b) Linear model

(c) Modified linear

(d) Jacobs' model

technology. There are four main stages (primitive food-gathering and hunting, herding, agriculture, and urbanization), linked by three processes (the domestication of animals, the permanent cultivation of crop plants, and the trading of goods). The position of these stages and processes with respect to the time line at the left reflects the pattern of archeological evidence.

Archeologists are increasingly divided over the ways in which urbanization fits into this sequence. Figure 12-5(b) shows a conventional "linear" view in which urbanization is a late stage in the developmental sequence, dependent on the buildup of a food surplus as a result of the increasing production by agricultural communities. More generally, herding is regarded as an incidental side-shoot, contributing little to the sequence. [See Figure 12-5(c).] But is this the correct order? Town planner Jane Jacobs has entered the fray with a controversial book called *The Economy of Cities.* She emphasizes (1) the increasing evidence of highly specialized and long-range trade (e.g., in obsidian axes) among human populations in the basic food-gathering and hunting stage and (2) the increasingly early dates assigned to cities.

Utilizing this evidence, Jacobs questions both the assumptions of the conventional "linear" view. In her own model [Figure 12-5(d)] urbanization is an early response to trade and exchange, and permanent agriculture a by-product of the food needs and hybridizing environment of the city. There are other scholars who question the assumption that cities originated primarily for *economic* reasons. Leading urban historian Lewis Mumford cites documentary evidence from the cities of ancient Egypt which suggests that they were founded as centers of royal or priestly power. The view of cities as *Zwinburg* (control centers) rather than marketing or manufacturing centers may be a correct interpretation of their role in the preindustrial Near East. However, by the time cities appeared in the eastern Mediterranean area, during the third millennium B.C., their role as control centers was becoming inextricably entwined with their role as centers of interregional trade.

The Location of Urban Hearths   If we leave alone the muddy waters of when and how cities began, we are left with the question of where they began. Unfortunately, our notions of the spatial distribution of early cities necessarily reflect the concentration of archeological activity. The patterns observed depend in part on where archeologists have chosen to look for evidence of urban centers and on fortuitous factors like the durability of building materials and the preservation of foundations. Most successful cities have experienced so many cycles of building and rebuilding on the same sites that traces of their early outlines are difficult to establish.

We know from the available evidence that urban development appears to have begun in four major river valleys: (1) the land lying between the Tigris and Euphrates rivers in the Mesopotamian area of the Middle East, (2) the valley of the Nile in Egypt, (3) the area near the Indus system in western India, and (4) the Hwang Ho (Yellow River) Valley in northern China. Table 12-1 gives the location and dates of the main areas of early urban cultures.

Prehistorians Childe and Wittfogel have seen a special link between the emergence of an urban civilization and the practice of large-scale irrigation. According to their *hydraulic hypothesis*, the environmental problems posed by the agricultural development of large, seasonally flooded river valleys could only be solved by integrating the collective efforts of many small communities. The construction and maintenance of large-scale irrigation works—dams, reservoirs, feeder canals—demanded the mobilization of large quantities of labor. Fair allocation of water between competing communities also called for the presence of a supervising authority which could plan the distribution of water use over both space and time.

Although the early emergence of Mesopotamia and Egypt as urban centers lends support to the hydraulic hypothesis, more recent work suggests the story may be a much more complex one. In a detailed study of another area of irrigation agriculture, the North China plain, geographer Paul Wheatley finds the emergence of urban forms in the second millennium B.C. is linked to the growth of ceremonial centers. As we saw in Chapter 11, the Chinese city is richly symbolic with its form and spatial organization directly planned to reflect prevailing ideas about the universe. [Look again at Figure 11-8(c).] Since this symbolic role, found in all the other nuclear urban areas, appears to predate the growth of large-scale irrigation, it may well be that urban forms helped to encourage large-scale irrigation rather than being dependent on them.

Table 12-1　Main urban nuclear areas[a]

Type	Zone	Location	Early Urban Culture	Representative City
PRIMARY NUCLEAR AREAS	Fertile Crescent (Middle East)	Mesopotamia (Tigris-Euphrates Valley)	Sumerian (2700 B.C.)	Ur, Uruk
		Nile Valley	Egyptian (3000 B.C.)	Memphis, Thebes
		Indus Valley	Indus (2500 B.C.)	Mohenjo-daro, Harappa
	East Asia	Hwang Valley	Shang (1300 B.C.)	Anyang
	America	Yucatan Peninsula	Mayan (A.D. 500)	Palenque, Tikal
		Central Mexico	Aztec (A.D. 1400)	Tenochtitlán
		Peru	Inca (A.D. 1500)	Cuzco
	West Africa	Niger Valley	Yoruba (A.D. 1300)	Ife
SECONDARY NUCLEAR AREAS	Southern Europe	Aegean Islands and Peninsula	Aegean (2000 B.C.)	Knossos Mycenae
		Italian Peninsula	Etruscan (400 B.C.)	Felsina, Rome
	South and East Asia	Mekong Valley	Khmer (A.D. 1100)	Angkor
		Japan	Yanato (A.D. 600)	Naniwa
		Central Burma	Pyu (A.D. 700)	Sri Ksetra
		Ceylon	Sinhalese (A.D. 1000)	Polonnaruva

[a]The dates are indicative of the middle period of the culture. The division between primary and secondary nuclear areas is based on Paul Wheatley, *The Pivot of the Four Quarters* (Edinburgh University Press, 1971).

Urbanized hearth populations of over 125 people/km² (325 people/mi²) existed in limited sections of agricultural districts as long ago as 4500 B.C. By the beginning of the Christian Era, the earth's total human population had grown to around 300 million, roughly doubling since the beginning of urbanization. At this time the main lines of the earth's present pattern of population were beginning to be blocked out.

The overwhelming majority of the world's population was located in three huge clusters. Probably the largest concentration of people was in the Indian subcontinent, which had over 40 percent of the estimated world population. The second largest concentration was that part of China within the Han Empire; perhaps 25 percent of the world's population was concentrated primarily in the delta plains of the Hwang Ho. Outside these two largest clusters lay the ancient Roman Empire, extending from Western Europe and the Mediterranean to the Middle East, and including the long-established population concentrations of the Nile Valley and Syria. These three areas contained well over four-fifths of the world's population. In the fertile alluvial parts of these areas, the population density reached levels in excess of 1000 people/km². Outside these areas, population continued to be thinly spread over the land surface, and only a few high-density pockets (e.g., in central Mexico) broke up the pattern of primitive agricultural settlements and food-collecting groups.

In settling questions of regional origins the main problem geographers face is lack of evidence. But as the evidence on the recent historical period builds up, the challenge shifts to making a comprehensive and convincing story of known but perplexing events. Each of the centers of urban culture we have recognized grew in population during the Christian Era, and each merits special study. Here, however, we choose to follow the history of only one of them, Western Europe.

The reason for selecting Western Europe rather than China, for example, is less chauvinistic than it may appear. The growth and spread of European culture is well documented; it provided, in fact, a pattern for much of the non-European world. In some areas like the Americas and Australia the pattern has been followed closely; in others like China, it has had scarcely any effect; in yet others like Africa and South Asia, we still await the final results of contact with the west.

12-2
The Question of Spread

Internal Organization: The European Hearth   In looking at the geographical spread of Western European culture, we begin with its internal crystallization. Then we move on to its overseas transfer.

Between the beginning of the Christian Era and A.D. 1500, the world's population probably doubled to around 500 million. This increase was particularly noticeable in the world's third major population concentration—the area of the former Roman Empire. Here the largest gains were registered in the more recently settled areas of Western and East-central Europe, areas which were to emerge in the later stages of the period as modern states like France, Britain, and Poland. For Europe, the improved information on population during the historical period allows us to see much more clearly the pattern of human organization. We can, for example, trace the westward spread of urban institutions in the Roman Empire (Figure 12-6). For the largest city, Rome, housing densities within the known city limits indicate a maximal population approaching 200,000 in A.D. 200. Below Rome extended an urban hierarchy that had the same general form we shall see later in characteristic modern city systems. (See Chapter 15.) The collapse of the Roman Empire and the reduction in the scale of political and commercial organization from a subcontinental to a local level was followed by a breakup of urban and regional links.

With the slow revival of trade in early medieval Europe came the emergence of a fine mesh of small inland cities whose sites were often chosen because of their good defensive positions (e.g., the *bastides* of southwest France). Populations remained surprisingly small. For example, by A.D. 1450, Nuremburg, Germany, an important inland town, still had a population of only 20,000. Even London, which by A.D. 1350 had regained the level of population it had reached in Roman times, contained only around 40,000 inhabitants. Below these principal cities came a regular hierarchy of smaller cities, not unlike that which exists in the modern world.

Figure 12-6. The spread of urban organization across Europe.
The earliest European cities occur in the extreme southeast corner of the continent, where cities like Knossos in Crete or Mycenae in southern Greece were established by around 2000 B.C. Major extension of the city-building in Europe accompanied the spread of the Roman Empire. The generalized contours show cities spreading north and west from the Aegean over an 1100-year period of history. [From N. J. G. *Annals of the Association of American Geographers* **59** (1969), p. 148, Fig. 6.]

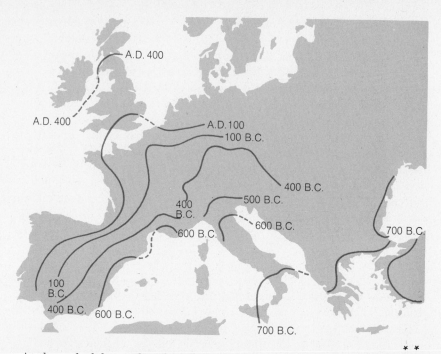

At the end of the medieval period, western and central Europe were firmly structured into an organized regional system of cities. At the top of the hierarchy stood the trading cities of emergent industrial areas like London, Flanders, Lombardy, and Catalonia. Below was a network of smaller inland cities like that in Figure 12-7, often playing an important role in trade and administration. The network of cities was growing in two ways. First, by spatial expansion through colonization and the establishment of new cities in Eastern Europe. Second, by the growth of smaller cities around fast-growing centers like Venice and Genoa, both booming as trade with the Levant increased.

Outside Europe lay other city hierarchies that were also highly differentiated and expanding. In the rest of the Old World the main areas of urban civilization were in eastern China and northern India. In the New World only central Mexico and the Peruvian valleys were urbanized, and these areas had much smaller populations than the Old World hearths of urban culture. Of the five hearths in existence in 1500, the European one was to experience the most significant expansion in the next 400 years of world history. Three of the four remaining hearths came directly under European influence in that period; only the Chinese hearth remained untouched by the major spatial reorientation of world trade produced by the growth of the European center.

Europe Overseas: Rim Settlements   The first phase of European overseas expansions, that of transoceanic *rim* or coastal, *settlement,* lasted from the

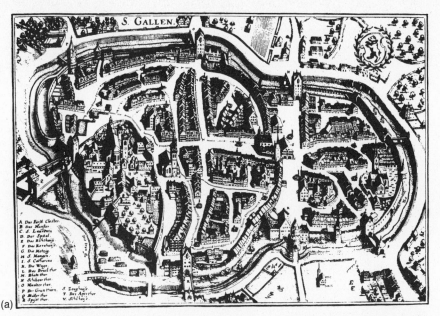

(a)

"Medieval city"

(b)

Figure 12-7. The growth of the West European city.

St. Gallen is a city in northeast Switzerland. Although the Romans colonized this part of Europe, the founding of St. Gallen is ecclesiastical in origin and later in date. It developed around a Benedictine abbey founded in the early seventh century. For the next four centuries this was to be the most famous educational institution north of the Alps, and it remained an important seat of learning throughout the medieval period. Along with the growth of the abbey came increased political and economic importance for the city. It was walled in the tenth century, became a free city in 1304, and joined the Swiss federation in 1454. An early seventeenth-century print (a) still shows the medieval core of the city with its walls and gateways. But in 1600 its population was probably still below 5000 people. Growth in the period since then to its present 80,000 has been associated with its role as a commercial center for the surrounding area (the canton of St. Gallen), its long-established textile industry, and new industries such as glass and metalworking. With this growth the city has sprawled well outside its original walls and now extends east–west along the Steinach Valley. The photo (b) shows the medieval core within the modern city today; forested valleyside slopes north and south of the city show as very dark areas. In comparing (a) and (b) note that north is to the right of the print but to the top of the photo. [From (a) H. Boesch *et al.*, *Villes Suisses a Vol d'Oiseau* (Kümmerly & Frey, Berne, 1963), pp. 212, 216. (b) Courtesy of Swiss National Tourist Office.]

original Age of Discovery in the fifteenth century to the early part of the nineteenth century. Different European peoples took the lead at different times—the Spanish and Portuguese earlier, the French, English, and Dutch later—in establishing settlements along the coastal rims of the Americas, Africa, and southern Asia. The settlements were of three main kinds: trading stations, plantations, and colonies of farm families. All clung, limpet-like, to the edge of the continents.

Coastal Trading Stations   Small trading posts were established widely on the coasts of India and southern China. Ports like Goa, Madras, and Canton served at times as points of exchange through which the products of the two great Asian urban civilizations of India and China could be brought to the growing urban markets of Western Europe. Trade was mainly in luxury articles such as spices, tea, and hand-crafted products like silks. The number of European settlers at any one time was quite small compared with the size of the indigenous Asian population, and only in India was it possible to exercise any real political control over the large hinterland areas that served the ports. Counterparts to the Indian ports, trading stations, were established on a minor scale in Malaysia, Indonesia, and East and West Africa.

Tropical and Subtropical Plantations   Plantations were originally established to grow sugar and spices, but the range of crops was later expanded to include the production of various foodstuffs (coffee, cacao, bananas, etc.). The earliest plantation settlements were generally on ocean islands like Madeira, offshore islands like Zanzibar off the east coast of Africa, or coastal strips like the Baixada Fluminense around Rio de Janeiro in Brazil. Inland extenstions of these settlements appeared largely in the nineteenth century. Such plantations demanded intensive labor, and European settlers primarily occupied only organizational roles, while the non-European population provided the field labor. When the indigenous population was unable to supply workers, slave laborers were brought in from other areas. (See Figure 12-8.) The current population mix in tropical America, East Africa, and parts of Malaysia and Australasia is largely a legacy of tropical plantation settlements.

Midlatitude Farm-Family Settlements   The third type of settlement associated with the period of European expansion is the farm-family colony of European migrants in the middle latitudes. This type included the settlements, mainly of English- and French-speaking groups, on the northeastern seaboard of North America, as well as later settlements in Australia and New Zealand. These settlements were in marked contrast to the tropical plantations because of their dependence on an influx of Europeans. Moreover, their agricultural products were destined for the local market rather than for export to Europe. Different groups from various parts of Europe (Swedes, Germans, Irish, etc.) brought to their overseas settlement some of the distinctive char-

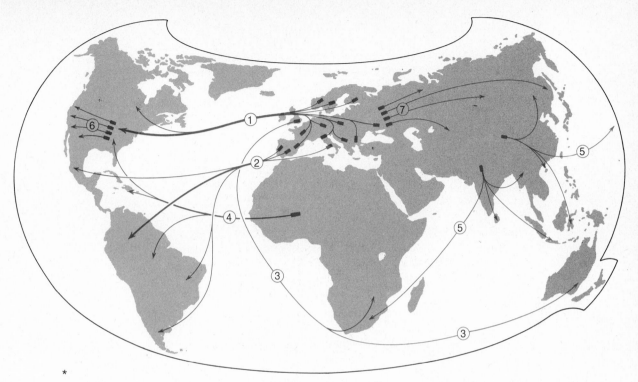

(1) Migration from all parts of Europe to North America
(2) Migration from southern Europe to Latin America
(3) Migration from Britain to Africa and Australasia
(4) Shipment of African slaves to the Americas
(5) Indian and Chinese movements
(6) Westward colonization in Anglo-America
(7) Eastward colonization in Russia

acteristics of their home areas. The layout of farms, villages and towns, the cropping patterns, and the farming technology often reflected traditions and practices in the original homelands. Even today the different ethnic backgrounds of farmers can be traced in the design of their farm buildings; the Pennsylvania barn shows just such a rich variability and has been carefully mapped by cultural geographers.

Europe Overseas: Continental Penetration The second phase of European expansion, that of *continental penetration*, began early in the nineteenth century and lasted until about World War I. This phase was accelerated by rapid industrialization in the European hearth, the development of transport innovations like the railroad, the growing overseas migration of Europeans, and a rapid increase in the rate of exploitation and trade in non-European resources. Its chief impacts on the distribution of population were the springing up of industrial cities in the midlatitude colonies and the inland penetration of the agricultural frontier as the rich, midcontinental grassland zones were exploited for grain or stock production.

Figure 12-8. Intercontinental migration.
The map shows the main currents of intercontinental migration since the beginning of the sixteenth century. Only the main flows are shown, and in a highly generalized form. [From W. S. and E. S. Woytinsky, *World Population and Production* (Twentieth Century Fund, New York, 1953), p. 68, Fig. 27. Copyright © 1953 by the Twentieth Century Fund.]

Midlatitude Grassland Settlements   The nineteenth century witnessed the occupation of the prairies and the pampas in the Americas, the veld in Africa, and the Murray-Darling and Canterbury plains in Australasia. The pattern and timing of settlement was greatly affected by technical innovations like the railroad, refrigeration, and barbed wire. Railroads made it cheap to move agricultural products to ports, refrigeration made it possible to preserve meat for long-distance shipping, and barbed wire resulted in the fencing of open rangelands.

If you look at historian W. P. Webb's book, *The Great Plains*, you will find an excellent example of how different groups responded to the resource mix provided by this midlatitude grassland area. The pre-Columbian Plains Indian, the sixteenth-century Spaniard, the cattleman in the 1840s, and the wheat farmer in the 1880s all experimented with the Great Plains environment. In the withered hedgerows (a pre-barbed-wire experiment at fencing, British style), in the bankruptcies that followed long droughts, and in the blowing topsoil, we see the aftermath of the experiments that failed. Each cultural group saw in the Great Plains environment different possibilities, linked to the current technology and the group's cultural background.

Meanwhile, on Europe's eastern continental border the Russian state was expanding its settlement of the steppe grasslands at a comparable pace. In the tropics the demand for plantation products accelerated, and the movement of non-European peoples, first African slaves and later indentured Indians and Chinese, into plantation areas like the Caribbean continued. Trading contacts with the Orient increased and intensified as western countries extended their political control through both treaties and military occupation. More European ports were established on the coasts of China.

Mining and Mineral "Rushes"   The mid- and late-nineteenth-century gold discoveries brought a rapid rise in both the white and the non-white population of mining areas. In some areas with an environment favorable to agriculture, mineral strikes provided the trigger which set off continued migration and long-term settlement. California provides the most well known example. The discovery of gold at Sutter's Mill on the American River in 1848 led to a rush to establish claims. "Forty-niners" poured into California from the east and from many distant parts of the world. In the next twelve years California's population leapt from around 26,000 to approaching 400,000.

The American experience was paralleled by Australia. The 1851 gold strikes at Ballarat and Bendigo, in Victoria, brought in 250,000 prospectors in the next five years. (See Figure 12-9.) By 1855 more people lived in Victoria than had lived in the whole of Australia before the discoveries. But in marginal environments like the subarctic zone the population and settlement that followed mining did not last. In 1898 some 30,000 prospectors moved down the Yukon River to the new gold strikes; today the total population of the whole Yukon territory is only half that number.

Figure 12-9. European overseas migration.
Gold rushes in the middle of the nineteenth century brought major increments of population to Australia. These contemporary prints show police checking mining licenses on the Ballarat, Victoria, gold field in the 1850s. [Print from News and Information Bureau, Department of the Interior, Canberra, Australia.]

The search for gold in the nineteenth centery was followed by a search for oil in the Middle East in the early decades of this century. Mining strikes and oil exploration continue to be part of a phase of settlement expansion but the emphasis is now on shifts in capital investment (e.g., drilling rigs) rather than major population movements.

Europe Overseas: The Geographic Legacy   We might say that a third phase of economic consolidation but political withdrawal began around World War I and appears to be continuing today. It was marked by a shift of economic power from the original European hearth to the United States and Soviet Russia. In addition there was a political withdrawal of formal European control of much of Africa and Asia, signaling a virtual end to the British, French, and Dutch overseas empires. Population movements from Europe to midlatitude countries like the United States, Australia, and Argentina have continued, and they have not been balanced by the counter flow of black population to Western Europe. Despite lessening political control, the presence of European capital, culture, and means of communication in much of Africa and Southwest and South Asia remains a fact of life. In Latin America Europe's economic role has been largely replaced by the United States. The rise of Japan as a prime industrial and trading power and the increasing role of China are bringing new waves of regional expansion based on East Asian hearths.

What was the effect of the 500 years of European expansion? On the spatial organization of the world community, the effect was enormous. For one thing, the expansion involved a transcontinental movement of around 95 million people. Over two-thirds of these were Europeans moving to temperate latitudes (notably to the United States). Another 20 percent were Africans forcibly transported from their homelands, especially to the American tropics and subtropics. The remaining fraction of over 10 percent were Asiatics. The pattern of Asian population movements is more diffused because there are considerable Asiatic populations growing in parts of Africa, in the Caribbean, and in parts of the Pacific.

Along with the changes in population went a massive exchange and mixing of the world's crops. Table 12-2 gives a list of just some of the leading crop plants now used by human beings and their probable areas of origin. As we saw earlier in this chapter (see especially Figure 12-4), the precise origins of agriculture occurred so far back in human history that we can only make reasoned guesses on their actual location. What is clear is that 500 years of European expansion turned the pre-Columbian pattern of crop distribution upside down. American crops like the potato and tomato were to become regular farm crops in Europe while Old World crops like coffee and wheat were to become major crops in the Americas. Indeed, your own backyard is now likely to contain a variety of plants unquestionably richer and more diverse than seemed possible to our more continent-bound forebears of the pre-Columbian period.

Along with the interchange of population and of crops went a fundamental reorganization of wealth. Figure 12-10 shows in a very generalized form the map of world income today in comparison with that of population. The twin-peaked North Atlantic center symbolizes the extreme degree of financial control exercised by institutions like Wall Street, London's "City," Paris's "Bourse," or the Zurich banks in the organization of much of the world's resource development. Certainly this dominance may now be past its peak with the emergence of a separate Russian center and the promise of new centers of financial power in the Middle East, Japan—and later China. Nonetheless, much of that control remains and the inequity between world population and world income that persists today is partly a reflection of the superimposition of a world urban system—centered first on Western Europe and then on a combined and widened North Atlantic core—upon much of the remainder of the world.

Table 12-2   Sixty of man's leading crop plants and their probable areas of origin[a]

GROUP 1: BEVERAGES AND DRUGS	
Cacao	Orinoco basin, S. America
Coffees	E. Africa
Opium poppy	S.W. Asia
Quinoa	Andean America
Teas	S.E. Asia
Tobacco	Plate basin, S. America

GROUP 2: ORNAMENTALS	
Bougainvillea	E. Brazil
Dahlia	Mexico
Marigold	Mexico
(Tagetes)	Tropics

GROUP 3: ROOT CROPS	
Cassava	Tropical America
Potatoes	Andes
Sweet potatoes	Meso-America
Taro	S.E. Asia (New Guinea)

GROUP 4: GRAINS	
Amaranths	Meso-America
Barleys	S.W. Asia
Maize	Meso-America
Oats	Near East
Rice	S.E. Asia
Rye	Asia Minor
Sorghums	E. Africa
Wheats	S.W. Asia

GROUP 5: SUGARS	
Sugar cane	S.E. Asia (New Guinea)

GROUP 6: FIBERS AND OIL PLANTS	
Cotton	Tropical America (Caribbean and Ecuador)
Flax	Mediterranean basin
Peanut	E. South America
Sunflower	N. America

GROUP 7: FORAGE PLANTS	
Alfalfa	S.W. Asia
Bluegrass	S.E. Europe
Cowpea	E. Africa

GROUP 8: VEGETABLES	
Beets	Mediterranean basin
Broad bean	E. Africa and S.W. Asia
Cabbage	Mediterranean
Carrot	S.W. Asia
Cucumber	India
Gourds	Tropics
Kidney bean	Meso-America
Lima bean	Tropical America
Pea	S.W. Asia
Red peppers	Tropical America
Rhubarb	China
Scarlet runner	Meso-America
Soybean	China
Squashes	Meso-America
Tomato	Andean America

GROUP 9: FRUITS	
Apple	Caucasus
Avocado	Meso-America
Banana	Malaysia
Citrus fruits	S.E. Asia
Coconut	S.E. Asia
Date Palm	W. India
Fig	S.W. Asia
Grapes	Turkestan
Mango	S.E. Asia
Melon	E. Africa and S. Asia
Papayas	Tropical America
Peach	China
Pear	Caucasus
Pineapple	E. South America
Plum	S.E. Europe
Quince	S.W. Asia
Strawberry	Americas
Watermelon	S. and E. Africa

[a]There are many hundreds of plant species used by man. This highly selective list is designed to illustrate some of the types of plants used and their probable areas of origin. Cultivated plants are very difficult to classify botanically because of the amount of hybridization and the ancestry of only a few has been fully worked out. The source areas should therefore be regarded as reasoned guesses from incomplete evidence in most cases. I am grateful to Professor Jonathan Sauer of the University of California at Los Angeles for commenting on this list, which was largely derived from Edgar Anderson, *Plants, Man, and Life* (Melrose, London, 1954), Chap. X and C. D. Darlington, *Chromosome Botany and the Origins of Cultivated Plants* (George Allen & Unwin, London, 1963).

Geographers have sought to build a spatial model of the diverse pattern of European overseas expansion. We pick up their attempts when we look more closely at models of migration in Chapter 17 and at the subject of economic growth in Chapter 21.

In the first part of this chapter we looked at the origins of cultural regions on the global scale. In the second part we took one culture—that of Western Europe—and traced its spread. In this third and final section we reduce the scale of our investigations again and look at the question of the persistence of

12-3
The Question of Persistence

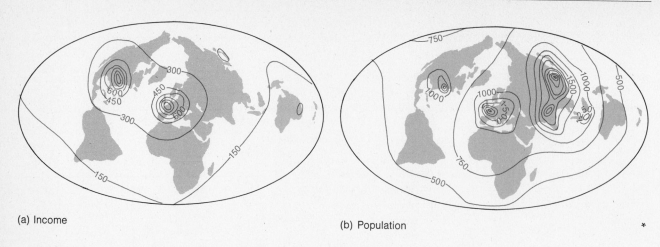

(a) Income

(b) Population

Figure 12-10. The legacy of European spatial organization.
These generalized maps show the pattern of world income and world population in the early 1960s. Note the similarities and differences between the two. [From W. Warntz, *Macrogeography and Income Fronts* (Regional Science Research Institute, Philadelphia, 1965), pp. 92, 111, Figs. 19, 24.]

cultural elements in one of the areas of European overseas settlement, the United States.

The Pre-Columbian Pattern European settlement of North America did not start in an empty land. The area north of the Rio Grande contained a population of more than 1 million American Indians in A.D. 1500. As we saw in Section 12-1, they had migrated southward from East Asia via the Alaskan bridge in the late glacial period. By the time of the first European contact the native American population had differentiated into a very complex variety of separate subcultures each with its own language and traditions.

Five broad regional groups can be distinguished, those of the eastern forests, the Plains, the northwest coast, the California-intermountain area, and the Southwest. Each region was characterized by differences in farming or hunting systems, in settlement patterns, in languages, and in organizational structures. Examples of three of the five groups are shown in Figure 12-11. The Sioux were part of the Plains Indian culture group. Originally farmers and hunters in the Mississippi valley forests, they were forced further west onto the grasslands as part of the general relocation of tribal areas that followed European occupation of the East Coast. The introduction of horses from the Spaniards moving north from Mexico also allowed a more mobile, hunting economy to be built up. The second example, the Navajo, were part of the southwest culture group. Originally a nomadic hunting people, they settled as northern neighbors of the sedentary Pueblo Indians ("pueblo" is the Spanish word for village) of Arizona and New Mexico and learned to raise corn and weave cotton. Again, the introduction of horses and sheep by the Spaniards modified the economic pattern of the Navajo by emphasizing the grazing and wool-processing side of their culture. Finally, the Florida Indians were part of the Eastern Forest Tribes. Tribal divisions in this area are less

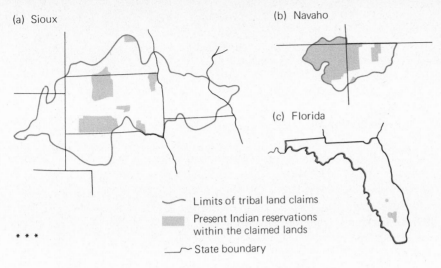

(a) Sioux

(b) Navaho

(c) Florida

Limits of tribal land claims

Present Indian reservations within the claimed lands

* * *

State boundary

Figure 12-11. American Indian culture areas.
The detailed culture regions of pre-Columbian America are not known with certainty. Many Indian tribes were mobile and were displaced from their original homelands by European colonists. The maps above show three from the several hundred distinct North American tribes: the Sioux in the northern Great Plains, the Navaho of the southwest, and the Florida of the southeast. Only the major reserves can be shown at this scale. [Tribal land claims are from a map compiled by S. H. Hilliard, *Annals of the Association of American Geographers,* Map Supplement No. 16 (1972).]

well recorded but the groups probably included the Seminole and relatives of the Creek Indians. In addition to hunting and fishing, the people of this area lived by raising such crops as corn, beans, squash and tobacco. Settlements were both permanent and complex in structure, often with houses built around a central court.

If we look back at the Slobodkin model in Figure 7-10 (page 154) we see the range of outcomes for two populations competing for the same resources. Given the advanced technology and organization of the European settlers it is not surprising that competition for the North American land resources lead to a displacement of Indians by the newcomers. White men also brought morbid allies in new diseases (measles, smallpox, tuberculosis, and influenza) which killed thousands of Indians. New cultural innovations (like liquor) were scarcely less disruptive. Forest clearance, enclosure of grassland, decimation of animal populations all changed the ecological balance against which Indian culture had evolved. Even though the Indian populations are now moving back toward the level of four centuries ago and reservations form extensive tracts for some tribes, the pre-Columbian pattern has been erased for major areas of the continent. The Navajo and the Sioux with numbers variously estimated between 50,000 and 100,000 are two of the largest remaining tribes; the Florida Indians number only a few thousand.

Sequences of Post-Columbian Migration Waves   The population that replaced the Indian came predominantly from Europe and Africa. A survey of the period since 1607 (the date of the first permanent English settlement, at Jamestown, Virginia) shows a series of five distinct *sequent waves* of immigration, each associated with a particular set of migration sources. Between 1607 and 1700 there was an initial wave of English and Welsh, along with a small number of African slaves. Numbers of Dutch, Swedes, and Germans were relatively few during this phase. The period from 1700 to 1775 brought increased migration from the earlier sources and an influx from Germanic and Scotch-Irish sources as well. The years 1820 to 1870 saw increased numbers arriving from northwest Europe (especially Britain, Ireland, Holland, and Germany), but the influx of Africans came to a halt. The 1870s saw the arrival of a vanguard from southern European countries, plus some Asians, Canadians, and Latin Americans. In the half century from 1870 to 1920, the period of "The Great Deluge," there was a massive increase in the number of immigrants and a widening of the source areas to include eastern and southern Europe and Scandinavia. Since 1920, there has been a drop in the number of immigrants, but the pattern of sources has remained wide and there has been a steady rise in the percentage of immigrants coming from Latin America.

While data for the early period is fragmentary, immigration in the period since independence is well documented. Figure 12-12 shows the pattern of numbers and changing origins over the last 150 years. The timing of the various waves of migration is reflected in the original areas of rural concentrations of migrants from different countries. The "early-wave" Scotch-Irish (the term used to describe both immigrants from Scotland and the Protestant areas of northern Ireland) settled in a belt running west from New England through the Appalachians into the near Midwest, while the "late-wave" Scandinavians were concentrated in the upper Midwest around Minnesota.

Figure 12-12. Sequences of migration waves.

The graph on the left shows the total number of migrants entering the United States over a 150-year period. The graph on the right shows the changing composition of the migrant stream. The curves indicate the different source areas dominant in the middle of the nineteenth century, the early twentieth century, and at present. Note that the graph describes the percentage of migrants from each source area and not the absolute number.

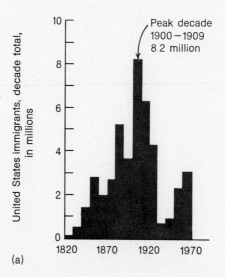

(a)

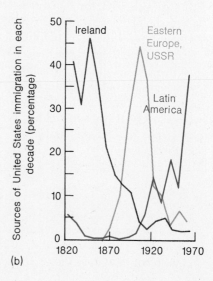

(b)

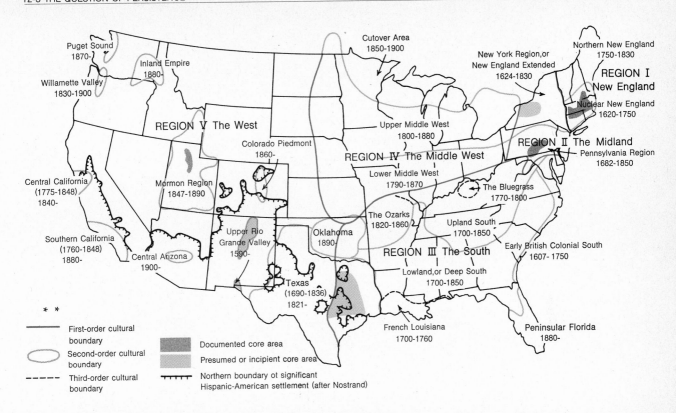

The initial concentrations of African blacks in the South, of Mexicans in the southwest, and of Italians in the cities of the northeast and parts of California are all well known. In each case, the primary pattern of settlement and the original balance of urban and rural concentrations has been muddied and altered by subsequent internal migration, particularly to the growing metropolitan areas.

Cultural Regions of the United States How do geographers make sense of the cultural patchwork that has emerged from these waves of immigration, settlement, and subsequent population growth? Many regional systems have been devised to describe the United States mosaic. Pennsylvania geographer Wilbur Zelinsky's system is shown in Figure 12-13. Zelinsky uses a fivefold classification. His first division, *New England* (Region I), was largely shaped by English migrations over the period from 1620 to 1830, with the development of settlement in northern New England lagging more than a century behind that of the southern nuclear area. The *Midland* (Region II), south of New England, was settled slightly later (between 1624 and 1850), by a wider

Figure 12-13. Cultural regions of the conterminous United States.
The map shows Wilbur Zelinsky's division of the United States into five major cultural regions and various subregions. The interregional boundaries vary in importance, and the status of three regions (Texas, Oklahoma, and peninsular Florida) is uncertain. The dates refer to the approximate limits of settlement and the emergence of a distinctive regional character. [From W.Zelinsky, *The Cultural Geography of the United States* (Prentice-Hall, Englewood Cliffs, N.J., 1973), p. 118, Fig. 4.3.]

variety of migrants. To the English element were added important Rhineland and Scotch-Irish populations in Pennsylvania. In the New York region, Dutch and southern European migration was more important, as was in-migration from New England.

The most complex and diffuse of the three original hearths of culture on the eastern seaboard is the *South* (Region III). Beyond the narrow coastal strip of English plantations with their African slave population (settled before 1750), the South is divided into two major regions. Each of these has important subregions: Louisiana (in the Deep South) and the Ozarks and Bluegrass country (in the Upper South). The triangular region of the *Middle West* (Region IV) has more definite boundaries. Settled largely in the century after 1790, it was strongly affected by the westward extension of two existing cultural areas—the Midland and New England. Other cultural elements were superimposed on the existing pattern by new waves of European migration (particularly from Germany and Scandinavia).

An attempt to piece together the cultural dependence of the later regions on the earlier ones and on outside sources is presented in Figure 12-14. Only the main lines of influence are shown. Beyond the four main regions that comprise the eastern half of the country stands the enigma of the *West* (Region V). Here Zelinsky chooses to isolate nine subareas with some claim to distinctive cultural identities and to leave the remainder of the West as something of a cultural vacuum. The archipelago of subregions includes those shaped by particular ethnic groups (e.g., the Mexican element in the upper Rio Grande Valley), by religious beliefs (the Mormon region, also described in Figure 11-13), and by resource-exploitation patterns (e.g., the Colorado Piedmont). Outside the five main regions lie three intriguing areas whose status and affiliation is uncertain. Texas and Oklahoma are distinctive subregions that run across the boundaries of major divisions, while peninsular Florida lies beyond the South yet is not part of it. Some of the links between these three areas and possible sources of cultural influence are shown in Figure 12-14.

Forces of Change   How long are the cultural divisions hammered out by nearly four centuries of immigration likely to persist? Urbanization, mass communication, and extreme social and geographic mobility all appear to be reducing regional differences. We shall look at these forces and their impact in Part Four of this book. In this chapter we have turned only a few pages in a rich library of research in cultural geography. Yet we have moved in time from eras that saw the earliest crystallizing of cultural differences to the brink of the modern age of mass culture in the United States. In our study, the major emphasis has been on cultural change through *migration*—the slow movement and mixing of peoples, each group carrying with it distinctive

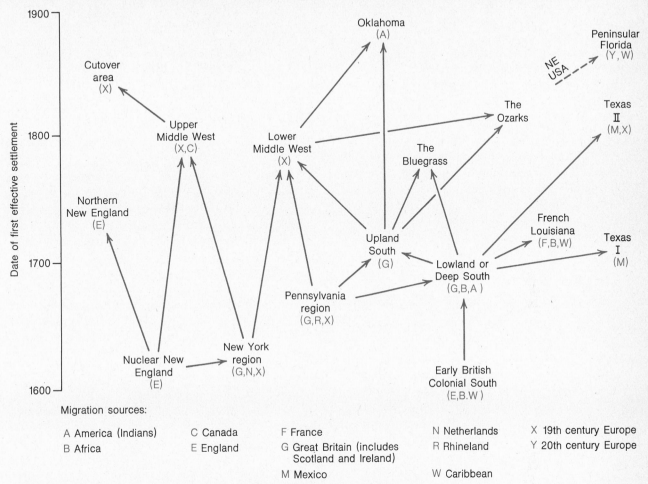

Figure 12-14. The origin of cultural regions.

The chart shows the main cultural links between the sixteen cultural regions in Zelinsky's model of the eastern United States. (See Figure 12-13.) The area's approximate geographic location is indicated on a north-to-south (left-to-right) horizontal scale. The date of the first effective settlement appears on the vertical scale. "Texas I" refers to Texas during its Spanish-American period (1690-1821) and Texas "II" to the area incorporated after 1821 into the United States. The arrows indicate major *internal* migrations and cultural contacts within the United States, and the letters indicate major *external* migrations from outside the United States.

patterns of language and behavior and other cultural freight. In the next chapter we continue to look at cultural transfers between regions, but the accent is on the *diffusion* of cultural elements by mass communication and the exchange of information.

Summary

1. Evidence of human groups on many of the earth's land masses is found by the end of the Pleistocene. Archeological evidence seems to point to the source of human origin in tropical Africa. World population during this early period prior to agricultural domestication has been estimated to have been about 5 million.

2. Agricultural hearths are areas in which early plant and animal domestication is believed to have taken place. The Sauer hypothesis locates hearths in South Asia and the northern Andes. More important is the effect of permanent agriculture on the spatial organization and density of the human population. Increased volume and dependability of food supplies led to a larger population and the opportunity to engage in nonagricultural activities. This also led to greater population densities in agglomerated settlements.

3. Urbanization is usually viewed as a final stage in the development of resource-organizing technology based on the food surplus from sedentary agriculture. The Jacobs hypothesis challenges this view and states that trade and exchange were responsible for early urbanization, and that city-building was the reason for the development of permanent agriculture. The urban historian, Lewis Mumford, considers cities to have been established as centers of royal or priestly power. The development of early great cities took place in major river valleys. By the early Christian Era the major population centers were India, China, and the Roman Empire. The earth's total population was then about 300 million.

4. Geographers interested in regional origins find that most evidence is available for the dispersal of European culture. Within Europe its spatial development can be traced from the early Roman urban hierarchy, through the medieval period, to the development of an increasingly organized regional system by the time of overseas settlement. This overseas migration by Europeans led to the establishment of three settlement types on the continental peripheries: coastal trading stations, tropical and subtropical plantations, and midlatitude farm-family settlements. European penetration of the continent interiors was generally delayed to the nineteenth century. In the twentieth century phase of European overseas activity there has been a consolidation of the cultural legacy, and a withdrawal from overseas empires.

5. A division of the United States into culture regions may be explained in terms of sequent migration waves. A chronological listing of sequent waves of immigrants beginning with the earliest makes possible a geographic description of the regional mosaic of the country. Wilbur Zelinsky defines five regions: New England, the Midland, the South, the Middle West, and the West. Persistence of these cultural divisions is being reduced through urbanization, mass communication, and social and geographical mobility.

Reflections

1. Where were the main hearths of agriculture thought to have been located? Why there?

2. List the crops grown on farms and gardens in your own locality. How many of these crops are native to the area, and how many are species introduced by crop exchanges between the Old and New World? (See Table 12-2 for a list of source areas for crops.) Select one crop plant and research its spatial spread.

3. Examine the different developmental sequences shown in Figure 12-5. Which do you think is most likely? Can you make a case for any alternative sequences?

4. Find out which Indian group occupied your own area when it was first settled by Europeans. What legacy did the group leave on the present landscape? (Students outside North America may wish to skip this question or, if appropriate, substitute an area of their own.)

5. Why are some immigrant groups in the United States concentrated in distinctive cultural areas, while others are widely dispersed throughout the land?

6. Collect the most recent data you can find on immigration into the United States. How far do the trends continue to follow those shown in Figure 12-12?

7. Review your understanding of the following concepts:

agricultural hearths   plantation settlements
Sauer's hypothesis   continental penetration
Jacob's hypothesis   sequent waves of
nodal hierarchies   migration
rim settlements   The Great Deluge
trading stations

One Step Further . . .

The literature describing human evolution and migration and the development of major cultural realms is a very rich one. A useful starting point, in which the growth and spread of population around the world is extensively treated from an ecological viewpoint, is

 Thomas, W. L., Jr., Ed., *Man's Role in Changing the Face of the Earth* (University of Chicago Press, Chicago, 1956).

The origins of the early food-producing centers are discussed in
 Isaac, E., *The Geography of Domestication* (Prentice-Hall, Englewood Cliffs, N.J., 1970) and
 Sauer, C. O., *Agricultural Origins and Dispersals* (American Geographical Society, New York, 1952).

Urbanization and its antecedents are discussed in
 Mumford, L., *The City in History: Its Origins, Its Transformation, and Its Prospects* (Harcourt Brace Jovanovich, New York, 1961) and
 Sjoberg, G., *The Preindustrial City* (The Free Press, New York, 1960).

Students should also look at the splendid accounts of the evolving geography of the Chinese city and the Western City in
 Wheatley, P., *The Pivot of the Four Quarters* (Aldine, Chicago, and Edinburgh University Press, Edinburgh, 1971) and

 Vance, J. E. Jr., *This Scene of Man: The Role and Structure of the City in the Geography of Western Civilization* (Harper & Row, New York, 1977).

The overseas expansion of Europe and the special case of the United States is described in terms of geographical models of expansion in
 Brown, R. H., *Historical Geography of the United States* (Harcourt Brace Jovanovich, New York, 1958) and
 Webb, W. P., *The Great Frontier* (University of Texas Press, Austin, Texas, 1964).

For a full treatment of the cultural regions shown in Figure 12-13, see
 Zelinsky, W., *Cultural Geography of the United States* (Prentice-Hall, Englewood Cliffs, N.J., 1973), pp. 110–134.

Much relevant research on the origins of cultural regions is published in archaeological and anthropological journals. A useful summary of this work is given from time to time in the "Geographical Record" section of the Geographical Review *(a quarterly).*

Because I know that time is always time
And place is always and only place
And what is actual is actual only for one time
And only for one place.

T. S. ELIOT
Ash Wednesday (1930)

chapter 13
Spatial Diffusion
Toward Regional Convergence

In 1905 the El Tor strain of cholera was first identified in the bodies of six Muslim pilgrims at a quarantine station outside Mecca. (El Tor was the name of the quarantine station.) In the 1930s the same strain was recognized as endemic in the Celebes, which has a largely Muslim population. For another 30 years there was little news of El Tor until, in 1961, it began to spread with devastating speed outward from the Celebes. By 1964 it had reached India (replacing the normal cholera strain endemic in the Ganges delta for centuries), and by the early 1970s it was pushing south into central Africa and west into Russia and Europe. (See Figure 13-1.) The seventh of the world's great cholera outbreaks was getting into its stride.

At about the same time an epidemic of a totally different kind was spreading from city to city. Hot pants were introduced as a style of western female dress in the spring fashion shows in Paris in 1970. In the autumn of that year boutiques from Sydney to San Francisco had caught on to the style, and by the spring of 1971 the first secretary in a conservative British university had been sent home for wearing the shorts to work. Now the style is just part of fashion history.

Waves similar to waves of cholera and hot pants have swept around the world in record time. Among the inconsequential waves were the brief Western passion for all things Japanese in the 1880s, and the late 1970s craze for skateboards among children on both sides of the Atlantic. Things as different as influenza epidemics and oral contraceptives, bank rate charges and computer data banks, Dutch elm disease and fire ants, have one thing in common. They originate in a few places and later spread over a much wider part of the world.

Why are geographers interested in such diverse things? Principally because their spread provides valuable clues to how information is exchanged between regions. Where are the centers of diffusion—and why? At what rates do *diffusion waves* travel, and

Figure 13-1. Spatial diffusion.
The map shows the spread of the El Tor strain of cholera from the Celebes during the decade 1961–1971. This strain of cholera appears to be *endemic* (i.e., permanently present) in the population of the Celebes. From this island it erupts from time to time to spread temporary epidemics in surrounding areas. When such epidemics reach major proportions and span several continents in the manner shown on this map, they are termed *pandemics.* In the fall of 1977 another major outbreak centered on the Middle East. (See press cutting.) Geographers are interested in the paths followed by epidemics and pandemics through the populations of human settlements because of the insights they yield on other spatial diffusion processes. [Data compiled from press reports and WHO bulletins. Press cutting courtesy of *Daily Telegraph*]

along what channels? Why do some waves die out rapidly and others persist? Some innovations may move slowly and quietly, like a tide lapping over mud flats. But rapid innovations are studied most frequently, not because of their intrinsic importance (indeed, they sometimes tend to be trivial), but because we can see the whole cycle of diffusion in a relatively brief time period.

The speed with which cultural changes now occur is clearly linked to rapid communication channels. In Chapter 12 we looked at the slow readjustments of the worldwide regional mosaic to the ponderous movements and migrations of human population. In this chapter we turn to the much more rapid changes that can occur through the diffusion of cultural elements, and we see how the ripples from information explosions in one part of the world find their way into another. In this chapter we try to present some of geographers' more recent research on

these topics and their use of computer models. (If you prefer to avoid these subjects, read only Sections 13-1 and 13-4.) Such models have far more than academic interest. If we wish to speed up the diffusion of certain cultural elements, like the adoption of family-planning methods, a knowledge of precisely how waves of change pass through a regional system may be of help. If we wish to halt or reduce the spread of other cultural patterns of behavior, like drug abuse, or protect certain very fragile cultures from being swamped by western civilization, this knowledge may be helpful there, too. In such cases the geographer, as he did in our case history of a beach in Chapter 1, contributes his spatial viewpoint as one way of gaining insight into a many-sided, multidisciplinary problem.

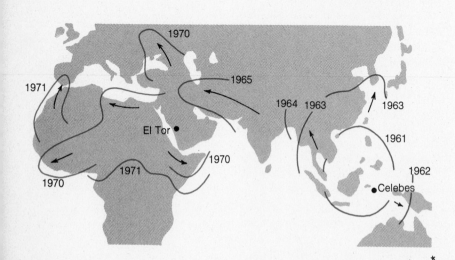

In Chapter 1 we saw how population spreads out over a beach. Again, in the last two chapters, we have come across other cases where a spatial distribution that occurs over time has a distinct diffusion pattern. In everyday language the term "diffusion" means simply to spread out, to disperse, or to intermingle; but for geographers and other scientists, it has acquired more precise meanings.

Types of Diffusion In geographic writing diffusion has two distinct meanings. *Expansion diffusion* is the process by which information, materials, and so on, spread from one place to another. In this expansion process the things being diffused remain, and often intensify, in the originating region; that is, new areas are added between two time periods (time t_1 and time t_2 are both located in a way that alters the spatial pattern as a whole). [See Figure 13-2(a).] A typical example would be the diffusion of an improved crop, such as the IR-8 strain of hybrid rice, described earlier in our discussion of the Green Revolution in Chapter 9 (Section 9.3).

Relocation diffusion is a similar process of spatial spread, but the things being diffused leave the areas where they originated as they move to new areas. The movement of the black population of the United States to the northern cities from the rural South could be viewed as such a relocation process, where members of a population at time t_1 change their location between time t_1 and time t_2. [See Figure 13-2(b).] In a similar manner, an epidemic may pass from one population to the next. Figure 13-2(c) illustrates the two processes and shows how they can be combined. The El Tor outbreak is an example of diffusion by both processes. The strain diffuses by relocation through some areas (e.g., as it did in Spain, where small outbreaks were recorded in 1971), but it also diffuses by expansion because it remains endemic in the Celebes. In this chapter we are discussing interregional interaction and are therefore mainly concerned with expansion processes. Relocation diffusion is treated more extensively in the discussion of regional growth models in Chapter 21.

Expansion diffusion occurs in two ways. *Contagious diffusion* depends on direct contact. It is in this way that contagious diseases like measles pass through a population, from person to person. This process is strongly influenced by distance because nearby individuals or regions have a much higher probability of contact than remote individuals or regions. Therefore, contagious diffusion tends to spread in a rather centrifugal manner from the source region outward. This is clearly shown in Kniffen's study of the spread of the covered bridge over the American cultural landscape, described in Figure 11-11.

Hierarchic diffusion describes transmission through a regular sequence of order, classes, or hierarchies. This process is typified by the diffusion of innovations (such as new styles in women's fashions or new consumer goods

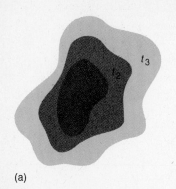

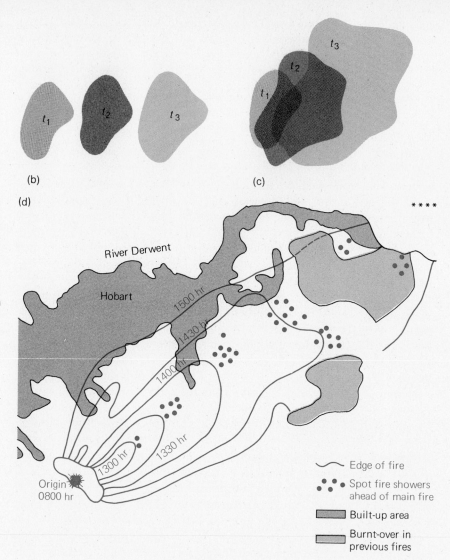

Figure 13-2. **Types of spatial diffusion.**

(a) Expansion diffusion. (b) Relocation diffusion. (c) Combined expansion and relocation processes. (d) Example of a combined expansion and relocation: the spread of the late summer Hobart bush fire in southern Tasmania on the early afternoon of February 7, 1967. The distance between the origin and the coast along the main axis of the fire's advance is about 14 km (9 mi). Spreads of twice the rate shown are common with grass fires. [From data by A. G. Arthur in M. C. R. Edgell, *Monash Publications in Geography,* No. 5 (1973), Fig. 2, p. 7.]

like TV) from large metropolitan centers to remote rural villages. Within socially structured populations, innovations may be adopted first on the upper level of the social hierarchy and trickle down to the lower levels. *Cascade diffusion* is reserved for processes always assumed to be downward, from large centers to smaller ones. When specifying a movement that may be either up or down a hierarchy, geographers generally prefer the term hierarchic diffusion. Figure 13-3 demonstrates how diffusion may begin at a lower point in a hierarchy, move slowly upward, and then expand rapidly. We might think of this as a "Beatles pattern." A musical style beginning in a provincial city (Liverpool) moves to the national capital (London), then on to other capitals throughout the world. Finally, it reaches the local music store in small towns thousands of miles from its point of origin.

Michigan geographer John Kolars has traced the growth of the Sierra Club as a hierarchic diffusion process. The club was founded in 1892 in San Francisco and a separate chapter established in 1906 in Los Angeles. For the next quarter of a century growth was confined to California but a New York center was set up in the 1930s and one in Chicago in the 1950s. With the leap in interest in environmental protection in the last two decades, the Sierra Club has flourished. As we should expect from Figure 13-3(d), this has been accompanied by many new branches being set up in smaller cities. Around Chicago seven new chapters were set up between 1963 and 1973.

Diffusion Waves Much geographic interest in diffusion studies stems from the work of the Swedish geographer Torsten Hägerstrand and his colleagues at the University of Lund. (See Figure 13-4.) Hägerstrand's *Spatial Diffusion as an Innovation Process*, originally published in Sweden in 1953, was concerned with the spread of several agricultural innovations, such as bovine tuberculosis controls and subsidies for the improvement of grazing, in an area of central Sweden. This book was the precursor of various practical studies, particularly in the United States.

In one of his early studies of a contagious diffusion process, Hägerstrand suggested a four-stage model for the passage of what he terms "innovation waves" *(innovations-forloppet)*, but which are more generally called diffusion waves. From maps of the diffusion of various innovations in Sweden, ranging from bus routes to agricultural methods, Hägerstrand drew a series of cross-sections to show the wave form in profile. Here we discuss the wave in profile and then the wave in time and space.

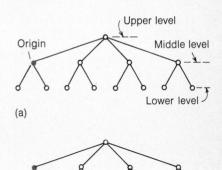

(a)

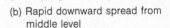

(b) Rapid downward spread from middle level

(c) Slow upward spread to upper level

(d) Rapid downward spread from upper level

Figure 13-3. Hierarchic diffusion. These diagrams show the spread of an innovation through a hierarchy. The innovation begins on a middle level (e.g., a small county town) and spreads rapidly down to a lower level (e.g., villages in its vicinity) but more slowly to an upper level (e.g., a regional capital). Once there, its downward spread is again rapid. Downward spread through a hierarchy is termed *cascade* diffusion. The map (e) illustrates the spread process shown in (a) through (d) for a hypothetical diffusion beginning on the west coast.

(e)

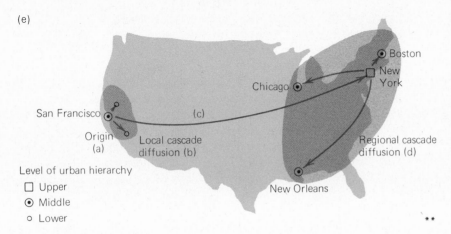

Figure 13-4. Torsten Hägerstrand. Born in central Sweden in 1916, Hägerstrand's doctoral work on spatial diffusion models provided a significant reinterpretation of work on an area of long-standing geographic interest. Under his leadership, geographers at Lund Universtiy have conducted innovative work into a wide area of population geography, computer application, and—most recently—space-time budgets. He is currently professor of geography at Lund and vice-president of the International Geographical Union. [Photo by Tony Philpott.]

Figure 13-5. Diffusion waves profile. The graph shows four main stages in the spread of an innovation by diffusion. The *innovation ratio* measures the proportion of a population accepting the item.

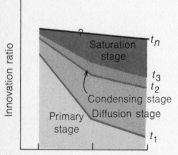

Distance from innovation center

The Wave in Profile Diffusion profiles can be broken into four types, each of which describes a distinct stage in the passage of an innovation through an area. Consider Figure 13-5, which shows the relationship between the rate of acceptance of an innovation and the distance from the original center of innovation. The first stage, or *primary stage*, marks the beginning of the diffusion process. Centers of adoption are established, and there is a strong contrast between these centers of innovation and remote areas. The *diffusion stage* signals the start of the actual diffusion process; there is a powerful centrifugal effect shown by the creation of new, rapidly growing centers of innovation in distant areas and by a reduction in the strong regional contrasts typical of the primary stage. In the *condensing stage* the relative increase in the number accepting an item is equal in all locations, regardless of their distance from the innovation center. The final, *saturation stage* is marked by a slowing and eventual cessation of the diffusion process. In this final stage the item being diffused has been accepted throughout the entire country so that there is very little regional variation.

Since Hägerstrand's original work, other Swedish geographers have carried out parallel studies to test the validity of this four-stage process. For instance, Gunnar Tornqvist has traced the spread of televisions in Sweden by observing the growth of TV ownership from 1956 to 1965. Using information obtained from 4000 Swedish post office districts, he demonstrated that television was introduced into Sweden relatively late, yet within 9 years about 70 percent of the country's households had bought their first set. Tornqvist's results broadly confirm Hägerstrand's analysis. The diffusion process slows down, thus indicating the beginning of the saturation phase, at the end of the study period.

The Wave in Time and Space More advanced work on the shape of diffusions in space and time has confirmed their essentially wavelike form. Figure 13-6 is based on American geographer Richard Morrill's work. By fitting generalized contour maps (called *trend* surface maps described on page 303) to the original Swedish data, he showed that a diffusion wave first has a limited height (reflecting a limited rate of acceptance). It increases in both height and extent, and then decreases in height but increases further in total area. The gradual weakening of the wave over time and space is both time-dependent (as the simultaneous slackening of acceptance rates shows) and space-dependent (as the effect when the innovation wave enters inhospitable territory, strikes barriers, or mingles with competing innovation waves shows). The nature of the medium through which the wave is traveling may cause it to speed up or slow down, and a wave traveling from one center of innovation will lose its identity when it meets a wave coming from another direction.

The exact form of wave may be difficult to spot when diffusion data are first plotted. There may be an apparently chaotic distribution of locations and dates. Geographers have experimented with mapping techniques designed to

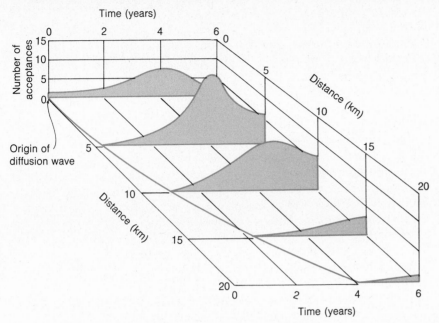

Time (years)

Number of acceptances

Origin of diffusion wave

Distance (km)

Distance (km)

Time (years)

Figure 13-6. **Diffusion waves in time and space.**

Waves of innovation change character with distance from the time and point of origin. In the case shown the maximum height of the wave (i.e., the largest number of new acceptances of the item being diffused) occurs at a point 5 km from the origin in space and 4 years from the origin in time. Although individual waves will vary, some moving very slowly and some very rapidly, a large number appear to follow the general shape shown here. [After R. L. Morrill, *Economic Geography* **46** (1970), p. 265, Fig. 12.]

filter out local variations so that the main form of the waves can be observed. For example, the spread of agrarian riots in Czarist Russia from 1905 to 1910 has a very spotty pattern. Using maps to filter out irrelevant data, two broad centers of unrest emerge: the southeastern Ukraine and the Baltic provinces. The location of the two centers is related to a high level of local tension caused by extreme contrasts in prosperity, the size of farms, and conditions of tenancy. Outward diffusion from the two cave areas follows different patterns. As Figure 13-7 indicates, rioting spread more rapidly along the Baltic Coast from the northern hearth, and in gentler, ripplelike movements from the southern hearth. The intersection of waves from other centers may create complicated patterns and make data difficult to interpret.

TREND-SURFACE MAPS

Geographers use trend-surface maps as a device for separating *regional trends* (regular patterns extending over the whole area under study) from *local anomalies* (irregular or spotty variations from the general trend with no regular pattern). Thus trend-surface maps are like filters which cut out "short-wave" irregularities but allow "long-wave" regularities to pass through. Thus the contours in Figure 13-7 show the general form of the spread of riots in Russia but cut out confusing local variations.

From his empirical studies Hägerstrand went on to suggest how a general operational model of the process of diffusion could be built. We shall look first at how the first and simplest of his models was constructed.

13-2
The Basic Hägerstrand Model

Contact Fields   If we take any of the examples of spatial diffusion in the past few pages, we see that the probability that an innovation will spread is related to distance. Distance can be measured in simple geographic terms, as when we measure the number of meters between trees affected with Dutch elm disease around the campus. Alternatively, distance can be measured in terms of a

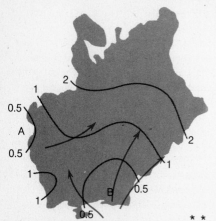

Figure 13-7. Multiple waves.
The arrows indicate the spread of agrarian riots in Russia from 1905 to 1907, an example of diffusion from multiple centers. This is termed *polynuclear* diffusion in contrast to *mononuclear* diffusion (where diffusion waves originate from a single source). [From K. R. Cox and G. Demko, *East Lakes Geography* **3** (1967), p. 11, Fig. 2. Reprinted with permission.]

hierarchy: E.g., the lower-level centers in Figure 13-3 are two "steps" away from the upper-level center. Let us take the first method and use it to measure the spread of information through a human population.

Let us begin by making the simple assumption that the probability of contact between any two people (or groups, or regions) will get weaker the farther apart they are. If we call one person the *sender*, we can say that the probability of any other person receiving a message from the sender is inversely proportional to the distance between them. Near the sender the probability of contact will be strong, but it will become progressively weaker as the distance from the source increases. The exact form of this decline with distance is difficult to judge, but the evidence on telephone calls indicates that it may be *exponential*. That is, it may fall off steeply at first but then ever more slowly. (Check back to Section 7-1, if you wish to refresh your memory on exponential curves.) Thus, we expect the volume of calls to fall off in the ratio 80, 40, 20, 10, 5, and so on with the first, second, third, fourth, fifth kilometers. This is, of course, an idealized decline and actual patterns will be less regular. Geographers term this spatial pattern a *contact field*, drawing their language from the use of gravitational and magnetic "fields" in physics.

In models of cascade diffusion we can retain the exponential contact field but replace geographic distance with economic distance between cities in an urban hierarchy or with social distance in a social-class hierarchy. Distance may not be symmetric in the hierarchic cases; for example, population migration up the hierarchy (from small to large towns) may be easier than migration down the hierarchy. This implies that the socioeconomic distance between levels depends on the direction of movement.

The contact fields in epidemics may be very complex. For instance, studies of measles indicate that the probabilities of contact (and thus infection) *within* a given group like a family or the students at an elementary school may be random. However, the probability of contact *between* such groups may be exponentially related to distance in the way we have already described. For example, research on southwest England showed that the probability of measles outbreaks in an area immediately adjacent to an area that had already reported cases was about 1 in 8. With further distance from the infected area the probabilities of infection fell steadily to about 1 in 30.

Mean Information Fields   How can we translate the general idea of a contact field into an operational model that can be used to predict future patterns of diffusion? Hägerstrand considered the problem in his early research, and he formulated various models to simulate diffusion processes. Figure 13-8 illustrates how he used probabilities of contact to determine a *mean information field* (MIF), that is, an area, or "field," in which contacts could occur. Superimposing of the circular field shown in cross section in Figure 13-8(a), on a square grid of 25 cells, enabled him to assign each cell a probability of being contacted. As Figure 13-8(b) indicates, the probability (P)

of contact for the central cells is very high, in fact, over 40 percent (P = 0.4432). For the corner cells at the greatest distance from the center, the probability of contact is less than 1 percent (P = 0.0096).

To make the grid operational we add together the probabilities assigned to the MIF cells. Thus, the upper left cell is assigned the first 96 digits within the range 0 to 95; the next cell in the top row has a higher probability of contact (P = 0.0140) and is assigned the next 140 digits within the range 96 to 235, and so on. Continuing the process gives the last cell the digits 9903–9999, to make a total of 10,000 for the complete MIF [Figure 13-8(c)]. As we shall see shortly, these numbers are important in "steering" messages through our simple distribution of population.

Rules of the Hägerstrand Model   We can present the basic structure of the Hägerstrand simulation model in terms of formal rules. The rules given here refer only to the simplest version. They can be relaxed to allow modifications and improvements.

1. We assume that the area over which the diffusion takes place consists of a uniform plain divided into a regular set of cells with an even population distribution of one person per cell.
2. Time intervals are discrete units of equal duration (with the origin of the diffusion set at time t_0). Each interval is termed a generation.
3. Cells with a message (termed "sources" or "transmitters") are specified or "seeded" for time t_0. For instance, a single cell may be given the original message. This provides the starting conditions for the diffusion.
4. Source cells transmit information once in each discrete time period.
5. Transmission is by contact between two cells only; no general or mass media diffusion is considered.
6. The probability of other cells receiving the information from a source cell is related to the distance between them.
7. Adoption takes place after a single message has been received. A cell receives a message in time generation t_x from the source cells and, in line with rule 4, transmits the message from time t_{x+1} onward.
8. Messages received by cells that have already adopted the item are considered redundant and have no effect on the situation.
9. Messages received by cells outside the boundaries of the study area are considered lost and have no effect on the situation.
10. In each time interval a mean information field (MIF) is centered over *each* source cell in turn.
11. The location of a cell within the MIF to which a message will be transmitted by the source cell is determined randomly, or by chance.
12. Diffusion can be terminated at any stage. However, once each cell within the boundaries of the study area has received the message, there will be no further change in the situation and the diffusion process will be complete.

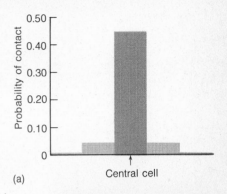

(a)

(b)   Mean information

0.0096	0.0140	0.0168	0.0140	0.0096
0.0140	0.0301	0.0547	0.0301	0.0140
0.0168	0.0547	0.4432	0.0547	0.0168
0.0140	0.0301	0.0547	0.0301	0.0140
0.0096	0.0140	0.0168	0.0140	0.0096

0–95	96–235	236–403	404–543	544–639
640–779	780–1080	1081–1627	1628–1928	1929–2068
2069–2236	2237–2783	2784–7215	7216–7762	7763–7930
7931–8070	8071–8371	8372–8918	8919–9219	9220–9359
9360–9455	9456–9595	9596–9763	9764–9903	9904–9999

(c)

Figure 13-8.   Mean information fields in the Hägerstrand model of diffusion. The probability of contact with distance (a) is superimposed on a square 25-cell grid (b). The probabilities for all the cells in the grid are summed to give (c) a mean information field.

Figure 13-9.   Simulated diffusion.
The opening stages of the Hägerstrand model are illustrated by a mean information field. The numbers refer to contacts determined by drawing random numbers.. When contacts are *internal* (i.e., with the cell on which the MIF is centered), a circle is added in that cell.

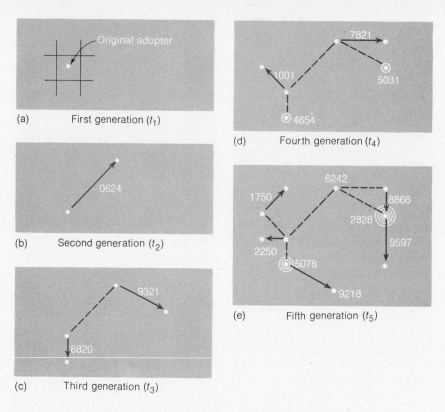

(a)   First generation (t_1)

(b)   Second generation (t_2)

(c)   Third generation (t_3)

(d)   Fourth generation (t_4)

(e)   Fifth generation (t_5)

Using the Model   The key to putting this model into use is in rules 10 and 11. In each time interval the MIF is placed over *each* source cell so that the center cell of the grid corresponds with the source cell. A random number between 0000 and 9999 is drawn and used to direct the message, following rules 4 to 6. *Random numbers* are sets of numbers drawn purely by "chance" (e.g., by rolling a dice). They can be taken from published tables of random numbers, or generated on a computer, or, for small problems, drawn from a hat. We show this process in Figure 13-9. In the first generation the number 0624 is drawn from a table of random numbers and a message is passed to a cell that lies to the northeast of the original adopter, located in the source cell.

Figure 13-9 goes on to present the first few stages in the diffusion process. In each generation the MIF is recentered in turn over each cell that has the message. Because the Hägerstrand model uses a random mechanism, each experiment or trial produces a slightly different geographic pattern. If we ran thousands of such trials (using a computer), we would find that the sum of all the different results matched the probability distribution in the original MIF; that is, we should arrive back at our starting distribution. In order to reap the benefits of the model, it should be applied not to simple, predictable diffusions whose end result is known, but to complicated, unpredictable diffusions whose end result is in doubt.

If we think about the rules of the basic Hägerstrand model, we can see that they represent a considerable simplification of reality. Areas where diffusions take place are not uniform plains with evenly spread populations; innovations are not adopted the instant a message about them is received; information is not passed solely by contacts between pairs of people; and so on. Hägerstrand was fully aware of these complications, and he used his basic model to provide a logical framework for more realistic versions of the diffusion process. Hägerstrand's later variants of his model contain significant modifications. Others have been added by American researchers.

Some of the modifications introduced into the original model are minor technical improvements in its structure. For instance, regular square cells can be reshaped to fit other regular divisions (hexagonal units have been adopted in some versions), but irregular areas present more of a problem. Adaptation of the contagious diffusion model to cascade processes involves substituting a hierarchy of settlements for an isotropic plain. Probabilities must be assigned to the links between the settlements rather than to cells.

Abandoning the Uniform Plain Some of the modifications can be simply made. Let us relax rule 1 and assume that the distribution of population is not uniform and that there are a variable number of people within each cell. The probability of contact is then a function both of the distance between the source and destination cells and of the number of people in each cell. Thus, we can multiply the population in each cell by its original contact probability to find a joint product. The ratio between the joint product of any cell and the sum of the joint products for all 25 cells in the MIF gives us a new contact probability based on both population and distance. (See the marginal discussion of weighting contact probabilities.) We have to buy this added realism at the cost of some tedious arithmetic, particularly because the new probabilities must be recomputed each time we move the MIF grid. On the other hand, such computations can be readily done by a computer.

Although this procedure may seem complicated, we are simply putting *back* into the model the geographic reality which the original assumption of a uniform plain took out. If we were concerned with understanding the spread of cultural artifact (e.g., TV ownership) through a region, one of our first concerns would be the distribution of population and thus of potential purchasers for the product.

Varying the Resistance to Innovations In discussing the impact of religious beliefs in Chapter 11 we noted their importance in insulating a group against

13-3
Modifying The Hägerstrand Model

WEIGHTING CONTACT PROBABILITIES

If we assume that the probability of contact in a diffusion model is a function of both the distance between the source and destination cells and the number of people in each cell, then we can estimate that

$$C_i'' = \frac{C_i' N_i}{\sum_{i=1}^{25} C_i' N_i}$$

where C_i'' = the joint probability of contact with the ith cell based on the MIF and population,

C_i' = the original probability of contact with the ith cell based on the 25-cell MIF,

N_i = the number of people in the ith cell, and

$\sum_{i=1}^{25}$ = the summation of all $C'N$ values for the 25 cells within the MIF, including the ith cell.

These revised values for the probability of contact (C'') must be recomputed each time the MIF grid is moved to allow for spatial variations in the population.

change. One of the examples we used was the persistence of some seventeenth-century cultural traits in the present-day Amish communities in the United States. Can the model be adapted to incorporate factors of this kind?

Well, it can be if we relax another of the original rules, in this case rule 7. The statement that adoption of an item takes place as soon as a message is received by the destination cell is an oversimplification. From research on agricultural innovations we know that there is generally a small group of people who are "early innovators" and another small group of "laggards"; the majority of a population adopts an innovation after the early innovators and before the laggards. In the case of spatial diffusions of population over a territory, this implies that settlements are established sporadically at first. The sporadic phase of settlement is followed by a period when everyone gets on the bandwagon, and eventually by a period of restricted settlement as the number of suitable unsettled locations in the territory diminishes. In the case of spatial diffusions of an innovation throughout a population, there are regional variations in the time of acceptance of the new item or way of doing things. For example, Figure 13-10 shows extreme regional variations in the time of adoption of tuberculin-tested (TT) milk by farmers in the counties of England and Wales. In some southern counties, 50 percent of the milk sold off farms in 1950 was TT milk; the proportion of TT milk in the milk sold in the far southwest had reached this level eight years before. Another example of varying resistance to change is illustrated in Figure 13-11.

We can approximate the symmetric course of the diffusion process by S-shaped curves. (See the discussion on p. 310 of innovations and logistic curves.) Standardized *resistance curves* of this type were used by Hägerstrand to take into account resistance to innovations. After one message, the probability of acceptance was very low (0.0067); after two messages, it rose to nearly one-third (0.300); and after three, to nearly three-quarters (0.700). From then on the rate of acceptance fell again. The probability of acceptance

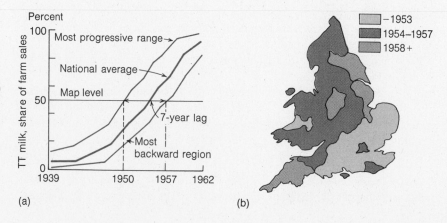

Figure 13-10.   Resistance to change. Regional variations in adoption rates are illustrated by (a) the diffusion of tuberculin-tested (TT) milk on farms in England and Wales. The map (b) shows the year by which each county had achieved 50 percent TT milk production. [Data from Milk Marketing Board. After G. E. Jones, *Journal of Agricultural Economics* **15** (1963), pp. 489–490, Figs. 6, 7A.]

(a)

(b)

Figure 13-11. Social factors in resistance to innovations.
One practical outcome of geographers' research on spatial diffusion is a concern with ways of speeding up desirable innovations (family-planning clinics) and slowing down or halting undesirable innovations (disease epidemics). The spread of family planning in Benares, India, illustrates the former concern. [Photographs by (a) Michaud and (b) Paolo Koch, Rapho Guillumette.]

rose slowly after four messages (to 0.933), and still more slowly thereafter. After five messages, even the worst laggards had accepted the item, and the rate of acceptance was 1.000. Like changing probabilities of contact, varying rates of acceptance (or resistance) can be readily incorporated into a computer simulation of a diffusion process. And so, if we have a community highly

INNOVATIONS AND LOGISTIC CURVES

The resistance of a population to adopting an innovation usually follows an S-shaped curve. [See diagram (a).]

This curve can be approximated by a *logistic distribution* given by the equation

$$P = \frac{u}{1 + e^{(a - bt)}}$$

where P = the proportion of the population adopting an innovation,

u = the upper limit of the proportion of adopters,

t = time,

a = the value of P when t is zero,

b = a constant determining the rate at which P increases with t, and

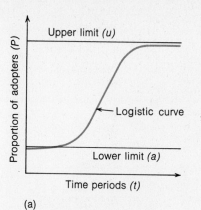

(a)

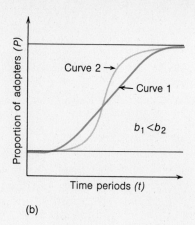

(b)

e = the base (2.718) of the natural system of logarithms

Thus, with u = 90 percent, a = 5.0, and b = 1.0, the proportion of adopters will be 4 percent at t = 2, 28 percent at t = 4, 66 percent at t = 6, 85 percent at t = 8, and so on. As diagram (b) shows, the constant b has a critical effect on the form of the innovation curve. Low b values de-

scribe smooth innovation curves (curve 1), whereas higher b values describe rates of acceptance that have a slow initial buildup, explode rapidly in a middle period, and enter a final period of slow consolidation (curve 2). [See P. R. Gould, *Spatial Diffusion* (American Association of Geographers, Commission on College Geography, Resource Paper 4, Washington, D.C., 1969).]

---

resistant to change, like the Amish, we can increase the number of messages sent to any appropriate large number. If a community were wholly resistant to change, the number of messages needed would be infinite!

Adding Boundaries and Barriers   In the original model, messages moving outside the boundaries of the study area were considered lost and had no effect on the situation (rule 9). In later models a boundary zone over half the width of the MIF grid was created so diffusion could proceed by way of these external source ·cells. More important modifications were involved in the introduction of internal barriers that act as a drag on the diffusion process. Like the other modifications we have discussed, such barriers allow observed variations in both the natural environment and the cultural mosaic to be incorporated into the model.

At the University of Michigan, Richard Yuill programmed the Häger-

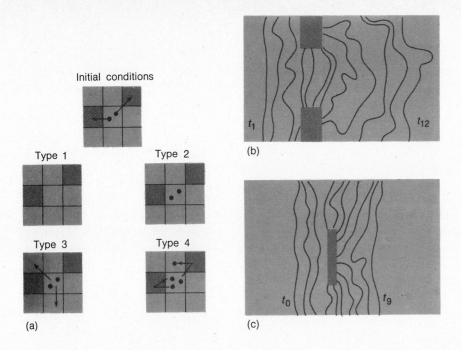

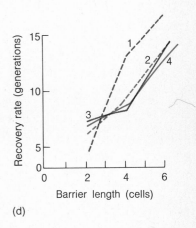

(a)

(b)

(c)

(d)

strand model to stimulate the effect of four types of barriers on the diffusion of information through a matrix of 540 cells within a 9-cell MIF. Figure 13-12(a) shows the 9-cell grid with the barrier cells indicated. Four types of barrier cells that provide a decreasing amount of drag are considered: a *superabsorbing barrier* that absorbs the message but destroys the transmitters; an *absorbing barrier* that absorbs the message but does not affect the transmitters; a *reflecting barrier* that does not absorb the message but allows the transmitter to transmit a new message in the same time period (see the arrows in the figure); a *direct reflecting barrier* that does not absorb the message but deflects it to the available cell nearest to the transmitter.

Each situation was programmed separately and the results plotted. Figure 13-12(b) shows the advance of a linear diffusion wave through an opening in a barrier. The time taken for the original line of the wave to reform determines the *recovery rate*. Varying types of barriers and gaps of varying widths were investigated. In the example shown, the line of the wavefront has recovered by about the eleventh generation (time t_{11}). Another type of barrier is presented in Figure 13-12(c). Here the diffusion wave passes around the barrier and reforms after about nine generations. The recovery rate of a wavefront is directly related to both the type and the length of the barrier it encounters; the curve for a superabsorbing barrier is quite different from the curves for the other three types of barriers. [See Figure 13-12(c).]

Yuill's work expands and develops modifications already begun by Häger-

Figure 13-12. Barriers and diffusion waves.

Four types of barrier cells (a) are used in this simulation model. In (b) diffusion waves pass through an opening in a bar barrier. In (c) diffusion waves pass around a bar barrier. The graph (d) shows the recovery rates around bar barriers constructed from the four different types of barrier cells. "Recovery rate" is the time taken for the straight line of the diffusion wave front to re-form. [From R. S. Yuill, *Mich. Inst. Univ. Comm. Math. Geogr. Disc. Papers,* No. 5 (1965), pp. 19, 25, 29.]

strand. The original model postulated what was, in effect, a row of absorbing cells around the preiphery of the study area. The internal barriers in Hägerstrand's model were represented by the lines between the cells. Such barriers could be adjusted to be absolutely effective (i.e., to allow no messages to get through) or 50 percent effective (i.e., to let one out of two messages cross the barrier). With such *permeable barriers* we can replicate a variety of environments. Thus, the original assumption of isotropic movement can be brought into line with known patterns. In other words, we can build *low-resistance corridors* into the model to allow faster diffusion in certain directions, and we can also build into the model *high-resistance buffers* to slow down diffusion across barriers.

To sum up, the basic Hägerstrand model can be easily modified to make it fit more closely to the realities of the geographic world. To the changes in population density and barriers discussed here, we can add such further refinements as variations in the "infectiousness" of the element being diffused.

13-4
Regional Diffusion Studies

Many of the applications of Hägerstrand's model stem from his own pioneering work in Sweden. Here we review two of the applications of the model to regions with contrasting environmental conditions. We look first at the spread of cultural attitudes (farmers' attitudes toward farm subsidies) and then at the spread of a cultural group (the Polynesians).

Farm Subsidies in Central Sweden   In the late 1920s the Swedish government introduced a scheme to persuade farmers to forego their traditional practice of allowing cattle to graze the open woodlands in summer. Grazing was proving to be a problem because it restricted the growth of young trees. To encourage fencing and improvements in pastureland, the government offered a subsidy. Figure 13-13 presents computer maps of the central part of Sweden and indicates areas where farmers accepted the subsidy during the years 1930 to 1932.

The maps indicate that in 1930 a few farmers accepted the subsidy in the western part of the region but there were scarcely any takers in the east. The next two years brought a rapid increase in the number of acceptors in the west but little change in the east. The sequence of maps suggests a spatial diffusion process in which distance is an important factor. To stimulate this process, Hägerstrand built a model using the 1928–1929 distribution of adopters as a starting point. The basic model was modified in two ways. First, the potential number of adopters (i.e., farmers) in each cell was added; second, barriers that were 100-percent and 50-percent permeable were added to simulate the long

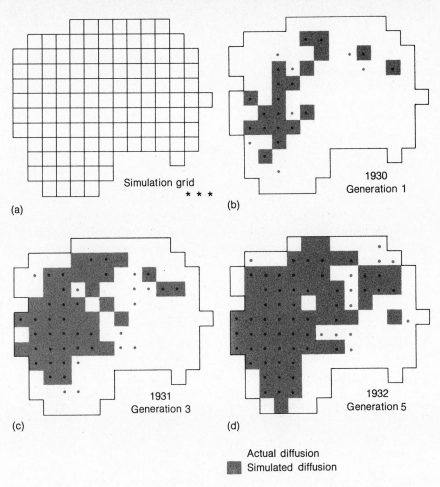

Figure 13-13. Simulating diffusion on a grid.
Here we see Hägerstrand's simulated diffusion and the actual diffusion of a decision by farmers in central Sweden to accept a farm subsidy. For simulation purposes the test area was approximated by a regular grid (a). The data for the three trial years shown in (b), (c), and (d) are part of more extended study. [After T. Hägerstrand, *Northw. Univ. Stud. Geogr.*, No. 13 (1967), pp. 17, 23, Figs. 6, 9.]

north–south lakes that lie across the region. Figure 13-13 compares the simulated diffusion process with the actual one. Because of the random element in the model, we should not expect the simulated pattern to match exactly the actual pattern. But the degree of matching is close, and both the general form of the expansion process and the location of the major clusters of adopters in the western areas are correct.

Kon-Tiki **Voyages in the Pacific**   When Thor Heyerdahl undertook his now historic voyage on the raft *Kon-Tiki* from the coast of Peru to the Tuamotu Islands, he was carrying out a single experiment: He was testing whether it was possible to cross the Pacific in such a craft. To analyze thoroughly the probabilities of contact between South America and different island groups

Figure 13-14.   Diffusion mechanisms.
This eastern Polynesian double canoe was
used in interisland voyaging. The settlement
of Polynesia by canoes of this kind is a
question hotly debated by anthropologists
and geographers. Where did the
Polynesians come from and how did they
move from island to island? Figure 13-15
shows how computer models of diffusion
may throw some light on both questions.
[Drawn by John Webber, artist on Captain
James Cook's third voyage. From A. Sharp,
Ancient Voyagers in the Pacific (Penguin,
London, 1957), Fig. 8.]

by this means would require too many voyages to be feasible. When direct
experiments prove too costly, too risky, or unlikely for other reasons, we may
be able to turn to computer simulation for answers to our questions. For
example, nearly 30 years ago, in Project Manhattan (the name for the atomic
bomb project), the atomic radiation from bombs was simulated mathematically. Heyerdahl's trans-Pacific migration exemplifies a spatial simulation
that can be approached by means of the Hägerstrand model.

The central issue to decide is how the Polynesians came to discover and
settle on the islands of the central Pacific. This question has recently attracted considerable attention from anthropologists, navigators, and geographers; but the sparsity of evidence has led to a clash between two schools of
thought. The first holds that the colonization process involved intentional
two-way voyages and hence a high degree of navigational skill on the part of
the Polynesians. (See Figure 13-14.) The alternative view was that colonization was largely accidental, by travelers drifting off course.

To test the probability of interisland contact as a result of accidental
drifting about, a group of investigators at London University (Levison, Ward,

and Webb) constructed a computer simulation model of the drift process. The stages in their computer program are shown in Figure 13-15. There are four main elements in the model: (1) the relative probabilities of wind strength and direction for each month, and of current strength and direction for each 5° square of latitude and longitude in the Pacific Ocean study area; (2) the positions of all the islands and land masses in the study area, together with their sighting radius; (3) the estimated distances that would be covered by ships given various combinations of wind and current strength; and (4) the relative probabilities of survival of ships during certain periods at sea. Sighting radius (the distance out to sea from which islands can be seen) was built into the model on the assumption that, once land had been sighted, a landfall could be made. During each daily cycle voyages are started from given hearth areas like the coast of Peru, and a simulated course is followed until it ends in either a landfall or the death of the voyagers. By simulating hundreds of voyages from each starting point, we can map the relative probability of contact with different island groups as a potential contact field.

The simulation program has already been run for various Pacific Island groups and for locations on the coasts of South America and New Zealand. Preliminary results indicate that the probabilities on interisland links from the accidental drift of ships differ from one area to another. Wind and current patterns create environmental boundaries which make drifting in certain directions highly unlikely. Some of these boundaries coincide with long-standing anthropological breaks in the geographic pattern of ethnicity and culture like that separating the Micronesian people of the Gilbert Islands and the Polynesian inhabitants of the Ellice Islands. Other low probabilities of contact coincide with important linguistic boundaries.

The computer model for this research simulates activities that cannot be observed at first hand and are too complex to be simulated by manual calculations. It confirms that certain existing population distributions are possible purely as a result of voyagers drifting off course. However, there remain certain hard cases, notably the Hawaiian Islands, whose settlement still remains a mystery.

We began this chapter with El Tor and skateboards and ended it with *Kon-Tiki*. Between, we have seen how the general notions of spatial diffusion can be simulated by probabilistic models—most of them developed from the work of Swedish geographers. These models help to throw some light on the process of diffusion by which past cultural changes have occurred. Modern mass communication media have made the power and significance of the forces of change we have studied immense. The TV antenna signifies the trend toward a global village where change no longer requires mass movements of people but spreads far more rapidly through the subtle osmosis of messages carried by the mass media. The long-term implications of the waves of innovation generated and reinforced by mass media for the persistence of the cultural variety of human beings on the planet Earth may be immense.

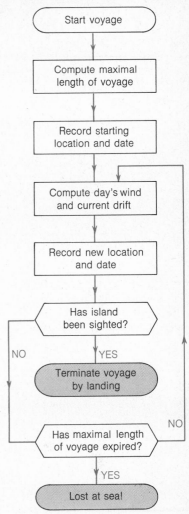

Figure 13-15. Polynesian voyaging. This flow chart shows the main elements in a computer program developed to simulate Pacific "drift" voyages. [After R. G. Ward, *et al.*, *Information Processing* **68** (1969), p. 1521, Fig. 1. Reproduced with permisision of North Holland Publ. Co., Amsterdam.]

Summary

1. The spread of culture between regions is studied by geographers as a diffusion process. Two types of diffusion are commonly recognized: contagious diffusion and hierarchial diffusion.

2. By studying a series of innovation profiles Häger-strand developed the concept of innovation waves, each containing a primary stage, diffusion stage, condensing stage, and saturation stage. Fitting mathematical trend surfaces to diffusion data shows that they may be regarded as an innovation wave through time and over space. Mathematical smoothing of wave forms has enabled geographers to identify diffusion centers and repeating patterns.

3. Hägerstrand's simplest model for simulating spatial diffusion is one in which the probability of contact between two persons, one being a sender, declines with distance exponentially. In this model diffusion processes may be simulated by specifying the probability of contact by use of a mean information field (MIF) in which random numbers are used to direct

messages. This basic simulation model assumes standard but unreal conditions; real world conditions require modifications to fit more complex variables. Specific examples indicate that resistance to innovation may be shown as an S-shaped resistance curve. Such curves may be represented mathematically as logistic curves.

4. Ways in which the basic Hägerstrand model may be relined are illustrated by the work of Yuill, who programmed four types of barriers to the diffusion of information. These are: superabsorbing barriers, absorbing barriers, reflecting barriers, and direct reflecting barriers. Those which eventually allow passage of information through are termed permeable barriers.

5. A number of empirical studies of regional diffusion in contrasting environmental situations indicate the utility of simulation models in studying cultural change. Typical of such studies is Levison, Ward and Webb's use of a modified Hägerstrand model to test alternative hypotheses about trans-Pacific migration.

Reflections

1. Gather data for your local area on (a) the location and (b) the date of foundation of any one sort of public institution (churches, banks, schools, colleges, and so on). Map the data and try to identify the kind of spatial diffusion processes that appear to be operating. Do contagious or hierarchic processes seem to be more important?

2. Imagine you are opening a new chain of motels or hamburger restaurants (e.g., Holiday Inns or McDonalds) in your state. Where would you try to locate the first five establishments? Why would you pick these locations? Compare your results with those of others in your class and identify any common locations you all wish (a) to adopt or (b) to avoid.

3. List the factors which affect the spread of family-planning information in a developing country like India. How many of these factors could you incorporate into the Hägerstrand model innovation diffusion?

4. Check how information spreads in a small group by planting a rumor (e.g., that your instructor has just become the father of twins) with *one* other member of the class. At the

beginning of the next class check (a) who now knows, (b) from whom the information was obtained, and (c) when the "telling" took place. Try to construct a tree like that in Figure 13-3, showing the way in which the rumor spread through the class.

5. Look carefully at the first four phases of Figure 13-9. What would the pattern of diffusion have looked like if the first seven random numbers had been 1920, 8520, 1567, 7803, 3223, 5059, and 2483?

6. Trace the loops in the simplified flow chart of Polynesian voyages in Figure 13-15. How accurate is such a model likely to be? How might this type of model be used to trace other cultural diffusion processes?

7. Review your understanding of the following concepts:

expansion diffusion	innovation profiles
relocation diffusion	contact fields
contagious diffusion	mean information fields (MIF)
hierarchic diffusion	barrier effects
random numbers	simulation models
adoption curves	

One Step Further ...

An excellent introductory review of theories of spatial diffusion and their use by geographers is given in

Gould, P., *Spatial Diffusion* (Association of American Geographers, Washington, D.C., 1969).

while a more mathematical approach is followed in

Haggett, P., A.D. Cliff, and A.E. Frey *Locational Analysis in Human Geography*, 2nd ed. (Arnold, London, 1977), Chap. 7.

The classic Swedish work on diffusion is by Torsten Hägerstrand. This has now been translated into English as

Hägerstrand, T., *Innovation Diffusion as a Spatial Process* (University of Chicago Press, Chicago, 1968).

Examples of more recent work on diffusion in the United States are

Bowden, L. W., *Diffusion of the Decision to Irrigate* (University of Chicago, Department of Geography, Research paper 97, Chicago, 1965) and

Berry, B. J., and F. Horton, *Geographic Perspectives on Urban Systems* (Prentice-Hall, Englewood Cliffs, N.J., 1970), pp. 419–434.

The "Polynesian drift" model discussed in the chapter and the application of diffusion models to the reconstruction of past cultural distributions is discussed in

Levison, M., R. G. Ward, and J. W. Webb, *The Settlement of Polynesia: A Computer Simulation* (University of Minnesota Press, Minneapolis, 1973), Chaps. 1 and 5.

A comprehensive review of diffusion theory, including its sociological and biophysical applications, is given in

Brown, L. A., *Diffusion Processes and Location* (Regional Science Research Institute, Philadelphia, Bibliography Series 4, 1968).

Although research in spatial diffusion is reported in the standard American geographic journals, it is worth paying particular attention to Swedish serials. The Lund Publications in Geography *(an occasional publication)* and Geografiska Annaler, Series B *(a quarterly)* are of special significance.

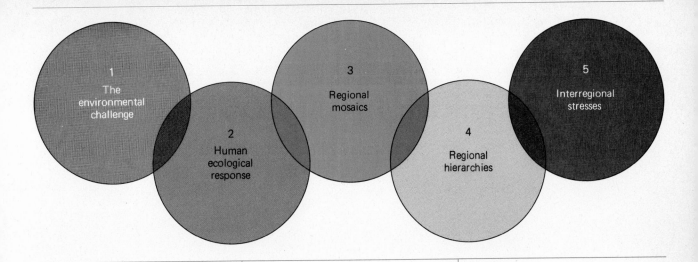

Part Four of the book turns from the study of regional contrasts—the mosaic of different cultural regions—to similarities between regions. We begin with the most powerful common force shaping regional structures in *Urbanization* (Chapter 14). We see how geographers are attempting to unravel the forces that cause more and more of the world's population to cluster into large cities. We look at the past and future trends in city growth. We find that cities and other population clusters grow not in a random fashion, but in a regular and rather orderly way. In *City Chains and Hierarchies* (Chapter 15) this delicate structure is dissected, and the beautiful symmetry and balance that underlies it

is exposed. People are shown to be a surprisingly orderly species, sharing with other populations an almost feudal structure, ranging from world city down to local village and farm. *Worlds Within the City* (Chapter 16) changes the focus to "close up" and looks at the geography of the city itself. The last two chapters (17 and 18) look to the *Worlds Beyond the City*. Cities are found to be intimately linked to the fortunes of surrounding rural areas, and the two together are found to form integrated city-regions. The mechanisms of this linkage and the way in which agricultural and industrial zones develop around cities is explored at a series of spatial levels. Finally, we examine the flows within a region and the delicate filigree of networks that binds city and region together into an interdependent whole.

part four
Regional Hierarchies

It isn't size that counts so much as the way things are arranged.

E. F. FORSTER
Howards End (1910)

chapter 14
Urbanization

If you had to think of a definition of a city, what would it be? Let us begin with a very simple one. "A *city* is a large number of people living together at very high densities in a compact swarm." Although there are many other significant aspects of cities, let us begin by just concentrating on density. Perhaps the easiest way to visualize what our definition describes is to think of a major rock or pop festival. During the summer weekends these may draw crowds of 100,000 or more to green fields around North America and Western Europe. Although newspapers and promoters vary in their estimates of the total attending, it is possible to utilize aerial photographs to measure the densities per square kilometer of these Woodstock-style events. If we put the densities, conservatively, at about 150,000 per square kilometer, then we need only some routine geometry (and a vivid imagination) to compute that, at these densities, the total world population could be corralled in a ring with a radius of only 96 km (59 mi). If we

packed people together at the densities encountered on the New York subway, we could get everyone into a ring with a radius of 12 km (7.5 mi).

The notion of such a seething swarm of humanity might repel you, and the ways in which such a swarm would be fed and watered, let alone sanitized, are wholly unknown. Actually, the world's biggest high-density clusters like Tokyo and New York are several orders of magnitude smaller than the hypothetical world city we have just described. Even so, these *megacities* of around 10 million or so people demand the most complex of life-support and communication systems and pose some of our largest social- and economic-control problems.

In this chapter and the four that follow we shall be looking at the city as a dominating force in the organization of the human population. Geographers must try to unravel the forces that lead to this dominance. Why do populations cluster into cities, and what keeps them from clustering even more strongly? What are the trends over time in this regard, and how urbanized is the world becoming? We turn, in the second half of the chapter, from questions of urbanization at the global level to an individual country. How did the United States get its urban pattern? What are the current trends—toward more urbanization, or less? We can only sketch in the most basic elements of the

answers to these questions in this book. For those of you who go on to more advanced geography courses, urban geography will form one of the most important and interesting areas of further study.

14-1
World Urbanization

Cities are the world's most crowded places. In New York City, about 55,000 people live on each square kilometer; in Montreal and Moscow, the corresponding population densities are about 52,000 and 49,000. These densities are several orders of magnitude higher than the overall densities for the countries that contain these cities. The United States has, on the average, less than 25 people living on each square kilometer. For Canada and Soviet Russia the national averages are lower still—about 2 and 11 persons per square kilometer, respectively.

Trends in Urbanization   Whatever the precise yardstick we use to measure a city, or urban place, the measurements show that the proportion of the world's urban dwellers is on the increase. If we use a threshold population of 20,000 for our definition of an urban place, then in 1800 only about one in forty (2½ percent) of the world's people were urbanites. By 1970 the proportion had jumped to about one in four (25 percent), and it is expected to reach one in two (50 percent) by the year 2000.

Trends in the rate of urbanization vary from country to country. For example, the United States is well ahead of the world as a whole. In 1800, about 5 percent of its population were classifiable as urban (using the same 20,000 threshold to define an urban place). By 1970 the proportion of urban dwellers in the population had risen to 70 percent, and by the end of the century it is expected to reach about 80 percent.

Can we detect any general pattern in the urbanization rates for different countries? Figure 14-1 shows a typical curve for a western country. This S-shaped *urbanization curve* has the same kind of logistic form which we have observed in other curves earlier in this book. (See Section 13-3.) Slow rates of growth in the early part of the nineteenth century are followed by sharp rises in the growth rate in the second half of the century and a progressive slowing down thereafter. During the most rapid period of growth, the key factor in population change was migration from rural to urban areas. Not only were birth rates lower in the cities, but the higher risk of epidemics and degenerative diseases there meant that death rates were higher too. Even when the figures are modified to take differences in the age structure of the populations in the two areas into account, London death rates in 1900 were one-third higher than rates in the surrounding rural counties.

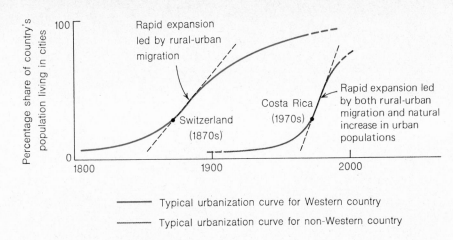

Percentage share of country's population living in cities

100

Rapid expansion led by rural-urban migration

Switzerland (1870s)

Costa Rica (1970s)

Rapid expansion led by both rural-urban migration and natural increase in urban populations

0

1800   1900   2000

——— Typical urbanization curve for Western country

——— Typical urbanization curve for non-Western country

Figure 14-1.   Urbanization curves over time.

On the left in the graph is an idealized S-curve for urbanization in a western country. Geographers are uncertain how reliable a guide this is for projecting the growth of cities in developing countries today, where natural population increase plays a more important role.

It would be tempting to regard the S-shaped western curve as a model for urbanization rates in the developing countries of the world. As Figure 14-1 shows, however, these countries do differ from the already-industrialized countries in two important respects. First, the process of industrialization not only started later, but is proceeding much more rapidly in the developing countries. The figure for the average annual gain in urban population for 34 countries in Africa, Asia, and Latin America over the last two decades is around 4½ percent. By contrast, nine European countries, during *their* period of *fastest* growth (for most, the latter half of the nineteenth century), had average gains just above 2 percent. Curiously, the United States, Canada, and Australia, which were hit by huge waves of European immigration, had urbanization rates closer to those of today's developing countries.

There is, however, a second difference between the two urbanization curves in Figure 14-1. Nineteenth-century urbanization was essentially "migration-led"; that is, the great proportion of the new urban dwellers were from rural areas. This is less true in today's developing countries. Although the popular images of the rural poor streaming into the shantytowns on the edges of Latin American or African cities are correct, the relative share of migration in the growth of urban populations is smaller in proportion to natural births. In Switzerland in the late nineteenth century about 70 percent of the urban growth was due to rural–urban migration. In Costa Rica, with a roughly similar proportion of urban dwellers in its population now (as compared to our Swiss example), such migration accounts directly for only about 20 percent of urban growth. In much of the developing world today the modern city is more healthy than were the cities of Europe and North America a century ago. The birth rates remain higher and the deaths rates lower. In short, many of the new urbanites are predominantly home-grown.

Changing Spatial Patterns   In the nineteenth century the growth of big cities was a feature of midlatitude countries. In 1900 the four leading cities

Figure 14-2. The world's fastest-growing cities.

The percentage figures indicate anticipated growth in the next decade. Note that all the cities lie in the less-developed countries. [Data from *Development Forum* February 1976, cited in L. Broom and P. Selznick, *Sociology,* 6th ed. (Harper & Row, New York, 1977), p. 487.]

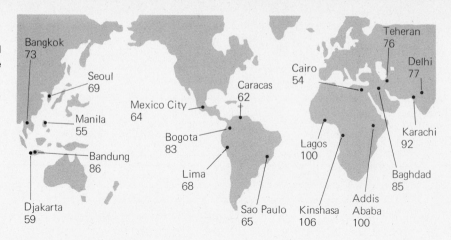

were London, Paris, New York, and Shanghai. By the early 1920s there were just 24 "millionaire" cities (i.e., cities with 1 million people or more) and only one of these lay within the tropics. In the present decade the number of millionaire cities has increased more than sixfold. One in ten of the world's people now lives in these huge cities, many of which are now located in the tropics.

One useful way of describing the geographical location of the millionaire cities is to look at their latitude. In 1920 the most poleward city was Russia's old capital Leningrad (latitude 60°N), and the most equatorward was Bombay (19°N) on the west coast of India. A half-century later, Leningrad remains the most poleward city but other millionaire cities extend right down to the equator: Singapore (1°N), with over 2 million people, is one of Asia's fastest-growing cities.

If we now calculate the *average* latitude (ignoring whether the city lies in the northern or southern hemisphere), the equatorward shift is clear. In the 1920s the average latitude for millionaire cities was between 44° and 45°, about the latitude of Minneapolis in the United States or Toronto in Canada. The average has now moved a further 10° equatorward, to the latitude of Los Angeles or Atlanta. Judging by current rates of growth (see Figure 14-2) the shift is likely to continue.

14-2
Dynamics of Urbanization

Why do cities grow? Can we identify the forces fueling the present rapid growth of world cities? If we leave aside growth by the natural increase of existing urban populations, we are left with the puzzle of why people increasingly concentrate in clusters so dense that the environment cannot support them harmoniously. Here we try and answer the question in two halves, looking first at why people leave the land ("push" factors) and then at why they cluster in cities ("pull" factors). Together these changes lead to major changes in the major job sectors of the whole economy.

Urbanization: Push Factors ("away from the land") For most of the world's history the human population was essentially rural. The "normal" unit of settlement was the small agricultural community, producing (at least in good seasons) the food needed to survive into the next year. As we saw in Chapter 7, over long periods of global history the number of people in such communities was fairly stable, linked to the carrying capacity of the community's land area. Now, one in ten of the world's population lives in cities of a million people or more, and the proportion of people on the land in rural communities is steadily falling.

We can see something of the reasons for leaving the land by considering the recent urban history of the United States. Today fewer than 1 in every 25 Americans lives on a farm; as recently as 1929 the number was 1 in 4! Over the last 50 years, as over the last 200, the rural areas have been net *exporters* of people. Higher birth rates in farm communities have produced a surplus of young people who sought jobs outside agriculture.

We can recognize three important factors in "pushing" populations off the land. First, if we disallow the period of the frontier expansion, then land is a fixed resource. As the number and size of farm families increased the extra hands proved valuable in clearing, weeding, and harvesting. Quickly, however, a point was reached where additional inputs of labor brought diminishing extra outputs from crops, i.e., in economic terms the law of diminishing returns had set in. If all the children had stayed down on the farm, there would have been greater crowding, with further reductions in the productivity of each man-hour. A cycle of crowding, farm subdivision, and rural poverty is a situation common in parts of rural Asia, Africa, or Latin America (see the discussion of shifting cultivation in Chapter 8). But in the United States the surplus of farm population was augmented by a second critical factor, technology. Mechanical progress in terms of the tractor, the combine harvester, or the cotton picker was supplemented by selective crop and livestock breeding and improved cultivation through rotational practices, irrigation, and fertilizers. As a result productivity on the farm has increased at a rate even higher than in manufacturing. The farm population needed to produce a given total of feed or fiber has been greatly reduced.

The long-run decline in agricultural population has been enhanced by a third factor, taste. As the living standards of the American population have risen in real terms, so the proportion of income spent on food products has declined. Investigations show that, internationally, once a certain level of living has been attained the proportion of additional income in the family budget spent on basic food declines. The parity ratio (the ratio of prices farmers get for their goods compared to prices they pay) has been generally moving against agriculture since at least 1900. Despite great cyclic fluctuations and occasional boom years, agriculture in many developed countries has become a problem for government, not because it produces so little but because it produces so much. Low farm prices have been countered with various forms of subsidies leading to curious anomalies like butter mountains or restrictions in farm acreage. Low prices are reflected in the fact that ten of the 25 million people in the United States who are below the poverty line live in rural areas—notably in Appalachia, the Deep South, and along the Mexican border. Canada faces a similar though less pronounced rural poverty problem in parts of the Maritime Provinces and Quebec.

To sum up: The factors pushing population *away* from the rural toward the urban areas appear to be fourfold: (a) a high rural population birthrate, (b) improved agricultural technology reducing the need for farm labor, (c) a swing away from foodstuffs toward other goods as income increases, and (d) relatively low and uncertain prices for farm goods. The relative strength of each factor will vary with different geographic situations: factor (a) is more important in developing countries, (d) more important in highly developed countries.

Urbanization: Pull Factors ("into the city")   Cities which grew just because life on the land was unattractive would resemble gigantic refugee camps. In fact, there are very positive factors pulling people into the cities.

Agglomeraiton Economics   The major benefits gained from high-density crowding are the so-called agglomeration economies. *Agglomeration economies* are the savings that can be made by serving an increasingly large market distributed over a small, compact geographic area. Economies of scale make production costs for each unit low, while the short distance separating buyer and seller in cities cuts back the costs of transporting goods. Clearly production efficiency is directly related to how much is produced. A large output commonly results in lower unit costs because machinery and plants are used more fully and labor can become more specialized. Research and development costs and fixed costs can be spread over more units of production.

Economist Adam Smith's dictum that "specialization depends on the size of the market" is clearly proven in the modern city. You need go no further than the Yellow Pages of New York's or London's telephone directories to be aware of the intricate degree of specialization the large metropolitan city permits.

Geographers have dissected the links between extreme specialization and extreme accessibility by examining the flows of information within a city. The persistence of highly specialized office complexes like New York's Wall Street and London's "City" district depends on face-to-face contacts as well as other forms of information flow and exchange. Each individual's activity is dependent on ready access to others, and the actions of the whole group form a complex, or knot, of specialized activities.

Urban Multipliers: Basic and Nonbasic   The export base of a city, like that of a country, is the activity that lets it sell goods or services or investments beyond its immediate boundaries. Although cities don't have precise boundaries like states, we can nonetheless think of them as having a balance of trade—importing and exporting. Of course, the distinction between the export sector and the rest of the activity of the city is not always clear-cut. Nonetheless, a useful distinction can be drawn between *basic* activities like manufacturing aircraft engines (to be "exported" to aircraft assembly plants in other cities) and those *nonbasic* activities like baking bread (to be consumed in the city itself). The *urban base ratio* is used by geographers to describe the ratio of jobs in "export" sectors of the city economy to the total population. If a city of 60,000 people has 10,000 jobs in its export sector, the urban base ratio is 1:6.

Let us look at the effects of an increase in the export base in a highly simplified situation. Figure 14-3 shows a small Minnesota city, whose export base is iron-ore processing. The ore is not consumed by the local community directly. But the inflow of funds from exporting the processed ore helps balance the "import" of food, gasoline, and so on, enabling the city to survive as an economic unit even though it is not self-sufficient. Let us suppose that a new market opens up, bringing an increase of 100 jobs in this basic export sector. What will be the effect on the local city?

A typical cycle of events is shown in Figure 14-4. First, as we have noted, the new market creates a number of entirely new jobs in the local area. If the average family size is four, we can say that every 100 jobs will lead to 400 more people in the household sector. Since these people will demand a set of service facilities—schools, churches, shops, hospitals—the households will, in turn, create a set of service jobs. As a rough estimate we adopt one service job for every ten members of the household sector. Moreover, the workers in the service jobs will have families and create a second, smaller cycle of more

Figure 14-3. Resource processing and urban growth.
The new town of Silver Bay (background) has grown up in this wilderness area of northern Minnesota in direct response to the location of an iron ore processing plant (foreground). Low-grade iron ore is hauled 70 km (43 mi) by railroad from the Mesabi Range to this coastal site on Lake Superior. After processing, the ore is shipped a further 1000 km (over 600 mi) to steel-making plants around the Great Lakes. [Photography courtesy of Reserve Mining Company.]

Figure 14-4. The impact of changes in the basic sector on urban growth. An increase in jobs in the basic sector of the urban economy has a multiplicative effect on other sectors within the city. In this highly simplified model every basic job creates an increase of four in the population of the household sector (i.e., the worker plus a hypothetical family). So the household multiplier (h) has a value of four. Every increase of ten people in the household sector creates one extra job in the service sector (for teachers, gas-station attendants, doctors, etc.). So the service multiplier (s) has a value of one-tenth. Each person in the service sector is also assumed to have a hypothetical family, which also demands services. And so on! The calculation of the total population is shown in (b). Since there is usually a lag between the establishment of jobs, the arrival of the workers' families and the provision of services, the actual distribution of jobs and households will be as in (c).

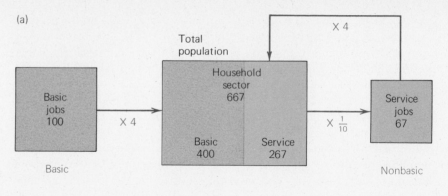

(a)

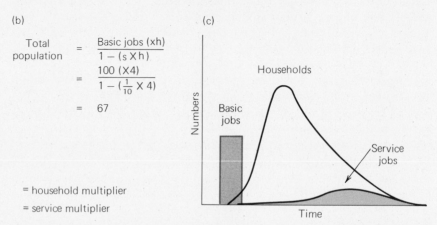

(b)

$$\text{Total population} = \frac{\text{Basic jobs } (xh)}{1 - (s \times h)}$$

$$= \frac{100 \, (\times 4)}{1 - (\frac{1}{10} \times 4)}$$

$$= 67$$

= household multiplier
= service multiplier

(c)

households and more service jobs. As Figure 14-4 indicates, these diminishing cycles can be repeated any number of times. If you follow the calculations through, you will find that the 100 new mining jobs eventually cause an increase of 667 people in the household sector, plus a further 67 jobs in the service sector.

This model of the domino effect of one activity on another was developed by an economist named I. S. Lowry and is generally termed a *Lowry model*. Two checks may be built into the model to modulate the cycles and thus get a "truer" picture. First, we can assume that land for residential development cannot be available in the same zone as the mine, and therefore housing must be located elsewhere. Second, we can stipulate a minimum size for service sectors. For instance, we can stipulate that a hospital may be created only when the population in the household sector exceeds an appropriate threshold level. Otherwise, the local population will have to go elsewhere for medical treatment.

Urbanization and Sector Changes The impact of the new jobs may not be confined to the household and service sectors of the economy. New jobs in mining may lead to associated industrial activities involving the refining, smelting, or processing of mineral ore, or the manufacture of equipment for the mining process. This will cause a proportional drop in jobs in the *primary sector* (a term used to describe jobs in agriculture, mining, fishing, etc.) and a relative increase in jobs in the manufacturing, or *secondary* sector. Further growth will lead to an expansion in the wide range of service jobs (in shops, education, hospitals, transport, etc.) in the *tertiary* sector. Large cities usually have a larger proportion of their work force in the tertiary sector. In the last half-century, the most rapidly growing job sector has been jobs in research and administration, the so-called *quaternary* sector. "Office" and "research lab" jobs account for an increasing share of employment opportunities in major metropolitan areas. One way of distinguishing among the four sectors is to take a given natural product and consider which jobs are related to it. For example, timber production gives jobs to a chainsaw gang (in the primary sector), a lathe operator in a furniture-manufacturing plant (in the secondary sector), a furniture-store operator (in the tertiary sector), and a researcher in wood technology (in the quaternary sector).

Figure 14-5 shows the job implications of urbanization at the national level. Sweden's excellent population records allow us to follow the trend from

(a) Sweden since 1750

(b) Sweden since 1850

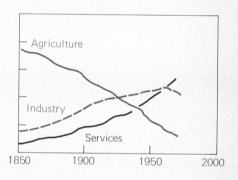

(c) General sector model

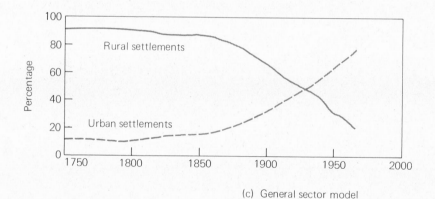

Figure 14-5. Sector model of economic growth.

The pattern of urban–rural settlement in Sweden (a) is compared with the changing balance of jobs in the three main sectors of economic activity (b) and with a general model of sector change (c). [Data from *The Biography of a People: Past and Future Population Change in Sweden.* (Royal Ministry for Foreign Affairs, Stockholm, 1975), Table A5.]

rural to urban settlements since 1750. Since 1850 we can also see the shift in jobs with agriculture declining and industry rising steadily until it reaches a peak in the 1960s. Since then it's started to decline and, as the graph shows, service employment has now taken over as the major job sector in Sweden.

Geographers see the Swedish changes as part of a *sector shift* model. Figure 14-5(c) shows the general form of the model as a series of curves, and you should take some minutes out to look carefully at this chart and see what it means. Note in particular the division into three stages termed *preindustrial*, *industrial*, and *postindustrial*. Most western countries would, on the basis of this graph, still be in the industrial stage while most Third World countries would be in the preindustrial stage.

But do countries necessarily have to go through these three stages? Canadian geographer Maurice Yeates thinks not. The evidence for Canada is that it has moved from the first stage straight through to the third. Primary activities (agriculture, mining, trapping, forestry, etc.) were very important in the first century of growth but industry has never been a dominant employer. Industrial employment reached its peak (around 23 percent of the workforce) in the 1960s and is now going down. But service employment has long been important. At the turn of the century it employed around 40 percent and its share is now approaching 70 percent. So Canada, unlike the United States, seems to have missed the industrial stage.

At the city level, whether or not the extraction of a given resource will trigger a chain of associated industrial growth depends on a great variety of factors. In many of the coal mining areas of northwest Europe that attracted large populations in the early nineteenth century, the major cities have continued to expand long after the original mining activity has ceased to be significant. The extraction of resources located in central, highly accessible zones (that may already have a dense population) is much more likely to trigger prolonged urban-industrial growth than the extraction of similar resources set in remote, peripheral locations.

14-3
Urban Implosion on the Large Scale

Geographers are especially interested in urbanization as a spatial process—a force shaping the distribution of the world's population map. Here we can see two powerful movements going in opposite directions. One is an *implosive* force bringing cities closer together; the other an *explosive* force pushing out the boundaries of the individual city. We shall look at each in turn.

Shrinking Travel Times Between Cities   The role of changing accessibility is an aspect of the urbanization process of great interest to geographers. For agglomeration economies to occur, markets have to be not only large but highly accessible. Over the last two centuries, in which time the average level of world urbanization has increased tenfold, overall travel times and transport costs have *fallen* in real terms by an even greater amount. New transport technology—whether it is a railroad, Telex, shuttle jet, or monorail—tends first to connect key cities, thereby increasing their locational advantage in

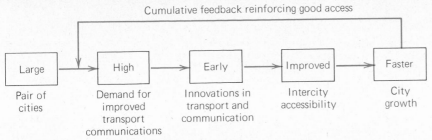

Figure 14-6. **Cumulative linkage advantages.**
The set of transport and communication advantages that help important cities retain their importance. This upward cycle is matched by a downward cycle of cumulative deprivation for unimportant pairs. (To find it, substitute the words "small," "low," "late," "retarded," and "slower" in the five boxes. Note that "retarded" may be relative rather than absolute, but the impact on locational advantage remains a negative one.

relation to important cities. (See Figure 14-6.) We can illustrate this "ratchet effect" by measuring changes in the average cost of movement between centers of different sizes. If we use time as a proxy for cost, between the years 1850 and 1900 the distance from New York to California was cut from 24 days (3 days by railroad plus 21 days by overland coach) to 4 days (by direct train). The next half-century reduced this distance to 8 hours by DC-6, and to 5 hours by jet. (See Figure 14-7.) Thus the relative cost of the trip (in hours) has been reduced by over one-half since the 1930s.

Yet these changes have not been uniform and the bigger cities gain more from improvements in services over time. As a result, the cost in time of traveling between them is reduced, and they move closer together in relative terms.

We can borrow a term from astronomer Fred Hoyle and describe the effects of this differential spatial shrinkage as an urban *implosion* —that is, the inverse of an explosion. The ways in which large cities converge on each other is shown by the diagram in Figure 14-8. Note that since large cities converge

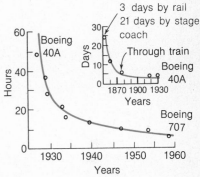

Figure 14-7. **Shrinking worlds.**
The graphs show changes in the average amount of time needed to go from New York City to Los Angeles since 1850, using the fastest means of public transport. [From R. E. G. Davies, *History of the World's Airlines* (Oxford University Press, London, 1964), p. 509, Fig. 91.]

Figure 14-8. **Urban implosions.**
Transport innovations are usually first introduced between pairs of large cities between which important flows of people and goods already occur. Such innovations make links between such pairs easier, setting off a cycle of increased contacts and flows. As Figure 14-9 shows, the net effect is to bring larger cities relatively closer to each other. Meanwhile smaller centers become relatively (and sometimes absolutely) more remote. This figure illustrates this process of "implosion." Note the convergent movement of the three largest cities (the size of the spheres is proportional to the size of the cities) in comparison with the divergent movement of the three smaller ones.

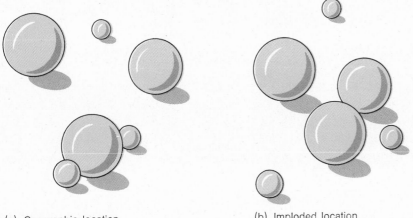

(a) Geographic location

(b) Imploded location

rapidly and the smaller cities converge less rapidly, the larger and smaller cities diverge in relative terms. The speed and self-reinforcing nature of this spatial implosion are principal factors in the current rapid growth of large cities. At Bristol University Philip Forer has traced the implosion of New Zealand cities over the last thirty years. Basing his research on changing air travel times, he was able to map New Zealand's changing structure; as larger cities were dragged closer together, smaller cities pushed outward into less central locations. For the South Pacific as a whole Forer's time maps show a curious pattern in which the largest Australian city (Sydney) is closer to the largest New Zealand city (Auckland) than are many small New Zealand cities (Figure 14-9.)

Figure 14-9.   Time maps of urban implosions.

(a) A conventional geographic map of New Zealand shows the location of major airports. (b) A time map shows the "distance" in minutes between the same urban airports. The map is based on travel times and service frequencies in 1970. Note the way in which the three leading cities—Auckland, Wellington, and Christchurch—are clustered near the center of the map. In contrast, the smaller city of Tauranga is displaced toward the periphery. (c) A time map shows the relations between the main New Zealand airports and other major urban airports in the South Pacific. It too is based on air travel times and service frequencies in 1970. Note how large numbers of high-speed flights between Sydney and the main New Zealand cities draw these locations together. These unfamiliar time maps are produced by a computer analysis of a mass of information on flight times and frequencies between all the pairs of locations shown on the map. [From P. Forer, *Changes in the Spatial Structure of the New Zealand Internal Airline Network,* doctoral dissertation, Bristol University.]

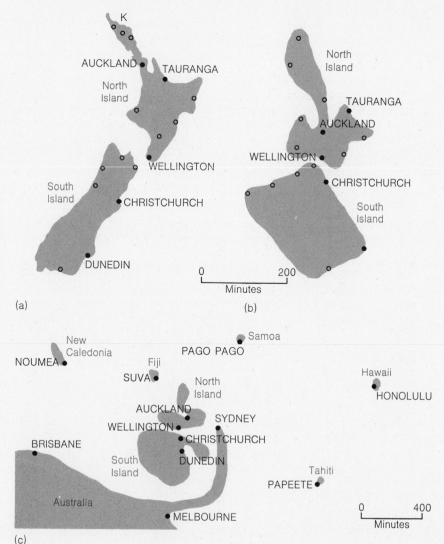

MAPPING CITIES IN TIME

How to plot maps showing the "correct" position of places in terms of time-space remains an unresolved problem in geography. We can show this by taking a simple example. Suppose we have four towns (p, q, r, and s) which are separated by travel times in hours. From p to q is 4 hours, from p to r is 1 hour, and so on as below (assuming travel time is the same in both directions):

$p \leftrightarrow q$	4 hours	$q \leftrightarrow r$	2 hours
$p \leftrightarrow r$	1 hour	$q \leftrightarrow s$	6 hours
$p \leftrightarrow s$	1 hour	$r \leftrightarrow s$	3 hours

If you settle down with a pencil and paper to try and draw this with a scale (say, 1 hour = 1 cm on the map) you'll have a frustrating time! Although we can easily map any pair of towns, the third and fourth simply refuse to fit on the paper in the correct position.

Although there is no precise solution to this mapping problem, there are ways of making a "best estimate" of the towns' location. One graphical method suggested by geographer Waldo Tobler is outlined in the diagrams and summarized in six steps:

Step 1 Locate the towns arbitrarily on the paper as a set of points labelled p, q, r, and s. In (a) we've chosen a square, but it could be any arrangement.

Step 2 Draw straight lines through each pair of points.

Step 3 On each line, center a segment with its length proportional to the given separation-time (say, 4 cm for the 4-hour distance between towns p and q). [See diagram (b).]

Step 4 Draw vectors (marked by arrows) from each point to the *ends* of the segments drawn in Step 3. [See diagram (c).]

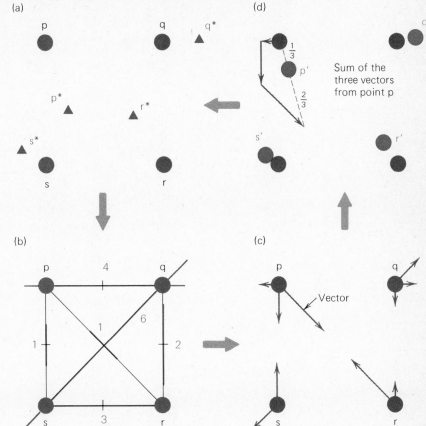

Sum of the three vectors from point p

Vector

Step 5 Average the vectors drawn in Step 4 to give a new set of locations labeled p^1, q^1, r^1, and s^1. The method of averaging is shown in diagram (a) for the first of the towns, and you'll find tracing paper useful for doing this. Since there are three towns the average will be one-third of the length of the sum of the displacements.

Step 6 If no points have moved in Step 5, stop. Otherwise use the new locations (p^1, q^1, r^1, and s^1) to start a new cycle starting at Step 2 and going around to Step 6 again.

For interest, the final stopping places are shown by triangles in (a) and marked p*, q*, r*, and s*. If you try the graphical method described

above using tracing paper, you'll find it tedious but rather satisfying. Notice how at each cycle the moves in position become less and less until no significant shifts are noted. For large numbers of places like those shown in Figure 14-9, the graphical method is far too tedious and a computer method must be used instead. The maps which result provide an accurate way of showing, as nearly as possible, how places are located in time-space. In principle we can substitute costs in dollars for time if this is more appropriate to the geographic study at hand. For a fuller description see P. Haggett, *et al., Locational Analysis in Human Geography* (Arnold, London 1977) p. 326–328.

The changing spatial relations of cities have given a new interest to ways of mapping change. Thanks to the computer, it is now possible to calculate and represent on maps the location of places in a "time" or "cost" space. (See the discussion on the preceding page.)

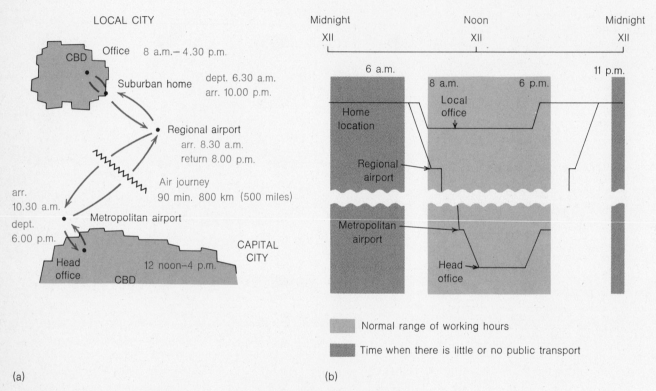

(a)   (b)

Figure 14-10.   Local and regional contact patterns.

(a) Two typical daily contact cycles for a suburban-living executive: first, the regular daily cycle between home and local office, with a short commuting period (about 40 minutes by automobile), and second, the occasional visit to the head office in the capital city, with a long commuting period (about 12 hours, using an automobile, plane, and taxi) and a short (4-hour) conference period. These contact patterns may be plotted on space-time diagrams, as in (b). In this diagram the horizontal lines indicate a static location, and the diagonal lines show travel between locations. The more a line diverges from the horizontal, the greater the speed of travel. Using detailed travel diaries, a group of Swedish researchers led by Torsten Hägerstrand and Gunnar Tornqvist have been able to show the importance of contact possibilities between Swedish cities in the growth of the quaternary sector. Clearly, increased transport speeds allow lower-density living patterns within the daily commuting range and an increased meshing of cities within the occasional daily contact range.

Contactability: Its Impact on Corporate Organization   Another approach to the urbanization process has been taken by a group of Swedish geographers headed by Torsten Hägerstrand and Gunnar Tornqvist. They note the growing importance of the quaternary sector in the urban growth pattern of the present century. Despite the growth of telecommunications of all kinds, the need for direct person-to-person contacts is critical in this sector. In research planning, for example, a large number of people may need to come together in informal groups. Note, however, that such groups are likely to meet occasionally and irregularly (rather than on a fixed daily basis) and that the membership of the group may be flexible.

In this situation the ease of contact between urban areas may become critical. Using detailed diaries of personal movements of business excutives, Swedish geographers have been able (a) to track down the degree of interdependence of different industries and types of activity in different geographic locations and (b) to rank Swedish cities on a *contactability scale*. To do this they use journeys made by private automobile or the fastest form of public transport. (See Figure 14-10.) Only journeys made between 6 a.m. and 11 p.m. are counted, and a "contact" is defined as a meeting of not less than four hours during the "normal" working day (8 a.m. to 6 p.m.). When journeys are weighted to include the cost of contacts, a very detailed regional picture can be built up. If we give the "most contactable" city—the capital, Stockholm—a score of 100, we can then rate the other cities by comparing them to the capital. Some smaller cities in the vicinity of Stockholm have scores in the 80s and 90s. Gothenburg and Malmo, the second- and third-largest cities, have scores of 78 and 70, respectively. The "least contactable" city is the iron-mining city of Kiruna north of the Arctic Circle, which has a score of only 36.

These results are useful in picking out the likely centers of further urban growth in the quaternary sector. Planned changes in transport services (e.g., the introduction of new airline schedules or the opening of new highway links) can be checked for their effect on the contactability rating of different cities. Which cities will benefit? Which will lose out? By repeating studies over time, the changing access of one city to another can be plotted and the course of the urban implosion charted and projected into the future.

One of the most striking impacts of changes in contactability has come through business reorganization. As industrial corporations have grown in size, so the control operations of management have become fewer but larger. (See the discussion in Section 22-4.) Figure 14-11 shows a hypothetical sequence in which the successive takeover of companies leads to centralized head offices. But the ecology of corporate management requires a steady inflow of energy in the form of information, and offices tend to cluster in just those central locations where information is richest: in the administrative and financial capitals of a country, or in its major cities.

Empirical evidence on the extent to which corporate headquarters have been separated from the remaining activities has been collected by Berkeley

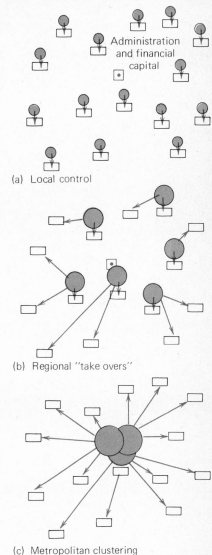

(a) Local control

(b) Regional "take overs"

(c) Metropolitan clustering of control centers

Figure 14-11.   Centralization of large offices.

As industrial corporations have grown so the head offices (circles) have become fewer and more concentrated. Control (arrows) of production plants (blocks) has shifted from local to more distant centers. The critical need of large corporations for information has tended to cause clusters of head offices to form in the major metropolitan centers.

geographer Allan Pred. In the United States the leading 500 industrial corporations show a highly concentrated pattern: in the middle 1970s one out of every three had its headquarters in New York City. Of the remaining corporations, half were concentrated in eight other major regional capitals—Chicago, Los Angeles, Cleveland, Philadelphia, Pittsburgh, Detroit, San Francisco, and Minneapolis-St. Paul. For Britain the situation was even more concentrated. London was the headquarters of two out of three of that country's 500 leading industrial companies. Comparison of the second-ranking centers shows a similar two-to-one contrast between the pattern in the two countries: Chicago had 11 percent of the headquarters, while Birmingham, England, had only 5 percent. For Canada the headquarters concentrated in either Toronto or Montreal (roughly in equal proportions but with Toronto growing faster) with Vancouver trailing in third place.

Concentration into high-contact centers is accelerated by the continuing need for the direct exchange of information on a face-to-face basis. The increasing importance of government as a source of major orders or as a body whose legislation affects a corporation's prosperity also points to a capital-city location.

14-4
Urban Explosion on the Small Scale

To understand the second major impact of transport innovations on urbanization, we must recall that people evolved biologically in an environment with a regular cycle of lightness and darkness, and most basic human activities—feeding, sleeping, working, procreating—have a daily rhythm. Not only is the basic biological clockwork that controls human activity important to agricultural communities, but even on the campus, the same daily rhythms can be shown (Figure 14-12.)

The Daily Contact Zone of the City   How do these time rhythms affect spatial organization? Basically, they operate through the dominant need of most of the human species for a fixed, or relatively fixed, home base. This home base may vary from an apartment in New York City to a sampan in Hong Kong, but it has the same essential economic, social, and biological role. At the very least the home base provides a place for a stock of household goods. It may also provide a site for the production and rearing of children, and a retreat for the satisfaction of sexual and emotional needs. The need to return regularly to a home base puts a severe constraint on the number of hours that can be devoted to travel to work, and therefore on the distance that separates the elements in any household economy. Even when the adult male breaks away from this pattern and works away from the home base for long periods, the constraint usually remains for females, children, and older folk in the household.

Over what distance does this constraint operate? Before Stephenson's *Rocket* rolled along the rails in October 1829, the only regular ways of moving over land were by manpower or horsepower. Horses and carriages could attain a speed of over 50 km/hour (over 30 mi/hour) over short distances, although in practice the rate of travel was reduced by poor or congested roads to less than a quarter of this figure. How fast or far man moved on his own legs is more difficult to estimate. Modern measurements of our walking speeds on city sidewalks indicate that the average adult pedestrian walks at 5.5 km, or 3.4 mi, an hour. The values vary with age and sex. Adolescents gallop along at a steady 6.4 km, or 4 mi, but mothers with young children walk only 2.6 km/hour (1.6 mi/hour). A slope of only 6° cuts speeds by over a fifth; a slope of 12° cuts speeds in half.

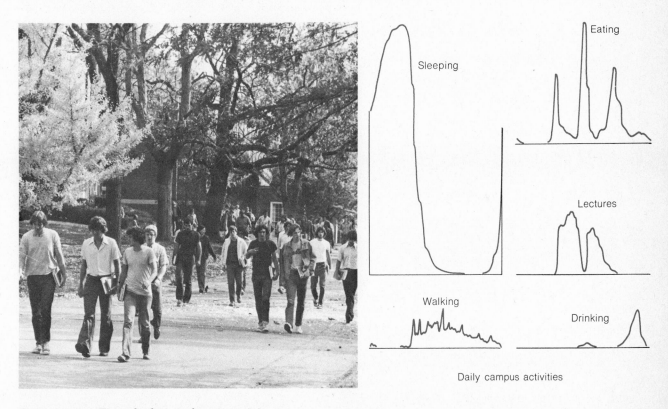

Figure 14-12. Time rhythms in human activity.
Activity patterns for a sample of students at Reading University, England. The concentration of drinking reflects English licensing laws. [From A. Szalai, Ed., *The Use of Time* (Mouton, The Hague, 1970), p. 736, Fig. 5.1-11A; and L. March, *Architectural Design* **41**(1971), p. 302, Fig. 14. Photo by Vannucci, De Wys, Inc.]

It is not surprising that most communities in the prerailway period were quite compact, usually with a diameter of not more than a kilometer or two. The average walking time from one end of town to another would thus be about 11 to 22 minutes. This compactness was directly related to the need for communication between different parts of the city and to the limitation imposed by walking or carriage speeds. The communication links between home and work, merchant and scribe, banker and businessman, magistrate and prostitute all had to be spatially short if transactions were to be completed within an acceptably small part of the day.

Urban Sprawl   The technological revolutions in land transport during the last 150 years have successively reduced the need for compact cities. As steam locomotive speeds increased, and the electric streetcar, the omnibus, the suburban electric railway, and the private automobile succeeded each other, so the links between home and workplace widened. (See Figure 14-13.) Living above the shop or workshop was replaced by daily commuting over increasingly long distances. Chicago's daily urban system is now reaching out 160 km (100 mi) from the downtown. Today all but 5 percent of the U.S. population lives within the bounds of such daily urban systems. (See Section 14-5.)

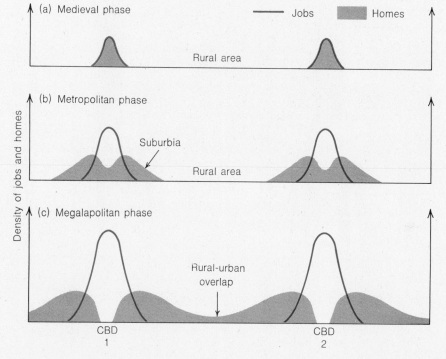

Figure 14-13.   A simplified, three-stage model of the sprawling expansion of two cities.

In the first stage (a) jobs and homes overlap in a small, high-density cluster (like the medieval walled city in Figure 16-13). (b) Transport innovations allow the links between jobs and homes to be loosened. Densities in the central areas of cities appear to fall as less land is available for homes and long-distance commuting creates a daily tidal flow of workers into and out of the city. (c) Continuation of this process leads to further evacuation of the central areas of cities, the destruction of the intervening rural land, and suburban sprawl. Corridors of cities develop like those along the German Ruhr and the Boston–Washington axis.

We can see the dramatic effects of these changes in the exploding size of London (in Figure 14-14). In 1800 London, Europe's largest city, could still be crossed on foot in little more than an hour. Even at its widest, the built-up area was only 10 km (about 6 mi) across. As the sequence of maps shows, this area expanded so rapidly that by 1914 its diameter had reached 35 km (about 22 mi), and it approached 70 km (about 43 mi) by 1958. Since then it has shown few changes due to a vigorous policy of protecting further extension into rural areas (the green belt policy).

London is modest in size by international standards. Los Angeles (Figure 14-15) already exceeds 80 km (50 mi), and Greater Tokyo 120 km (75 mi) in diameter. If we accept Oxford University geographer Jean Gottmann's idea that the whole network of cities along the east coast between Boston and Washington, D.C., is being fused by factors such as shuttle jets into one vast urban area, then we have a world city of mammoth dimensions, over 600 km (373 mi) in diameter. Gottmann terms the city complex thus created a *megalopolis.* This Greek word, which literally means "great city," was first used to refer to a city being planned on large lines in the Peloponnesus area of ancient Greece. The Boston–Washington axis is only one of a number of elongated chains of urban settlements that show similar trends. As Figure 14-16 shows, in each case cities have the advantage of a central set of transport arteries (road, rail, and air) along which massive flows on information, people, energy, and freight moves.

The tendency for such transport links to forge urban centers into a megalopolis is already well established. Some geographers argue that by the year A.D. 2000 the existence of three principal megalopolises—sometimes

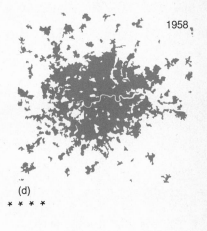

1958

(d)

* * * *

Figure 14-14. Expanding daily contact zones within a city.
The maps show the growth of London over a period of 108 years of progressively faster transport. [From D. J. Sinclair, in K. M. Clayton, Ed., *Guide to London Excursions* (Twentieth International IGU Conference, London, 1964), p. 12, Fig. 9.]

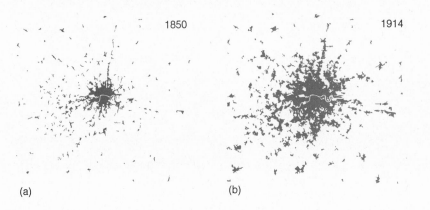

1850

1914

(a)   (b)

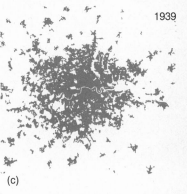

1939

(c)

Figure 14-15.   Urban sprawl.
Los Angeles is often seen as the classic example of urban sprawl and the prototype of the twentieth-century city. It is now the third-largest city in the United States, with an urbanized area that sprawls outside its legal limits to include over 7 million inhabitants in the metropolitan area. Although equal in population to Chicago, it has twice its area, and it is nearly ten times as large as San Francisco. This low-density, automobile-dependent metropolis has had the fastest growth of any major city in North America. In 1870 it had a population of only 5728 inhabitants; since then, the population has grown to more than 100 times that figure. This photo shows us the view across the Los Angeles basin to the San Gabriel Mountains, with the civic center and high-rise buildings around the CBD in the middle distance. [Photo by William A. Garnett.]

termed *Boswash, Chipitts,* and *Sansan*—will have become more evident. By that date three gargantuan metropolises are expected to contain roughly one-half of the total U.S. population. If they are correct then Boswash, extending from Boston to Washington, should have a population of around 80 million; Chipitts, the lakeshore strip from Chicago to Pittsburgh, should have around 40 million people; and Sansan, stretching from Santa Barbara to San Diego, should have around 20 million. The three megalopolises would probably contain a large fraction of the scientifically most advanced and most prosperous segments of world population; even today, the smallest of the megalopolises (Sansan) has a larger total income than all but a dozen of the world's nations.

However, the 1970s have seen some slowing up in the trend toward big-city coalescence. In some regions strong counter forces are beginning to emerge and the situation is locally rather complex. It is to this pattern of regional variability that we turn in Section 14-5.

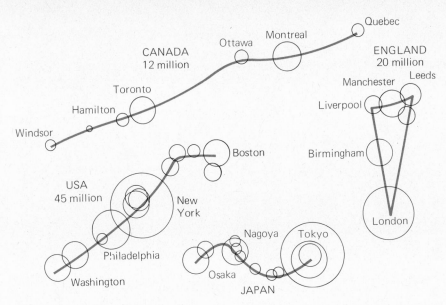

Figure 14-16. Urban agglomerations. Five of the world's major elongated urban agglomerations drawn to the same scale. Circles are proportional to city population size. [From M. Yeates, *Main Street* (Macmillan of Canada, Toronto, 1975), Fig. 1.7, p. 34.]

Crossing the Daily Contact Zone Finding our way across the daily contact zone of an ever-expanding city poses acute problems. The journey to work saps our energy and steals our money and time as we move along a congested route to and from the campus or office. Ways of "solving" the transportation problem of the city are frequently put forward. But such improvements in transportation often bring short-lived benefits. Indeed, some actually increase congestion over the longer term. Why should this paradox occur?

Some insights into the whole process of urban growth may be gained from looking closely at the problem. Consider a typical route into a city center as shown in the first map in Figure 14-17. The area used by commuters traveling in along the highway (1) forms a "teardrop" shaped catchment area, not unlike a river basin. Movements along the highway may be measured in terms of the daily travel time for each trip (the cost of each journey) and the volume of trips each day (the number of journeys). These are plotted as position (1) on the graph. Since in some senses travel is an economic commodity, we can add to the graph the economists' conventional supply and demand curves (S_1 and D_1) indicating that the number of trips and their cost are likely to be related. If the time into the city is short, we may be tempted to go in more often, say, for an evening meal or weekend shopping; if the time is long, we may stay home or shop locally.

The second map in Figure 14-17 shows a major highway improvement. A new multilane freeway is built in place of the old highway. Travel times are reduced, and we expect that there will be some small increase in the number of trips from the old users in the existing catchment area (2). In terms of the diagram the supply curve for the old users has shifted downward to position

Figure 14-17.   Spatial impact of an improved highway link.
In the diagram the area beyond the acceptable commuting time is shaded. S and D describe supply and demand curves. [From R. D. MacKinnon, in L. S. Bourne, *et al.*, Eds. *Urban Futures for Central Canada* (University of Toronto Press, Toronto, 1974), p. 238, Fig. 13.1.]

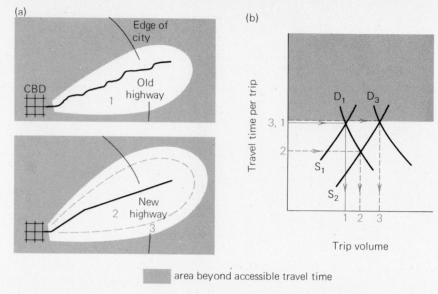

area beyond accessible travel time

S_2 with the time of trips reduced and the volume of trips increased (as shown by position 2 on the vertical and horizontal axes, respectively).

We might expect this position to be maintained only for a short period. New drivers will learn of the improved highway and shift from their old routes. Others may locate their homes to take advantage of the new highway. So in the longer run we would expect the catchment area to be enlarged as shown by area (3) on the map. The limits of enlargement would be determined by the longest acceptable daily trip time. In terms of the diagram the new supply curve S_2 will be intersected by a new demand curve D_3 relating to the increased demand by new users who are driving the improved highway (position 3 on the vertical and horizontal axes).

Congestion is built up again partly because the old users now make more trips and partly because new users augment the flow. In some senses the "solution" has failed, but in another sense it hasn't, because the new highway is now performing more effectively than the old did in terms of the *total* load it is carrying. We must also add the benefits from the other previously used routes, which may now be less congested. Altogether the impact of highway improvements may be complex, and the evaluation may be very different when viewed from an individual rather than from the city in total. In the growing, sprawling city, transport "improvements" are not likely to bring too many improvements to the individual, yet the city itself gains—and so goes on growing.

14-5
American Urbanization

Urbanization is such a complex spatial process that we can only hope to understand it region by region. In this introductory book we only have space to look at one and choose the country familiar to most readers, the United States. For those readers in other countries, the books listed in *One Step*

Further . . . (p. 349) should allow you to build up a similar regional picture for your own area of interest.

Phases of Urban Growth Harvard geographer Brian Berry recognizes a four-stage process in the emergence of cities in the United States (Figure 14-18): mercantile, industrial, heartland–periphery, and decentralized.

The first stage recognized was a *mercantile phase* beginning with the growth of Atlantic seaboard towns in the eighteenth century. Such towns were generally deepwater ports serving as nuclei of communication and ex-

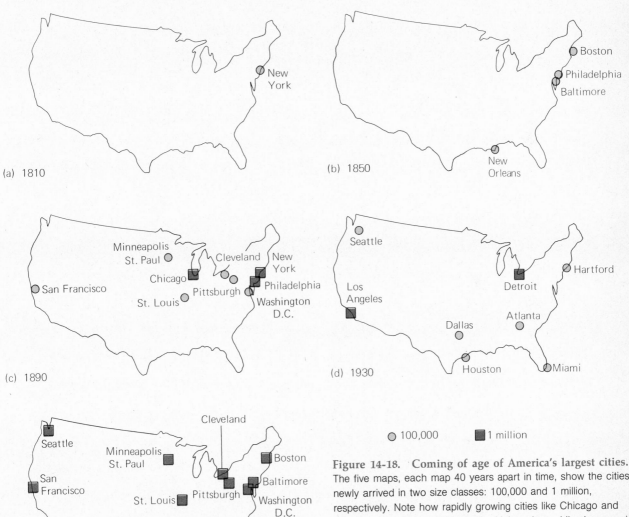

(a) 1810

(b) 1850

(c) 1890

(d) 1930

(e) 1970

○ 100,000   ■ 1 million

Figure 14-18. Coming of age of America's largest cities. The five maps, each map 40 years apart in time, show the cities newly arrived in two size classes: 100,000 and 1 million, respectively. Note how rapidly growing cities like Chicago and Los Angeles jump straight into the higher class while slow-growing cities like Boston take more than a century to move from one class to another. Figures refer to metropolitan populations for the 20 most populous cities in the United States judged by their 1970s size.

port centers for agricultural hinterlands that produced staples for the world's markets. The fact that the hinterlands of Boston, Philadelphia, and Charleston were physically more limited than those of New York, and the relative separation of New Orleans from the main domestic market, allowed New York City—with its middle location along the Atlantic strip and its good internal communications—to move into a dominant position that it retained in succeeding decades. The increase and spread of population inland from these coastal cities followed natural corridors, reinforced by later canal and railroad links, toward the heart of the agricultural-processing regions. With such expansion came the growth of a second generation of inland rail and processing centers like Cincinnati, Chicago, and St. Louis.

A subsequent *industrial phase*, dating from around 1840 to 1850, took place because of the rapid expansion of manufacturing. The growing demand for iron, and later steel, thrust into locational prominence those areas with (1) appropriate resource combinations (iron ores and coal) and (2) central cities already established during the mercantile phase. Buffalo, Cleveland, Detroit, and Pittsburgh shared these advantages, while other peripheral locations with natural resources (but without ready access to a market) did not. The industrial phase further strengthened the position of New York City and saw the emergence of a major heavy-industry "ridge" running westward into the heartland of the United States.

This heartland is usually delineated as the area within the Boston-Washington-St. Louis-Chicago rectangle. It had the initial advantages of excellent agricultural resources and a strategic location with respect to mineral resources. We could characterize the period after 1870 as a *heartland–periphery phase*, when contrasts between this core region and other parts of the United States were strengthened. The processes of circular and cumulative causation that increased the wealth of the core area are part of the same processes of regional evolution that brought wealth to the European cultural hearth (Section 12-2). We can regard even the spatial changes outside the heartland—that is, the emergence of new peripheral centers in the Far West, the South, and Texas—as direct responses to the needs of the heartland. The resource demands of the peripheral areas fostered regional specialization there, but, at least until World War II, this peripheral growth was essentially dependent on central growth.

Countervailing Forces Since around 1950 we can detect a *decentralized phase*. In this phase the location of amenity resources (e.g., a sunny climate or unpolluted environment) has become more important. These resources have stimulated interregional movements of population, bringing rapid urban growth to Arizona and the southwest. They have been responsible for the intraregional rise of small- and medium-sized urban centers with above-average housing, schools, amenities, and so forth. And they have affected local populations by encouraging the suburbanization of manufacturing. These population shifts away from established urban centers are partly related to a rapid overall rise in real incomes. More wealth and leisure have made natural environments (high mountains, persistent sunshine, clear water, extensive forests) a pervasive new influence in the location of population. Of course, in interpreting these signs, we should recall that the total number of people who can respond to this desire is still small and that the area into which the affluent population is moving is largely in the western, southwestern, and southern parts of the country.

Among the first group affected was the retirement population. To these retirees have been added workers employed in research and development projects whose location is not greatly dependent on either natural resources or urban markets. Both groups are likely to be more significant elements in the makeup of the end-of-century populations than now. Already research has created important new centers in areas that were previously lagging behind the rest of the United States (e.g., Huntsville in Alabama) or that were sparsely populated (e.g., Los Alamos in New Mexico). Urban growth in states like Texas and Florida has also vastly increased.

This transference of affluent groups from the outer fringes of suburbia (the "stockbroker belt") to distant amenity-rich parts of the nation outside the metropolises is one of the latest phenomena in a spatial sorting process. This transfer can be illustrated by the spread of a prime medium of communication, television. For black and white TV, stations were opened in a hierarchic sequence from a few major cities in the early 1940s down to the smallest in the late 1950s. The spatial distribution of receivers followed a similar pattern; there were fewer TVs per capita in the remote interurban areas. [See Figure 14-19(c).]

Color TV has a different spatial pattern. Color television receivers were distributed widely in new settlement areas. [See Figure 14-19(d).] The startling contrast between the two TV distributions is an indicator of an important swing toward amenity-resource areas by the country's wealthier population. This movement may well intensify prior to A.D. 2000. If the U.S. experience is repeated in other areas of the world with high living standards, then some major relocations of population may be in line. The Mediterranean and Alpine parts of Western Europe will probably become increasingly attractive relative to other areas, particularly as barriers to the movements of people and their capital are reduced in an enlarged Common Market.

The shifts in growth locations within the United States over the last two centuries may be partly traced to the urban population's changing definitions of natural resources. In the agricultural period the most valued resource was agricultural land, which relied on basic distributions of climate, water supply, and soil. With industrialization the location of mineral resources—particularly coal—became a dominant factor in growth. During the growth of the heartland and peripheral areas, location and good communications were stressed; the current emphasis on leisure activities makes amenity resources an important locational factor. As different amenity resources become important, they provide new directions for the expansion process.

Figure 14-19.   Forces changing American urban regions.
The traditional contrasts between urban and rural areas is shown in (a) and (b). Comparison of TV distributions for black and color suggests some new forces on the American urban scene are beginning to emerge, drawing population away form the major metropolitan areas that have been magnets for so long. [From B. J. L. Berry, *Area* **1** (1970), p. 46, Fig. 1.]

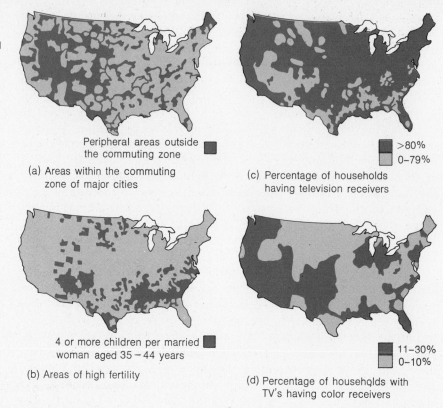

Peripheral areas outside the commuting zone ■

(a) Areas within the commuting zone of major cities

(c) Percentage of households having television receivers
■ >80%
  0–79%

4 or more children per married woman aged 35 – 44 years ■

(b) Areas of high fertility

(d) Percentage of households with TV's having color receivers
■ 11–30%
  0–10%

Limits to Urban Growth   In the United States there is evidence of a recent shift in economic vitality away from the very big cities toward the medium- and smaller-size cities. It is also clear from Figure 14-1 that the rates of urbanization have slowed down in the western, industrialized world. Indeed, in some western countries an equilibrium position of low overall population growth and very little continuing urban growth has been reached. Some major cities, like London, show little overall change in population and even absolute declines. Many major cities, like New York, have acute financial and social problems.

Can we then talk about some limits to the eventual size of cities? We've already seen that a major factor in urban growth is the *centripetal* force of agglomeration economies. But these advantages do not accrue indefinitely as size increases. Lower production costs may be outweighed by increased transport costs as urban areas grow to such a size that raw materials must be imported and finished goods exported over greater distances. Cities may become increasingly congested, internal transport costs rise, and public health dangers from infectious diseases or antisocial behavior increase. Let us take again the familiar example of the United States. As Table 14-1 indicates, most big cities have severe social problems—soaring crime rates, drug abuse,

Table 14-1   Some implications of city size[a]

Characteristic	10,000 Inhabitants	100,000 Inhabitants	1,000,000 Inhabitants
Median income	90	100	120
Crime rates			
Murder	37	100	310
Rape	38	100	260
Auto theft	30	100	320
Air pollution	82	100	155

[a]United States data for the 1960s standardized to give characteristics of cities of 100,000 an index value of 100. Note that median income is probably underestimated due to the "suburbanization" effect (the fact that those who earn large incomes in the city may reside outside its legal or statistical boundaries).

poverty, and increasing pollution. These are the *centrifugal* forces that tend to scatter population away from the cities. In general the problems appear to get worse the larger the city. But, as the table also shows, the rewards (as measured by median income) also increase with city size.

Urbanization continues to increase so long as the benefits to be gained from crowding exceed the costs. As the cost of overcoming spatial separation has fallen, so the relative strength of centripetal forces has grown. Water pipelines, supertankers, and refrigerated ships are symbols of a continuing transport revolution which enables centralized population agglomerations to draw on widely distributed resources; equally important are the sewage lines and disposal services which allow a city to dispose of its massive daily burden of wastes. It is difficult to measure the exact balance of costs of cities, but the probable form of the cost curves as the size of the cities increases is given in Figure 14-20. This figure suggests that there is a threshold over which further increases in size are uneconomic and growth will slow down.

At any one time, one particular bottleneck may be critical. For the eighteenth-century European city, the water supply and infectious diseases were significant barriers to growth; for today's cities, crime, energy and finance may be higher on the agenda. As we saw in Section 8-4, geographers view cities as large and complex ecosystems. As in all such systems, life is dependent on the inflow of resources and the outflow of wastes. Over the last few centuries that flow has been at very high levels, and cities have grown accordingly. But what of the future? We shall look at this question again in Chapter 16.

Figure 14-20.   Urban economies. The general cost curve for all activities suggests that middle-sized cities may be more efficient than either large cities or small towns. This does not necessarily mean that they are preferable, however, as comparable curves for welfare or satisfaction levels show. [From R. L. Morrill, *The Spatial Organization of Society* (Wadsworth Publishing, Calif., 1970), p. 157, Fig. 8.01. Copyright © 1970 by Wadsworth Publishing Co., Belmont, Calif. Reprinted with permission.]

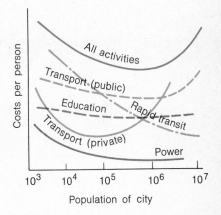

Summary

1. The proportion of the world's population living in cities has constantly increased over the period since 1700 for which reasonably accurate figures are available.

2. Rates of urbanization, however, are not the same all over the world. Western countries typically show an S-curve for urbanization rates. Urbanization rates in developing countries show more rapid growth, with a population increase based not only on migration but also on natural increase.

3. Present rapid city growth is based primarily on (a) the displacement of jobs from the rural sector and (b) the

benefits derived from the agglomeration economies in cities. These are essentially savings derived from having large markets and labor forces concentrated in small geographic areas.

4. Jobs may be divided into four sectors: primary, secondary, tertiary, and quaternary. In urbanizing countries the proportion of jobs changes systematically over time with the proportion of primary-sector jobs falling. The sector model divides the growth process into preindustrial, industrial, and post-industrial stages.

5. The ratio of jobs in the export sectors of a city to its total population is the urban base ratio. Increased employment in the export sector leads to consequential growth in the service and household sectors. The Lowry model attempts to explain the domino effect one activity has on another.

6. Accessibility of markets is an important part of the urbanization process. The major transportation innovations of the last two centuries have increased the locational advantage of the large urban centers as new transport technology is first used to connect these centers. The resultant convergence of large cities is termed an urban implosion. Plotting the changing accessibility of urban centers may aid in determining the course of urban implosion.

7. Improvements in transportation have also led to geographical expansion of the city. Slow transport made compact cities a necessity, but high-speed movement has led to the joining of some urban centers into what geographers term a megalopolis.

8. Study of the United States and other western countries suggests that very large cities are now growing less slowly; some are experiencing an absolute population decline. Limitations to city size are based on centrifugal forces such as public service problems, rising transport and tax costs, antisocial behavior, and the like, which favor decentralization of jobs and homes from cities.

Reflections

1. Why do cities grow? List (a) the benefits and (b) the costs of high-density crowding of the human population into large cities. Is the balance between the two changing? If so, in what direction is it shifting?

2. What is going on in Figure 14-4? Substitute alternative multipliers for those in the diagram (say, 5 for 4, and $\frac{1}{10}$ for $\frac{1}{8}$), and recalculate the effect of 100 new jobs in the basic sector, ignoring the effects of the constraints. How would you improve this model to make it more realistic?

3. Gather information on the proportion of jobs in the primary, secondary, and tertiary sectors for your own country. Do you consider that it is now in the preindustrial, industrial, or postindustrial phase?

4. How does the biological need to sleep for around eight hours a day affect the organization of cities? What would happen if we needed sleep once every six hours, or once every six days?

5. The growth of very large cities in western countries has now slowed. List factors you think might be important in explaining this. Compare these with those chosen by the rest of your class.

6. Review your understanding of the following concepts:

urbanization curves
agglomeration economies
basic-nonbasic jobs
sector model
centripetal and centrifugal forces

urban base ratios
urban implosion
primary, secondary, and tertiary sectors

One Step Further . . .

For a general introduction to urbanization and the problems it poses, see

Scientific American, *Cities* (Knopf, New York, 1966), esp. Chaps. 1 and 2, and

Berry, B. J. L., *The Human Consequences of Urbanization* (Macmillian, London, 1973), esp. Chaps. 2 and 3.

Excellent brief case studies of seven of the world's largest cities, together with studies of the urban experience in individual countries, are given in

Hall, P. G., *The World Cities* (Weidenfeld & Nicholson, London, 1966), and

Jones, R., Ed., *Essays on World Urbanization* (Philip, London, 1975).

Modern work on the evolution of urban systems which emphasize the role of contact patterns and the location of corporate headquarters are

Bourne, I. S. and J. W. Simmons, eds., *Systems of Cities* (Oxford University Press, Oxford, 1978), and

Pred, A., *City Systems in Advanced Economies* (Hutchinson, London, 1977).

The classic geographic study of urban growth in the Boston–Washington corridor is

Gottmann, J., *Megalopolis* (MIT Press, Cambridge, Mass., 1964).

while a comparable study for the Windsor–Quebec corridor in Canada is

Yeates, M., *Main Street* (Macmillan, Toronto, 1975).

Urban geography now dominates many geographic journals, and you will find something of relevance and interest in most issues. Keep a special eye on the book review section to check the growing literature in this area. Those really enthusiastic about urban studies can keep up-to-date with journals such as Urban Studies *(quarterly) and the* Journal of the American Institute of Planners *(monthly).*

A night flight over any densely populated part of the world presents a unique opportunity to see the intricate spatial structure of the human ant's nest. Aboard an intercontinental jet flying at an altitude of 11,000 m (36,000 ft), our concern with cultural complexity slips away. Large cities are visible only as faint clusters of lights spaced many kilometers apart; as the plane loses height, and smaller settlements and isolated farms come into view, sporadic pinpoints of light appear. At night the earth viewed from above looks much like the heavens from the earth. The great galaxies visible to the naked eye dissolve into a host of stars when we look at them through a telescope.

Geographers have long been fascinated by the galactic patterns of human settlement. What forms do they take? Are their forms random and chaotic—or can patterns and formative processes be discerned? If there are regularities, what lies behind them? We look here at some of the answers to these questions, at the models of settlement geographers have built, and how we can use these models to predict changes and to plan more efficient and attractive patterns of settlement.

chapter 15
City Chains and Hierarchies

15-1
Defining Urban Settlements

To answer questions about settlements, we must first puzzle out how to define them. We must try to find ways of describing them and comparing the characteristics of settlements in one region with those of another. Perhaps the simplest solution is to begin by asking how big human settlements are. For if we can define their size, we can go on to compare their magnitudes and relate them to other findings.

Figure 15-1. Difficulties in defining cities.

The census limits of cities rarely coincide with their actual built-up area (shaded). Matched boundaries are rare; underbounded cities, the most common.

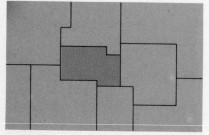

(a) Matched boundaries

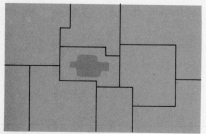

(b) Overbounded city

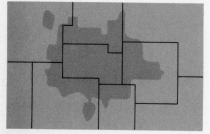

(c) Underbounded city

Questions of Size   Let us consider the various definitions of urban settlements. The definitions used in legal and administrative documents will tell us precisely what we mean by Topeka, Kansas, or Melbourne, Australia. Unfortunately, the legal and administrative borders of cities are often a historical or constitutional legacy. Typically, the legal city has fixed boundaries that survive long after urban development has exceeded those bounds. Thus the legal city is often "underbounded" [Figure 15-1(c)]. Parts of the urban area may remain legally outside the city though but share a common boundary with it. Beverly Hills, completely surrounded by the city of Los Angeles, is a case in point. In England, some boroughs still have a municipal status that is a legacy of their former importance and is out of line with their present small size. The number of inhabitants an area must have to be considered urban also varies from country to country. In Iceland places with a few hundred people are termed urban, whereas in the Netherlands a population of 20,000 is needed.

A second approach to defining urban settlements is to ignore the legal boundaries and try to define each settlement in terms of its physical structure. For example, we might define a settlement on the basis of continuous housing, or population above a certain density, or the intensity of traffic. But there are difficulties here too. What do we mean by "continuous" housing, and what happens when different definitions don't all give the same answer? Figure 15-2 presents some different definitions of New York based on both its legal boundaries and its physical structure. Note that New York City itself (Manhattan, Staten Island, Brooklyn, Queens, and the Bronx) is only a small part of the continuous urban sprawl that is greater New York.

The "mismatch" between the legal and the physical city becomes vitally important when the legal city, with its static or declining population and limited tax base, has to provide public services like transport and police for the millions of commuters who cross its legal boundaries to work each day. As the discrepancy between the legal and economic boundaries of the city becomes worse, the pressure for some form of revenue sharing or boundary adjustment grows. This discrepancy also affects our ability to answer even the simplest questions about the size of the city. To take an extreme case, the "legal city" of Sydney, Australia, in 1975 had a population of only 58,000 while the "built-up area" of Sydney had a population of 2,898,000. This difference of around 50 times in size is unusual, but important enough to show that definition of settlements is a matter of concern.

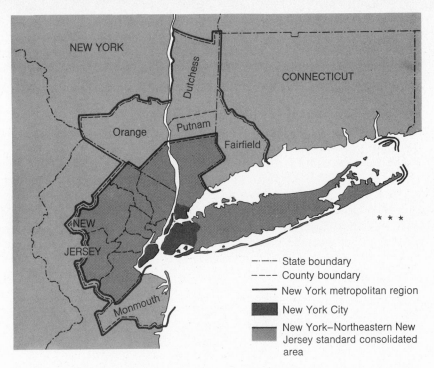

Figure 15-2. Varying definitions of a metropolis.
The map shows three alternative boundaries for New York, none of which coincides exactly with the limits of the built-up area. Part of New York City's financial problems stem from the mismatch between its area and the built-up area of the city. As industry and high- and middle-income residents have moved out to the suburbs, so the tax base of the city has been eroded.

State boundary
County boundary
New York metropolitan region
New York City
New York–Northeastern New Jersey standard consolidated area

Some Possible Solutions As a result of this problem, international, and indeed intranational, definitions of urban settlements are being standardized. One definition of world metropolitan areas, by demographer Kingsley Davis, runs to 12 pages, including 2 pages on difficult cases. In the United States the concept of a Standard Metropolitan Statistical Area (SMSA) was introduced in 1960 so that metropolitan areas could be defined realistically by using three criteria. First, a population criterion: Each SMSA must include one central city with 50,000 or more inhabitants. Special rules allow contiguous cities (i.e., those directly adjoining each other) and nearby cities (within 32 km, or 20 mi, of each other) to be combined. Second, the metropolitan character of an area is taken into account. At least 75 percent of the labor force of the county must be employed by nonagricultural industries. Other criteria for SMSAs relate to population density, the contiguity of townships, and ratios between the nonagricultural labor forces of the counties making up the unit. Finally, the integration of the areas that constitute the SMSA is considered. Counties are integrated with the county containing a central city if 15 percent of the workers living in the county commute to the city, or if 25 percent of the workers in the county live in the city. This measure of integration can be supplemented by other measures based on the market area for newspaper subscriptions, retail trade, public transport, and the like.

Despite their apparent comprehensiveness, the SMSA definitions have still not fully solved the problem of urban boundaries. Improved definitions using county blocks and commuting data continue to be sought by geographers interested in international comparison between cities.

Not only metropolitan areas but small towns and villages as well can be difficult to define. The smaller settlements, however, unlike urban areas, are usually overbounded [Figure 15-1(b)]. Similar sets of dull, but necessary, rules must be worked out for these cases too.

Measuring Patterns of Settlement If we mark the location of cities, towns, and villages on a map, we can see the overall pattern of settlement. Let us look at some examples. Figure 15-3 shows four sample areas from different parts of the United States; each square has an area of 5000 km^2 (2000 mi^2), and each dot represents an urban settlement as listed in the United States Census. We can distinguish among the four areas in two ways. First, they differ in density (i.e., in the number of urban settlements per square kilometer). For example, the North Dakota area has only 16 towns (3.2 towns/1000 km^2) while the Ohio area has 98 (19.0 towns/1000 km^2). Mea-

Figure 15-3. Patterns of settlement.
The diagrams show the density and spacing of towns in four sample 5000-km^2 (1930-mi^2) areas in the northern United States. The classification of the spatial patterns of "clustered," "random," or "scattered" is determined by the value of the nearest-neighbor index, R. (See marginal discussion.) [From L. J. King, *Tijdschrift voor Economishe en Sociale Geografie* **53** (1962), pp. 4–6, Fig. 3-7.]

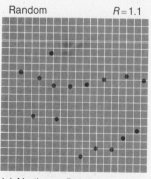

Random   R = 1.1

(a) Northwest Dakota
3.2 per 1000 km^2

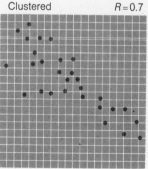

Clustered   R = 0.7

(b) Southern Washington
5.8 per 1000 km^2

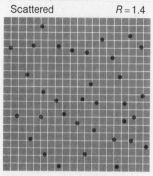

Scattered   R = 1.4

(c) Western Minnesota
6.4 per 1000 km^2

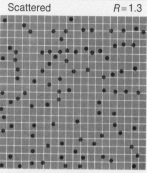

Scattered   R = 1.3

(d) Northwest Ohio
19.0 per 1000 km^2

✳ ✳ ✳ ✳

THE NEAREST-NEIGHBOR INDEX

Assume a spatial distribution of towns as in Figure 15-3. Using measures first developed by plant ecologists, geographers can define a *spacing index* by comparing the observed pattern of settlements in an area with a theoretical random distribution; that is,

$$R = \frac{D_{obs}}{D_{exp}}$$

where
R = the nearest-neighbor index of spacing,

D_{obs} = the average of the observed distances between each town and its nearest neighbor in kilometers, and

D_{exp} = the expected average difference between each town and its nearest neighbor in kilometers.

The expected average distance is given by

$$D_{exp} = \frac{1}{2\sqrt{A}}$$

where A = the density of towns per km^2. Thus, in an area with an ob- served nearest-neighbor distance of 3.46 km (2.15 mi) and an observed density of 0.0243 towns/km^2, the nearest-neighbor index of spacing (R) will be 1.08. Values of 1.00 indicate a random pattern. Dispersed or scattered patterns of settlement have values greater than unity, and clustered patterns of settlement have values less than unity. Nearest-neighbor indices are discussed at greater length in L. J. King, *Statistical Analysis in Geography*, (Prentice-Hall, Englewood Cliffs, N.J., 1969), Chap. 5.

surement is no problem, and it is a simple matter to arrange the four areas in a sequence of increasing density, as in Figure 15-3.

The patterns of settlement also differ in a second characteristic which is less easily measured. Compare the patterns of the Washington and Minnesota areas in Figure 15-3. Both have similar densities (5.8 and 6.4 towns per 1000 km^2, respectively) but strikingly different arrangements of the towns in space. The Washington towns are clustered, whereas the Minnesota towns are scattered. To measure this second property, geographers have adopted a spacing index (see the above discussion of the nearest-neighbor index) that enables them to rank patterns of settlement along a scale from "highly clustered" to "highly dispersed." The values of the index range from a theoretical low of zero, when all the settlements are concentrated at a single point, to a maximal value (2.15) when the pattern of settlement is that of a triangular lattice.

Most of the patterns of settlement examined thus far would have values on this scale between around 0.5 and 1.5. These values hover around the value we would assign to a group of randomly generated points (1.0), thereby implying that there are no strong pattern-forming forces deciding how settlements are arranged. In relatively uniform environments the index values for settlements drift toward 2.15 at the uniform end of the spacing scale; conversely, in environments with greater contrasts in population density the values drift toward the clustered end (0). With this index geographers can compare patterns of settlement with different spacing and estimate the probable amount of environmental influence on the location of settlements.

15-2
Settlements as Chains

Once we find a commonly accepted method of defining cities, then analysis of their comparative size and importance can begin. One of the first steps in such analysis is to arrange them in order of population size. Table 15-1 ranks the 20 largest cities in three areas of decreasing size: the world, the United States, and the state of Texas. At first sight, this looks like a dull collection of statistics. But look at Figure 15-4. This plots the size of each of these "top 20" cities against its rank. Geographers have repeated this process for large and small areas; in each case they have looked for a repetitive pattern in the array of sizes. Have any rules been discovered? What do they tell us about the way such "chains" of city sizes are linked together?

A Rule for the Size Distribution of Settlements   Although some nineteenth-century investigators sought for patterns, one of the first to actually find a significant one was a German geographer, Felix Auerbach, in 1913. He noticed that if we arrange settlements in order of size (1st, 2nd, 3rd, 4th, ..., nth), the population sizes for some regions are related. Auerbach found the

Table 15-1   The twenty largest urban settlements for areas of different magnitude[a]

World metropolitan areas		United States metropolitan areas		Texas metropolitan areas	
New York	14.26	New York City	9.63	Dallas–Fort Worth	2.50
Tokyo	11.28	Chicago	6.97	Houston	2.22
Shanghai	10.82	Los Angeles	6.93	San Antonio	0.98
Mexico City	10.47	Philadelphia	4.81	El Paso	0.41
Paris	9.86	Detroit	4.43	Austin	0.39
Buenos Aires	8.44	Boston	3.92	Beaumont–Port	
Moscow	7.73	San Francisco	3.14	Arthur–Orange	0.34
Peking	7.57	Washington, D.C.	3.01	Corpus Christi	0.30
London	7.17	Nassau–		McAllen–Pharr–	
Los Angeles–		Suffolk	2.62	Edinburg	0.22
Long Beach	7.03	Dallas–Fort Worth	2.50	Lubbock	0.20
Calcutta	7.03	St. Louis	2.37	Brownsville–Harlingen	
Chicago	6.98	Pittsburgh	2.33	–San Benito	0.19
Bombay	5.97	Houston	2.22	Waco	0.18
São Paulo	5.92	Baltimore	2.14	Galveston–Texas	
Cairo	5.72	Minneapolis	2.01	City	0.17
Seoul	5.43	Newark	2.01	Amarillo	0.17
Philadelphia	4.82	Cleveland	1.98	Wichita Falls	0.15
Leningrad	4.37	Atlanta	1.78	Abilene	0.14
Tientsin	4.28	Anaheim–Santa		Texarkana	0.11
Rio de Janeiro	4.25	Ana	1.66	Odessa	0.11
		San Diego	1.52	Tyler	0.10
				Sherman–Denison	0.08
				Midland	0.08

[a]The figures refer to the estimated population in millions for the mid-1970s according to the United Nations *Demographic Yearbook* and the United States Census Bureau. Notice that the world series figures for the United States cities may be greater than that of the national series since they are based on a different definition of metropolitan areas.

Actual population size (in millions)

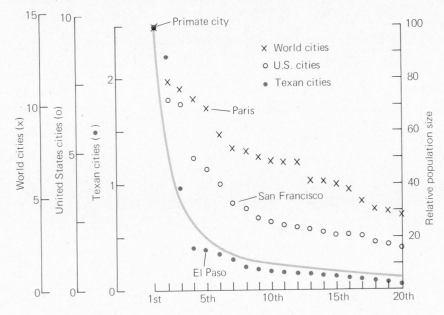

Figure 15-4. City chains.
The twenty largest cities in three areas (see Table 15-1) are arranged on a population rank-size diagram. The largest city in all three series—the "primate city"—has a relative population size of 100. Actual population sizes are shown on the vertical scales to the left. The continuous line is an idealized rank-size curve.

THE RANK-SIZE RULE

Assume a set of cities ranked according to size from the largest (l) downward. In its simplest form, the rank-size rule states that the population of a given city tends to be equal to the population of the largest city divided by the rank of the given city; that is,

$$P_r = \frac{P_l}{R}$$

where P_r = the population of the rth city,

P_l = the population of the largest city, and

R = the rank of the rth city in the set.

This basic formula is often modified by a constant (b) to allow variations from the strick rank-size rule—for example,

$$P_r = \frac{P_l}{R^b}$$

Thus, when the largest city has a population of 1,000,000 and $b = 0.5$ (a low-angled slope in Figure 15-5), we should expect the population of the fourth-largest city to be 500,000. If b is raised to 2.0, then we should expect the population of the fourth city to be smaller—that is, only 62,500. For an extensive critical study of rank-size rules in a specific regional context, see C. D. Harris, *Cities of the Soviet Union* (Rand McNally, Skokie, Ill., 1970).

simplest relationship to be that the population of the nth city was 1/n the size of the largest city's population. Thus the fourth ranking city was found to have approximately ¼ the population of the largest. This inverse relationship between the population of a city, and its rank within a set of cities is termed the *rank-size rule*. (See the marginal discussion of the rank-size rule on this page.) If we apply this rule to the United States and look at Table 15-1, we should expect Chicago (ranked second) to have one-half of the population of New York City (ranked first). Thus, in the mid-1970s, Chicago should have had a population of 4.82 million; in fact, its population was more than this (6.97 million). In earlier synthesis, there is a closer correspondence with Auerbach's ideas. The match with the ideal rank-size rule (the curve in Figure 15-4), is better for the Texan cities, but rather poor for the world metropolitan areas.

We can more easily compare the fit between distributions of real cities and the idealized distributions predicted by the rule if we make the axes on which we plot the cities' size and rank nonlinear. The 20 hypothetical cities in Figure 15-5(a) all conform exactly to the rank-size rule and make an awkward, J-shaped curve. If we transform the values on the axes to logarithmic scale, the curve becomes a straight line, as in Figure 15-5(b). The simple rank-size relationships for the United States and Texas can also be described by a straight line, with a slope of 45° to the horizontal; that is, they have the form predicted by the rank-size rule and shown in Figure 15-5(b). Yet other lines that are equally regular but have *different* slopes have been found. For

Figure 15-5. Hypothetical relations between city size and rank.
Twenty cities arranged in an idealized rank–size pattern are plotted on (a) a graph with arithmetic axes and (b) a graph with logarithmic axes. Figure (c) shows three alternative patterns of city sizes. Figure (d) shows the evolution of idealized city chains as population size increases through three time periods. (See also Figure 15-16.)

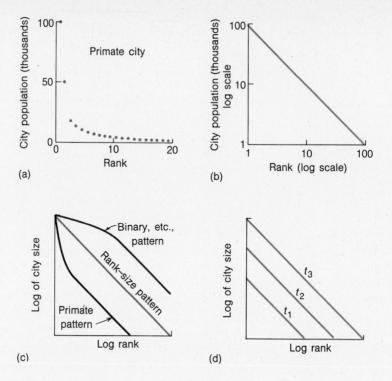

example, the curve for the cities of Switzerland in 1960 has a much gentler slope, and that for urban areas in India in 1921 has a much steeper one. Gentle slopes imply that the decrease in a city's population with rank is extremely slow; steep slopes imply a sharp falloff in size with rank.

Regional Variations in the Rule   Since the size of any one city appears to be linked to the size of all other settlements in a region, we can regard the whole set of cities as forming an interlinked chain running from the largest city to the smallest. These are termed *rank–size chains.* When considering these rank–size chains, some geographers have found it more meaningful to subdivide the distribution of settlements into distinct segments. Australia has a distinct pattern of urban areas which when plotted on a graph showed a flat upper section and a steeper lower section, with the critical break at a population of 75,000. This convex Australian pattern is in contrast to the concave distribution of Soviet cities, in which a steep upper section is followed by a flatter lower section. Here the break between the two sections comes much higher than in the Australian pattern; it happens at a population level of around 500,000. In Figure 15-5(c), we distinguish between these two patterns of settlement. The Australian pattern is dominated by a few large cities that are roughly equal in size to a "tail" of smaller cities conforming to the rank–size rule. Conversely, in the Russian pattern the falloff in population among the first few cities below Moscow is sharper than the falloff predicted by the simple rank–size rule. We call this a *primate* pattern to indicate the dominant role of the first, or primate, city. (The term primate is taken from ecclesiastical language where the primate of a church is its superior bishop.)

When historical census figures are available, geographers can trace the changes in rank–size relationships over time. If the total population of a region grew and its cities remained distributed in a simple rank–size sequence, we should expect it to change through time periods t_1 to t_3 in the way described in Figure 15-5(d). The figures for the United States over one and a half centuries are rather stable. The curves for 1790 to the middle of this century are generally parallel, and there is some evidence of increasing regularity (manifested by the straightening of the curves over time), as we shall see later in Figure 15-18. The statistics for Sweden over the same period reveal a reverse effect; here an S-shaped curve is retained and even accentuated by an overall growth in the population. Figure 15-6 shows the pattern followed by Israel in relation to the expected development of city-size distributions.

The Logic of Rank–Size Chains   Enough evidence is now available to prove that regular rank–size chains are recognizable for settlements in many types of regions during different time periods. Why are many towns and cities arranged in this regular fashion?

Geographers are not the only ones to be puzzled by these size distributions. Rank–size rules are not confined to human settlements. Similar distributions have been observed by botanists studying the number of plant species, and by linguists studying the frequency with which different words are used in our speech. The pervasiveness of this type of distribution has led organization theorist Herbert Simon to postulate rank–size rules as the equilibrium slope of a general growth process. We can visualize this process as one in which each unit, such as a city, initially has a random size and thereafter grows in an exponential manner that is proportional to that initial random size. (See Section 7-1, especially page 143.) Simon points out that extremely general processes of this kind tend to produce distributions that approximate a regular rank–size form.

Simon's hypothesis has been translated into urban terms by Brian Berry. Berry studied the rank–size distribution of towns with populations of 20,000 or more in 38 countries. Of all the countries, only 13 had rank–size distributions like the ones postulated in the Simon model. These countries were among the largest in the group (e.g., the United States), had a long history of urbanization (e.g., India), and were economically and politically complex (e.g., South Africa). By contrast, 15 countries had primate distributions in which one or more large cities dominated the size distribution and were much larger than we might have expected them to be on the basis of rank–size rules.

In contrast to regular rank–size distributions, primate distributions appear

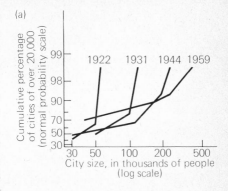

(a)

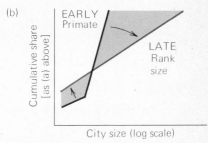

(b)

Figure 15-6.   Changes in city-size distribution over time.

(a) Cumulative share of the population in different classes of city size in the area now occupied by Israel (formerly Palestine) over the period from 1922 to 1959. (b) Hypothetical pattern of change from a primate to a rank–size distribution of city size with urbanization. [Data from G. Bell, *Ekistics* **13** (1962), Fig. 1, p. 103.]

to be products of urbanization processes in countries that are smaller than average, have a short history of urbanization, and have simple economic and political structures. Hence, primate distributions typify the impact of a few rather strong forces. For instance, the impact of imperial status on the large cities in Austria, the Netherlands, or Portugal, each the hub of former empires, is certainly potent. Another powerful force is the superimposition of outside influences on an existing hierarchy. Examples would be the institution of a dual economy (such as a peasant and plantation sector in Sri Lanka) or the influence of a westernized city such as Bangkok on the Thai system of cities.

Simon's model has two principal advantages: It introduces time into the rank–size model by taking into account an urban system's history of development, and it emphasizes the effects of numerous small forces in producing regular structures. Variations from the rank–size model may be caused by the distorting effect of a few powerful forces.

15-3
The Christaller Central-Place Model

Two-dimensional patterns of settlement as shown on maps of population distribution invite questions similar to those provoked by one-dimensional patterns. Is any order discernible? If so, what forces lie behind it? Although this problem had been stated by German geographers in the nineteenth century and some tentative hypotheses were made, the main breakthrough did not occur until 1933, when Walter Christaller published his now-famous doctoral dissertation on *Central Places of Southern Germany*. (See Figure 15-7.) Christaller's works had little impact in Germany at the time, and it was not until their introduction into the United States in the 1940s and 1950s that their value was realized. Since then Christaller's ideas have been verified, extended, and challenged.

Central Places   The terminology of Christaller's model is straightforward. *Central places* are broadly synonymous with towns that serve as centers for regional communities by providing them with *central goods* like tractors and *central services* like hospital treatment. Central places vary in importance. Higher-order centers stock a wide array of goods and services; lower-order centers stock a smaller range of goods and services—that is, some limited part of the range offered by the higher center. *Complementary regions* are areas served by a central place. Those for the higher-order centers are large and overlap the small complementary regions of the lower-order centers.

Schools provide a good example of a central-place organization. The local elementary school provides a lower-order center (to use Christaller's terms) which serves a small part of a city or a single rural community. There is a large number of such schools in any state, and they teach children drawn from only a few square miles (i.e., they have small complementary regions). Above the elementary schools come the higher-order services provided by the junior high schools, the high schools, and colleges of various kinds. As we move higher up the educational ladder, the number of centers becomes smaller and their complementary regions become larger. At the top of the ladder stands the state university, sometimes a single institution serving students drawn from the whole state, its complementary region. Education is just one range of central goods and central services that give character to central-place organizations and help to distinguish the central-place functions of one settlement from those of another.

Christaller defined the *centrality* of an urban center as the ratio between all the services provided there (for both its own residents and for visitors from its complementary region) and the services needed just for its own residents. Towns with high centrality supplied many services per resident, and those with low centrality few services per resident. Christaller found that, the number of telephones offers a useful indicator of the range of central goods available in a town. Using telephone data, he defined the centrality of a town as equal to the number of telephones in the town, minus the town's population multiplied by the average number of telephones per population in the town's complementary region. A town of 25,000 with 5,000 telephones in a region with only 1 telephone for every 50 people would have an index of 5,000–25,000 ($\frac{1}{50}$), or 4,500. Thus the index basically measures the difference between the expected level of services (i.e., that needed by a town to serve its *own* inhabitants) and the level of services actually measured within the center.

Later researchers have revised Christaller's terminology to include two simple concepts. The first is a *market-size threshold*, below which a place will be unable to supply a market good. That is, below the threshold, sales will be too few for firms to earn acceptable profits. The second concept added was that of the *range of a central good*, the limits of the market area for the good (Figure 15-8). The market area's lower limit is determined by the threshold market size, and its upper limit is defined by the distance beyond which the central place no longer is able to sell the good. If we assume that travel is equally easy in all directions, the range of a good will be a perfect circle. This circle is the outer limit of a *demand cone* in which the quantity of a central good consumed decreases with distance from the central place because of increased transport costs.

Complementary Regions   Given a circular demand cone for central goods, Christaller demonstrated that a group of similar central places will have hexagonal, complementary regions with the central places arranged in a regular triangular lattice. Figure 15-9 depicts the stages by which such a pattern might emerge as population colonized a new area and central places were established. (Cf. Section 15-5.) The final hexagonal patterns follow directly from five simplified assumptions.

1.   An unbounded isotropic plain with a homogeneous distribution of purchasing power. (Imagine farms uniformly distributed over a flat plain that

Figure 15-7.   Walter Christaller.
The German geographer Walter Christaller (1893–1969) was something of an "odd man out" among the scholars of his generation. His thesis on the structure of settlements in southern Germany was one of the cornerstones on which central-place theory was to grow. His ideas owed something to the German settlement geographer Robert Gradmann and to the locational theorists whom we shall meet in the next two chapters (Johann von Thünen, J. G. Kohl, and—Christaller's old teacher—Alfred Weber). His ideas of model-building were out of harmony with the prevailing geographic ideas of his time, and he did not hold a university post. It was not until the 1950s that Christaller's ideas were widely recognized in the English-speaking world. [Photo from Inter Nations.]

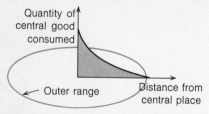

Figure 15-8. Idealized demand zones in the Christaller model.
With uniform transport costs, the demand for central goods falls with distance from the central place, and the market range (the area within which the goods will be bought) forms a circle.

has the same fertility everywhere. On this flat plain, travel costs are the same in any direction).

2. Central goods to be purchased from the nearest central place.
3. All parts of the plain to be served by a central place; that is, the complementary areas must completely fill the plain.
4. Consumer movement to be minimized.
5. No excess profits to be earned by any central place.

The hexagons result from our attempt to pack as many circular demand cones as possible onto the plain. If we require all parts of the plain to be served by a central place (assumption 3), the circles will overlap. But because of our second assumption, that consumers will shop at the closest central place, the areas of overlap will be bisected. A perfect competitive situation will be achieved only when the plain is served by the maximal number of central places, offering identical central goods at identical prices to hexagonal complementary regions of identical size. Only this arrangement ensures that consumers travel the least distance to central places.

Central-Place Hierarchies Christaller was able to account for the varying levels of central places within a settlement hierarchy by varying the size of the complementary regions, as in Figure 15-10. He discusses three cases.

The first is a *market-optimizing* case, in which the supply of goods from central places is as near as possible to the places supplied. A higher-order central place will serve *two* of its lower-order neighbors. It may do this by serving only two of its six equidistant nearest neighbors and thus having an asymmetric complementary region. Alternatively, a higher-order central place may share the same neighbors with two others, for instance, competing neighbors. Note in Figure 15-10(a) how settlement 2 lies on the edge of three complementary regions (those of centers 1, 3, and 4). This arrangement is termed a $K = 3$ system, where K refers to the number of places served, that is, the central place plus two nearest neighbors or the central place plus one-third of each of its six nearest neighbors.

The second case involves a *traffic-optimizing* situation in which the boundaries of the complementary regions are rearranged to allow a more efficient

Figure 15-9. Hexagon formation.
The overlapping of circular demand cones, along with the close packing of centers, gives a network of hexagonal territories. (Compare with Figure 15-14.)

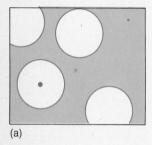

(a)

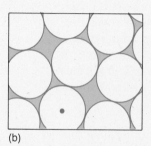

(b)

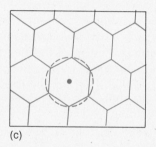

(c)

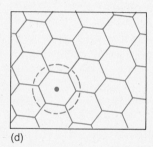

(d)

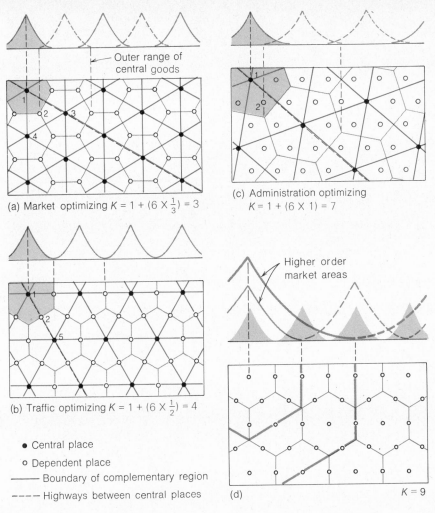

(a) Market optimizing $K = 1 + (6 \times \frac{1}{3}) = 3$

(b) Traffic optimizing $K = 1 + (6 \times \frac{1}{2}) = 4$

(c) Administration optimizing $K = 1 + (6 \times 1) = 7$

(d) $K = 9$

- ● Central place
- ○ Dependent place
- —— Boundary of complementary region
- ---- Highways between central places

Figure 15-10. Alternative principles of organization in the Christaller model.

Settlements can be partitioned in one of three basic ways: (a), (b), or (c), by enlarging and rotating the hexagonal cells. Note the way in which the K number depends on whether a dependent place is shared with three other places ($K = 3$), two other places ($K = 4$), or unshared ($K = 7$). The hexagonal cells can be grouped hierarchically to give tiers of higher-order centers. For example (d) shows higher-order centers in a traffic-optimizing ($K = 4$) hierarchy. Note the way in which lower-order centers "nest" within the market areas of higher-order centers in a manner reminiscent of sets of Russian dolls.

highway pattern than in the first case. As Figure 15-10(b) shows, as many places as possible now lie on traffic routes between the larger towns; for example, the direct route from center 1 to 5 goes right through center 2. This situation is represented by the $K = 4$ hierarchy, where a higher-order place serves three adjacent lower-order places. It may do this by dominating three of its six nearest neighbors or by sharing them with another central place of the same order.

The third case Christaller discusses is an *administration-optimizing* situation, in which there is a clearcut separation of the higher-order place and its neighboring lower-order centers. That is, each lower-order center falls clearly within the trade area of a single central place; in Figure 15-10(c), for example, center 2 falls within the area of center 1. Such arrangements are likely to be economically and politically more stable than divided settlements. This relationship produces the $K = 7$ hierarchy.

All three cases assume that relationships established for one level (e.g., between villages and small towns) will also apply to other and higher levels (e.g., between small towns and larger cities). They are usually called *fixed-K hierarchies* because the same fixed relationships hold at *all* levels of the settlement hierarchy. This means that we can expand each of Christaller's three central-place variants by building higher and higher levels on top of the basic framework. Consider the situation in Figure 15-10(d), where a second and third $K = 4$ central place is superimposed on the first. As we add each successive upper level, the size of the hexagonal regions increases and the number of places is reduced by a quarter. Thus if a region had 2000 central places on the lowest level, it would have 500 on the next level, and 125 on the next higher level again. If we start at the top, we can put this a simpler way. In an idealized $K = 4$ school system with three levels or tiers, 1 junior college would draw students from 4 high schools, each of which drew children from 4 elementary schools (i.e., 16 in all). For each of the three cases considered by Christaller, typical sequences would be 1, 3, 9, 27 for the $K = 3$ network, 1, 4, 16, 64 for the $K = 4$ network, and 1, 7, 49, 343 for the $K = 7$ network.

Southern Germany as a Test Area   As we have noted, Christaller developed his basic ideas using southern Germany as his original test area. The theoretical distribution of the status and location of towns in this area, according to his market-optimizing principle, is summarized in Table 15-2. Christaller postulated seven levels of the hierarchy from the level of the hamlet (Figure 15-11) to that of the city, each level showing an increase in the area of the complementary region. Approximate populations have been added to the table by extrapolating from Christaller's detailed work on southern Germany.

Table 15-2 Status of towns in Christaller's system[a]

Type of town	Order	Approximate population	Distance from other towns		Service area	
			km	mi	km^2	mi^2
Landstadt (L)	Upper	500,000	187	116	35,000	13,514
Provinzstadt (P)		100,000	109	68	11,650	4,498
Gaustadt (G)		30,000	63	29	3,880	1,498
Bezirkstadt (B)		10,000	36	22	1,243	478
Kreisstadt (Kr)		4,000	21	13	414	160
Amtsort (A)		2,000	13	8	140	54
Marktort (M)	Lower	1,000	7	4	47	18

[a]Values are based on a study of southern Germany (Figure 15-12).
SOURCE: R. E. Dickinson, *City and Region* (Humanities, New York, and Routledge & Kegan Paul, London, 1964), p. 76.

Figure 15-11. South Germany.

South Germany was used by Christaller as a test area for his central-place model. Settlements like this second-tier Amtsort (at the bottom of the photo) were seen as part of an intricate structure linking lower-order rural settlements to higher-order urban settlements. [Photograph courtesy of the German Information Center.]

On the upper level of the hierarchy are the *Landstadt* cities with populations of around 500,000—Munich, Frankfurt, Stuttgart, and Nuremburg, together with the border cities of Zürich in Switzerland and Strasbourg in France. The lowest market center at the base of the hierarchy has a service radius of a little more than 3 km (2 mi). The application of the Christaller's theoretical model to southern Germany is shown in Figure 15-12.

Despite the general agreement between the model and reality, Christaller found several specialized centers—mining towns, border towns, and so on—that deviated from the general pattern. The resources in a particular region or subregion may cause a general increase in the density of settlement, resulting in a closer spacing of centers.

In the pages available here, we've only been able to discuss Christaller's main ideas. From very simple assumptions the model can be further extended and refined to give a more realistic representation of the real-world complexities of human settlement. It is to these we now turn.

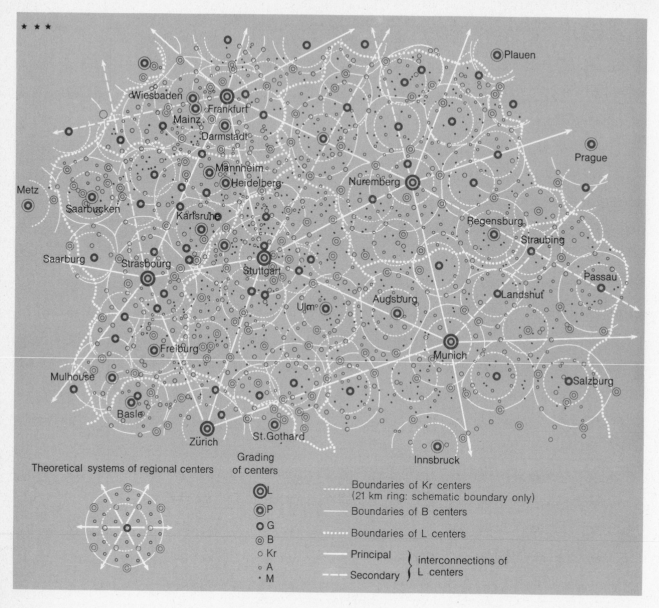

Plauen

Wiesbaden
Frankfurt
Mainz
Darmstadt

Prague

Mannheim
Heidelberg

Metz

Nuremberg

Saarbucken

Regensburg

Straubing

Saarburg

Strasbourg

Karlsruhe

Passau

Stuttgart

Ulm

Augsburg

Landshut

Freiburg

Munich

Mulhouse

Salzburg

Basle

Zürich   St.Gothard

Innsbruck

Theoretical systems of regional centers

Grading
of centers

⊚ L
⊙ P
○ G
◉ B
○ Kr
○ A
· M

---- Boundaries of Kr centers
      (21 km ring: schematic boundary only)
—— Boundaries of B centers
····· Boundaries of L centers
—— Principal } interconnections of
--- Secondary } L centers

Figure 15-12. South Germany.
Distribution of cities, towns, and villages in southern Germany shown by means of
seven-level hierarchy. (See Table 15-2.) The map also shows the boundaries of the
complementary regions about the four highest levels of centers (drawn as ellipses or
circles) and the main routes interconnecting the highest-level centers (drawn as straight
lines). The boundary of the area is simply the limit of Christaller's study area; border
cities like Zürich and Plauen lock into a continuing, continent-wide hierarchic system.

Since its publication the Christaller model has provoked two main reactions from fellow geographers. First, there are those who have accepted the general argument of his model. Their reaction has been to extend and refine it. Second, there are those who found the Christaller model too rigid and static. They have reacted by trying to build alternative models with a stronger time dimension and a closer correspondence to actual settlement history. In the last two sections of this chapter we look at these two approaches.

Lösch's Modifications The prime theoretical extension of the Christaller model was created by a fellow German, August Lösch (1906–1945), in his *Die raumliche Ordnung der Wirtschaft* published shortly before his death. He clarified the ways in which spatial demand cones are derived and verified the optimal hexagonal shapes of complementary regions where the population served was uniformly distributed. However, Lösch's main contribution was to extend the notion of fixed-K hierarchies.

Lösch took all the hexagonal networks in Figure 15-10 and extended them to higher orders by superimposing them on a common central place. This common central place is the hub of the settlement system, its single most important city dominating the trade and services in the whole of the surrounding region. Each of the networks was then rotated about this common central city until as many as possible of the higher-order services coincided in the same centers. Such an arrangement insures that the sum of the minimal distances between settlements is small and that not only shipments, but transport lines, are reduced to a minimum.

You can envisage this process by imagining that the fixed $K = 3$ network is drawn on a map. The $K = 4$ network is now drawn on an overlay of transparent tracing paper and pinned to the $K = 3$ map by a single thumbtack through the common central place. By rotating the overlay, many major places on both the $K = 3$ and the $K = 4$ paper are made to coincide. For example, if we have a $K = 3$ school system and a $K = 4$ hospital system, we try to rotate the overlay so that the high school and the doctor's hospital both coincide in the *same* locations rather than being split between two. Lösch went on to add the $K = 7$ and still higher K networks to the map, always trying to get as many services as possible to overlap in the same locations.

A simplified version of his final result is shown in Figure 15-13. It shows that the resulting central-place system changed with distance away from the common central hub and was arranged, like a wagon wheel, with alternating

Figure 15-13. The Löschian landscape.
City-rich and city-poor sectors in the Löschian landscape. (a) Twelve sectors. (b) Centers with the largest number of functions. (c) An enlargement of a pair of adjacent sectors to show the underlying regular hexagonal pattern; the size of the dot is proportional to the number of functions. [From A. Lösch, *The Economics of Location* (Yale University Press, New Haven, Conn., and Fischer Verlag, Stuttgart, 1954), p. 127, Fig. 32.]

15-4

Extensions of the Christaller Model

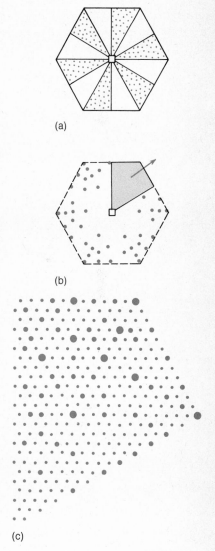

(a)

(b)

(c)

sectors. Twelve sectors are produced, six with many production sites and six with few (called by Lösch "city-rich" and "city-poor" sectors). In Figure 15-13 the metropolitan center is the center of 150 separate fields.

Thus, using the same basic hexagonal unit and the same K concept as Christaller, Lösch evolved a markedly different hierarchy. Christaller's hierarchy consists of several fixed tiers in which all places in a particular tier have the same size and function and all higher-order places perform all the functions of the smaller central places. In contrast, the Löschian hierarchy is far less rigid. It consists of a nearly continuous sequence of centers rather than distinct tiers. So settlements of the same size need not have the same function (e.g., a center serving seven settlements may be either a $K = 7$ central place or a center where both a $K = 3$ and $K = 4$ central place coincide), and larger places need not perform all the functions of the smaller central places.

Lösch's model represents a logical extension of the Christaller model. It is based on the same hexagonal unit and hence suffers from the same rigidity, but it yields a relationship between the size and function of central places that is continuous rather than stepped, and therefore more in accord with the observed distributions described in Section 15-2.

Periodic Variations The permanent provision of central goods implies a high and continuous level of demand. In most peasant societies central goods are provided by markets that are not open every day, but only once every few days on a regular basis. Although periodic markets are now only a small element in the central-place structure of western society and generally sell only agricultural products, they continue to be important in most peasant societies and are vital to the exchange structure of two-thirds of the world.

The relevance of Christaller's scheme to periodic markets is demonstrated by the central-place network of rural China. Figure 15-14 shows a portion of the Szechwan Province, southeast of Chengtu, which is simplified to a basic $K = 3$ system of nesting. [Cf. Figure 15-10(a).] Two levels of the hierarchy are shown: an upper level (Chung-ho-chen) and a lower level (Hsin-tien-tzu) which has a market area about one-third the size. Periodic markets are superimposed on the system as indicated in Figure 15-14. A 10-day cycle is usually divided into three units of 3 days. No business is transacted on the tenth day. With synchronized cycles, central goods can be circulated around several markets on a regular schedule and firms can accumulate enough trade to remain profitable. Hence, a central-place system is maintained by *rotating* rather than *fixed* central-place functions.

Market cycles vary widely in length. In tropical Africa, the market week varies from 3 to 7 days. For example, Yorubaland in western Nigeria works on an interlocking system of 4-day circuits. Generally the higher the population and per capita income, the greater the total trade and the shorter the length of the cycle. Where demand for goods is high enough, the market opens every day (and thus is a permenent central-place function); where demand is low enough, the cycles become so long that a service ceases to be provided within the region.

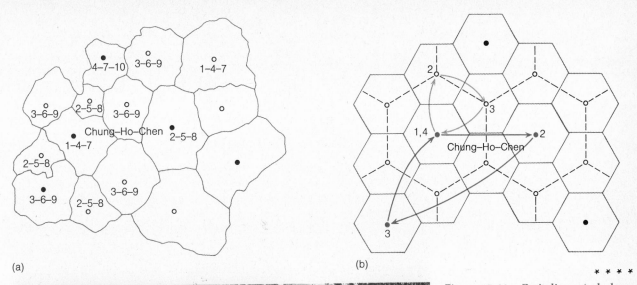

(a)

(b)

* * * *

(c)

Figure 15-14. Periodic central places.
(a) A map of rural centers in the Chinese provinces of Szechwan, showing the days on which markets are held. (b) The rural centers as part of a $K = 3$ Christaller network. The cyclic movements of traders around a market ring are examples of the space–time meshing of central-place functions. [After G. W. Skinner, *Journal of Asian Studies* **34** (1964), pp. 25, 26, Figs. 3, 4. Photo (c) by Ralph Mandol, DPI.]

Historical research reveals instances of space–time interlocking in medieval Europe, where cattle might be sold or cloth exchanged at large spring and autumn fairs. In the modern international economy the time period may widen to years and the region broaden to involve all the major capitals of the world. The World's Fairs and the Olympic Games, with their 4-year cycle, could be viewed as extreme extensions of the periodic case of the Christaller model.

15-5
Alternatives and Applications

Not all geographers are happy with the Christaller model, even in its more refined forms. They regard it as essentially a special case in two important ways. First, it is a special case in *conception* in that it describes a closed system where change can occur only from the bottom upward (i.e., with increased rural productivity in the lowest level of settlement leading to an enlarged hierarchy and therefore more higher-order centers). Within that closed system, pure competition for space is allowed. Second, it is a special case in *reality* since it emerged from the study of a particular area (southern Germany, a midcontinental location) with a particular history of settlement in which the "feudal" organization of agriculture had played a notable part. The Christaller model gives insights to settlement growth only in geographical areas that are rather uniform in character, economically isolated, and historically stratified.

The Vance Mercantile Model   Geographers who accept the above criticisms as valid do not see how the Christaller model can help in studying settlement patterns where the main forces of change are from the outside (i.e., the system is open rather than closed). In these patterns the hierarchy may evolve from the top downward with large seaboard cities, like those on the east coast of North America in the nineteenth century, acting as centers of innovation for external commercial forces.

Berkeley geographer James Vance's book *The Merchant's World* (1970) represents one of the first major attempts to challenge and augment the Christaller model of settlement structure. Vance approaches the problem through the eyes of the historical geographer, following the twisting path by which actual settlements have evolved—particularly the mercantile cities of America's east coast. There city growth had begun, in a sense, from the top down (the reverse of the Christaller model). Thus Boston was from the start of settlement the point of attachment, the economic hinge, through which the staple goods of New England (timber and fish) were concentrated and shipped back to Europe. If you look at the five charts in Figure 15-15, you can follow the way in which the settlement pattern in the Vance model emerges. Note the establishment of depots for staple collection, the antecedent pattern of land division (refer back to Figure 2-7) and the emergence of major wholesaling centers in the interior. Finally comes central-place infilling in the last phase of the model. Since Vance's ideas stress the importance of trade at each stage, the model is termed a *mercantile* model after the Italian word meaning "merchant."

Vance's model applies not just to North America but makes an equally coherent story for lands like Australia where the "limpet" cities of Brisbane, Sydney, Melbourne, Adelaide, and Perth were the attachment points from which the central-place development later partially emerged. Thus the attraction of the model is that it introduces a dynamic element into settlement models and supplements, rather than supplants, the earlier work. It also enriches our understanding of those long-settled areas to which the original

Vance's mercantile model

Based on exogenic forces
introducing basic structure

Christaller's central-place model

Based on "agriculturalism" with endogenic
sorting-and-ordering to begin with

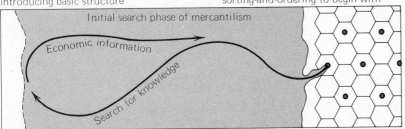

Initial search phase of mercantilism

Economic information

Search for knowledge

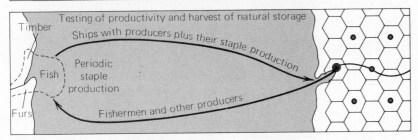

Testing of productivity and harvest of natural storage

Ships with producers plus their staple production

Timber

Fish

Furs

Periodic
staple
production

Fishermen and other producers

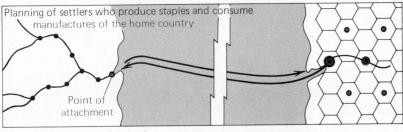

Planning of settlers who produce staples and consume
manufactures of the home country

Point of
attachment

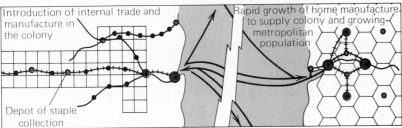

Introduction of internal trade and
manufacture in
the colony

Rapid growth of home manufacture
to supply colony and growing
metropolitan
population

Depot of staple
collection

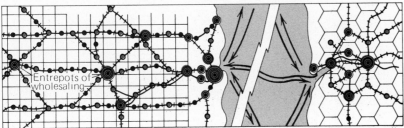

Entrepots of
wholesaling

Mercantile model with domination by
internal trade (that is, with emergence of
central-place model infilling)

Central-place model with a mercantile
model overlay (that is, the accentuation
of importance of cities with the best
developed external ties)

**Figure 15-15. The Vance mercantile
model of settlement evolution.**
Five stages in the evolution of settlement
under the mercantile model (left) and the
central-place model (right). The term
exogenic refers to external, and *endogenic*
to internal forces. The term *entrepot* is the
French word for storehouse and means a
commercial center for the import and export,
collection and distribution of goods. [From
J. E. Vance, Jr., *The Merchant's World*
(Prentice-Hall, Englewood Cliffs, N.J., ©
1970), Figure 18, p. 151. Reprinted by
permission of Prentice-Hall, Inc., Englewood
Cliffs, N.J.]

model best applies, accentuating the role of those coastal cities with the best-developed overseas trade links.

Postscript: Toward Applications   We have left until last the important question of the applications of a settlement model. How can geographers use ideas of city chains and central-place structure? Well, we can use these ideas in two ways. First, the relationship between the size and pattern of settlements summed up by the rank–size rule is stable enough over time for us to use it to project future patterns of settlement sizes. Consider, for example, the regular progression of curves for the United States in Figure 15-16, and try to think what forces we would have to bring in to distort or disturb the pattern. Of course, it is only the whole system of cities that is stable. Individual cities may have widely varying paths of growth. Compare, for example, the constant lead position of New York to the faltering pattern of a city like Savannah, Georgia. Still greater contrasts separate the rapid rise of Los Angeles with the dropout pattern of Hudson, N.Y.

  A second use of the central-place model is in regional planning. The settlement hierarchy in a newly settled area often tends to move from a primate to a rank–size form as population increases and the separate settlements are more closely integrated with each other within the region. For instance, if we were to design such a system for the settlement of the middle-west plateau of Brazil around the new primate city of Brasilia, the later readjustment of the hierarchy could be anticipated and investment in infrastructure (roads, power stations, schools, hospitals, etc.) adjusted accordingly. Central-place theories also have played a role in the designing of hierarchies of shopping and service

Figure 15-16. Hierarchic evolution. The diagrams show changes in the role of individual cities with a hierarchy. The regular growth of the U.S. urban system over 150 years contrasts with the varying growth trajectories of individual cities. [From C. H. Madden, *Economic Development and Cultural Change* (University of Chicago Press, Chicago, 1956), Vol. 4, p. 239, Fig. 1. Copyright © 1956 by The University of Chicago.]

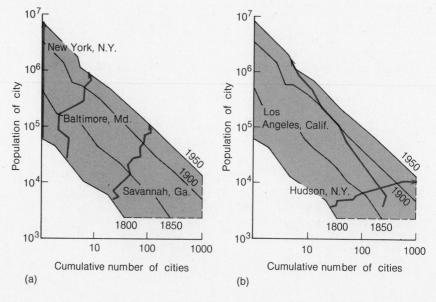

BECKMANN'S HIERARCHIC MODEL
The idea of integrating the concepts of one-dimensional city chains and two-dimensional city hierarchies has been proposed by mathematician Martin Beckmann [*Economic Development and Cultural Change* 6 (1958), pp. 243-248]. In his model, the population of a city of a given order is given by

$$P_r = \frac{LC_1K^{r-1}}{(1-L)^r}$$

where P_r = the population of a central place with order r in the hierarchy,

L = the proportion of the total population served by a central place located in that central place,

C_1 = the rural population served by the place with the lowest order in the hierarchy, and

K = the number of places of the next lowest order $(r-1)$ served by places of order r.

Thus, in a Christaller hierarchy with $k = 4$, $L = 0.5$, and $C_1 = 100$, a second-order central place will have a population of 800, a third-order place a population of 6400, a fourth-order one a population of 51,200, and so on. The more orders there are in the hierarchy, the more closely city sizes will conform to a continuous size distribution. Moreover, if L is small relative to 1, then the product of the rank of a place and its population size will approximate a constant, as the rank–size rule requires.

centers within cities. Increasing use is also being made of hierarchic concepts in the design of key service sectors like hospital systems.

In beginning this chapter we looked first at the simple chain models of settlements in one dimension and went on to consider hierarchic models in two dimensions. Both types of models are closely linked. Indeed, mathematician Martin Beckmann has shown that the more levels there are in a system the closer the array of city sizes will come to a continuous distribution. By contrast, regions with only a few units in the hierarchy will have sharply stepped rank–size distributions. (See the above discussion of the Beckmann model.) Rank–size distributions must be logical byproducts of the central-place system.

Like Alice in our opening quotation the geographer finds the world laid out like a chessboard. In this chapter we have been able to cover only the simplest and most basic moves in the complicated chess game by which cities "capture" and organize one another's territories into a kind of feudal hierarchy of metropolis, city, and village. Like all models our hexagonal chess set is an oversimplification of reality. To follow the actual "moves" by which any individual city develops would tax the skill of a Bobby Fischer.

Summary

1. Settlements are difficult to define accurately in statistical terms. Legal and administrative definitions of a city may seriously under- or overestimate its real population through so-called bounding problems.

Standardized definitions of larger settlements in the United States are based on criteria for the Standard Metropolitan Statistical Area (SMSA). Geographers are attempting to work out accurate city definitions based on functional economic areas in order to make comparisons between countries.

2. A spacing index for settlement patterns, the nearest neighbor measure, has been adopted by geographers to measure spatial patterns. Values of this index range on a scale from highly clustered to highly dispersed.

3. If settlements are ordered by size within regions, there is a relationship between a settlement's size and its rank. This regularity has led to the development of a rank–size rule. This rule appears to fit well only in some regions and distinctive variations may be seen when rank–size distributions are plotted on logarithmic graphs. The number of cities on the upper slope of the distribution form patterns termed primate, binary, and so forth.

4. Research into the nature of order in settlement patterns has led to the development of a central-place model. Central places provide an array of goods and services for complementary regions. The size of the array locates a central place within a settlement hierarchy. Recent research has introduced the con-

cepts of threshold and range of a central good into the model. In model form the range of a good is a circular outer boundary to a demand cone. Assuming equal economic and geographic conditions, the overlapping demand cones for a group of central places will take the form of a hierarchial hexagonal pattern. This pattern will minimize consumers' average travel to central places.

5. Various hierarchical levels of central places are made possible by complementary regions of varying sizes. The German geographer Christaller dealt with three special cases: the market-optimizing, the traffic-optimizing, and the administration-optimizing. Each has a specific system. Extensions of the fixed-K hierarchy were made by Lösch, who created a modified central-place model in which the size and function relationship is not stepped but continuous.

6. Central-place concepts may also be applied to explain the periodic markets of peasant societies. In this case, the central function circulates around the settlement on a regular schedule.

7. The historical study of population increase over time has led to the development of alternative models, such as Vance's mercantile model. This allows bridges to be built between the static theoretical models and the actual historical record of urban growth.

Reflections

1. What is the size of your own community? Look very carefully at your answer to see how it is affected by the bounding problem illustrated in Figure 15-1. Have you over- or underestimated your community's real size? How might you make your estimate more accurate?

2. Gather data for the size of cities in your state or province. Plot their position on a population rank–size diagram like that in Figure 15-4. How closely does the resulting distribution correspond to the one predicted by the rank–size rule? Suggest reasons for any departures from this rule you observe.

3. Why do so many small towns in western countries appear to be losing their central-place functions? Can you suggest any ways in which their decline might be arrested?

4. What periodic central functions are still found in western countries? List any you can find for your own area, and try to map their tracks. Would you consider college football an appropriate example?

5. Debate the relevance of the very formal settlement models presented in this chapter. Do they (a) fail to reflect the true complexity of city settlements or (b) provide a unique insight into their structure?

6. To what extent does the distribution of schools and colleges in your own city, state, or province have a regular spatial order? Identify some of the factors that you think might be "disturbing" this order?

7. Review your understanding of the following concepts:

bounding problems	primate patterns	market-size thresholds	the Vance
SMSAs	central places	fixed-K hierarchies	mercantile model
rank–size rule	complementary regions	periodic market cycles	

One Step Further . . .

Reviews of the basic theories of settlement hierarchies and their structures are provided in most texts on human geography. See, for example,

Abler, R., J. S. Adams, and P. Gould, *Spatial Organization* (Prentice-Hall, Englewood Cliffs, N.J., 1971), Chap. 10, and

Haggett, P., A. D. Cliff, and A. E. Frey, *Locational Analysis in Human Geography*, 2nd ed. (Arnold, London, 1977), Chaps. 4 and 5.

The classic work in central-place theory is Walter Christaller's study of southern Germany, published in 1933. It should certainly be dipped into and is available now in translation. See

Christaller, W., *Central Places in Southern Germany* (Prentice-Hall, Englewood Cliffs, N.J., 1966, transl.).

Another classic German work that has more ideas in its footnotes than many books have in their text is

Lösch, A., *The Economics of Location* (Yale University Press, New Haven, Conn., 1954).

Some modern theoretical departures are authoritatively presented in

Berry, B. L., *Geography of Market Centers and Retail Distribution* (Prentice-Hall, Englewood Cliffs, N.J., 1967), Chap. 4, and

Beavon, K. S., *Central Place Theory* (Longmans, London, 1977).

For a critical approach to settlement theory with emphasis on the historical evidence and dynamic models, the outstanding book is

Vance, J. E., Jr., *The Merchant's World: The Geography of Wholesaling* (Prentice-Hall, Englewood Cliffs, N.J., 1970).

Current research is reported in the major geographic journals. Look especially at the University of Chicago Department of Geography Research Papers *(published occasionally) for applications of central-place concepts.*

> The crowd, and buzz, and murmurings of this great hive, the city.
>
> ABRAHAM COWLEY
> *The Mistress* (1647)

chapter 16

Worlds Within the City

Cities as we know them are not like the dots on the map of the previous chapter. They are places of immense variety, throbbing or quiet, dangerous or secure, impressive or repellant. Times Square and Washington Square in New York City, or Soho Square and Grosvenor Square in London may be within a few city blocks of each other. Yet in other senses they are worlds apart. So here again, as in so many places in the book, we have to put on fresh lenses and change from a focus on cities like small specks on the global surface (''cities in space'') to a focus on the worlds within the city itself (''space within the city'').

Studies of the internal geography of the city is one of the fastest-growing areas within geography so here we are only able to touch on some aspects of geographers' work. We begin with the pattern familiar to most readers and look at the spatial structure of the western city, paying special attention to a city which has played a crucial part in studies of urban geography, Chicago. Second, we remain with the western city but look at it through the eyes of its residents. What image do they hold of their city? Third, we change our focus and try to put the western city in context. How typical is it of other cities around the world, differing as they do in their age and cultural origins? Finally, we look back at cities both western and nonwestern, as homes for an increasing share of the world's population. What problems do cities have and what are their prospects?

16-1
Spatial Patterns Within the Western City

The basic geometry of the North American city is well known to most readers. At the center is the downtown shopping district, the banks and offices, the hotels and theaters. Surrounding this is an area of rundown housing, mixed with some industry. Beyond this we run into modest residential areas, typically houses and apartments for office and blue-collar workers. Still further out come the family homes of middle-class suburbanites, thinning out until we come to the golf courses and the rolling acres of the very rich.

In this section we shall probe the familiar pattern to try and see what factors lie behind it. We shall look at the value of land within the city, the use to which it is put, and the population density it supports. Secondly, we shall take Chicago as an illustration of a western city and examine its pattern in relation to existing models of urban structure.

The Geometry of Land Values One measure of the value an urban community places on different sites within the city is provided by land values. As an example, Figure 16-1 shows a three-dimensional representation of land values in Topeka, Kansas. Despite minor variations in peripheral areas, the dominating elements in this and in most western cities are the extremely high values attached to land in the central part of the city and the generally steep decline of land values with distance from the center. The high point in values is found somewhere near the center of the central business district (CBD). The CBD is usually marked by tall buildings, an extremely high daytime population, and high traffic densities, and geographers have worked out various ways of mapping its precise location. (See the discussion on the next page about delimiting the CBD.)

If we take a series of such cities, we can draw a general picture of average land values (Figure 16-2). The value of land is highest in the city center and decreases toward the periphery, but the pattern is modified by two additional elements: main traffic arteries and intersections of main arteries with secondary centers at regular distances from the CBD. When we superimpose these three effects on one another on a *three-dimensional model*, the result is a conic hill whose flanks are disturbed by ridges, depressions, and small peaks. This land-value surface directly reflects the different accessibility of parts of the city and shows where the most intense competition for space occurs (and hence where the higher land values occur).

Land Values and Land Use What will the effect of these variations in land values be on the distribution of land uses within the city? Let us assume that a city wishes to establish a new university somewhere within its limits. If the university is constructed near the city center, it will be most accessible by public transport to all students, but it will be taking up valuable land that could be leased at high rents to commercial firms. Conversely, if we choose a green field on the edge of the city as the site of the campus, the land values there will be low and the university grounds can be more spacious. This

Figure 16-1.   Land values in an urban community.

This three-dimensional map shows land values in the city of Topeka, Kansas. Note the extremely high values in the city center and the smaller secondary centers. [From D. S. Knos, *Distribution and Land Values in Topeka, Kansas* (University of Kansas, Bureau of Business and Economic Research, Lawrence, Kansas, 1962), Fig. 2.]

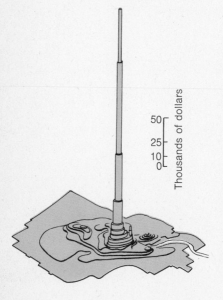

DELIMITING THE CENTRAL BUSINESS DISTRICT (CBD)

The central parts of western cities are marked by high buildings, high land values, and distinctive types of commercial activities (e.g., major department stores, corporation headquarters, specialized banks). To map the exact extent of this central business district (CBD), geographers have devised a number of specific measures. One of the most commonly used is that by Vance and Murphy. This needs the following information on floor area for each building:

(A) ground plan area (e.g., as shown on a large-scale map);

(B) total floor area on all levels (i.e., including all the upper and lower basement levels as well as the ground floor);

(C) total floor area devoted to specialized central business activities. These activities *include* stores, entertainment, banking and insurance but *exclude* residences, schools, parks, churches, wholesaling, vacant premises, and storage.

From this information two measures can be constructed: first, a *height index* (C/A) and second, an *intensity ratio* (C/B × 100). To take an example: assume a building with a ground plan of 100 m × 200 m (i.e., A = 20,000 m^2) which has a total floor area of 78,000 m^2 (= B) on four floors, and of this 45,000 m^2 (= C) is devoted to central business activities. Then its height index will be 45,000 ÷ 20,000, = 2.25, and its intensity ratio (45,000 ÷ 78,000) × 100 = 58 percent.

For mapping purposes it is more convenient to sum the areas for all buildings on a given city block, thereby giving a single block index. Where this height index has a value of 1.00 or more *and* the intensity ratio is 50 percent or greater, then a city block is potentially part of the CBD. The line is usually drawn by starting at the city's peak land value intersection and adding in *contiguous* "CBD blocks" (i.e., those with *both* measures at or above the critical levels). The exact nature of the area delimited will depend on the precise definition of a specialized *central* business activity. Lists of such activities with more refined modifications of the Vance-Murphy index are given in M. H. Yeates and B. J. Garner, *The North American City*, 2nd ed. (Harper & Row, New York, 1976) Chap. 12.

advantage, however, will be offset by the school's eccentric location. Most students will have to travel farther, some right across the city from the far side.

All types of urban land use entail this type of quandary. As businesses move farther from the city center, they gain from the cheaper land prices or rents; but they are increasingly divorced from the center of their potential market and stand to lose money from increased transport costs. We can describe this trade-off between land prices and transport costs by a series of nonintersecting lines, termed by the economist *bid-price curves* [Figure 16-3(a)]. Each line represents the rent values that exactly balance the increased transport costs due to locating away from the city center. If we assume that transport costs increase regularly with distance, then the bid-price curves will be parallel straight lines sloping downward with distance from the city center. Note that the lower lines within the family of parallel lines represent lower rent levels and therefore are always preferred over higher lines.

By superimposing bid-price curves on the actual rent curve for a given city, as in Figure 16-3(b), we can determine both the point where the best trade-off is reached and the corresponding rent level. The best location will be where the actual rent curve just grazes the lowest possible bid-price curve. At this

Figure 16-2.   A general pattern of city land values.

This idealized land-value surface emphasizes the high value ridges that run outward from the CBD along the main arterial roads. Secondary centers on these roads cause local peaks.

Figure 16-3.   City land-use models.

For any urban land use—be it a cemetery, a motel, or a college—a location represents a trade-off between the convenience of being near the accessible city center and the high rent levels we must pay to locate there. If we describe our assessment of these two conflicting goals by a family of bid-price curves (a), we can compare these with the actual rent levels (colored line). Note how the equilibrium position (A_1) indicating the best trade-off (and thus the best location) may shift (to A_2 or A_3) as the city grows and land values rise (in time periods t_1 through t_3). Thus certain types of urban land use (e.g., single family homes) may be squeezed out toward the margin of the city as it grows.

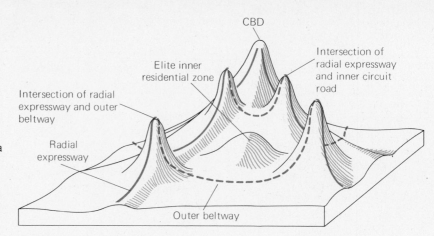

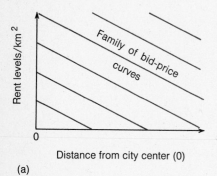

(a)

point (A_1) the values of both the actual rent curve and the bid-price curve are equal. If we think of this diagram as three dimensional, the bid price curves become a series of cones centered on the CBD, and we see that A_1 is not a point but a ring of possible locations around the city center.

As a city grows larger, land values increase, especially at the center, so the land-value curve becomes both higher and more concave, as in Figure 16-3(c). Hence, the ring of optimal locations for land-using activities is forced outward from near the city center (A_1 in time period t_1) to successively more peripheral locations at A_2 and A_3. We can trace this effect in the displacement of activities like manufacturing from inner city to suburban locations.

Each type of land-use activity in a city has a distinct pattern of bid-price curves. Those activities which gain greatly from locating "where the action is" near the city center will have steep curves. Theaters, insurance brokers, publishers, are all examples of activities which depend on a high degree of contactability and need accessible locations. Conversely, other activities may not be greatly affected by their location but may be very anxious to avoid high rentals: These will have bid-price curves with a gentle slope. Activities with steep bid-price curves will be able to cling to the steep slopes in land values around the CBD; others are gently angled and find a foothold only in the remoter parts of the city. Figure 16-4(a) illustrates hypothetical curves for

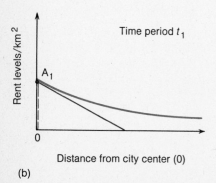

(b)

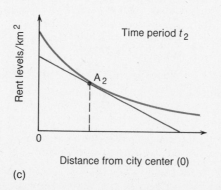

(c)

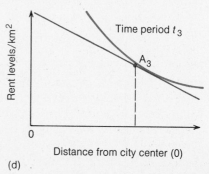

(d)

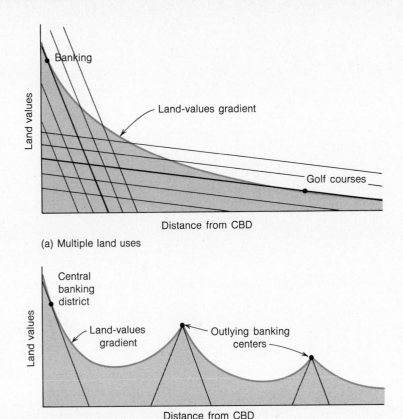

(a) Multiple land uses

(b) Multiple centers for single land use

Figure 16-4. Multiple land uses. An extension of the arguments in Figure 16-3 to two types of land use. (a) Characteristic bid-price curves for an intensive high-rent land use (banking) and an extensive low-rent land use (golf courses). Note how the equilibrium position for the first is in the downtown area, while that for the second is on the outskirts of the city. (b) This shows how a varied land-values gradient with secondary peaks may allow banking to find a series of equilibrium positions at varying distances from the CBD.

banking and insurance offices (which typically cling tenaciously to downtown locations) and golf courses, which are usually located on the fringes of cities and sensitive to rises in land values. Note that the cross section shown here is a very simple one; if we add outlying secondary centers to the figure we can see that banking may be carried on there too, albeit on a smaller scale. [See Figure 16-4(b).]

Land-Use Mosaics, Chicago Style Land use in urban areas is a major research area for the urban geographer. Probably the most thoroughly dissected city is Chicago, where social scientists like Robert Park and E. W. Burgess began a trail of research in the 1920s that is still being followed a half-century later by geographers like Harold Mayer and Brian Berry. If we take an aerial photograph of Chicago from the south, looking toward the CBD in the Loop (Figure 16-5), we are viewing a city whose structure has formed a touchstone for studies of large metropolises elsewhere.

Can we make any sense of the complex mosaic of land uses shown by Chicago? Figure 16-6 shows one guide to the pattern provided by geographers. Look first at the top row of the diagram. This shows an idealized

Figure 16-5. Chicago.
Poet Carl Sandburg's "City of the Big Shoulders" has had a special fascination for geographers. From the 1920s on, its spatial structure has been studied and dissected by a series of distinguished researchers. The key to its semicircular structure is the CBD, whose high-rise buildings form the famous "Loop" area (named from the loop of elevated railroad tracks that encircle it). The zones and sectors that flare out from the CBD are summarized in Figure 16-6. The transportation lines that are so prominent in the photograph have had an important effect on the city. Chicago grew up in the 1830s at the strategic southwest corner of Lake Michigan, and the canal connecting the Chicago River to the Illinois River (constructed in 1848) brought a rapid rise in the city's role as a freight-hauling center. In the same year, the first railroad link was completed and the stage was set for a period of remarkable economic growth linked to Chicago's dominant position on the expanding railroad network of the Midwest. Its population grew around 30,000 in 1850 to past the 1 million mark by 1890. Today, with over 7 million people in its metropolitan area, Chicago is America's second largest city and the tenth-largest city in the world. To meet the vast transit needs of the metropolis, an intricate system of urban highways like the Dan Ryan Expressway (in the foreground) has been constructed. [Photography by William A. Garnett.]

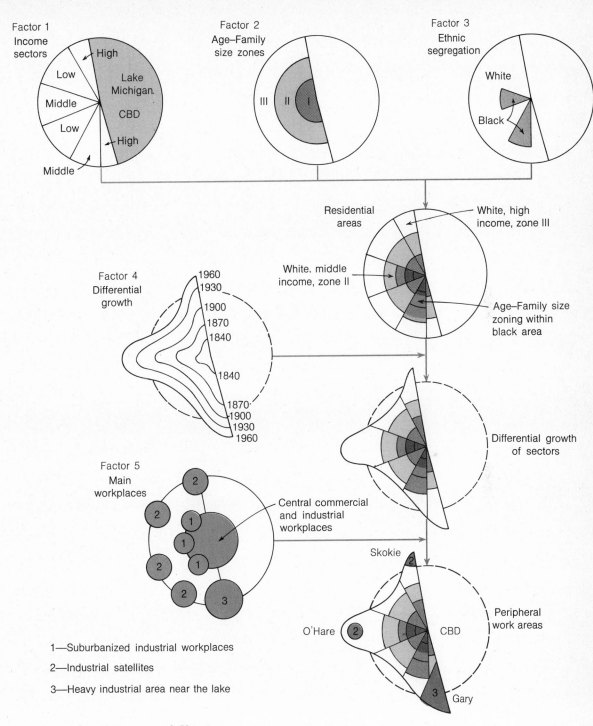

Figure 16-6. The regional structure of Chicago.

Much of the variety within Chicago's semicircular urban structure (itself a product of the lakeshore location) can be explained by the combined effect of five factors. The arrows show the ways in which factors are cross-bred together in this simplified model of the city. For a description of each factor and its impact on the city, see the text discussion. [From P. H. Rees, University of Chicago, unpublished master's thesis, 1968.]

Chicago as a circle centered on the CBD with Lake Michigan to the right. The CBD is the key feature, marked out by a score of factors (its high land values, its soaring daytime population, its high buildings, its age, and so on). Around this CBD the city's housing has developed as a series of wedge-shaped *income sectors* (upper left) which fan out as we move outward. A drive north or south from the CBD along the lakeshore will take you through much higher-income housing than a drive to, say, the northwest. Once housing of a particular type had been established in a sector, it tended to persist as new housing of similar quality was built. Thus the city expanded outward in this wedge-shaped fashion.

A second factor is a series of ring-shaped *age and family-size zones* centered on the CBD. The three zones shown mark a regular progression from the apartment-dwelling population of Zone I (typically, people in their twenties or an older generation in their fifties or beyond) to the suburban family-home population of Zone III (typically, families with young children). On the right of the top row is the third factor, *ethnic segregation.* This shows the presence of two distinctive black wedges to the south and west, in the inner part of an otherwise white city.

If you follow the arrows from the first three factors, they lead to a fourth diagram of Chicago. This is made by overlapping the first three factors to give a mixture of wedges and zones. Thus we get a white, high-income, Zone III in the north of Chicago (the Evanston area). Note how the two black zones distort the simple model: Each black zone is itself subdivided into three zones related to age and family size, and each is an area of relatively low income despite its occurring in middle-income sectors in terms of our first diagram (upper left). Indeed, an area like the black south side of Chicago is in some ways a detailed replica of the whole city, finely divided into a complex mosaic of income, age, and family-size units.

Of course, Chicago is not the simple ring shown in the upper half of Figure 16-6. So we introduce a more accurate map of the city showing its *differential growth* between 1840 and 1960. This fourth factor is also blended into the model. Note how the zones within the various sectors are now displaced in our map showing differential growth of sectors. Finally, a fifth factor, the *main workplaces,* can be added to the model: The main commercial center plus three types of outlying industrial area are shown. Since to add all these to the model would be too complicated for our final diagram of the city (lower right), we include just some examples of the way in which industry helps to shape the land-use mosaic of the city. Two industrial satellites are shown, the Skokie area in the north and the area near O'Hare airport in the west. South of the city the heavy industrial area of Gary is also shown.

The patterns shown in Figure 16-6 show how various types of spatial segregation interact in shaping the mosaic of land use within a city like Chicago. Spatial segregation affects not only the clustering of social groups as in the black ghetto on the south side of Chicago or the Italian quarter to the northwest; it also plays a role in decisions to group certain land uses, like light industry, in particular sections of the city.

Electronics have shrunk world space in the 1970s to the size of a TV screen. Satellites and television allow each of us city-bound dwellers an unprecedented opportunity to peer into our neighbor's backyard. On an average day's TV program we find items which range from hotel fires in São Paulo, to floods in the Australian outback, to a nature film shot in the Canadian tundra. Never before have our senses been stretched over the whole globe in this way.

But what about our immediate neighbors? How do we view the urban world that is right around us? The evidence at hand from geographers and psychologists suggests that we retain primitive, twisted, and biased pictures of our local "world" that may be far removed from the "real" world described in the maps in the preceding section. So we look at some of the early results in an important and rapidly growing area of geographic research on images of the western city.

Mental Maps of Los Angeles Consider the way we look at the city in which we live. How much of it do we know well? How much of it do we not know at all? Is the geographic view of the city from the central areas the same as that from the suburbs? If not, then how does it differ? Figure 16-7 presents three maps of Los Angeles that show the perception of the city by people residing in three district areas within it. A sample of respondents in each area was asked to sketch maps of the city based on their own travels within the city and the contacts they had made. Their combined views are summarized in the maps. Those living in Boyle Heights, a poor inner section of the city near downtown Los Angeles, have a quite limited view of the city dominated by the nearby city hall, railroad station, and bus terminal [Figure 16-7(a)]. Suburban residents of Northridge in the San Fernando Valley had a limited but spatially more extensive view than Boyle Heights residents [Figure 16-7(b)]. They had an extensive perception of their own valley and its facilities, but little real familiarity with the main city, beyond the Santa Monica Mountains. The respondents around the University of California campus at Westwood, an

16-2
The Image of the Western City

Figure 16-7. Images of a city. The maps show Los Angeles as seen by a cross section of its residents from three contrasting districts (marked by a star). Boyle Heights (a) is a largely black neighborhood near the downtown; Northridge (b) is a suburban residential community in the San Fernando Valley; and Westwood (c) is a high-class housing area near the UCLA campus. Each cell indicates a different section of the city; tints indicate the proportion of those interviewed who were familiar with that section of the city. Unshaded cells were unfamiliar and lie outside the residents' area of perception. [Data by R. Dannenbrink, Los Angeles City Planning Commission. From *National Academy of Sciences Publication No. 1498* (1967), pp. 107–112, Figs. 2-4.]

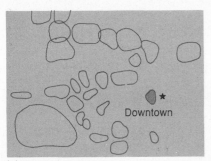

(a) Boyle Heights

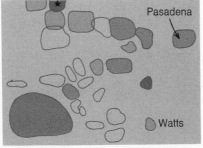

(b) Northridge

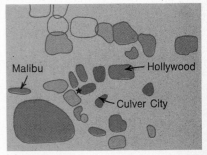

(c) Westwood

Familiarity (percent)

upper-class sector just west of Beverly Hills, had a wide-ranging and detailed image of almost the whole Los Angeles metropolitan area [Figure 16-7(c)].

This kind of research indicates something about the warped mental maps we have of cities. Groups from higher-income areas have a wider view of the city, and more educated groups have a more accurate view of the city. On the other hand, this research fails to explain why certain parts of the city environment are commonly known, or why some cities produce more positive and memorable mental images than others. One day in Cincinnati may leave a person with a clearer mental map of the city than one day in Kansas City, for example. Why?

Mental Maps of Boston Some clues to the answer to this question have been provided in *The Image of the City* which describes work carried on by Kevin Lynch at the Massachusetts Institute of Technology. Residents of three contrasting North American cities (Boston, Los Angeles, and Jersey City) were interviewed and asked to sketch a map of their city, to provide descriptions of several trips through the city, and to list and comment upon the parts of the city they felt were the most distinctive. With these documents Lynch pieced together the public image of each city held by its inhabitants.

There were considerable, and interesting, variations in the responses of individuals linked to their age, sex, length of residence, area of residence, and so on. But enough common ground was found to allow some citywide generalizations. Lynch organized the common elements of the mental maps into five types of spatial phenomena. The results for Boston (Figure 16-8) are shown in Figure 16-9.

The five types of elements can be defined as follows. *Paths* [Figure 16-9(a)] are the channels along which we customarily, occasionally, or potentially move within the city. They range from streets to canals and are the reference lines we use to arrange other elements. *Edges* [Figure 16-9(b)] are linear breaks in the continuity of the city. They may be shorelines, railroad tracks, or barriers to movement. The Charles River in Boston and the lakefront in Chicago have all the abrupt barrier qualities of an edge. *Nodes* [Figure 16-9(c)] are focal points within the city. They are commonly road junctions or meeting places. Louisburg Square in Boston or Times Square in Manhattan are typical nodes. *Districts* [Figure 16-9(d)] are medium-to-large sections of the city that we can mentally enter "inside of" and that have some common identifying character. Beacon Hill or South End in Boston are typical districts. Finally, *landmarks* [Figure 16-9(e)] are also reference points but much smaller in size than nodes. A landmark is usually a simple physical object: a building, a store, a mountain. It may be memorable for its beauty or its ugliness. The gold dome of Boston's state house or Nelson's Column in London illustrate the role of landmarks in giving structure to our image of a city.

Waterfront   CBD   State House   Beacon Hill

Charles River

Figure 16-8. Images of Boston.
An aerial view of the city of Boston looking across the Charles River. But how do the residents of the city see it? What landmarks, markers, and districts do they recognize and use to orient themselves? Figure 16-9 suggests some of the answers from research by Kevin Lynch. [Photograph from Rotkin, P. F. I.]

The relative strength and richness of these five spatial elements give coherence and character to a city. Cities with strong elements may be interesting environments to live in, despite dilapidation or deterioration. San Francisco, Cincinnati, New Orleans, Montreal are North American cities that come into this first category. Cities with weak elements may be formless, monotonous, and lacking in character. You may like to provide your own candidates for this type of city.

Lynch's approach has been repeated in many cities around the world on a variety of scales. Dutch geographers have emphasized that although regular street structures help us to find our way easily, difficulties in orienting ourselves arise when the overall structure of a city is clear but the individual elements are too uniform to be individually distinguished. It appears that we construct mental maps more readily in an older European city (a baffling jumble of streets, but with distinctive elements to serve as locational clues to where we are) than in a newer North American city (a regular street pattern, but too many roads looking like each other).

Figure 16-9.   Elements in mental maps of Boston.

Five main elements in the image of Boston as seen by its residents. [From K. Lynch, *The Image of the City* (MIT Press, Cambridge, Mass., 1960), p. 147, Fig. 37. Copyright © 1960 by The Massachusetts Institute of Technology. Photo of Beacon Hill by Christopher S. Johnson, Stock, Boston.]

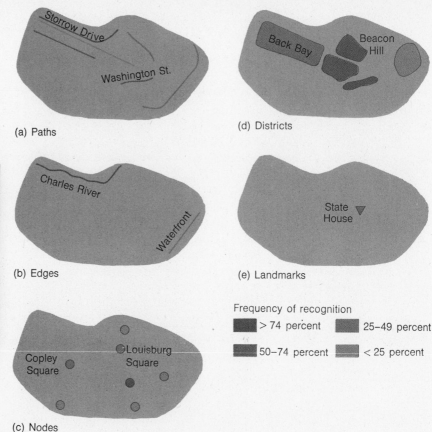

(a) Paths

(b) Edges

(c) Nodes

(d) Districts

(e) Landmarks

Frequency of recognition

> 74 percent   25–49 percent

50–74 percent   < 25 percent

The Persistence of Urban Images   But why are these images important? Our answer, as we saw in an earlier discussion of regional images (see Section 11-4), is that images underline action. We make decisions on the basis of what we believe to be true. This is clearly seen in the persistence of elite districts within the city. Now you do not have to live long in any western city to pick out the prestige areas to live. To reside in the Bel Air area of Los Angeles, the South Yarra area of Melbourne, or the Mayfair area of London is to have an address which has been associated with wealth and elegance for several decades.

The Beacon Hill area of Boston is typical of these elite districts. It lies on the western edge of the Shawmut Peninsula on which the city of Boston was established. Originally an area of merchants' housing close to the harbor wharves and centers of commerce and government, it became fashionable in the late eighteenth century. As the city grew over the following 200 years it retained its character and attraction, resistant to downtown encroachment and inner-city decay. For the well-to-do it remains an attractive and convenient location for weekday living, increasingly backed up by weekend "second homes" in rural New England.

Urban geographer Ron Johnston has tracked a similar persistence for the fashionable southeastern suburbs within the Australian city of Melbourne. He used as a definition of high social status a resident's inclusion in the annual *Who's Who in Australia*. Gridding the city into squares, Johnston found the greatest concentration of prestige addresses in the 1960s was in exactly the same square as a half-century earlier. This square lay in the suburb of Toorak. Like Beacon Hill, Toorak is within a short distance of the CBD. It contains large mansions which have been progressively subdivided as demand for this desirable space increased.

Johnston's study goes on to show that the social pattern in Melbourne is neither a static nor a simple one. The center of gravity of prestigious residences has moved outward though at a much slower rate than the outer boundaries of suburban living. Indeed, the convenience of inner-city living has caused a reversal of the suburban-flight trend common in most western cities. As commuting distances lengthened, so higher-income families have been moving back into inner-city areas. This movement is now happening in many western cities. In London, the process of upgrading inner-city housing is termed *gentrification*, from the English slang term "gentry" for the upper-middle class in that country.

So far in this chapter we have chosen to stress broad themes and common elements, the things that give western urbanization its overall character and western cities their general flavor. But each city is also unique and special. Cincinnati is not Columbus, Ohio; neither is Denver a Detroit. Outside the United States the pattern of diversity gets wider. Clearly the cultural mosaic of a Montreal, the vitality of a Hong Kong, the formality of a Vienna, or the poverty of a Calcutta cannot be squeezed into a simplified model. For as we saw in Chapter 12, cities emerged at different times, for different reasons, and from different cultural backgrounds. A model based on the western city alone needs to be checked against a more general framework.

16-3
Cross-City Comparisons

Structure, Process, and Stage in City Evolution   In this section we shall try to put the western city in a broader context and see how it fits into the more diverse world picture. To do so we are going to borrow some ideas from physical geography that we first met in Chapter 6 (see Section 6-2). If you look back you'll notice the photograph of an important physical geographer, William Morris Davis. Now Davis is important because he developed a way of analyzing the structure of the world's land surface, its terrain. Davis argued that a land form was a product of three sets of circumstances, its original *structure*, the erosive and constructive *processes* which had worked on it, and the *stage* that its evolution had reached.

To geographers, cities have something in common with the constructional land forms studied by Davis. As we saw in the case of Chicago (Figure 16-6),

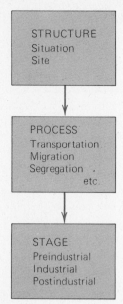

Figure 16-10. Elements in the comparison of cities.

Use of the notions of structure, process, and stage in evaluating the geographic structure of a city.

many cities grow outward ring by ring, something like a tree. A building boom every 20 years or so pushes out the edge, new territory is annexed, and gaps nearer the center are filled in. Driving into the city, the urban geographer can often read the brick, concrete, and stucco landscape much like geologists read a drill core of ocean sediments.

Figure 16-10 sketches a model of city evolution on Davisian lines. We shall follow the sequence shown there, taking illustrations under each heading.

Structure: Situation and Site   The starting point for urban comparisons is their physical geography. If we begin with the big cities of North America, then differences in physical setting is striking. Montreal began on an island in the St. Lawrence while Miami started on a low, sheltered beach ridge on the Florida coast. Pittsburgh grew up where the Allegheny and Monongahela rivers meet to form the Ohio, while San Francisco rose around a partly sea-filled depression along the California coast. Some cities are located in dramatic, strikingly beautiful locations with strong terrain differences—Vancouver in British Columbia, Rio de Janeiro in Brazil, Sydney in Australia. In other cities it may be hard to spot any differences at all in elevation.

Geographers usually consider the location of a city by two scales. *Situation* refers to the general regional location, while *site* refers to the characteristics of the actual location where the city began and the terrain over which it spread. Figure 16-11 illustrates the distinction between the two terms for an African city. Cape Town is the legislative capital of South Africa, with a population of 850,000 people. Its foundation in 1652 by Jan van Riebeeck of the Dutch East India Company reflected its ideal situation as a halfway station (supplying water and fresh food) for ships sailing on the 22,000 km (14,000 mi) journey between Holland and its overseas empire in the Dutch East Indies (now Indonesia). But Cape Town also had a remarkable site. Table Bay provided a safe harbour on a generally inhospitable coast, while Table Mountain rising to 1090 m (3570 ft) had a dramatic impact on the future shape of the city. As Figure 16-11(b) shows, outward growth around the original settlement (now the CBD) has been split into distinct eastern and western suburbs. Notice that although the absolute character of situation and site may be unchanged, their interpretation may. The opening of the Suez Canal reduced the situational significance of Cape Town as a way station. Likewise, civil engineering works have modified the site value by allowing part of Table Bay to be filled in order to extend the commercial sector of the city.

Process: Sorting Mechanisms Within the City   Once a city has been established, then various growth processes may be set in motion. We have already met two of these. First, we saw in Section 14-2 how any urban activity sparked off a train of related effects for housing and service activities. Second, we saw in Section 14-3 how changes in urban transport allowed a city to grow outward at various speeds and in distinctive ways. (You may like to look back

Figure 16-11. Cape Town: situation and site in city location.
Cape Town on the southeastern corner of Africa was founded in the seventeenth century. It had the advantage of a unique situation at the halfway point between Europe and the East Indies, combined with local site advantages. An early eighteenth-century print is compared with a contemporary air photo. Note the way in which urban spread from the old center has been guided by the terrain contrasts. [From P. Kolben, *The Present State of the Cape of Good Hope,* (London, 1731). Photo by Wide World Photos.]

Figure 16-12. General circulation model of city migrations.

Contrasts in the movements of black and white migrants into and out of an American city. Percentage figures (shown) are for the east coast city of Baltimore and are estimated from U.S. census figures for the 1950s and 1960s. All figures in the lefthand charts sum to 100, while those in the right show net differences. Thus 35 percent of all black movements were from the rest of the nation to the city, while 12 percent moved in a counter direction, giving a net cityward movement of 23. In absolute terms, the number of black movements is only about one-tenth of the white. This figure is based on data given in a penetrating analysis of the Baltimore situation by geographer Sherry Olson in *Baltimore* (Ballinger, Cambridge, Mass., 1976).

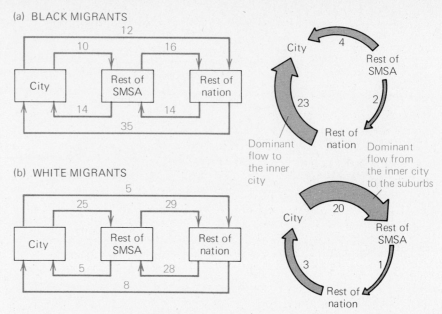

again at Figure 14-13 and Figure 14-14 to revise these ideas.) Since these have already been covered, we shall concentrate here on a third group: the social processes that sort out population into distinct patterns within the city.

Consider the situation shown in Figure 16-12, which charts recent flows of black and white population for the American east coast city of Baltimore. Note that in the case of the black population the dominant movement is an inward one, bringing new migrants from the southern part of the United States into the central part of the city. In contrast, the circulation of white migrants is more complex: It is dominated by an *outward* movement from the city to the suburbs. Exchanges between the suburbs and the rest of the United States are roughly in balance.

The trends shown in Figure 16-12 have together led to a continuing social and ethnic differentiation of inner-city districts. In the central residential areas nearest the downtown, employment opportunities are low and population is on the decline. Nevertheless, these districts continue to serve as reception areas for migrants from poor rural areas. The international migrants from Ireland and Italy have been succeeded in the middle part of this century by blacks from the southern states, whites from Appalachia, and a substantial Mexican population in the southwestern cities. The proportion of black people in the cities is rising and, if we assume that present city limits will not change, will probably continue to rise. Washington, D.C., now about 65 percent black, is expected to be 75 percent black by the end of the century. By the same date Cleveland, Ohio, and Newark, New Jersey, will be over 60 percent black, while six other cities (Baltimore, Chicago, Detroit, New York, Philadelphia, and St. Louis) will be 50 percent black. In the suburbs surrounding these cities the black population is expected to expand, but it is unlikely to exceed 25 percent. The highest increases are projected for the San Francisco–Oakland suburbs.

Stage: From Pre- to Postindustrial Cities   In considering the general urbanization process, we noted the way in which countries could be divided into preindustrial, industrial, and postindustrial phases. (See Figure 14-5.) This division provides a useful basis for discussing the third element in our model, the role of time.

The Preindustrial City   The preindustrial city is one which serves a population in which most of the jobs are still in agriculture. Such a situation was typical of most countries before the nineteenth century and remains true for a significant number of Third World countries today. So it is not surprising to find a great variety in preindustrial cities in both time and space: A medieval European town like Ghent is clearly different in detailed structure from Mecca in Arabia or Mandalay in Burma. Nontheless, urbanist Gideon Sjoberg has argued that all share in common certain characteristics which mark them off from the urban-industrial centers of the last two centuries. What are these characteristics?

The first set of common features is structural. Most preindustrial cities were sharply set off from the surrounding area by a defensive wall. (See Figure 16-13.) Access into the city was controlled by a number of gates. Within the city, buildings were highly concentrated with a narrow, irregular street pattern. With the exception of public buildings (churches, temples, palaces), housing was one- or two-story.

The second set of characteristics is functional. Overall, the preindustrial city was largely concerned with marketing, commerce, and craft industries, with secondary religious and administrative functions. The various functions were intermingled so that homes, workshops, street markets, and shops intermingled. There was, however, zoning in the sense that workers in a particular craft were concentrated in a particular street or district. So the Kansari brassworkers have concentrated in one small district of northern Calcutta. Frequently these loosely defined cells might take on added significance if the trade was linked to membership in particular ethnic or religious groups.

Figure 16-14 depicts the largest black city in Africa, Ibadan, in the western zone of Nigeria. The city can be divided into distinct zones, each with a separate ethnic origin and a different culture. The largest zone is the original core area of the Yoruba people. It is characterized by densely packed compounds, each housing a large and extended family unit. It has a high population density and an intricate system of special markets. On the northern edge of Yoruba town lies a small zone of Hausa people from northern Nigeria. They fulfill a special role in marketing cattle from the north. In the third zone is the European community associated with the original British colonial administration. It now has houses of banking and commerce, medical and

Figure 16-13. Preindustrial cities. Three examples of preindustrial cities from different time periods and cultural contexts. (a) A planned, walled city built on a fortified hilltop site, Carcassone in the southern part of France. (b) Bristol, in southwest England, retained its fortified gates and walls into the seventeenth century. This section from a 1673 map shows two hubs of its economic life: first, the busy port with its important links with Africa and North America through the slave and tobacco trade; second, the streets focusing on the market cross where markets for the rich agricultural area that surrounded the city were held. Note the signs of prosperity in the large merchant houses and numerous medieval churches. (c) The concentration of the upper classes in the central areas of cities is still maintained in many Third World cities. Popayan in western Colombia shows a pattern of social areas, rich near the center, poor on the periphery, that is the opposite to the pattern in most western cities today. [(a) Photo by Ray Delvert; (b) from Jacobus Millerd, *An Exact Delineation of the Famous City of Bristoll,* 1673; (c) after H. H. Whiteford in H. Carter, *The Study of Urban Geography,* 2nd ed. (Arnold, London, 1975), p. 187, Fig. 9.3.]

(a)

(b)

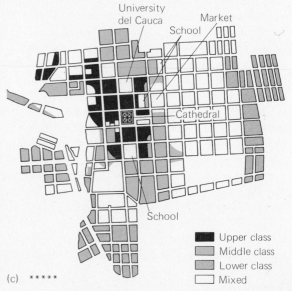

(c) *****

University del Cauca

School

Market

Cathedral

School

Upper class
Middle class
Lower class
Mixed

educational services, and low-density suburban housing. In addition to the three broad zones there are areas dominated by Lebanese and Syrian traders; and within the Yoruba area there are important differences in neighborhoods related to the origin, skills, and time of arrival of family groups. The spatial form of the city, its population density, and its economic organization are inseparable from the cultural diversity of its inhabitants. Each ethnic group forms part of a social and spatial mosaic whose components touch but do not penetrate one another.

Overall, there appears to be a complex patchwork within the preindustrial city rather than steep gradients. Densities of population were generally high, and the compact size of the city allowed ready access between the various districts. As to the rich, they tended to live centrally near the high-activity areas like the central market places.

The Industrial City We have already seen in the first half of this chapter something of the structure of major industrial cities. If we take a single western city and map its changing population density over time, we find that the population tends to spread out like a slowly melting ice-cream cone, covering ever wider areas but at ever lower densities. Figure 16-15 shows the pattern of population change for Chicago during a 100-year period by a series of population-density gradients that steadily decline in steepness from around 1860 to 1920. The change over the last half-century has been less spectacular but has included a market reduction in density at the center of the city.

How typical of the spatial structure of cities is Chicago? Cross-cultural checks on population declines with distance from the city center have been intensively studied, and investigators have reached some general conclusions about the shape of these slopes. For example, economist Colin Clark studied

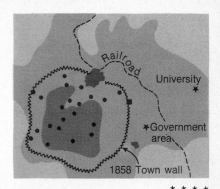

- ● Yoruba market
- ■ Hausa market
- Core region pre-1850
- Western banking and commercial area
- Boundary of built-up areas in 1960s

Figure 16-14. Ethnic and cultural divisions within a city.
The map shows the main regions within the Nigerian city of Ibadan, which has a population of approximately 700,000. [From A. L. Mabogunje, *Urbanization in Nigeria* (University of London Press, London, and Africana, New York, 1968), p. 206, Fig. 26.]

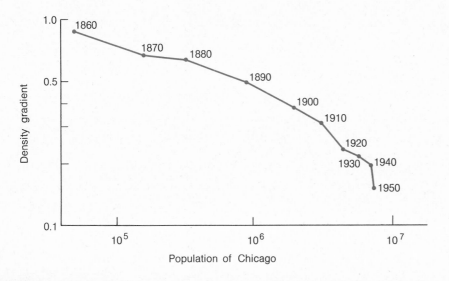

Figure 16-15. Changing urban density gradients over time.
The average slope of Chicago's population density away from the city center was steep in 1860 but has declined steadily as the city expanded. Compare this graph with the map of Chicago's outward spread in Figure 16-6. [From P. H. Rees, University of Chicago, unpublished master's thesis, 1968.]

URBAN DENSITY FUNCTIONS

Study of scores of urban areas throughout the world led Colin Clark to suggest a general model for the decline of urban population densities away from the CBD. He proposed a *negative exponential* form in which population decreases at a decreasing rate with distance; that is,

$$Z_d = Z_0 e^{-bd}$$

where Z_d = the population density at distance d from the CBD;

Z_0 = a constant indicating the extrapolated population density at distance zero, that is, at the center of the city;

e = the base of the natural logarithms (2.718);

b = a constant indicating the rate of decrease of population density with distance; and

d = the variable distance.

Thus with a central density of 1000 people/km^2 at the center and b = -1.0, we should expect a density of 368 people/km^2 at 1 km from the CBD, 135 people/km^2 at 2 km, 50 people/km^2 at 3 km, and so on. A comparison of Z_0 and b values allows us to compare different urban structures easily. Clark's work has been extended by others to allow more complex density functions to be matched and compared. [See M. H. Yeates and B. J. Garner, *The North American City* (Harper & Row, New York, 1971), Chap. 10.]

population-density gradients for a group of 36 cities from Los Angeles to Budapest from 1807 to 1950. He found that all the curves could be described as *negatively exponential*—that is, as decreasing sharply at first with distance from the center and then getting progressively flatter. (See the marginal discussion of urban density functions.) In western cities the decline in population density is reflected in a familiar sequence of housing types, with high-rise apartment blocks near the CBD and low-density housing on semirural tracts of land on the suburban periphery.

Extension of these cross-cultural studies to nonwestern cities shows that we still have a lot to learn about the way cities evolve. For example, one striking feature of an Indian city like Calcutta is the continued increase in densities near the center and the stability of the density gradient with urban growth. The falling degree of compactness and crowding that characterizes the growth of western cities is not seen in Calcutta, where both tend to remain constant. Given the same increase in population, the periphery of the nonwestern city expands less than the periphery of the western city. Figure 16-16 summarizes the variations in both time and space that characterize the two types of city. As the transport revolution which shaped the western city makes its impact elsewhere, more cities may conform to this pattern.

The Postindustrial City What kinds of forms will replace the present urban-industrial center? If we follow the argument of Figure 14-5, then we should expect jobs to be heavily concentrated in the tertiary and quaternary sectors. This means that most jobs in the city would be in offices and research laboratories; such jobs would be highly dependent on massive inputs of information and on the availability of skilled people.

One indicator of the postindustrial city is the distribution of those highly qualified people who form the work force of the quaternary sector. For example one of the highest concentrations of Ph. D.s in the United States is not in the large industrial cities, or academic New England or Southern California, but in North Carolina. There an area known as the Research Triangle has grown up based on three nearby university towns: Raleigh, the state capital; Chapel Hill, the attractive state university town; and Durham, home of Duke University. Since 1956 active promotion has brought to the area some 25 research laboratories for industrial corporations and federal organizations with 12,000 new jobs. Newcomers include IBM, Burroughs Wellcome, Monsanto, and the Environmental Protection Agency. The attractive environment, the lack of big-city problems, and the scale economies of common servicing arrangements suggest that the Triangle is likely to retain its distinctive role as a quaternary complex.

New capital cities, such as Australia's Canberra, form a major example of postindustrial cities. They are characterized by low population densities, small differences in living standards and social differentiation, a reliance on personal transport (the automobile) rather than public transport, and a high level of recreational amenities. It is too early to know, however, if such cities

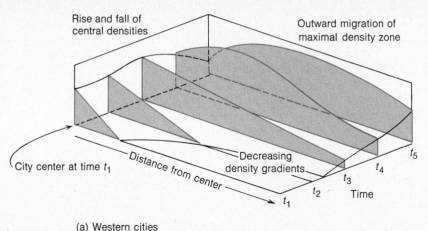

Rise and fall of central densities

Outward migration of maximal density zone

Decreasing density gradients

City center at time t_1

Distance from center

t_1 t_2 t_3 t_4 t_5

Time

(a) Western cities

Figure 16-16. Western and nonwestern cities.

Cross-cultural contrasts are evident in these urban-density gradients for a sample of western and nonwestern cities. [From B. J. L. Berry *et al., Geographical Review* **53** (1963), p. 403. Reprinted with permission.]

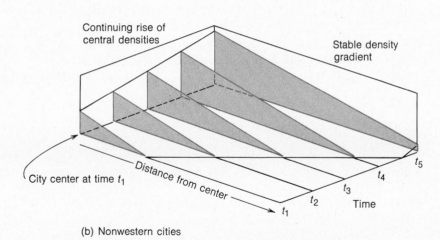

Continuing rise of central densities

Stable density gradient

City center at time t_1

Distance from center

t_1 t_2 t_3 t_4 t_5

Time

(b) Nonwestern cities

are merely the favored byproducts of the industrial city. A move toward a more general pattern of postindustrial urbanization could pose enormous problems in terms of land use and employment opportunities. Mechanization and automation in manufacturing industry have drastically reduced the need for labor. We might expect similar changes in the tertiary sector in the decades ahead as the impact of new technologies (e.g., that of silicon chips) is felt.

When representatives from 132 nations assembled in Vancouver in June 1976 to convene HABITAT, the United Nations Conference on Human Settlements, their focus was on the city as an international problem area. As we saw in Chapter 14, unprecedentedly high rates of population growth and

16-4
The City as a Problem Area

massive rural to urban migration is causing critical urban problems around the world. We can divide these problems into those arising from *inward* migration to the city, those from *outward* spread of the urban area, and those from the operation of the city as an *ecological* unit.

Problems of Inward Migration: Uncontrolled Settlements   Rapid population growth puts stress on the housing stocks of a city. Much of the very rapid urban growth in Third World countries occurs through *squatter* settlements. These uncontrolled settlements often lie around the periphery of the built-up area, and are made up of temporary buildings (usually built by the squatters themselves) with few public services. Their names vary from country to country: In Latin America they may be called *ranchos* or *favelas*; in Asia, *bustees* or *kampongs*; in Africa, *bidonvilles* or *shantytowns*. Attempts by United Nations agencies to assess the share of such squatter settlements in the total city populations suggest they make up from 10 to 80 percent. For example, about a quarter of Rio de Janeiro's population is housed in this way. (See Figure 16-17.)

Attitudes toward the squatter settlements have changed in ways that remind us of the changing attitude toward shifting cultivation in the tropics (discussed in Section 8-2): This was first regarded as harmful, later recognized

Figure 16-17.   Uncontrolled settlements around a Brazilian city. Sharp contrasts between the *favela* in the foreground and the apartment blocks in the background. Favelas are roughly constructed of boards or tin plate, flout most of the city zoning laws, have few or no public facilities (water, sewers, telephones, etc.), and are usually built on land whose ownership is undetermined. Shantytowns of this type are a problem for urban government but also represent a logical, low-cost response to the problems of new migrants to the city. In Rio de Janeiro, over 1 million people live in shantytowns of the type shown. [Photo by Wide World Photos.]

as a logical reaction to tropical ecology. Much the same shift has occurred over the squatter settlements. As late as 1970, they were described in official reports as a "spreading fungus" and "excessively squalid and deprived." Today a more tolerant attitude prevails as the shantytowns are recognized as performing an important transitional role in urban evolution.

Squatter settlements provide six functions of major importance. They act as reception centers for migrants. They provide housing within the means of the very poor. They provide a variety of small-scale employment. Their social and communal structure provides a cushion for residents during times of unemployment and other periods of difficulty. They encourage self-help in improving the standards of the houses. Finally, they provide a location within range of possible workplaces within the ctiy.

Detailed studies of the squatter settlements show they range enormously in the standards of housing, amenities, and access. For most the standards of facilities are very poor, and the risks of public health problems are correspondingly high. But they provide for some newcomers steps on the urban escalator that probably could not be provided in other ways. Clearance of shantytowns and rehousing the population in more distant areas, as has recently happened in Manila, has generally been expensive for the city government and has pushed the displaced population ever further from job opportunities near the city center.

Ghetto Formation   Uncontrolled squatter settlements are largely a Third World problem. Inward migration of the rural poor in the western world has led to a spatially different though socially similar phenomenon, the ghetto. The term *ghetto* originally described Jewish areas within the medieval cities of Eastern and Southern Europe. In such cities the Jews lived apart from the rest of the community and in some cases even had a wall separating their area from the rest of the city.

In terms of the modern American city the term ghetto is more generally used to describe a confined part of the city occupied by a distinct ethnic or cultural group. Thus it is typified by the distribution of blacks, Puerto Ricans, and Mexican Americans in many cities. Note that the ghetto is held in place by two sets of forces: the internal cohesion of the ethnic group, and the external "walls" placed around it by the rest of the urban community. Internal cohesion may be common links of language or culture, and the role of the ghetto community as a transition point for new migrants coming into the city. Although the boundaries around the ghetto are not physical, they may be real enough in terms of the cost of housing or the willingness of families to sell to ghetto occupants. In most ghettos the exclusion process is complex: Everybody excludes everybody else.

Typically the ghetto limits are coincident with areas of urban deprivation. Personal income tends to be low, housing crowded, unemployment higher, health poorer, and delinquency rates higher. In St. Louis, no less than 87 percent of the poverty areas of the city are occupied by nonwhites; in Seattle

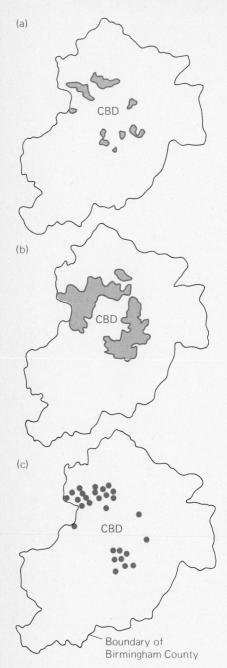

(a)

CBD

(b)

CBD

(c)

CBD

Boundary of
Birmingham County

Figure 16-18. Population change in the inner city.
Maps show for the city of Birmingham in central England the areas where the percentage share of New Commonwealth population was 10 percent or higher in (a) the 1961 census and (b) the 1971 census. Map (c) shows the location of elementary schools (to age 11 years) where colored pupils make up half or more of the enrollment. Note the concentration of immigrant population in the ring of old, inner-city housing around the central business district (CBD). The main sources of New Commonwealth migration are the Caribbean and South Asian countries. [After R. I. Woods, *Trans. Inst. Brit. Geogr.*, New Series **2** (1977), p. 476, Fig. 1.]

the corresponding figure is 60 percent. A similar relationship exists in some European cities. (See Figure 16-18.) But as population density builds up, the pressure to break out of the ghetto also rises. Typically the process of spatial extension is a jerky one—a city block or school district switching rather quickly from one group's occupation to another. But at a more general spatial scale we can see how the expanding ghetto follows a definite wedge structure.

In a large American city there may be a variety of ethnic nuclei. Thus Baltimore not only has major black areas but other smaller groups—Lumbee Indians, Appalachian populations, Chinese, Jamaicans, and Cubans. Boston has well-defined Italian and Irish districts. Study of the growth of these ethnic areas usually shows that, even if the pattern is discontinuous, an outward wedge seems to develop. Everyone gradually increases the size of their slice of pie, although the slices may be less well defined the further from the pie's center.

This process of neighborhood change may be described in terms of a *tipping mechanism.* When the level of black house ownership or school enrollment reaches a critical level (say, 15 percent) then there is (a) rapid acceleration in the outward movement of whites and (b) a refusal of whites to buy property in that area. The combination of both factors may together cause very rapid change. (See Figure 16-19.) Studies of census tracts in Chicago show that up to 75 percent of dwellings may shift to black ownership over three years once the tipping point is reached. This represents the upper range of "white flight" behavior, and turnover will generally be substantially less depending on the age-composition of the population and the condition of the housing stock. No single value for the tipping point exists, and some communities are much readier than others to coexist at particular mixture levels.

Although we have discussed the tipping phenomenon in terms of ethnic groups, the same kind of models may apply to other groups. In Belfast, Northern Ireland, the Catholic–Protestant dispute has also led to separation on a street-by-street basis. We shall see later in the book (Section 20-3) cases where the sorting process on racial grounds has been formalized, as through the *apartheid* policy of South Africa's nationalist government.

Problems of Outward Migration   We've noted in this chapter that the streams on inward migration were associated with an outward spread of population at lower densities. Two of the problems that this has led to are those of city government and sustaining the CBD.

City Government   We illustrate the problem of governing the rapidly expanding city by looking at Toronto. (See Figure 16-20.) As Canada's largest city and commercial capital the city has gone through wave after wave of growth in the last century, building and rebuilding its central core and sprawling outward in a profusion of new factories and homes on the margins of the built-up area. The maps show the rapid expansion of the urbanized area. Growth brought with it a series of problems in terms of such items as highway planning, housing shortages, lack of adequate utilities such as water and sewerage. These conditions led the Ontario legislature to create the Municipality of Metropolitan Toronto ("Metro") in 1954. This was modified in 1966 by the merging of the municipalities into six units—the city of Toronto, and five "outer" boroughs shown on the map. Metro has brought a series of successes to the government of this prosperous and attractive city. New highways, an extended subway system, housing for the aged, flood-control schemes, and major developments in recreational areas.

But the city has continued to boom. As Figure 16-20 shows, the tidal wave of urban Toronto has now swept well outside the 1954 Metro boundaries. While the 1970 Metro population was 2.0 million, the urbanized area was 2.35 million. If we extend the boundary to include the area within an hour's travel of the CBD (a rough approximation to the outer limits of the daily commuting zone) then the population leaps to 3.5 million. Just how far out to draw this boundary of any expanded Metro will be a problem for the next decade.

In comparison to many other city areas, Toronto's revision of city government has been a progressive and successful one. In others the financial problems caused by loss of tax-paying industry, shopping, and homes have

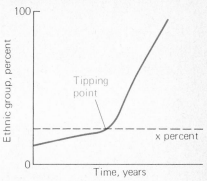

Figure 16-19.   Tipping-point mechanism.
Study of the expansion of the ghetto areas of many western cities suggests that a critical level (x) may be reached at which there is a sudden switch from, say, white to black occupation. The level x is not a fixed percentage but varies from city to city.

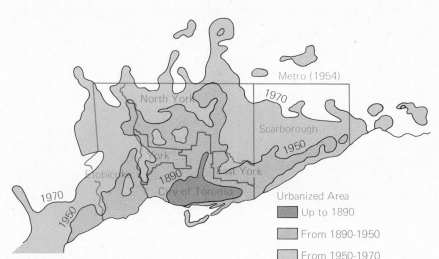

Figure 16-20.   Governing the expanding metropolis.
The rapid growth of the Canadian city of Toronto led to the amalgamation of existing communities into a unified metropolitan area (Metro) in 1954. At that time the Metro boundary contained the greater part of the urbanized area. Over the next 20 years the urban tide has swept well beyond this boundary. See also p. 577 for a photo of Toronto. [After J. W. Simmons and L. S. Bourne, in L. Gentilcore, Ed., *Ontario* (University of Toronto Press, Toronto, 1972), p. 92, Fig. 5.5).]

meant that inner-city government has been left to cope with major problems on an insufficient and sometimes dwindling tax base.

Sustaining the CBD   Along with the outward sprawl of housing has gone a decentralization of industrial and commercial activity. Symbolic is the out-of-town shopping mall with its major department stores, its massive parking lot, and its location near one of the superhighways. As suburban development has pushed further from the city, so the ring of shopping plazas has become larger and more diverse. Originally built to cater for convenience and week-end shopping (with an emphasis on food and clothes), the newer centers include more specialized shops.

As this trend has gathered momentum, so the role of shops within the CBD has declined. For example in Chicago the last 30 years have seen the daily number of shoppers in the CBD fall by nearly one-half. Since the late 1960s insurance giants like Prudential and John Hancock and major retailers like Sears Roebuck and Montgomery Ward have been investing heavily in downtown centers, attempting to bring shoppers (as well as jobs and homes) back to the central areas. Private and public investment in the downtown has also led to very substantial schemes aimed at restoring the flagging centers of American cities.

Another approach to sustaining central activity has come through transport innovation. For example, commuters to San Francisco's CBD face special problems. The very broken topography of the San Francisco Bay metropolitan area (3.1 million people and the United States' sixth most populous SMSA) means bridge crossings between the west and east sides. The year 1972 saw the opening of a new 120 km (75 mi) rail facility for commuters, the Bay Area Rapid Transit District (BART) system (Figure 16-21). This aims to provide a high-speed commuter service and to divert commuters from cars to trains and to overcome the congestion for cars crossing the bridges into San Francisco from the East Bay Area. Trains are very frequent on the new system during the peak rush hours and are scheduled by computer to run at average speeds of 72 km per hour (45 mph), nearly twice as fast as average commuter speeds by automobile. The system includes a number of interesting innovations in the design of the trains, the stations, and control, and set new standards in luxury.

Despite the imaginative scheme, BART has proved only a partial success. Tactical problems relating to fare levels and commuter parking at stations remain to be solved, and much of the original system is still unbuilt. Perhaps the biggest strategic problem relates to CBD employment. As the Bay Area has grown, so satellite centers have emerged outside San Francisco itself, which now has only 36 percent of the region's jobs. So, much of the region's commuting problem lies in cross-movements within the metropolitan area rather than simple suburbs-to-CBD transit. Few of those commuters in the outer Bay Area use public transport, and those who do tend to use the more flexible bus services. Thus the BART system shares, like all fixed-rail systems,

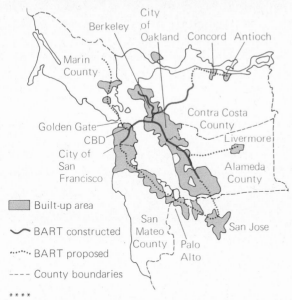

City
of
Berkeley
Oakland Concord Antioch

Marin
County

Contra Costa
County

Golden Gate
CBD
Livermore

City of
San
Francisco

Alameda
County

Built-up area

BART constructed

BART proposed

County boundaries

San
Mateo
County

San
Jose

Palo
Alto

****

Figure 16-21. Rapid transit solutions to city commuting problems.

The physical configuration of San Francisco Bay poses special problems for commuters crossing the bridges to get to the CBD area. The Bay Area Rapid Transit (BART) System opened in 1972 represents one attempt at a solution. Earlier studies envisaged most of the inner Bay Area counties being served by the system but San Mateo withdrew from the scheme in the 1960s and Marin County, joined to San Francisco by the Golden Gate Bridge, was considered too difficult to serve under existing conditions. [Photo courtesy of BART.]

the fundamental geographic difficulty that it must focus on the high-intensity traffic moving into downtown. The more jobs decentralize, the harder it is to run fixed-rail systems, except by massive public subsidy.

Epilogue: The City as a Complex System   As with the natural ecosystems discussed in the first two parts of this book, so the city ecosystem demands flows of energy and nutrients to keep it alive. Geographer Sherry Olson has described how metropolitan Baltimore—2 million people, three-quarters of a million dwellings, and 2,000 factories are packed onto 3,540 km² (2,200 mi²)—is sustained by flows. The 150,000 people who move into and out of the city to work each weekday and the 2 million internal trips are supplemented by the trucks, ships, and planes coming in and out along freeways, harbor lanes, and airways. To this must be added the 250 million gallons of water pumped in each day, and draining out along the sewers; the flows of electricity and gas; the 5 television and 26 radio stations and the millions of telephone calls binding the life of the city together. Cleaning the city itself can be viewed as a special kind of ecosystem.

The complex worlds within the city are local reflections of the city's role in relation to the outside world. It is to that outside world that we turn in the next two chapters.

Summary

1. In the western city transportation is a controlling factor in the development of land-use zones. The central business district (CBD) of a city is most accessible and generally has the highest land values; peripheral areas decline rapidly in value with distance. Land value surfaces may be modified by the transport route structure. Decreasing land prices away from the CBD are generally balanced by increased transport costs. Location of least-cost points may be found through the use of bid-price curves. Such curves can be determined for each type of land use in a city.

2. The pattern of land use in western cities can frequently be seen as a combination of sectors and rings. Models developed for the city of Chicago on this basis have been widely used by urban geographers in studying other western cities.

3. The structure of the city is filtered through the perception of its residents to form a series of mental maps or images. Mental maps can be broken down into significant elements (e.g., paths, nodes, districts, and landmarks). Images may have a powerful impact on the city, as in the persistence of prestige areas within a city.

4. Comparison of cities may be carried out in terms of a structure, process, and stage model borrowed from physical geography.

5. Structure involves the notion of the location of the city, in terms of its regional situation and its local site. These represent the platform from which the city develops.

6. Process covers the operation of those expansion and sorting mechanisms by which the city grows and is differentiated. Social processes involve the separation of groups of different cultural, ethnic, or income bands. In extreme cases groups may be separated into ghettos which grow spatially in a step-by-step fashion through a tipping mechanism.

7. Stage divides cities into preindustrial, industrial, and postindustrial. Each stage is characterized by distinctive spatial patterns of land use, economic role, and social ecology.

8. Problems faced by cities are of three main kinds: those arising from inward migration, from outward migration, and from the operation of the city as an ecosystem.

Reflections

1. Gather data on the distribution of different types of land use within your local urban community. (Maps are usually available from the city planning offices.) How far do the patterns you find (a) resemble or (b) differ from the patterns described in Section 16-1? Suggest some reasons for the differences, if any.

2. Is your local city's central business district (CBD) growing, stabilizing, or declining in importance? Give reasons for your opinion. What evidence would you need to assemble to test your views?

3. Choose any one city in a country other than your own. Using the resources of your college library, try to build up a picture of how the city evolved. How far does it fit a structure, process, stage model?

4. Select a major metropolitan area and look at its boundaries. How would you reorganize these boundaries to make them more appropriate to the problems faced by the city?

5. Review the set of urban problems described in Section 16-4. Which do you think is the most serious for your own major city? Why? Compare your answer with those of others in your class.

6. Renew your understanding of the following concepts:

bid-price curves	ghetto
CBDs	mental maps
ring and zone structures	site and situation
structure, process, and stage	squatter settlements
tipping mechanism	gentrification

One Step Further . . .

There are a number of very good introductions to the study of urban geography. See, for example,

Carter, H. *The Study of Urban Geography*, 2nd ed. (Arnold, London, 1975).

Bourne, L. S., Ed. *The Internal Structure of the City*, 2nd ed. (Oxford University Press, New York, 1978).

For a general study of North American cities and the special problems of the ghetto, see

Yeates, M. and B. J. Garner, *The North American City*, 2nd ed. (Harper & Row, New York, 1976).

Rose, H. M., *The Black Ghetto: A Spatial Behavioral Perspective* (McGraw-Hill, New York, 1971).

You should also look through the pages of a magnificent urban atlas

Abler, R., Ed., *A Comparative Atlas of America's Great Cities* (University of Minnesota Press, Minneapolis, 1976).

and the several vignettes of individual metropolitan areas which were written to accompany the atlas. For study of cities in a contrasting cultural environment, see

McGee, T. G., *The Southeast Asian City* (Bell, London, 1967).

Environmental problems in the city are discussed in

Berry, B. J. L. and F. E. Horton, *Urban Environmental Management* (Prentice-Hall, Englewood Cliffs, N.J., 1974).

For urban journals, see those recommended at the end of Chapter 14.

chapter 17
Worlds Beyond the City, I
Agricultural Zones and Industrial Centers

Where does a city end? Think about the answer by recalling the last time you drove along a highway out of a major city. At one mile out from the CBD the landscape was probably still urban, dominated by a built-up environment of houses and streets. At ten miles out the landscape was probably semirural; but in this suburban range there would still be massive daily commuting to the city. At fifty miles out the landscape may have been truly rural; but the car radio would still carry, faintly, the voice of city radio stations, the farmers would still purchase city newspapers from their local general store, and their wives would still frequent the city to carry out their major shopping.

Cities, like old soldiers, don't die—but simply fade away. Their spatial dominance lessens as we move further from their centers but is never entirely eliminated. Some minute proportions of a city's mail or telephone calls will reach out even to the far side of the world; its billboard advertising ("Only 1050 miles to Harry's Place") may stretch across a continent. It is this city-centered world that we look at in this chapter. We begin by looking at the effect of the city on how the rural land around it is used. We then go on to look at the parallel impact of the city on the processing of natural resources and study the seesaw of forces that shape the spatial pattern of industry. In both these sections we shall be concerned with the locational theory that geographers have built up to explain the city-oriented world. Locational theory is one of the three major elements in geographers' research, and we shall be asking questions about why certain activities are located where they are. Since some of the answers turn out to be complicated, you may need to move rather slowly through this and the following chapter.

407

17-1
Thünen and Land-Use Zoning

The earliest attempt to correlate land-use patterns with the spatial relationship of a city to its surrounding region was made by a German, Johann Heinrich von Thünen (Figure 17-1). In his classic work on the location of agricultural land-use zones, *Der Isolierte Staat in Beziehung auf Landwirtschaft*, first published in 1826, he not only laid the foundation for refined analysis of the location of agriculture, but stimulated interest in a much broader area of locational analysis. In 1810 Thünen, at the age of 27, acquired his own agricultural estate, Tellow, near the town of Rostock in Mecklenburg on the Baltic coast of Germany. For the next 40 years, until his death in 1850, he supervised the cultivation of the Tellow estate and amassed a remarkable set of records and accounts that provided the empirical basis for his published theories.

The Isolated State   Thünen's isolated state is modeled on the agricultural patterns of nineteenth-century Mecklenburg. The basic form of the land-use patterns he envisaged is shown in Figure 17-2, and the characteristics of each of the zones are presented in Table 17-1. The patterns are a series of concentric shells ranging from narrow bands of intensive farming and forest to a broad band of extensive agriculture and ranching to an outer "waste."

To understand the form of these spatial patterns, we need to review the six assumed conditions that dictated it:

1. the existence of an isolated state cut off from the rest of the world;
2. the domination of this state by a large city that served as the sole urban market;
3. the setting of the city in a broad, featureless plain that was equal everywhere in fertility and in which the ease of movement was the same everywhere so that production and transport costs were the same everywhere;
4. the supplying of the city by farmers who shipped agricultural goods there in return for industrial produce;
5. the transport of farm produce by the farmer himself, who hauled it to the central market along a close, dense trail of converging roads of equal quality at a cost directly proportional to the distance covered; and
6. the maximizing of profits by all farmers, who automatically adjust the output of crops to meet the needs of the central market.

These assumptions are, of course, unrepresentative of actual conditions either in the early 1800s or, indeed, now. Why, then, did Thünen make them? To understand this, we have to go back to our initial discussion of the role of models in science (Section 1-5). The objective of models is to simplify the real world in order to understand some of its characteristics. Thünen's model permits us to do just this by sorting out some of the key factors (albeit in a simplified form) that caused land-use rings to form.

Given these assumptions, Thünen was able to demonstrate that rural land values would decline away from the central city in the same way urban land

Figure 17-1.   A pioneer locational theorist.
Johann Heinrich von Thünen (1783–1850) laid the foundations for agricultural location theory in a book published in 1826, called *The Isolated State*. Basing his ideas on observations from his own farms, Thünen extended them to encompass world wide patterns of land use variation. Although agriculture has changed dramatically in the last 150 years, his models still give useful insights into the geographic patterns observable today. [Photo courtesy of The Mansell Collection.]

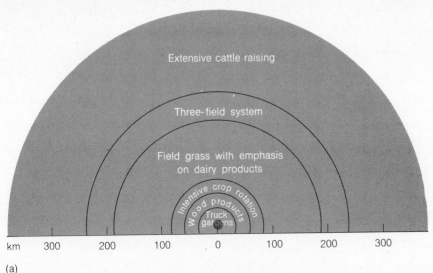

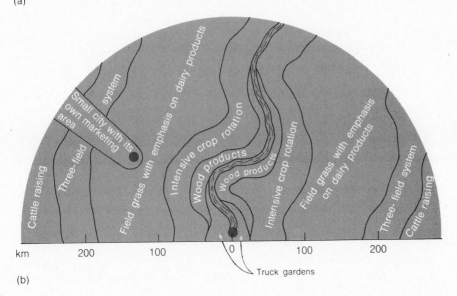

Figure 17-2. Land-use zones in Thünen's Isolated State (1826).
(a) Thünen's original land-use rings.
(b) Thünen's rings modified by a navigable river bringing cheaper transport costs. The distance figures were added to Thünen's original model by L. Wailbel in 1933, in his *Probleme der Landwirtschafts-geographie.*

values do, though at much lower rates and with gentler slopes. (See Figure 17-3.) Like each urban land use, each agricultural land use has a characteristic set of bid-price curves and finds an appropriate location with respect to distance from the city. Thünen stated his model in terms somewhat different from those of urban models (see the discussion on page 411 of zoning mechanisms in the Thünen model), but the processes by which eventual land uses are determined are identical. A product that is bulky (i.e., has a large tonnage per unit of area) or is difficult to transport will have steep bid-price curves and will be quite sensitive to displacement from the market. Conversely, one that is lighter (i.e., has a low tonnage per unit of area) or easy to transport will be

Table 17-1   Thünen's land-use rings

Zone	Percentage of state area	Relative distance from central city	Land-use	Major product marketed	Production system
0	<0.1	−0.1	Urban-industrial	Manufactured goods	Urban trade center of state; near iron and coal mines
1	1	0.1–0.6	Intensive agriculture	Milk, vegetables	Intensive dairying and trucking; heavy manuring; no fallow period
2	3	0.6–3.5	Forest	Firewood	Sustained-yield forestry
3a	3	3.6–4.6		Rye, potatoes	6-year intensive crop rotation: rye (2), potatoes (1), clover (1), barley (1), vetch (1); no fallow period; cattle stall-fed in winter
3b	30	4.7–34	Extensive agriculture	Rye, animal products	7-year rotation system: field grass with an emphasis on dairy products; pasture (3), rye (1), barley (1), oats (1), fallow period (1)
3c	25	34–44		Rye, animal products	3-field system: rye, etc. (1), pasture (1), fallow period (1)
4	38	45–100	Ranching	Animal products	Mainly extensive stock-raising; some rye for on-farm consumption
5	—	Beyond 100	Waste	None	None

SOURCE: P. Haggett, *et al.*, *Locational Analysis in Human Geography* (Wiley, New York, and Arnold, London, 1977), p. 205, Table 6-4.

Figure 17-3.   Ring formation in the Thünen model.

Figure (a) shows how the bid-price curves for four types of land use have slopes of different steepness. (To understand this figure more easily, you may find it useful to check back to Figure 16-13 where we first looked at the idea of bid-price curves.) Use of land for dairying has a steeper curve than that for wheat farming, indicating that dairying is a more intensive way of using the land and one which stands more to gain from locating near the city. But dairying is itself displaced by land used for residential housing on the fringe of the city. The broken lines in (a) mark the breaks where one type of land use "outbids" the other in terms of the rent levels it can afford to pay for a convenient location near the city center at X. If you follow these broken lines down to Figure (b), you will see that four distinct land use zones are formed. ("Waste" indicates land not in use because it lies too remote from the city.) If you complete the arcs of the circles shown in (b), a series of land use rings are formed around the city.

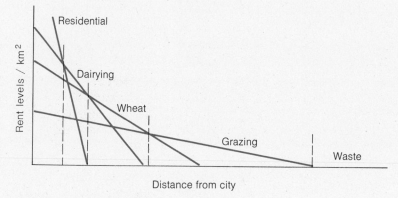

(a) Optimal land uses

(b) Land-use rings

ZONING MECHANISMS IN THE THÜNEN MODEL

Thünen regarded locational rent (*Bodenrente*) as the key factor sorting the uniform area of his isolated state into distinct land-use zones. The location is given by

$$L = Y(P - C) - YD(F)$$

where L = the locational rent (in $/km^2),

Y = the crop yield (in tons/km^2),

P = the market price of the crop (in $/ton),

C = the production cost of the crop (in $/ton),

D = the distance to the central market (in km), and

F = the transport rate (in $/ton/km)

Thus, for a crop yielding 1000 tons/km^2, fetching $100/ton at the central market, and costing $50/ton to produce and $1/ton/km to haul, the locational rent at the city center would be $50,000/km^2, at 10 km distant it would be only $40,000/km^2, and at 20 km distant it would be down to $30,000/km^2. Beyond 50 km production would be at a loss. The competition of two crops (i and j) for the same area depends on their yield (Y) and relative profitability ($P - C$). When the condition

$$1 < \frac{Y(P - C)_i}{Y(P - C)_j} < \frac{Y_i}{Y_j}$$

obtains for crops i and j, they form two distinct spatial zones; crop i dominates a circular area adjacent to the city, and crop j occupies a ring-shaped zone immediately outside it. The symbol $<$ means "is less than." Any other relation of the terms in the equation above results in the two crops' being reversed so that j occupies the inner ring, *or* one crop dominating all available land to the complete exclusion of the other, *or* both crops being grown side by side with no spatially differentiated zoning. [See E. S. Dunn, *The Location of Agricultural Production* (University of Florida Press, Gainesville, Fla., 1954), Chap. 1.]

less sensitive to displacement. As Figure 17-3 indicates, land-use boundaries occur at the intersections of bid-price curves. Land use in the remoter areas is adapted to take into account the poor accessibility. A smaller tonnage of a given crop may be grown on each unit of area by using extensive farming methods; for instance, long fallow periods may be used to restore the land's fertility rather than fertilizers. Alternatively, a product may be shipped in a more compressed form (e.g., as cheese rather than liquid milk), or animals may be used to "concentrate" a crop (as when hogs raised on corn are marketed in place of corn).

We can extend the basic Thünen model to explain situations quite unlike the ones it was first proposed to explain. Although the original study was of ring formation around a single node, this is merely one geometric case. If we substitute a linear market for a central one, the model still explains the formation of zones, but in straight parallel bands. Land-use zones along a coastal strip or transport axis are also common variants of the conventional Thünen rings. Several alternatives of the ringed model were discussed by Thünen himself. By introducing a navigable river on which transport is speedier and costs are only one-tenth that of land transport, a minor market center with its own trading area, and spatial variations in the productivity of the plain, he was able to explain considerable variations in land-use patterns. [See Figure 17-2(b).] Once we allow these kinds of variations and add to them the wider rings that the technological advances in transport make possible, then the Thünen model can provide insights into contemporary patterns of land use on larger spatial scales.

City-Centered Zoning in the United States   Attempts to compare the Thünen model with the real world are hindered by the difficulties geographers have had in drawing up unambiguous definitions of land use. Different crops are grown in different environmental zones, so the Thünen rings are disturbed by other zones related to the ecology of crops rather than the accessibility of cities. Geographers have tended, therefore, to use population density as a substitute for land use and to look for gradual changes in density instead of sharp discontinuities between distinct land-use zones.

Demographer Donald Bogue has investigated how population distributed itself around 67 major cities in the United States. Using the census figures for counties, he analyzed the changes in population density with respect to distance as far as 800 km (500 mi) from the city.

Bogue's general conclusion is that the main metropolitan centers dominate the spatial arrangement of population in the United States. If we take the average population density at various locations and plot it against the distance to the nearest city, we find a rapid falloff; if, however, we transform both the axes of the graph to a logarithmic form, the decrease appears as a simple linear rate of decay (Figure 17-4). For example, 40 km (25 mi) outside the city the density exceeds 500 people/km^2 (1250 people/mi^2); 400 km (250 mi) out, the density is only 10 people/km^2 (25 people/mi^2). The detailed pattern of the decay curves reflects the size of the central city. Large metropolitan cities with over half a million people have densities much higher than smaller cities at similar distances. Farther out, these differences diminish.

Figure 17-4.   Density gradients around cities.
Variations in population density with distance from the nearest large city (metropolis) are shown for three regions of the United States in 1940. Both axes of the graphs are logarithmic to allow details of the middle parts of the density curves to be shown more clearly. This also allows easier comparison among the three regions. If plotted on arithmetic scales, all the curves would appear to be L-shaped, emphasizing only the dramatic drop in population density within the first few miles of distance from the metropolis. [From D. J. Bogue, *Structure of the Metropolitan Community* (Scripps Population Institute, Ann Arbor, Mich., 1949), p. 58.]

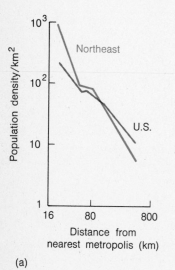

(a)

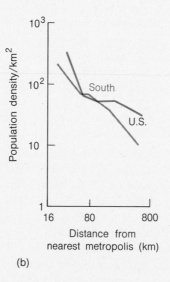

(b)

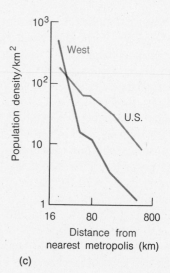

(c)

We can also detect strong regional differences within the United States. The northeast has curves that are higher and steeper, reflecting its dense network of cities separated by rural pockets of relatively low population density such as northern New England and Appalachia. The south has lower population densities and greatly irregular curves, reflecting its fewer and smaller cities and a more uneven pattern of rural population. In the west there is a sharp decline in population density with distance from the city. [See Figure 17-4(c).]

We can relate these differences in population density more directly to land use by sorting the population into employment categories. Bogue showed that for farm populations there was a gentle decline in density from the metropolitan centers out to about 150 km (93 mi) and a sharp decline at 500 km (300 mi). Again, the national trend concealed strong regional contrasts. The density in the south, for example, changed little with distance. Industrial land use as approximated by industrial employment was much more dependent on distance. Employment declines very sharply with distance from the major metropolises, but the decline is arrested around 50–100 km (30–60 mi) out before dropping sharply again. The brief halt in the decline probably represents a concentration of specialized manufacturing towns at this distance.

One imporatant feature of Bogue's study, which has been picked up in later work, is that population density depends not only on distance from the metropolis but also on the direction in which other cities lie. If we divide the area around the city into wedge-shaped sectors, then sectors with routes running to other cities represent ridges of high population density compared with those not containing such routes.

Similar research in different countries but on comparable geographic scales indicates that the findings for the United States are reasonably typical. Population density the world round is sensitive to both the distance from and the direction to the centers of cities.

Zoning in Farm Communities   Geographic studies of land-use zones at the micro level of the village and farm (Figure 17-5) have been done as well as studies of subcontinental zoning. Here, too, the effort required to use an area of land is going to increase with the distance from the center of the community. If we take a small community, the individual family farm, the time taken to reach the most distant fields will be greater than the time needed to visit the home paddock adjacent to the farmstead. By tracking individual farmers from dawn to dusk, comprehensive diaries of their movements in space over time can be compiled. What do these show? In Holland arable plots only 0.5 km (.3 mi) from farmsteads receive about 400 man-hours of care per hectare annually. At a distance of 2 km (1.2mi), the care level drops to 300 man-hours; at 5 km (3.1 mi), it dwindles to only 150 man-hours. We can convert such figures into costs by adding information on the jobs carried out in these movements between farmstead and field. In Punjab villages in

Figure 17-5. Agricultural land use.
The detailed pattern of crops at the field and farm level provides evidence from which the models of land use zoning described in this chapter are tested. Distinctive zoning at the micro-level, such as these fields in Guatemala, is repeated at larger geographic scales all the way up to world scale. [Photograph by David Haas.]

Pakistan the cost of ploughing increases by around 5 percent, the cost of hauling manure rises 10–25 percent, and the cost of transporting crops rises 15–32 percent every time we move an additional half-kilometer from the village.

How rapidly costs increase with distance is a matter for debate; different rates have been reported in different studies. What does seem clear is that locational adjustments begin to occur at distances as low as 1 km (.6 mi) from the community center and that the costs of operation rise sharply beyond about 3 or 4 km (2 or 2.5 mi). We can see the form of these locational adjustments in the distinctive zones of land use around villages. Figure 17-6 shows the sequence of crops extending a distance of 8 km (5 mi) from the Sicilian village of Canicatti. Note that the growing of olives and vines falls off rapidly with distance; beyond 4 km (2.5 mi) the open, arable land is cultivated mostly for wheat and barley. Some clue to the "sorting" process by which the decision is made to grow certain crops on certain locations is provided by the average number of man-days per hectare expended. This energy expenditure begins at 52 man-days in fields near the village but declines to less than 40 in remote fields at distances of 8 km (5 mi) or more.

Geographer Mansell Prothero has described a similar zoning around villages in northern Nigeria. He distinguishes four zones. The first is an inner garden zone with close interplanting, a continuous sequence of crops, and intensive care. A second zone at 0.8 to 1.2 km (.5 to .7 mi) out is continuously used (mainly planted with Guinea corn, cotton, tobacco, and groundnuts) and fertilized. A third zone with an outer boundary at 1.6 km (1 mi) is used for rotation farming; that is, the land is cultivated for 3 to 4 years and then allowed to return to bush for at least 5 years to get back its fertility. Finally comes the fourth zone of heavy bush. Within this zone are isolated clearings, in which the three-zone sequence of the main villages is reproduced.

Studies similar to those in Sicily and Nigeria reveal a variety of responses by farmers to distance. Sometimes the reponse leads to sharp land-use zoning, as in Figure 17-7; in other instances farmers may grow the same crops over a wide range of distances but plant them in different combinations and give them varying degrees of care. Moreover, the simple symmetry of the zones is usually interrupted by local variations in terrain, soils, patterns of ownership, and the like.

Zoning at the International Level   During the nineteenth century, a major urban-industrial nucleus emerged in Western Europe and eastern North America. This nucleus may be thought to play a similar role in the international economy to that played by the single city in our Thünen's isolated state. Geographer Richard Peet has looked at this notion for one country in the "nuclear" area, the United Kingdom. (See Figure 17-8.) Over the period from 1800 to the outbreak of World War I in 1914, the British population increased fourfold. Over the same period living standards increased and food consumption per capita rose. Where did this food come from?

Figure 17-8 shows part of the answer. The broad pattern followed is an outward shift of the source areas from home production to imports from

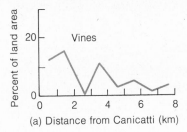

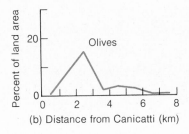

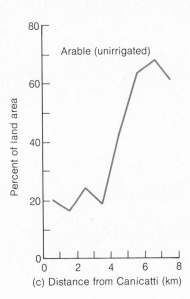

Figure 17-6.   Distance and rural land use.

The diagrams show changing crop patterns with increasing distance from the Sicilian village of Canicatti. [From M. D. I. Chisholm, *Rural Settlement and Land Use* (Hutchinson, London, 1966), p. 57, Table 6.]

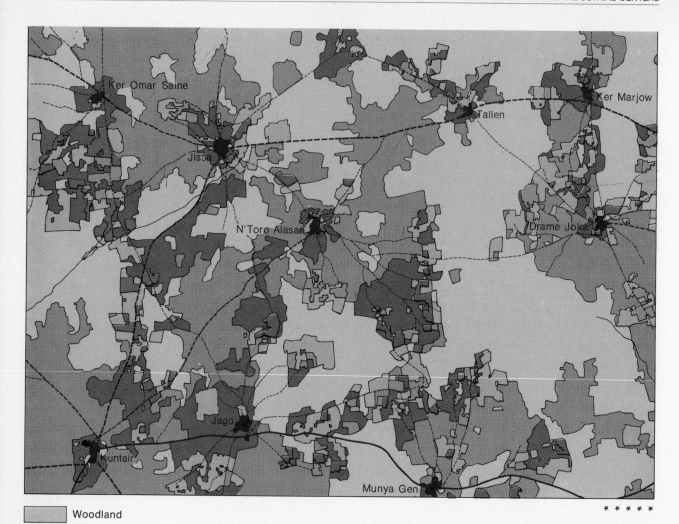

* * * * *

	Woodland
	Fallow bush
	Grass
	Groundnuts
	Groundnuts, sorghums, and millets
	Sorghums, millets, and other food crops
	Villages

Figure 17-7. Rural land-use banding in the African tropics.
The map shows areas of permanent and shifting cultivation in the Kuntair area Gambia, West Africa. Note the relations of the different zones to the location of the nine villages named: The total distance across the map is about 5 km (3 mi). Woodland occurs as a residual zone at the margins of the cultivated land and along boundary areas between villages. Fallow bush is woodland which has previously been used for cultivation, and where the land is now lying idle while recovering its fertility. Groundnuts (peanuts) are an important cash crop, while sorghum and millet (both grain crops) are consumed locally. These crop areas (see color key) tend to be located on the woodland margins in newly cleared patches of land with relatively high fertility. [From the Directorate of Overseas Surveys, Gambia, Land Use Sheet 3/111, 1:25,000, 1958.]

(a)

(b)

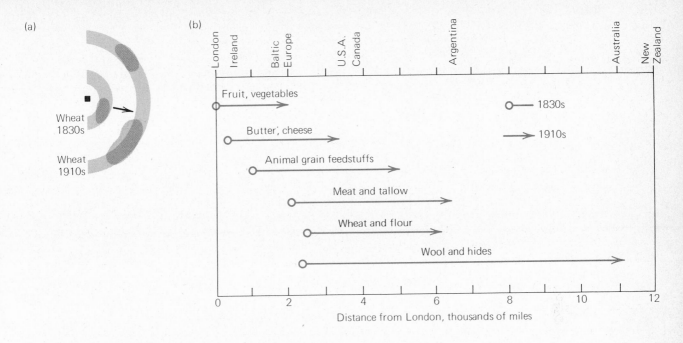

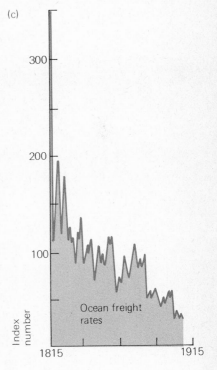

Figure 17-8. Spatial expansion of land-use zones.
(a) Outward movement of source of wheat for a central city to serve distant areas.
(b) Change in average distances from London over which various types of British agricultural imports moved between the 1830s and the 1910s. (c) Decline in freight rates on American exports as measured by a standard index (1830 = 100) in real cost terms. [Data from J. R. Peet, *Economic Geography* **45** (1969), p. 295, Table 1].

overseas areas as demand increased. If we take the case of wheat [Figure 17-8(b)] the average distance from which imports were drawn rose from 3,850 km (2, 400 mi) in the 1830s to 9,500 km (5,950 mi) by the outbreak of World War I in 1914. In the mid-nineteenth century wheat for England came from Prussia and the Black Sea ports. Russia rose to importance as a granary in the 1860s until supplanted by the wheatlands of North America in the 1870s. The wheat frontier moved westward into the Great Plains and had reached western Canada by the 1890s. Australia comes onto the scene as a major producer at the turn of the century.

The outward movement of the zones was partly related to increased demand at the center, partly to the overall downward trend in shipping costs as shown in Figure 17-8(c). Although the "rings" are greatly fragmented by continents and oceans, the dynamics of the expanding Thünen model remains visible.

17-2
Weber and Industrial Location

Figure 17-9. Least-cost locations. Simplified examples of the least costly location for a resource-processing facility are shown as a "tug-of-war" between the resources located at R and the markets located at M. Cost on the vertical axis is plotted against distance along a line between R and M on the horizontal axis. The three different cases shown are discussed in the text.

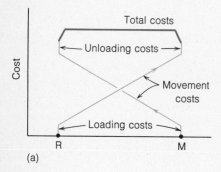

(a)

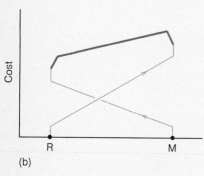

(b)

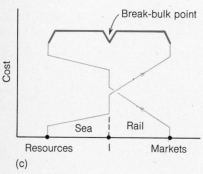

(c)

We have seen how cities, as centers of agricultural consumption, play a dominant role in shaping rural land uses. But cities are also the great consuming centers for all natural resources. What influence does the city as a marketplace have on the location of resource processing and thus on industrial patterns in the world beyond the city?

This question of the location of resource processing, with all its population-multiplying implications (Figure 14-4), has long attracted investigators. One of the simplest but most penetrating analyses of this subject came from the German spatial economist Alfred Weber, who published his original study of industrial location in 1909. Weber paid particular attention to the loss of weight or bulk involved in resource processing. He demonstrated that this weight loss played a significant part in the location of certain industries. Industries with manufacturing processes that involved a large weight loss were found to be *resource-oriented* (i.e., located near the natural resource that was to be processed). To convert wood to pulp and paper involves a 60-percent weight loss, and so pulp and paper mills tend to be located near major forested areas rather than in the heart of the large cities in which the bulk of their products is eventually consumed. Conversely, an industry like brewing is *market-oriented*, since its product is very weighty in relation to the malt, hops, and other materials that go into it. (Water is assumed to be generally available; hence a need for it is assumed not to affect location decisions.) This analysis is clearly greatly oversimplified, but much current theory, although different in terminology, sophistication, and method, follows Weber's basic line of thinking and is therefore termed *Weberian*. We first look at Weberian analysis in a single spatial dimension and then go on to more realistic two-dimensional analyses.

Location on a Line Let us assume that a single resource is supplied at location R and a single urban market for that resource is located at city M as in Figure 17-9. We further assume that transport costs increase regularly with distance from the supply point and that all other costs (labor, power, taxes, etc.) are fixed and equal everywhere. Figure 17-8(a) shows a simplified situation with a single resource supply point (R) and a single city (M) located on the horizontal axis; costs are marked on the vertical axis. If the loading costs are equal, and delivery costs from R to M are identical functions of distance, then the total transportation costs have the form shown by the heavy line. Transportation costs may be at a minimum at either the supply site or the market site. At all other possible locations on the line of haul between the two (RM), the transportation costs are equal, but higher, because they involve an additional loading or unloading charge.

The processing of resources commonly involves changes in their mass, weight, or value that affect unit transportation costs. The inequality in transit costs for raw and processed goods is illustrated in Figure 17-9(b) by the steeper slope of the delivery-cost curve from R. Transportation costs are lowest from the supply point, R. When transportation costs are higher for the

TERMS USED IN THE STUDY OF
INDUSTRIAL LOCATION

Agglomeration economies are savings
to the individual manufacturing
plant that come from operating in
the same location. These may come
from common use of specialist
servicing industries, financial services, or public utilities.

Heavy industries have (a) finished
products which have low value per
ton, (b) a high material index [see
below], and (c) a high tonnage of
materials used per worker.

Isodapanes are contours of total
transport costs.

Isotims are contours of transport
cost for a single element in the
manufacturing process.

Light industries have (a) finished
products which have high values
per ton, (b) a low material index
[see below], and (c) a low tonnage
of materials used per worker.

Market orientation is the tendency
for certain industries to locate near
their market. Typical is the brewing
industry which has to add a bulky
material (i.e., water) to the finished
product.

Material index is given by the total
weight of localized materials used
per product divided by the weight of

the product. Most industries have an
index greater than 1 and are
described as "weight-losing."

Resource orientation is the tendency
for certain industries to locate near
their source of localized raw
materials. Typical are the
mineral-processing industries where
a great deal of waste material in the
ore can be removed before the
refined material is shipped.

Ubiquitous materials are those which
can be found in any location and
therefore play a minor role in
locational decision-making.

---

finished product, the reverse situation obtains and the lowest-cost point
moves to location M. To make the situation a little more realistic, we can
introduce two zones (land and sea) with different transport costs. Because the
commodity being moved must go by two modes of transportation, the costs
of loading, unloading, and reloading at the intermediate point [I in Figure
17-9(c)] may be substantial. Thus all three points (R, I, and M) share the
status of *least-cost transport points.* By modifying some of the assumptions in
Figure 17-9 we could go on to introduce further and more complex least-cost
patterns.

We can see something of the tug of war between these supply and demand
points by considering the location of oil refineries. In the early period of oil
production (to 1920, approximately) there were advantages to be gained from
locating refineries on or near the oil fields themselves. Transport costs were
high and the demand was mainly for kerosene, with about half the crude oil
being dumped or burned as waste. As the advent of larger pipelines and
supertankers reduced crude-oil transport rates, and as the range of petroleum
products and the capacity of refineries to use more of the oil grew, it became
more advantageous to locate refineries near the point of consumption. The
vastly increased refinery capactiy on the seaboards of Western European
countries reflects both the changing economics of location in the industry and
the changing political situation. Market-based refineries can deal with a wider
selection of crude-oil suppliers, thus reducing the risk of supplies being cut off
from a single source; and they can be massive enough for the economies of
scale since they can draw on various small oil fields for supplies. The jump
from the simple diagram in Figure 17-9 to the complexities of refinery loca-

tion shows how large is the gap that more advanced locational theory must try to bridge.

Location in a Plane   We can move a small way forward if we change the focus of our analysis from a one-dimensional line to a two-dimensional plane. Figure 17-10(a) presents a simple situation in which we have two resource supply sites (R_1 and R_2) and a single market city M. If we begin by assuming equal transport costs per unit of weight, the costs from each of the three points can be represented by a series of equally spaced, concentric, and circular contours called *isotims*. Each isotim describes the locus of points about each source where delivery or procurement costs are equal. Here the isotims are circular about each of the three points because transport costs across the isotropic plane are everywhere the same. Total transportation costs can be computed by summing the values of intersecting isotims. Lines connecting points with equal total transportation costs are termed *isodapanes* and, like isotims, may be regarded as contours showing the cost terrain of a particular region. In Figure 17-10(a) the lowest point on the isodapane is equidistant from each of the three resource supply points.

If we relax the assumption of equal transport costs per unit of weight, then the isotims around each point may vary. Figure 17-10(b) shows a situation in which the movement costs from R_2 are twice those from the other two sites. This situation distorts the shape of the isodapane surface and displaces the point with the higher transport costs (i.e., toward R_2). Weber compared this displacement process to a set of scales that are weighted too heavily in one

Figure 17-10.   Isodapanes.
Contours of the total transportation costs on a plane. In these simplified situations there are two resources (R_1 and R_2), a single market (M), and a uniform plane. Transport costs around each of the three points (R_1, R_2, and M) are shown as concentric circles (*isotims*). Equal isotims about each point gives a midway compromise point of lowest transport cost in (a). Doubled costs from R_2 draw this compromise point toward it in (b).

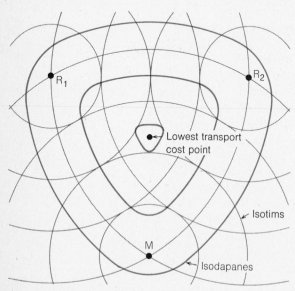

(a) Equal transport costs

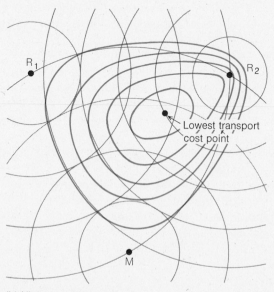

(b) Higher costs from R_2

scale pan. He suggested that this process might be illustrated by a simple physical model in which the pull of alternative sites is represented by physical weights attached to a circular disk by a system of cords and pulleys. Although clumsy in a practical sense, the idea of locational weights helps us to visualize the many competing forces involved in locational decisions.

Isodapane maps prove useful in practice, and a Swedish example of their use is given in Figure 17-11. This map shows the total costs of transporting pulp wood from various assumed manufacturing centers in southern Sweden. The map emphasizes the relatively advantageous location of the central lakes lowland and the coastal strip at its two ends. Costs increase sharply toward the interior and to the north. Of course, isodapane maps indicate only transport costs; to get a more complete picture we need to consider other costs not related to the transport of goods.

Space-Cost Curves So far we have analyzed the location of resource processing sites simply in terms of transport costs. This is an oversimplification, as Weber recognized; for location is affected by three other kinds of costs. First, there are labor and power costs that vary over space and are dependent on location. Second, we have costs that are largely independent of location. That is, there may be advantages to be gained from producers joining together to share the overhead costs of marketing and research or to encourage local suppliers to specialize in certain goods. Third, government legislation may cause spatial variations in costs through subsidies or taxation.

We can combine these nontransport costs with transport costs in locational analyses of resource processing. In Figure 17-12 space-cost curves are used to create a two-dimensional profile of the distribution of both types of costs. In the first case [Figure 17-12(a)] nontransport costs are considered uniform irrespective of location, whereas transport costs vary systematically about a central location (A). If we assume a horizontal demand curve at a common market price, then the area of profitable production forms a shallow, saucer-like depression where we can locate the point of least-cost production (A) and the spatial margins of profitability. When we allow factors such as labor or power costs to vary, the area of profitable production can be altered. Thus, we can have a second production outlier with low nontransport costs forming a geographically isolated pocket, as in Figure 17-12(b). Changes in the third group of nontransport factors, government intervention, are introduced in Figure 17-12(c). By placing heavy taxes on locations near A and subsidizing peripheral areas, we ensure considerable alterations in the spatial distribution of profitable locations.

These simple cross sections give some idea of how variations in nontransport costs can be introduced into locational analysis. The cross sections can be supplemented by contour maps whose cost elements are plotted on a plane rather than on a line. Again, these combinations are only first steps in unraveling the complexities of real-world locational patterns and intricate locational criteria.

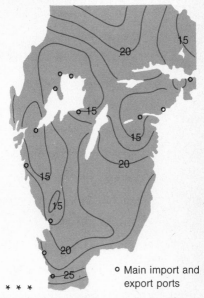

○ Main import and export ports

Figure 17-11. Isodapanes in Sweden. Contours of total transportation costs for paper-making in southern Sweden. The figures include the total cost of transporting pulpwood, coal, sulfur, limestone, and the paper itself in kroner per ton of output. The lowest-cost zone runs across the middle of the map and to the northwest. [From O. Lindberg, *Geographische Annalen* **35** (1952), p. 39, Fig. 20.]

Figure 17-12. The effect of nontransport costs on location. All three diagrams show a cross section through an economic landscape in which transport costs vary systematically with distance from a central location (a), but the basic (or nontransport) costs either (a) are uniform, (b) vary because of local production conditions, or (c) are modified by government intervention. In practice basic costs are also likely to vary over space, being lower (because of greater economies of scale) near the market center (A).

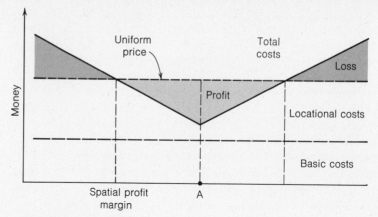

(a) Uniform basic costs

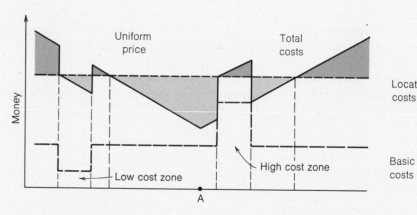

(b) Variable basic costs

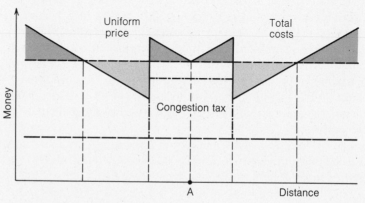

(c) Government intervention

We can remind ourselves of how sensitive to change are real-world resource processing locations by considering the decisions faced by a ski-tow operator in Pennsylvania during the 1960s (Figure 17-13). If he moved closer to the natural resource (i.e., areas of heavy and reliable winter snowfall), then he would have been further from his main urban markets—the skiers in Philadephia, Baltimore, and Washington. If he moved toward these sources of skiers, he would have had to locate in areas of lower and less reliable snowfall. He also had to keep an eye on other ski-tow operators and evaluate the benefits of locating near or far from them. Thus, in this resource processing situation, a subtle game was being played in which both nature and the individual processors had a role. Artificial snowmaking would vary this locational game still further. It is complications of this kind that modern locational theory, derived from Weber's theory but now vastly extended, is designed to explain.

Locational Change over Time   So far we have used Weberian analysis to study fixed locations. It can also be extended to industry changing over time.

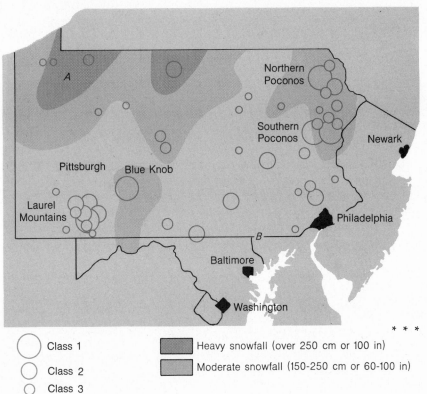

Figure 17-13.   Locational decisions in a resource-use framework.
This figure illustrates the problem of finding the best location for a ski resort, given variations in snowfall, major urban markets, and the competition of other ski resorts. The sizes of the circles indicate the sizes of the existing ski resorts. A ski resort at *A* in northwest Pennsylvania would be fine for snow conditions and lack of competition, but short on skiers. Location at *B* would have lots of potential skiers in nearby Baltimore and Philadelphia but be short on snow. [From J. V. Langsdale, Pennsylvania State University, unpublished master's thesis, 1968.]

* * *

⭘ Class 1

⭘ Class 2

○ Class 3

∘ Class 4

▨ Heavy snowfall (over 250 cm or 100 in)

▨ Moderate snowfall (150–250 cm or 60–100 in)

One of the most instructive examples of the changing locations of a manufacturing industry over time is provided by the iron and steel industry. Iron has been in human use for over 4000 years. Evidence from the eastern Mediterranean suggests that the earliest method of producing iron was the so-called direct process. Iron ore was mixed with limestone and charcoal and heated in a furnace for about a day. The resulting spongy mass of metal was passed through repeated cycles of hammering and reheating as it was forged into tools or weapons. This iron is what we would today call wrought iron.

Phase One The locational pattern associated with iron making by the direct process can be seen from medieval Europe. This first locational phase was essentially one of small-scale and widely dispersed production. [See Figure 17-14(a).] Small deposits of iron ore occur at or near the surface in a wide range of geological formations. Iron making was critically dependent on a large supply of wood, since every ten tons of iron smelted required about two acres of timber to be felled for charcoal. The progressive clearing of the woodlands (see the description in Chapter 10) throughout the Middle Ages was to make the presence of forests an ever-more critical factor in locating iron production. For example, by the sixteenth century the only two areas of England that could support major charcoal-using industries were the Forest of Dean in the southwest and the Wealden Forest in the southeast. A secondary locational was a regular supply of water to supply energy for the furnace "blast," using bellows, and for hammers for forging the iron. High transport costs ensured that iron was made largely to satisfy local demand, and so the scale of operation remained small. This first phase of iron making was replicated on the North American seaboard in the eighteenth century as New England and the middle Atlantic states also developed their own charcoal-based iron industries dependent on the eastern forest.

Phase Two The second locational phase in iron making is associated with larger-scale productions concentrated on the coalfields. [Figure 17-14(b).] By 1500 attempts were being made to replace charcoal as a fuel by coke, and to smelt the iron ore in larger blast furnaces. Successive improvements in the smelting technique using coal meant that by the early nineteenth century coal had largely replaced wood as the essential fuel. The location of iron making on a coalfield site now held many attractions. First, under the prevailing technology eight times as much coal as iron ore was needed to produce a given quantity of smelted iron so that, in Weberian terms, coal had a very high weight-loss index. It was clearly much more advantageous to take iron ore to the coal, rather than vice versa. Movement of iron ore was not always necessary for a second advantage of many coalfields was the presence of iron ores interbedded with the coal seams (the "black-band" ores). Older iron ore sources were often abandoned in favor of those located in or adjacent to the coalfields. Third, the coalfield regions in parts of Europe had long traditions of iron making associated with forging and reprocessing iron using

(a)

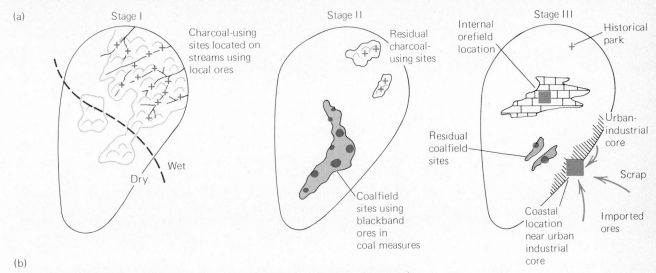

Stage I

Charcoal-using sites located on streams using local ores

Wet
Dry

Stage II

Residual charcoal-using sites

Coalfield sites using blackband ores in coal measures

Stage III

Internal orefield location

Historical park

Residual coalfield sites

Urban-industrial core

Scrap

Coastal location near urban industrial core

Imported ores

(b)

Figure 17-14.   Evolution of industrial location patterns.

(a) An idealized sequence of three stages in the growth of the iron and steel industry.
(b) An example of a Stage III coastal location for a modern iron and steel plant. Sparrows Point near Baltimore, site of a Bethlehem Steel plant using imported Venezualen and Labrador iron ores. [Photo courtesy of Bethlehem Steel Corporation.]

either charcoal (e.g., the Sheffield region of England) or coal itself (e.g., Midland England). These forces of locational change in favor of the coalfield areas were probably running at their height in the late eighteenth and early nineteenth century. The rapid growth in demand as the Industrial Revolution accelerated, was marked by the emergence of booming steel making areas such as those of Pittsburgh, around Sheffield in England, and around Liege in Belgium. Here production was increasingly in large-scale units with finished products being transported over longer and longer distances.

Phase Three   The third locational phase in iron making is the most complex. [See Figure 17-14(c).] It runs from the late nineteenth century through to the present and is closely bound to the continuously changing technology of steel manufacturing. It is marked by a redispersal of the industry away from the coalfield areas, and in particular toward (a) orefield sites and (b) coastal sites. Coalfields became less significant as the amount of coke needed in blast furnaces was steadily reduced: Today only about half as much coke is needed as iron ore. Exhaustion of the blackband iron ores in the coalfield areas added to this trend. As the attraction of coalfields declined, so other areas rose in importance. From 1878 a technical breakthrough in steel manufacturing (the Gilchrist–Thomas process) meant that the rich but hitherto unused limestone orefields could be used. Where orefields were located near major market areas (e.g., the Jurassic orefields of eastern France and Luxembourg), major iron and steel industries emerged. Where orefields were located in more remote areas without a local market (e.g., Schefferville on the Canadian shield, or Mauritania in Saharan Africa), the ore was exported. Growing international trade in iron ore has matched the decline in production from home orefields. Other than the USSR, all the major steel producers—the United States, Japan, West Germany, and the United Kingdom—are now all heavily dependent on imported ores. In consequence, coastal sites where materials (including imported ores) could be easily assembled have become important. In North America the Great Lakes complex of Cleveland, Detroit, Chicago, and Gary (in the United States) and Hamilton (in Canada) using first Lake Superior and later Labrador ores, overtook the Pittsburg area in steel production in the 1920s. On the middle Atlantic coast the Fairless and Sparrows Point sites draw on South American ores. Similar coastal concentration has occurred in Britain with the growth of the South Wales coastal plants.

The locational pattern of iron and steel production continues to be in a transitory state today. Plants are increasingly specialized, drawing on more distant and richer ores and producing materials for specialized world markets. Technical change continues to reduce the dependence on raw material location and increase the importance of the market, although with an increasing volume of scrap metal in circulation the major urban-industrial centers now serve a double role as both market and source area. Ore enrichment has allowed the same earlier raw material sources to be reevaluated, leading to the revival of iron ore production (e.g., Minnesota).

Thus the locations of the steel industry today represent a partly inherited pattern in which past locational forces as well as those of the present may be seen. With increasing size the huge investment in a single steel making plant may give it an inertia sufficient to maintain its location long after the forces which created it have weakened. Both the company (in terms of capital investment) and the government (in terms of the workforce) may have a strong vested interest in preserving such inertia.

In our discussion of both the Thünen and the Weber families of models we have stressed geographic distance from the city as a dominating theme. But geographic distance (in the sense of the length of routes in miles or kilometers) is often only a crude measure of the costs of movement. Consider Figure 17-15, which shows actual freight costs per ton from six ports in eastern New Guinea to other points within the territory. The costs form an intricate patchwork related to the volume and type of goods to be carried, the mode of transport (truck, aircraft, or ship), and the degree of competition, as well as the geographic distance to the six ports. The situation in New Guinea is by no means atypical. At any point in time the exact cost of moving resources from one part of the earth's surface to another is a function of a dozen or more different considerations. Think of the different rates quoted for a passenger fare from Boston to London—from scheduled air fares, through reduced air charter rates, to variable shipping rates.

Notwithstanding these complications, we can still make out some rough order in the relationship of costs and distance. First, we find that geographic distance does play a role in determining most rates. Other things being equal, a longer haul costs more than a shorter haul. There are of course exceptions. Within many countries the rate for sending a parcel varies with its weight but

17-3
Complexities of Space

Figure 17-15.   Spatial variations in transport costs.
Freight rates per ton are shown for general cargo from the six main ports in eastern New Guinea to a selection of inland and coastal locations, using the cheapest mode of transport available. Note the contrasts between the low unit costs but restricted operating area of coastal shipping (a) in contrast to the very high unit costs but flexible operating area of air freight (c). The fact that the road network into the mountainous interior is poorly developed is reflected in the cluster of high costs in (b). The six main ports are marked on each map by stars. [From H. C. Brookfield and D. Hart, *Melanesia* (Barnes & Noble, New York, and Methuen, London, 1971), p. 357, Fig. 14-9.]

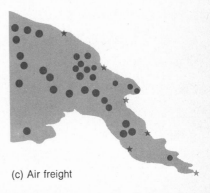

Freight rates ($/ton)

· <8   • 8–16   • 16–48
• 48–96   • 96–144   ● >144

(a) Coastal shipping

Wewak
Madang
Lae
Oro Bay
Port Moresby
Samarai

(b) Road

(c) Air freight

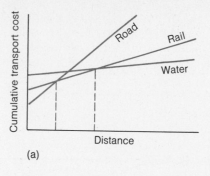

(a)

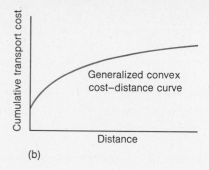

(b)

Figure 17-16. Transport costs and distance.
Some of the factors that distort a simple linear relationship between costs and distance. (a) Competition between modes of transport. (b) Length-of-haul economies. (c) Freight-rate zoning.

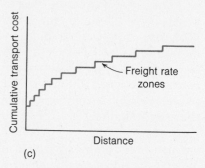

(c)

not with the distance it is sent. Similarly, companies may charge a uniform delivery rate throughout their sales area. Such blanket rates do not mean that costs do not increase with distance. The blanket rate subsidizes the longer-distance movements by charging more than is necessary for the shorter-distance movements.

Second, we find that total costs have two elements: a terminal, or handling, element (unrelated to distance) and a delivery, or haulage, element directly related to distance. These two cost elements may vary for different modes of transport, as Figure 17-16(a) shows. A more realistic curve for distance costs is convex and nonlinear, indicating that transport costs increase, but at a decreasing rate, with distance [Figure 17-16(b)]; that is, the cost of moving the first 10 km may be much higher than the cost of moving a similar 10-km stretch between 150 and 160 km away. We can add further realism to our model by breaking up the continuous cost curve in Figure 17-16(b) into a series of steps, each related to a particular level of freight charge, as in Figure 17-16(c). An extreme case of this kind of blanket zoning is provided by the postal service, which charges uniform rates over very large areas, regardless of the distance a letter or parcel is transported. The costs of transport may be absorbed by either the producer or the consumer. (See the discussion of spacial pricing on the next page.)

Thus it remains broadly true that the economic costs of connecting a city to agricultural zones and industrial-processing centers are a function of geo-

SPATIAL PRICING

Three alternative pricing strategies are commonly used to recover the costs involved in transporting a product.

1. *Source pricing.* The price is established at the production point and the customer pays the transfer costs of moving the product. This system is also termed *f.o.b.* (free-on-board) pricing. Customers may be charged freight rates on the basis of the actual distance covered [Figure 17-16(b)], blanket zones [Figure 17-16(c)], or a uniform *postage stamp* rate in which the charge levied is unrelated to the distance involved.

2. *Uniform delivered pricing.* The price is the same for all customers regardless of their location. The producers pay all the transport costs involved in shipping their product but recover this by taking the average transport costs into consideration when deciding on the price at which they offer the product. This system is also termed *c.i.f.* (cost-insurance-freight) pricing.

3. *Basing-point pricing.* All production of a given commodity is regarded as originating from a single point, a uniform price is established for all sources regardless of their location, and customers pay the transfer costs from the basing point

regardless of the actual location of the producer from which they purchased a product. The most famous case of basing-point pricing was the *Pittsburgh plus* system that operated for some time in the United States steel industry. Here customers were charged freight costs from Pittsburgh regardless of the location of the plant from which they actually purchased steel.

Pricing policies and their locational implications are discussed in D. M. Smith, *Industrial Location* (Wiley, New York, 1971), Chaps. 4, 5.

graphic distance. The twisting and blurring of freight rates, competition between modes of transport, and the like may be important for individual activities in particular locations. But the overall picture of a city organizing the space around it—but with its influence gradually fading as we move further away from it—remains. In the next chapter we go on to explore the implications of this city-centered organization for other aspects of the world beyond the city.

Summary

1. Early in the nineteenth century German locational theorist Thünen developed a theoretical model of geographic variations in land use. Starting with very simple assumptions, the model led to concentric "rings" of crops with agricultural land values sloping away from the central city. By relaxing these assumptions, environmental variables and transport links can be built into the Thünen model giving a closer match to real-world patterns.

2. Support for Thünen's interpretation has come from studies using different geographic scales. Using county census figures for the United States the demographer Bogue found that there is a rapid decrease in average population density with respect to distance from the city. Various curve forms indicate regional and employment category differences, with density also dependent on routes joining cities. Research at the village and farm level, and at the international level, shows the urban-industrial core of Western Europe and eastern North America acts much like the single city in Thünen's original isolated state.

3. The large amounts of natural resources used by urban centers influence the location of resource

processing and industry. The spatial economist Weber showed that industrial location is highly dependent on loss of weight or bulk in resource processing. Products of resource-oriented industries have large weight losses in processing, while market-oriented industries have weight and/or bulk gains in processing.

4. Plotting cost contours (isodapanes) allows the specification of least-cost locations in terms of transport costs. But locational analysis of resource processing is not restricted to transport costs only, but considers nontransport costs. This may be accomplished through the use of space-cost curves in which variations in nontransport costs can be introduced.

5. The impact of geographic distance on transport costs comes through a complex system of freight rates. Costs are usually convex over distance and made up of terminal and haulage components. Freight-rate fixing may be used by companies or governments to reinforce the locational advantages of particular producers.

Reflections

1. Examine the sequence of land-use zones in Thünen's isolated state (Figure 17-2). List the reasons why (a) truck farming is close to the city and (b) extensive cattle raising is carried out on the far periphery of the region. Do any of these factors affect location decisions in agriculture today?

2. Transport costs for the bulk movement of agricultural products continue to decline in real terms. What effect do you think this has on the changing importance of environmental resources (e.g., climate and soils) as a factor in the location of crop areas?

3. Give examples from your local area of any industrial concerns which appear to be (a) resource-oriented and (b) market-oriented. Is there an "in-between" category? How can the firms in this in-between category be fitted into a locational theory?

4. To what extent do you agree with the statement "Transport costs are a function of distance"? Why?

5. Assume that a major oil field is discovered in your community. Where do you think the oil would be refined, and why there?

6. Review your understanding of the following concepts:

land-use rings   Weberian analysis
the Thünen model   isotims
resource orientation   isodapanes
market orientation   space-cost curves
weight loss

One Step Further . . .

The classic theory of ring formation by J. H. von Thünen has been translated in
    Hall, P. G., Ed., *Von Thünen's Isolated State* (Pergamon, London, 1966).

Good accounts of its role in the structuring of land use in rural areas is given in
    Chisholm, M. D. I., *Rural Settlement and Land Use* (Hutchinson, London, 3rd ed. 1979), Chaps. 4 and 7.

Changing patterns of agricultural land use in intercity areas are described in
    Gregor, H. F., *Geography of Agriculture* (Prentice-Hall, Englewood Cliffs, N.J., 1970) and

Morgan, W. B., and R. C. Munton, *Agricultural Geography* (Methuen, London, 1971).

A modern view of industrial location is given in
    Smith, D. M., *Industrial Location: An Economic Geographical Analysis* (Wiley, New York, 1971) and
    Estall, R. C., and R. O. Buchanan, *Industrial Activity and Economic Geography*, 2nd ed. (Hutchinson, London, 1970)

and a useful selection of background readings is provided in
    Karaska, G. J., and D. F. Bramhall, Eds., *Locational Analysis for Manufacturing: A Selection of Readings* (M.I.T. Press, Cambridge, Mass., 1969).

For an excellent discussion of locational decision-making in the context of resource use and economic theory, see

    Lloyd, P. E. and P. Dicken, *Location in Space*, 2nd ed. (Harper & Row, New York, 1977).

Research on agricultural and industrial geography is generally reported in the regular geographic journals, especially Eco- nomic Geography *(a quarterly). Understanding developments in locational theory now demands considerable mathematical competence; browse through the* Journal of Regional Science *(a quarterly) to get some idea of what lies beyond the highly simplified models presented in this chapter.*

To arrive at a clear decision on these questions, let us take familiar examples, but set them out in geometrical fashion.

JOHANNES KEPLER
The Six-Cornered Snowflake (1611)

In this second chapter on the worlds beyond the city, we move away from agricultural zones and industrial centers to the lifelines that hold all three together. Transport links of all kinds—from TV transmissions to crude-oil supertankers—form an essential system by which cities and regions are maintained and grow.

This dependence of the city on links to the world beyond has roots which run back well beyond the modern period. Even in the most primitive of early urban settlements, trade and exchange were the distinctive features which al-lowed urbanization to emerge. What distinguishes primitive from modern systems of exchange is the intensity of exchange, its product range, and its spatial reach, rather than its essential nature. In this chapter we consider two aspects of this exchange. We look first at the questions of how and why exchanges between city and region take place and report on the regularities that geographers have discerned in these exchanges. Second, we examine the spatial structure of exchanges and the networks of routes that have been erected to facilitate them. As in Chapter 17 we link our discussion to the names of people whose concepts have thrown light on these geographic problems.

chapter 18

Worlds Beyond the City, II
Movements and Pathways

18-1
Newton and Intercity Flows

To raise the name of Sir Isaac Newton, the seventeenth-century English mathematician, in a geography text requires some explanation. Like many men of genius, his ideas in one field have been later found to be productive in others. Specifically, Newton's laws of gravitation have been found to throw light on geographers' understanding of the way in which flows occur between cities. This does not mean that you and I are swept along between cities like molecules in an "urban gravitational field." It does signify that the trillions of telephone messages, billions of freight-car journeys, or millions of aircraft movements that link the world's galaxy of settlements show a tendency, taken as a whole, to move in a way not unlike Newton's physical laws would predict.

Types of Flow   We have already mentioned three different kinds of intercity flows: telephone calls, freight cars, and aircraft movements. Just what sort of flows do we have in mind? We can draw a broad distinction between *transport* flows and *communication* flows. Transport involves the physical movement of something, be it people or freight, between two places. This movement can take place through a series of different transport modes, each with a specific set of advantages and disadvantages. These are summarized in Table 18-1. Thus if we wish to move large cargoes of a bulky commodity, then slow-moving barges are very cheap (on a per tonnage basis) and may be preferable. Conversely, aircraft make up for their very high costs per ton by their very great speed and freedom from the environmental barriers set by mountain, ocean, or icecap. The relative advantages of the various modes have not remained constant; indeed, only two of the five modes listed in Table 18-1 would have been available for intercity flow 150 years ago. The current trends in both passenger and freight flows are toward more emphasis on highways and airways.

More striking than the changes in individual transport modes has been the much faster increase in communication flows. Unlike transport, communication flows do not involve the physical movement of an element between places. Communication is the sharing of information. Although short-range communication is as old as man himself, most of the mass communication now flowing between cities is the product of the last two centuries of technology. The nineteenth century was characterized by the invention and rapid spread of "wire" communication systems. Thus Washington and Boston were the first two cities to be linked by a commercial telegraph line (1844); Europe and America were linked by submarine telegraph cable (1858). After Bell's Boston experiments in 1876, the telephone system spread slowly. By the century's end, Chicago and New York had been linked. The twentieth century saw breakthroughs in "wireless" communication systems—radio in the 1910s, television in the 1930s, satellite communications in the 1960s. Intercity communication using these new modes has been increasing exponentially at rates which make the expansion of world population look slow. For example, the volume of intercity telephone calls is now doubling every decade over most of the world.

Mode of transport	Principal technical advantages	Use
Railroads	Minimum resistance to movement, general flexibility, dependability, and safety	Bulk-commodity and general-cargo transport, intercity; of minimum value for short-haul traffic
Highways	Flexibility, especially of routes; speed and ease of movement in intraterminal and local service	Individual transport; also transport of merchandise and general cargo of medium size and quantity; pickup and delivery service; short-to-medium intercity transport; feeder service
Waterways	High productivity at low horse-power per ton	Slow-speed movement of bulk, and low-grade freight where waterways are available; general-cargo transport where speed is not a factor or where other means of transport are not available
Airways	High speed	Movement of any traffic where time is a factor—over medium and long distances; traffic with a high value in relation to its weight and bulk
Pipelines	Continuous flow; maximum dependability and safety	Transport of liquids where total and daily volume are high and continuity of delivery is required; potential future use in movement of suspended solids

Table 18-1 The relative advantages of different modes of transport

SOURCE: W. H. Hay, *An Introduction to Transportation Engineering* (Wiley, New York, 1961), p. 283, Table 8-5.

We have already commented in this section of the book on the way in which the changing costs of flows have led to two contrary spatial movements—an *implosion* of major cities at the intercity level, but an *explosion* of suburban sprawl at the intracity level. (Compare Figure 14-8 with Figure 14-15.) This drastic change in the geography of the world's leading cities is closely linked to the innovations in transport and communications reviewed above. New airline services or telex links tend to be first established between pairs of cities where the demand is greatest, so reinforcing the already commanding position of the leading cities.

Spatial Patterns of Flows   If we map the origin and destination of flows, we find that most moves are over a short distance. As an example let us take the hundreds of heavy trucks that roll down the freeways from Chicago bound for other parts of the United States. Most unload their contents a few kilometers from the city; relatively few move a long distance. In Figure 18-1(a) we plot the decrease in traffic with distance from Chicago out to about 650 km (400 mi). Notice that the diagram is drawn so that both the amount and the distance of the flow are plotted not on a linear scale, but on a logarithmic one.

Plots of similar flows on much larger geographic scales have a rather similar pattern. Figure 18-1(b) presents the pattern of rail-freight flows within the United States, which fall off regularly out to about 2,400 km (1500 mi). Figure 18-1(c) shows the pattern of world shipping out to a distance of 20,000 km (12,500 mi). Similar patterns can be found on geographic scales from the world level right down to the level of local kindergarten districts.

We can generalize about these *distance-decay*, or *distance lapse-rate*, *curves* by simply stating that the degree of spatial interaction (flows between regions) is inversely related to distance; that is, near regions interact more intensely than distant regions. The general form of this rule has been firmly established since the 1880s. However, the exact form of the relationship between distance and interaction has been difficult to pin down. On a graph with arithmetic scales, plotting distance against interaction produces a J-shaped curve in which flows decrease rapidly over shorter distances and more slowly over longer distances. Using logarithmic scales on both axes commonly yields an approximately linear relationship, and various alternative mathematical funtions can be used to describe such forms. The results of Swedish work on migration indicate that spatial interaction is inversely related to the square of the distance between *settlements*, but this is an approximation of variable empirical findings. (See the discussion of the distance-decay curves on page 437.)

The Gravity Model　More than a century ago observers of social interaction had noted that flows of migrants between cities appeared to be directly related to the size of the cities involved and inversely proportional to the distance separating them. By 1885 the British demographer E. G. Ravenstein had incorporated similar ideas into elementary "laws" of migration. Although

Figure 18-1. The decay in spatial interactions with distance.
(a) Truck trips around Chicago. (b) Railway shipments in the United States. (c) World ocean-going freight. [Data from M. Helvis and G. Zipf, adapted from P. Haggett, *Locational Analysis in Human Geography* (St. Martin's Press, New York, and Arnold, London, 1965), p. 34, Fig. 2-2.]

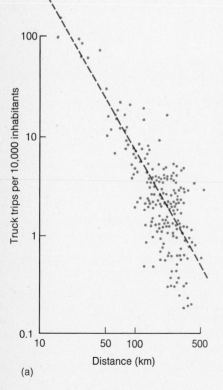

(a)

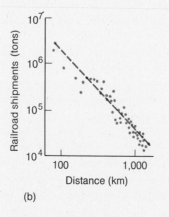

(b)

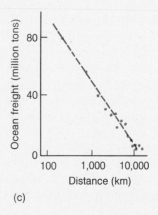

(c)

the specific term *gravity model* did not appear until the 1920s, it is clear that nineteenth-century workers were drawing on the relationships formalized by Sir Isaac Newton in his law of universal gravitation (1687), which states that two bodies in the universe attract each other in proportion to the product of their masses and inversely with the square of their distance. Gravitational concepts were specifically introduced by W. J. Reilly in 1929 in discussing the ways in which trade areas are formed. Reilly's ideas were subsequently expanded by researchers concerned with predicting flows in applied fields like highway design or retail marketing studies.

We can roughly estimate the size of flows between two regions by multiplying the mass of the two regions and dividing the result by the distance separating them. Thus a flow of 6 units would be produced by two regions with masses of 4 and 3 units, respectively, that are 2 units of distance apart. But what do we mean by "units" of mass and "units" of distance?

Mass has been equated with population size in many gravity studies. Information on population is easy to find, and we can readily estimate the size of most population clusters from census figures. However, population data may conceal significant differences between regions that affect the probability of spatial interaction, and the use of some system of weighting has been urged to take these differences into account. Economist Walter Isard has argued that just as the weights of molecules of different elements are unequal, so too the weights assigned to different groups of people should vary. Weights of 0.8 for population in the Deep South, 2.0 for population in the Far West, and 1.0 for population in other areas of the United States (reflecting regional differences in travel patterns) have been suggested. Multiplication of the population of each area by its mean per capita income is another possible way of refining our measuring stick for mass.

As we have already seen in Section 17-3, *distance* can be measured in several ways. The conventional measure in gravity models is simply the straight-line or cross-country distance between two points. In commuting studies, travel time rather than miles or kilometers may be an appropriate measure, as it may take as long to go a short way in urban areas as to go a long way in rural areas. When different forms of transport are available, distance may be measured in terms of the ease and cost of movement also. Fares (for people) and terminal costs and delivery charges (for goods) may be taken into account.

A simple example of the use of the gravity model to estimate flows between four cities is given in Figure 18-2. If you follow the sequence of maps carefully and refer to the caption to check on terms used, then you should see how simple the arithmetic is. You may be puzzled that two separate estimates of flow are given, one using distance [Figure 18-2(c)] and one using distance squared [Figure 18-2(d)]. Much of the work of geographers has been concerned with estimating just how distance should be measured and how it should be blended into the formulas. It turns out that, although distance squared is the better of the two, we have to use considerably more sophisticated formulas if we are to make estimates of acceptable accuracy. These lie outside the scope

DISTANCE-DECAY CURVES

Consider Figure 17-1, in which spatial interaction falls off with distance. One of the simplest and most common ways of describing curves that relate flows and distance is with a *Pareto function:*

$$F = aD^{-b}$$

where F = the flow, D = the distance, and a and b are constants. Geographers are especially interested in the value for the constant b. Low b values indicate a curve with a gentle slope with flows extending over a wide area, whereas high b values indicate a sharp decrease with distance so that flows are confined to a limited area. This formula was used extensively by a group of Swedish geographers in studies of migration between regions on a large variety of geographic scales as far back as the nineteenth century. Their findings showed b values going from as low as -0.4 to as high as -3.3. The mean value for all the studies was just below -2 (in fact, -1.94). This figure would suggest that

$$F = aD^{-2}$$

which we can rewrite as

$$F = a\frac{1}{D^2}$$

Apparently, spatial interaction falls off inversely with the square of distance. That is, the size of a flow at 20 km (12 mi) is likely to be only one-quarter of that at a 10-km (6-mi) distance. This *inverse-square* relationship is analogous to that used by physicists in estimating gravitational attraction.

Figure 18-2   A gravity model of flows between centers.

Population size (a) and distance (b) are used to estimate the spatial interaction between four centers. In (c) the actual distance between cities and in (d) the square of the distance between cities is used to estimate the flow. Note in (b) that D_{12} refers to the distance between city 1 and city 2, and so on. Since the formula in (c) and (d) is general and refers to flows (F) between *any* two pairs of cities, we use the general term F_{ij} to refer to flows between city *i* and city *j*. The same general terms *I* and *j* are used in the rest of the formula so that the population of the cities is P_i and P_j and the distance between them in D_{ij}. Check to see that you understand how the figures in (c) and (d) were achieved. For example, flow F_{14} in (d) is 4.0. This is given by multiplying the population of cities 1 and 4 and dividing this product by the square of the distance between them, i.e., $(64 \times 1) \div (4 \times 4)$ $= 4.0$.

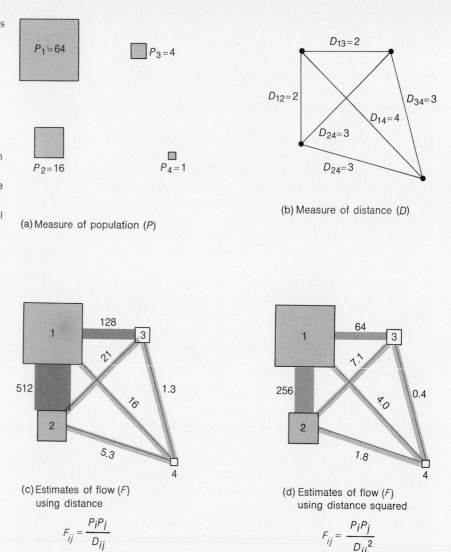

(a) Measure of population (P)

(b) Measure of distance (D)

(c) Estimates of flow (F)
    using distance

$$F_{ij} = \frac{P_i P_j}{D_{ij}}$$

(d) Estimates of flow (F)
    using distance squared

$$F_{ij} = \frac{P_i P_j}{D_{ij}^2}$$

of this book but will form an important part of most future courses you select in this field.

The Ullman Model   A different approach to the study of flows between regions is to ask why they occur. We can begin to answer this question by inverting it, or by trying to define the conditions under which flows would *not* occur. For example, if travel between the regions were expensive or if each region were highly self-sufficient, then we would expect rather little to be exchanged. Washington geographer Edward Ullman has systematized these

GRAVITY MODELS: SPATIAL INTERACTION

Assume two regions, region 1 and region 2, spatially separated by an intervening distance. Spatial interaction between the two regions can be estimated by a gravity model of the form

$$F_{12} = a \frac{M_1 M_2}{D_{12}^{\,b}}$$

where F_{12} = the flows between region 1 and region 2,

M_1, M_2 = the mass of the two regions (mass may be equated with population),

D_{12} = the distance between the two regions,

a = an empirical constant, and

b = a distance exponent (assumed to have a value of 2.0 in the original gravity model).

Thus, if we assume two cities of 1000 people each, 10 km (6 mi) apart, with $a = 1$ and $b = 2$, then the total flow between the two cities would be 10,000 units; at 20 km (12 mi) apart, the flow would fall to 2500 units. The units of flow are arbitrarily determined by the definitions of M and D. Values for the constants a and b can be estimated empirically by studying regional situations in which flows (F) as well as population (M) and distance (D) values are known. For an extension of gravity models see W. Isard, *Methods of Regional Analysis* (The M.I.T. Press, Cambridge, Mass., 1960), Chap. 11.

notions in a useful model of spatial interaction based on three factors: regional complementarity, intervening opportunity, and spatial transferability. We can illustrate Ullman's model using his study of interstate flows of lumber in the United States (Figure 18-3).

The first factor on which his model is based, *regional complementarity*, is a function of the resources available in any particular region. In order for regions to interact, there must be a supply or surplus of resources in one region and a demand or deficit in another. Thus in Figure 18-3 shipments of forest products from Washington to the southeast states are low partly because of the easy availability of forest products in each; conversely, flows of forest products to New York and Pennsylvania are heavy, despite the long distance, because of the high demand there and the small size of their own forest area.

Complementarity will, however, generate flows between pairs of regions only if no *intervening opportunity* for a flow occurs—that is, if there are no intervening regions in a position to serve as alternative sources of supply or demand. Seventy years ago little lumber moved from Washington to the northeast because the Great Lakes region provided an alternative and intervening source of supply.

The third factor in Ullman's model, *transferability*, refers to the possibility of moving a product. Transferability is a function of distance measured in real costs or time, as well as of the specific characteristics of the product. Table 18-2 outlines the relationship between the specific value of three types of lumber products (measured in dollars per ton) and the length of shipment. Local products may be substituted for products that are difficult to transfer in the same way that intervening areas of supply or demand substitute for more distant ones.

Figure 18-3.   Interregional freight flows.

Width of the arrows is used to show the volume of forest products moved by rail from Washington to other states in a single year. The row of arrows centered over the state of Washington itself (upper left) shows that most forest products moved very short distances. [From E. L. Ullman, in W. L. Thomas, Jr., Ed., *Man's Role in Changing the Face of the Earth* (University of Chicago Press, Chicago, 1956), p. 869, Fig. 162. Copyright ⓒ 1956 by the University of Chicago.]

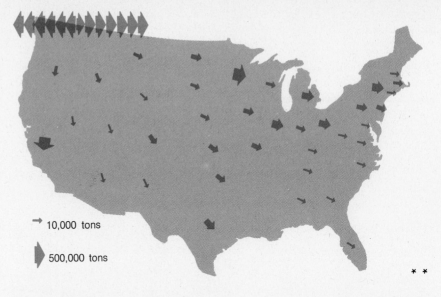

→ 10,000 tons

▶ 500,000 tons

Table 18-2   The relative transportability of three lumber products

	High-value veneer logs	Medium-value pulpwood	Low-value mine props
Value (dollars/ton)	150	20	5
Average length of longest haul by railroad (km)	640	32	8

SOURCE: W. A. Duerr, *Fundamentals of Forestry Economics* (Copyright ⓒ 1960 by McGraw-Hill, New York), p. 167, Table 21. Reproduced with permission.

18-2
Kohl and Intercity Networks

In the previous section we assumed that flows between cities and between a city and its region occurred across a uniform plane. For wave transmissions via radio and TV, this is broadly true. Other flows, however, are generally confined to a series of channels, or routes (Figure 18-4). Geographers see this delicate filigree of transport networks as the arterial and nervous system of regional organization, along which flow signals, freight, people, and all the other essential elements that allow the structure to be maintained. But do these networks have a characteristic spatial structure? And, if so, what controls it?

Networks as Regional Lifelines   Although transport systems form an essential and permanent feature of the economic landscape, the early locational theorists like Johann von Thünen and Alfred Weber had little to say about them. Yet as early as 1850 a German geographer, J. G. Kohl, created a series of branching networks to serve the settlements in his idealized city-region [Figure 18-5(a)]. His ideas were taken up by Walter Christaller nearly a century later in his own scheme for a system of cities [(Figure 18-5(b)]; since then, these ideas have been extended by other workers.

(a)

(b)

Figure 18-4. Contrasting modes of spatial interaction.
Each mode has distinct advantages for different types of movements. Check out the three shown—waterways, highways, and airways—against the characteristics listed in Table 18-1. [Photographs (a) from the Bettmann Archive, (b) Rotkin, P. F. I., and (c) courtesy of Port of N.Y. Authority.]

(c)

(a) Kohl 1850

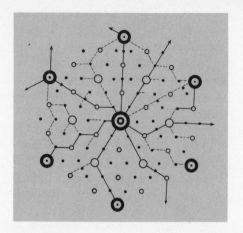

(b) Christaller 1933

Figure 18-5. Transport networks for theoretical settlement systems.

Three alternative schemes proposed between 1850 and 1956. Kohl's network (a) serves a Thünen-like isolated state. Only the upper half of the circular state is shown and the broken lines subdivided it into identical segments. The Christaller scheme (b) shows traffic routes in a $K = 3$ landscape. [Check back to Figure 15-10(a) to refresh your memory.] In this case the networks are not symmetrical about the central city: Compare the upper right quadrant (a wealthy region in which long-haul traffic prevails) with that of the lower left quadrant (a poor region in which short-haul traffic prevails). Note the contrasts between the straight routes of the former and the zigzag routes of the latter. Minor routes are shown by broken lines. Isard's network (c) has two centers each of which is surrounded by zones of decreasing population density. Boundaries of the complementary regions are shown as polygons which increase in size with distance from the two centers. [From P. Haggett and R. J. Chorley, *Network Analysis in Geography* (St. Martin's Press, New York, and Arnold, London, 1969), p. 125, Fig. 3-11.]

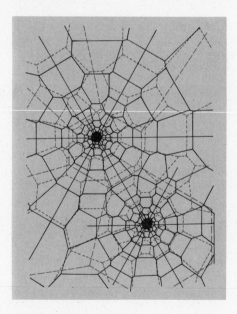

(c) Isard 1956

Some features of both the Kohl and Christaller schemes deserve notice. First, the transport networks are *hierarchic* in that they consist of a few heavily used channels and many lightly used feeders, or tributary channels. Like the city systems they serve, the segments of the transport system form an inverse distribution of size with frequency.

A second feature also is analogous to river systems, for the transport network has a branching structure in which the angle of branching is intimately related to the flow. The rule governing this phenomenon is familiar.

The angle of departure between the branch and the main stem is inversely related to the size of the branch. As the flow on the branch diminishes in relation to the main stem flow, the angle of departure becomes bigger. There is a precise interaction between the shape of the system and the work it has to do.

The number of major routes emerging from each city also has been the subject of research. For inland centers the most frequent number of key routes radiating from a city is six; few cities have less than three or more than eight. These figures are about what we would expect from our earlier discussion of the spacing and location of cities.

Mathematician Martin Beckmann has shown that if a region has a uniform population density and the costs of building a route are everywhere the same, the ideal transport system has a hexagonal honeycomb pattern [Figure 18-6(a)]. It is assumed in this system that both the origins and the destinations of flows are evenly spread over the region. Beckmann's system is a complicated one and involves some advanced theory that need not concern us here. What is interesting is that the basic honeycomb pattern can be modified to be more realistic.

For example, suppose we retain the idea of a uniform population but assume that we are concerned simply with linking this population with a single source (say, a central place in terms of Christaller's model). In this instance, all the destinations are uniformly spread over the whole region, but the origin is a single point. The best type of transport network is a symmetric honeycomb with holes placed in such a way that the system remains simply connected, as in Figure 18-6(b). By "simply connected" we mean there are no loops in the system and it still has a basic branching form like a tree. This treelike form is important because it helps us to bridge the gap between Beckmann's honeycombs and Kohl's branching patterns [Figure 18-5(a)]. The two schemes differ in two ways: Kohl's region is bounded (in fact, it is circular) and has a higher density of settlements in the center than at the periphery, whereas Beckmann's region is continuous (no boundary is specified) and the population density is uniform throughout. If we modify the Beckmann model by allowing a higher population density (and thus, smaller honeycombs) near the center and add a circular boundary, the model has a spatial form very similar to that of Kohl's network. One network is then simply a special case of the other. The missing link between the two is provided by Isard's landscape [Figure 18-5(c)] where a honeycomb system of transport links is developed about a pair of centers.

Can geographers find regions in which to test their models of transport networks? Certainly the number of areas of entirely new settlements is fairly small. The pattern of settlement in a Dutch polder (i.e., major area of land reclaimed from the sea) may be the nearest approximation we can find to the pattern that would result if a new, empty, and rather uniform landscape were settled all at once rather than slowly over a historical period. The scheme actually adopted there is rectangular, not unlike the pattern of roads laid out in the new territories of the United States during the 1800s. Other small

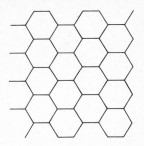

(a) Uniform origins

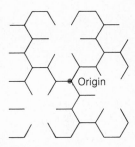

(b) Single origin

Figure 18-6. Optimal transport networks.

These networks assume a uniform population and homogenous regions. In (a) there are multiple origins and multiple terminals for flows. In (b) there are a single origin and multiple terminals.

agricultural areas like plantations also have been designed according to the principle of the Beckmann model.

Network design is important when a new system of roads is superimposed on existing ones. The interstate highway system in the United States and the new highway system of the United Kingdom typify the locational compromise that must be made between the cost of building networks compared with the costs of using them. We can illustrate this need for compromise by considering Figure 18-7, which presents a simple network designed to connect only five towns. If we design the network to minimize the costs to the user, we will make as many of the links as direct as possible. [See direct link AB in Figure 18-7(a).] But if we design the network to minimize building costs, we will have a different kind of pattern. In this second case the link AB will be much longer, but the *total* length of the network will be much smaller [Figure 18-7(b).] Comparisons with actual transport systems reveal an evolution from the first to the second type as flows increase. Look at a historical atlas, and compare the changing structure of the United States railroad network during different periods of its growth. Indeed, there is still a strong contrast between the sparse rail network in the west, where the costs to the builder were very important and the denser rail network in the east, where greater intercity flows made the costs to the user more significant.

Networks as Graphs   In our discussion of urbanization processes we saw how changes in the relative accessibility of cities had important repercussions on their relative growth. (Take another look at Section 14-1 to refresh your memory on this point.) So if we add a new transport link to a regional network, we should expect it to affect the relative accessibility of all the cities that are connected to it. Let us consider a specific example. Figure 18-8 shows the road network of the northwestern part of Ontario, Canada. Each of the 37 nodes represents either a settlement with at least 300 inhabitants or a main highway intersection. The road network, extending from Sudbury to Sault Sainte Marie, is connected to road systems in the remainder of Canada, but these external links are so few that we can treat the network as a closed system. Now let us suppose that seven new road links are proposed [links (a)

Figure 18-7.   The shortest networks connecting five urban centers.
Two alternative solutions are shown. In (a) the costs of the system to the user are most important and the interchanges are located within the cities. In (b) the cost of building the network is more important and the interchanges are located in nonurban areas. [After W. Bunge, *Lund Studies in Geography C*, No. 1 (1962), p. 183, Fig. 7-10.]

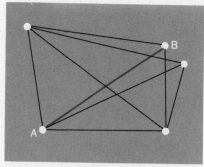

(a)   User optimum

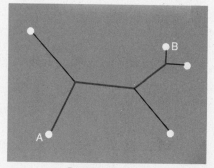

(b)   Builder optimum

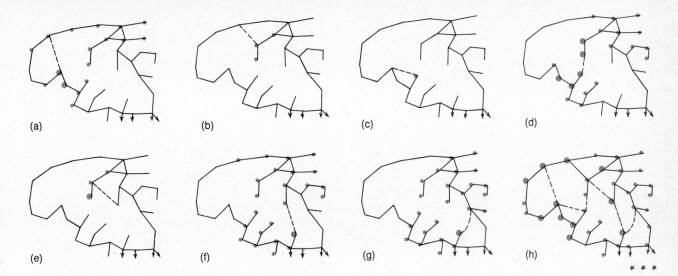

(a) (b) (c) (d)

(e) (f) (g) (h)

* * *

through (g) in Figure 18-8]. Which links will have the greatest impact on the total acessibility of points within the area to one another? Also, what local effect will each new link have on the *relative* accessibility of individual nodes within the network?

One way of answering these questions is to turn to a branch of mathematics that treats networks on the most primitive topological level. Topology is a branch of geometry that is concerned with the quality of "connectivity," that is, whether objects are or are not connected in some way. Its earliest and most famous geographic application is shown in Figure 18-9. Note that Euler was not concerned with distance or direction (the usual concern of geometry) but simply with whether a particular path through a network—the seven bridges connecting the different parts of the city of Königsberg—was possible or not. Euler's puzzling over this problem was to grow into *graph theory,* a branch of mathematics whose concepts are proving of increasing interest to geographers studying regional networks.

To use graph theory, we must reduce networks to graphs. This reduction involves throwing away a great deal of information about flows and characteristics of routes, but retaining the essential spatial factors of networks, nodes, and links. *Nodes* are the termination or intersection points of a graph. They may be assigned values denoting their location, size, the traffic they can handle, and so on. Depending on the varying scale of the analysis, nodes may be whole cities or street intersections. *Links* are connections or routes within a network. Links also can be assigned values relating to their location, length, size, and capacity. Some information on the connectivity of graphs can be obtained by measuring the *average path length.* Path lengths are determined by the number of steps (or "hops") between pairs of nodes, moving one link at a time along the shortest path through the network. (See the discussion of connectivity in graphs on page 447.)

Armed with this simple ruler, we can now go back to our Ontario road

Figure 18-8. The impact of new links on accessibility.

The maps show the impact of projected road links (a through g) on the accessibility of nodes in western Ontario, Canada. Points on the network which are expected to benefit by the improved connections are shown by dots: Large dots indicate major improvements, small dots only minor improvements. The final map (h) shows the combined effect of all new links. [After I. Burton, *Accessibility in Northern Ontario,* unpublished paper, 1963, Fig. 1.]

Figure 18-9. The Königsberg problem.
Sometimes apparently trivial spatial problems can have profoundly important implications. Mathematician Leonhard Euler (1707–1783) puzzled over why it was impossible for the citizens of the Prussian city to visit four areas (A, B, C, and D) and cross all its seven bridges (marked by colored dots) without recrossing at least one bridge. His later studies of the structure of networks laid the groundwork for a major branch of mathematics, graph theory, which has proved of importance in designing computer circuits and is of direct use to geographers analyzing the spatial structure of networks. [Drawing courtesy of The New York Public Library, The Astor, Lennox, and Tilden Foundations.]

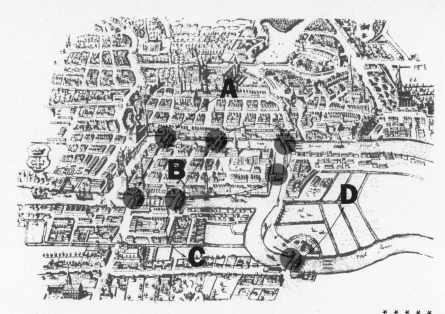

* * * * *

network and measure the impact of the proposed new links on accessibility within the system. For while each of the new links must bring some improvement and cut average path lengths in the network, some do this more effectively than others.

If we build the new links into the network one at a time, we can test just how much each one improves the connectivity of the system. Link (d) (from Folyet to Chapleau) does the most, cutting average path lengths in the network by 9.5 percent. Next best is link (a). Note that both (a) and (d) short-circuit the network by joining the northern and southern halves. Other links, like (c), are on the periphery of the network and their building has little overall effect.

How do individual cities benefit from the new links? This is measured by calculating the changes in the average path length for *each* city. Figure 18-8 shows the spatial pattern of improvements in connectivity, city by city, from each link proposed. Note that some links have a very localized benefit [(e.g., (c) and (e)]. In contrast, (d) brings the biggest gains in improved access to the center of the network; (g) brings less striking improvements, but spreads them rather equitably over almost all the eastern nodes. Some projects which are desirable on local grounds may not have the most beneficial effect on the whole network.

Graph theory provides only a first step in the analysis of transport systems. Links must be weighted to reflect the traffic they carry. We also must relax our implicit assumption that each link is just as costly to construct as all of the other ones. The tools needed for a full analysis of transport systems—cost-benefit ratios, spatial allocation models, and the like—are a bit complicated for an introductory course and are generally reserved for courses in transport geography. Graph theory does, however, serve to highlight the delicate

CONNECTIVITY IN GRAPHS

Consider a simple graph consisting of five nodes (A, B, C, D, and E) connected by five links (shown by solid lines).

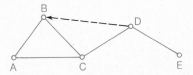

We can summarize the information in the graph in a connectivity matrix. Here the distance between pairs of nodes is expressed in terms of the number of intervening links along the shortest path connecting them, as follows.

To:					Row	Average path
A	B	C	D	E	sum	length
From:						
A 0	1	1	2	3	7	1.75
B 1	0	1	2	3	7	1.75
C 1	1	0	1	2	5	1.25
D 2	2	1	0	1	6	1.50
E 3	3	2	1	0	9	2.25
					Total 34	1.70

The row sum for each node provides a measure of its relative accessibility. Thus node C is the most accessible and node E is the least accessible. The grand total (termed the *dispersion value* of the graph) provides a measure of the graph's size in terms of all the paths within it. Dividing the row sums and the dispersion value by the number of positive values provides a measure of the average path length, which can be used in comparing one network with another.

The connectivity matrix in this simple case is symmetric about the diagonal of zeros because all five links are two-way. If we introduce a one-way link from D to B, it will disturb the symmetry. That is, DB will have a link value of only 1, but BD will have a link value of 2. The effect of the new one-way link is to slightly improve the network's connectivity. (The dispersion value falls to 33.) If we were to reverse the direction of the same link from B to D, the improvement in connectivity would be greater. For a further discussion of graphs including weighted as well as directed links, see K. J. Kansky, *Structure of Transportation Networks* (Department of Geography, Research Paper 84, University of Chicago, Chicago, Ill., 1963), Chap. 1.

---

balance between the urban system of cities and the regional transport net that links them and binds them together. Each new airline link, pipeline, highway, or shipping schedule tilts the balance of locational advantage this way and that, in a manner geographers must continue to monitor carefully if they are to be able to track—and eventually forecast (see Chapter 24)—changes in the world about them.

Disruptive Changes in Networks   Locational advantages may of course be reduced as well as increased by network change. Let us take one dramatic example.

On Sunday, January 5, 1975, at 9:30 P.M. the Australian city of Hobart was cut in two. The capital city of Tasmania, Hobart's population at the time was 145,000 spread along the eastern and western sides of the Derwent River (Figure 18-10). On the western shore lay the old city of Hobart with 104,000 people; on the eastern, the fast-growing suburb of Clarence with 40,000 people. Joining them across the kilometer-wide river was the four-lane highway over the Tasman Bridge joining the city to eastern shore suburbs and the airport.

The disaster which struck the bridge was so sudden and unexpected that a few cars ran on into the deep water. The *Lake Illawarra*, a freighter loaded with zinc concentrate, collided with the bridge pillars. The collision brought down a major section of the bridge decking, which fell onto the freighter, sinking it under the impact.

Although we have met natural disasters earlier in this book (see Section 6-3), this catastrophic change in transport networks has few parallels. Apart from the immediate disaster of loss of life, the longer-term effects were soon apparent. Approximately 34,000 persons usually crossed the bridge in each direction on a busy weekday. Although over a quarter of the population lived on the eastern shore, this was largely a dormitory area which had mushroomed since an earlier bridge had first opened in 1943. Over 90 percent of jobs in retailing and manufacturing, and all the entertainments lay on the western shore.

With the severing of a bridge link, two alternatives were open to eastern shore commuters. One was a 48 km (30 mi) circuitous route to an upstream bridge; another was to queue for a place on the few and now overloaded ferryboats. Both solutions had the effect of suddenly converting a few minutes' drive into a frustrating and expensive 60- to 90-minute journey. Its effect on both the social and economic geography of the city is being monitored by geographers at the University of Tasmania. Preliminary findings show how the impacts span a wide range of city life: from changing price levels for houses to shifts in patterns of crime. The two halves of the city were linked together again when the bridge reopened in 1978, and it will be fascinating to study the return to "precatastrophe" patterns.

Figure 18-10.   Impact of link removal. The Tasman bridge linking the two sides of the city of Hobart, Australia, was smashed by an ore-carrying ship in January 1975. Severing this vital link faced east-side commuters with a 48 km (30 mi) additional drive or long queues for ferry boats. [Photo by author.]

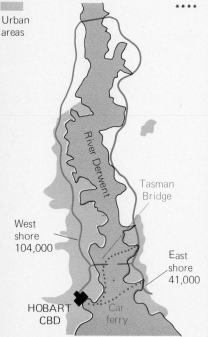

Urban areas

River Derwent

Tasman Bridge

West shore 104,000

East shore 41,000

HOBART CBD

Car ferry

The real world is made up of an immensely complex mosaic of regions. As we have seen, geographers have attempted to make sense of this by devising formal systems of regions. There is, however, no single or generally accepted set of world regions. Rather, there are alternatives, each with certain strengths and weaknesses. The most successful types of regional units appear to be those whose spatial boundaries coincide most closely with ecological or socioeconomic systems.

Many geographers have chosen as a basic unit a "city region." By a *city region* we mean the area that surrounds a human settlement and is tied to it in terms of its spatial organization. (These central settlements may be smaller than cities measured by conventional standards, but the term "city region" is arbitrarily extended to include smaller nodal regions when it is appropriate to do so.) The spatial ties between the city and its surrounding area are essentially movements of people, goods, finance, information, and influence.

The arguments for adopting the city region as the basic spatial unit are persuasive. A growing proportion of the world's population is concentrated in cities, and consequently human organization of the globe is increasingly city-centered. Cities form easily identifiable and mappable regional units. (See Figure 18-11.) Reasonably uniform statistical data have been available for many cities for the last century and a half. Further, city regions stress the comparability of different parts of the world, and thus encourage the search for general theories of human spatial organization. Finally, city regions are

18-3
The City Region as an Ecological Unit

Figure 18-11. Defining the city region.
Pinning down the limits of the city region is shown for the Australian town of Tamworth (23,000) in northern New South Wales. Tamworth (a) is the marketing center for a prosperous sheep-farming and wheat-growing area. If we map the area served by each of a city's professional establishments (doctors, lawyers, etc.) then we get a series of overlapping boundaries [shown hypothetically in (b)] from which we can identify an outer and inner range and a median line (c). This is done in (d) for Tamworth. Repeating the study for other sectors of city activity and plotting all the medians (e) allows a clear idea of the city region's limits to emerge. Note the important effect of the interstate boundary. [From J. Holmes and R. F. Pullinger, *Australian Geographer* (1973), p. 221, Fig. 8. Photo courtesy of Tamworth City Council.]

(a)

(b)
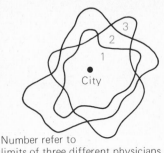
Number refer to limits of three different physicians

(c)
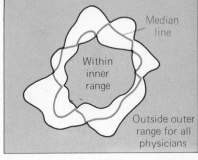

(d)
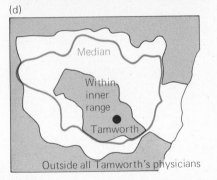

(e)
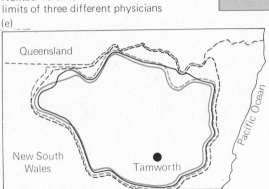

hierarchic. Like watersheds, they nest inside one another, and the city-region concept can be enlarged up to the world level or reduced down to the level of the smallest hamlet. We have already seen in Chapter 3 the emphasis geographers place on the ecosystem in studying the natural world. Can ecosystems be extended to include people? Cambridge geographer David Stoddart thinks they can. He sees several advantages in extending ecosystem concepts, which are usually confined to plant and animal communities, to human communities. Ecosystems bring us together with our physical environment into a single framework, encouraging a monistic rather than a dualistic view of a region. Ecosystems are studied in terms of structure and function, structural cohesion being seen as a logical response to the cycling of material and energy. Finally, ecosystems have certain features in common with other systems, and the networks used to construct models of those systems—say, by engineers or physicists—can be extended to ecosystems. Experiments in using electric circuits to construct models of biological systems are an example of how analogies might be developed.

Can we think of the city region as an ecosystem? Like watersheds, city regions need a constant flow of energy to maintain themselves. If we cut off the movement of people, freight, or funds into a city, it will stagnate; if we increase those flows, it will respond by growing in size. The city region is, like the watershed, in a state of balance with the forces that maintain and mold it. We have used the city region as a basic organizational unit in the last four chapters, but it is only one way of ordering the complexities of the real world. A complete integration of such regions with ecological and cultural regions still lies in the future.

Summary

1. Different modes of transport are competitive over different distances and different geographic environments and with different types of payloads. No single mode is optimal for more than a few purposes.

2. Study of the spatial pattern of transport flows shows that short-distance movements tend to dominate from the world scale down to the local scale. Distance decay or distance lapse-rate curves are J-shaped on arithmetic scales; logarithmic scales give a straight-line relationship.

3. Spatial interaction can be studied using a gravity model where mass equals either population or the size of an attraction value and distance is measured in one of several ways (e.g., distance, time, or cost.) Ullman developed a framework to examine spatial interaction based on regional complementarity, intervening opportunity, and spatial transferability.

4. Most intercity flows are confined within a fixed transport network. These networks have distinct spatial patterns which contribute to differences in accessibility. The model network schemes of Kohl and Christaller include both hierarchic conditions and branching structures. Beckmann's idealized regional model has a hexagonal honeycomb pattern which is based on uniform population density and building costs.

5. Addition or subtraction of transport links to regional networks raises questions of relative accessibility, solutions to which may be partially based on graph theory. This allows measurement of the effect of changing links on nodal accessibility.

6. City regions form an important basis for understanding the world's spatial organization. The use of the city region has grown out of the increasing city-centeredness of the earth's population and the non-unique aspects of such regions which allow for comparative studies.

Reflections

1. Define the three basic conditions needed for flows to occur in the Ullman model. Illustrate each, using flows into or out of your own state or province as an example.

2. What changes would you expect to occur in the flows between two cities if (a) both doubled in population or (b) the travel time between them was halved (e.g., by a bridge built to replace a slow ferry link)? Use Figure 18-2 to guide your calculations.

3. Sketch a map of the campus and the surrounding community and plot on it the paths followed by members of the class in getting to the building each day. Compare your results with Figure 18-5. How far does your network of paths resemble Kohl's branching tree? How are the paths of walkers, cyclists, bus riders, etc., different?

4. Gather data on any one pattern of spatial interaction (e.g., the number of guests in a local hotel, the number of visitors to a national park, or the number of students to a state university). Plot the observed number coming from each "source" area on the vertical axis of a graph and the expected number on the horizontal axis. For the expected number use a gravity model with the relevant population of each source area divided by its distance from the hotel, park, or university campus. Comment on your findings.

5. Consider the three modes of transport in Figure 18-4. Sketch typical networks for each mode and convert to a graph. What differences can you see in the connectivity of the three types of networks?

6. Review your understanding of the following concepts:

distance-decay curves

gravity models

regional comple-
mentarity

intervening oppor-
tunity

transferability

nodes and links

connectivity

average path length

city region

One Step Further . . .

For excellent but brief introductions to the topics covered in this chapter, see

Taaffe, E. J., and H. Gauthier, *Geography of Transportation* (Prentice-Hall, Englewood Cliffs, N.J., 1973), and

Hay, A., *Transport for the Space Economy* (Macmillan, London, 1973).

A wide range of geographic work is brought together in a set of readings,

Eliot, Hurst, M. E., Ed., *Transportation Geography* (McGraw-Hill, New York, 1974).

An authoritative review of both classical and modern work on interregional flows is provided in

Olsson, G., *Distance and Human Interaction* (Regional Science Research Institute, Philadelphia, 1965).

The spatial structure of transport networks and the locational principles that determine their form are discussed in

Haggett, P., and R. J. Chorley, *Network Analysis in Geography* (St. Martin's Press, New York, and Edward Arnold, London, 1969) and

Kansky, K. J., *Structure and Transportation Networks* (University of Chicago, Department of Geography, Research Paper 84, Chicago, 1963).

Research on both spatial interaction and the structure of networks is reported regularly in the major geographic journals. The more advanced quantitative work is often presented in Geographical Analysis *(a quarterly) and in the* Journal of Transport Economics and Policy *(also a quarterly).*

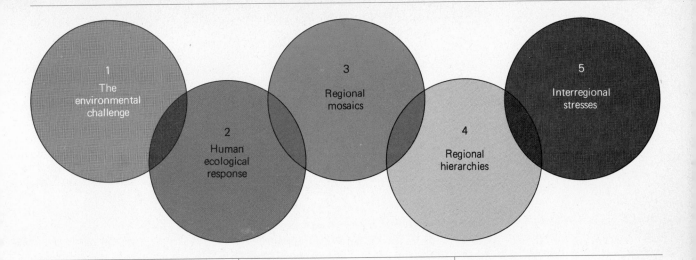

In Part Five of the book we look at the stresses that the regional structures we saw emerging in Parts Three and Four place on contacts *between* regions. *Territory and Conflict* (Chapter 19) examines the barriers people have set up to separate and delimit areas of ownership. We consider the geography of the modern nation-state, the most powerful of territorial organizations, and probe the causes of conflicts between such states. In *Coalition and Partition* (Chapter 20) this leads to cooperation between states and the common problems of dividing up the earth's last frontier—the offshore and ocean waters —and their own internal space. In *Inequalities Between States* (Chapter 21) we explore the vast differences in present prosperity and future prospects between the world's nation-states. Special attention is paid to trends over time and the disputed question of whether poorer countries are catching up with the rich, or whether inequalities are growing. Finally, in *Inequalities Within a State* (Chapter 22) we follow the problem down the spatial scale. We note how governments intervene to try to adjust imbalances in the regional mosaic of wealth and poverty, and consider the issues raised by regional planning and the role of geographic concepts in guiding spatial policy decisions. We close the section by looking back at the state as a spatial organization, and consider how far its role is now being challenged by such institutions as the multinational corporation.

part five

Interregional Stresses

My apple trees will never get across
And eat the cones under his pines, I tell him.
He only says, "Good fences make good neighbors."

ROBERT FROST

Mending Wall (1914)

So far in this book we have been concerned with ways in which the interactions between people and the environment lead to regional differences around the globe. The view taken has been a peaceful one. But the world we live in, the one which geographers have to describe, is not peaceful. Over all of our recorded history the story is one of conflict between the people occupying these regions—whether at the tribal level in earlier periods, or between superpowers today. Much of this conflict is concerned with disputes over territory.

Now the word "territory" has several alternative meanings in the English language, so it is useful at the outset to define its interpretation here. Territory is used in a legal sense to mean land belonging to a given sovereign state, such as the Northwest Territory of Canada. More specifically, it describes areas not yet granted full status in comparison with other parts of the state. For example, Alaska was a territory of the United States, with an appointed,

not an elected, governor, until it became the forty-ninth state in 1959.

Geographers, like biologists, use the term territory in a much more general sense, to indicate an area over which rights of ownership are exercised and which can be delimited or bounded in some way. We use the word "boundaries" to describe the limits of such territories. Sometimes ownership is formal and may be legally enforced. A householder may legally own the lot on which his house is built, and a nation-state may legally own its lands. At other times ownership may be unstable, and territories may be precariously held by displays of strength at the borders. Ornithologists studying the robin singing on the garden post at the edge of its territory and sociologists noting the gang member painting obscenities on a wall at the boundary of his turf are both recording displays of territorial ownership.

Boundaries may be marked by things as fleeting or trivial as birdsongs or graffiti or by boundaries as fixed and patrolled as the Berlin Wall. The earth's surface is laced with an

chapter 19
Territory and Conflict

intricate network of boundaries. In this chapter we look at how geographers interpret the idea of territoriality, noting differences and similarities with other biological creatures. Second, we look at the most powerful spatial unit of territory, the modern nation-state. Third, we ask why conflicts between states occur. Is this simply a matter of politics and history, or are there certain geographical conditions (those discussed in the earlier parts of this book) which make conflicts more or less likely?

19-1
Concepts of Territoriality

Territoriality is not a purely human trait. In this opening section of this chapter, we look at the evidence of this trait in other species and compare their behavior with the ways in which human beings stake out their "home range" on the earth's surface.

Animal Territories: Biological Evidence   In Chapter 1 we drew attention to the way in which groups on the beach spaced themselves out, creating, in effect, local "family territories." (See the discussion of interpersonal space, Section 1-2.) At many points in this book we have had to recall that, for all our cultural and technical uniqueness, we humans are still an animal species. To understand the intricate and formal way in which we organize our territory, we need to begin by looking over the species fence at the simpler territorial behavior of other animals. In doing so we shall need to draw on the findings of *ethology*, that branch of the study of animal behavior in a natural environment pioneered by European zoologists like Konrad Lorenz and Niko Tinbergen.

Many animal species "stake out" a specific area of space for their activities (e.g., feeding, breeding, or nesting). Territories will be defended very aggressively against other animals of the same species—sometimes by real fighting, but more often through highly ritualized "displays." Detailed maps of bird territories show distinctive spatial patterns (Figure 19-1). Note the distinction between discrete and overlapping territories for both isolated and gregarious species. Gregarious species of birds live in colonies, and the territories are those of the whole colony rather than the individual breeding pair. Figure 19-1(d) shows a special case, sea birds whose colonies may be confined to a few traditional nesting sites on islands and whose territories are overlapping areas of the sea.

While the fact that animals have territories is not in doubt, there is considerable debate over its meaning. Just why do territories occur? Two reasons appear the most likely. First, territories help to regulate population density and thus preserve an ecological balance with food supply. Those animals that cannot secure a territory for themselves are forced to migrate or starve. Second, territories ensure that the strongest members of a population (i.e., those able to obtain and to hold a territory) are the ones that breed and perpetuate the group. Because it forces out the weaker members of a population, ter-

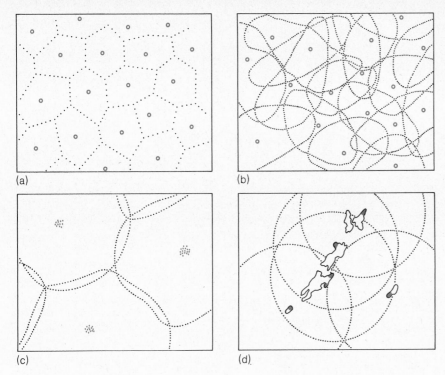

(a)

(b)

(c)

(d)

Figure 19-1. Territories of animal populations.

The maps shows four types of "home ranges" for territorial animals. (a) Solitary animals with sharply bounded and discrete territories. (b) Solitary animals with somewhat overlapping ranges. Core areas around a nesting site for bird populations will be discrete and defended. (c) Gregarious animals in colonies with slightly overlapping group territories. (d) Gregarious animals with highly overlapping ranges—in this case, seabird populations with island nesting sites. [From N. C. Wynne-Edwards, *Animal Dispersion in Relation to Social Behavior.* Copyright © by Oliver & Boyd, Ltd., Edinburgh.]

ritoriality may be an important mechanism in the natural selection process. Note that in the special case of sea bird colonies [Figure 19-1(d)], the competition is not for the marine feeding areas but for the few square inches of rock ledges in the traditional nesting areas.

Rural Territories: Median-Line Models   But what have the sea bird colonies to do with us? It would be as easy as it would be dangerous to draw facile comparisons between mice in cages and gangs in ghettos, or sororities in suburbia. In approaching the territorial behavior of the human animal, we choose to begin by looking at some points of similarity with the behavior of other animals, and then go on to review the differences.

Let us replace our sea birds with a scatter of pioneer farmsteads like those which existed during the colonial period of European overseas expansion. Each farm family clears the virgin forests around the homestead, tilling the most accessible land first and then gradually moving further afield as the family grows and more help is available. If we make simplifying assumptions which ensure that all families are the same size and have the same resources and that the land is everywhere of the same quality, what territorial pattern of farm boundaries will emerge?

Figure 19-2 shows the probable course of events. In the earliest stages each farm can expand in isolation and its territory forms a circle. The untamed

Figure 19-2.   The evolution of farm boundaries.

The maps show stages in the idealized settlement and colonization of a forest area. The method of drawing boundaries is as follows: First, lines join a given farmstead to each adjacent farmstead; second, each of these interfarm lines is bisected to obtain the median (or midpoint) of the line; third, from this median point a boundary line is drawn at right angles to produce a series of polygons. This type of territorial division is based on two assumptions—that there is a random scatter of farmsteads and that each farmer clears only the land nearest.

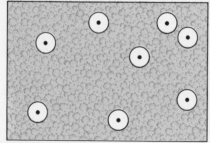
(a) Early phase

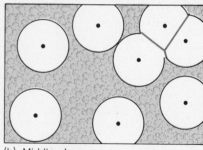
(b) Middle phase

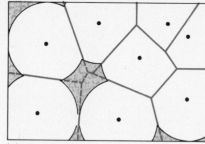
(c) Late phase

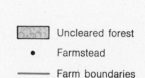
Uncleared forest
• Farmstead
── Farm boundaries

forest outside the homestead looks like pastry rolled out on a kitchen table, from which round biscuits have been cut. But as the closer farms expand, their boundaries meet and fence lines are staked halfway between them. In the later stages of farm expansion, only the last vestiges of forest remain and the farm boundaries look like those in Figure 19-2(c). Polygonal territories formed in this way are termed *Dirichlet polygons,* after the German mathematician who studied their geometric properties in the last century. They have the unique quality of containing within them areas that are nearer to the point around which they are constructed (in this case, the farmstead) than to any other points. Essentially, each side of the polygon is a median line, drawn at right angles to the line joining two farmsteads at its halfway point.

Clearly our two simple assumptions—that there is a random scatter of farmsteads and each farmer clears the land nearest to him—produce a territorial division of some complexity. It bears some resemblance to the patterns of villages and forests in Gambia, West Africa, that we met in our study of the Thünen model of regional divisions. (See Figure 17-7.) If the farmsteads had been arranged in a regular triangular lattice, then the farm boundaries would have been hexagonal, just like those of Christaller's complementary regions.

Thus we could regard some human territories simply as median-line divisions of space. Each animal, or gang, or farmer, or nation expands its territory outward until it meets its neighbors. The boundary is fixed at the halfway point between an individual and its neighbors. Unfortunately, this simple geometric view of how territories are formed ignores two significant complications. First, the animals, gangs, and so on may not be uniform but may vary in strength and agressiveness. Second, the space over which the distances are measured may be not simple and uniform (as in Figure 19-2) but complex and highly differentiated.

City Territories: Unequal Competition for Space   Recall that in Figure 19-2(c) the inequalities in farm areas come from the initial irregular location of the farms and the fact the farmers are assumed to be a homogenous group. Let us suppose, however, that the farmers were quite different—that some had large families and some had small ones, that some had more resources than others, were more agressive, and so on. Can geographers incorporate the effect of such differences into a territorial model?

One approach is to use the gravity models we met in Section 18-1. When we have two farmsteads of equal size, we expect the boundary to be exactly halfway between them, that is, at the median point. If they are of different sizes, however, we expect the boundary to be displaced away from the median point in the direction of the smaller farmstead. Just how big the displacement will be can be estimated from a gravity model. (See the marginal discussion of estimating boundaries by gravity models.)

Working with models of competition, regional economists have provided another perspective on how space can be partitioned into territories. Consider the position of the two sellers at center 1 and center 2 in Figure 19-3(a). Both produce homogenous goods and both have to pay the same freight charges, which are proportional to the linear distance the goods are shipped. The costs form an inverted cone about each center. These are shown as circular contours on the left of the diagram and as V-shaped cross sections on the right. The boundary (B) is located where the cones intersect; here the cost of goods from both centers is exactly equal.

In the first case the boundary is a straight line, so that a set of centers would form territories in the same way Dirichlet polygons were formed. But we can go beyond the simple polygon model. In the second case, the freight rates remain the same; but the production costs in center 2 are higher than those in center 1. The intersection of the two cones is now a curve, and the boundary forms a hyperbola [Figure 19-3(b)]. In the third case the position is reversed; that is, there are equal production costs but unequal freight rates. In this last case the boundary forms a circle, and the market area of seller 2 becomes an enclave within the much wider territory of seller 1. The circle is displaced eccentrically from the location of seller 1 [Figure 19-3(c)]. The formal structure of the model can incorporate even more complicated variations in both production and transport costs.

So far we have concentrated on certain rough analogies between the spatial form of animal and human territories. What about their purpose? Do human territories also have a role in the control of population density, regulating numbers by limiting the area from which resources are drawn? Certainly in some cultural groups farms cannot be subdivided if the number of potential

ESTIMATING BOUNDARIES BY GRAVITY MODELS

We can estimate the location of a boundary line between the market area of two centers by using a gravity model like those encountered in Section 18-1. Assume that we have two cities (city 1 and city 2), each with a specific market area (M_1 and M_2) and separated by distance D_{12}. We can estimate a breakpoint (B_2) in units of distance from the second city as

$$B_2 = \frac{D_{12}}{1 + \sqrt{\dfrac{M_1}{M_2}}}$$

If we now assume a simple case in which the two cities are 12 km (7.5 mi) apart and both have the same size market ($M_1 = M_2 = 10$), then

$$B_2 = \frac{12}{1 + \sqrt{\dfrac{10}{10}}} = 6 \text{ km (3.7 mi)}$$

That is, the boundary line occurs halfway between the two equal-sized centers. If we make the two cities unequal in size ($M_1 = 20$; $M_2 = 5$), then

$$B_2 = \frac{12}{1 + \sqrt{\dfrac{20}{5}}} = 4 \text{ km (2.5 mi)}$$

In this second case the boundary is displaced away from the halfway position in the direction of the smaller center.

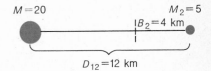

$M = 20$ $M_2 = 5$

$B_2 = 4$ km

$D_{12} = 12$ km

Figure 19-3 Competition for a market territory.

The diagram shows the hypothetical effect of variations in transport costs and production costs on the location of boundaries between retail centers. There are two centers (1 and 2) producing identical goods and located on a uniform plane where freight charges are proportional to the straight-line distance across the plane. The boundary line between the two market areas forms an *indifference curve* along which prices from both centers are equal and a potential consumer is indifferent to which center sells him the goods. These curves form hypercircles determined by the equation $P_1 + T_{1x}D_{1x} = P_2 + T_{2x}D_{2x}$, where P is the market price at the point of production, T_x is the freight rate between a production point and a given consumption point, x, and D_x is the distance between a production point and a given consumption point, x. Three alternative boundaries are shown here. The original market partition model was developed in the 1920s by economist F. A. Fetter and is known as Fetter's model. [From H. W. Richardson, *Regional Economics* (Praeger, New York, and Weidenfeld & Nicolson, London, 1969), p. 27, Fig. 2-3.]

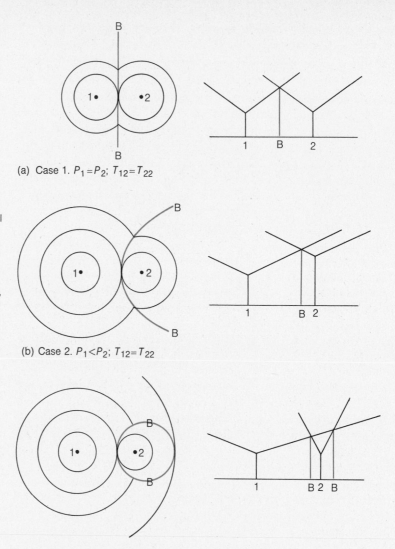

(a) Case 1. $P_1 = P_2$; $T_{12} = T_{22}$

(b) Case 2. $P_1 < P_2$; $T_{12} = T_{22}$

(c) Case 3. $P_1 = P_2$; $T_{12} < T_{22}$

farmers increases; the legal practice of *primogeniture*, by which the eldest son inherits property and the younger sons leave the estate to make their own fortunes, might, by a long stretch, be seen as ritual ecological behavior. The second reason for animal territories, survival of the fittest, may also have analogies in the division of market territories by business corporations, with small firms "going to the wall."

What about the ways in which animal and human territories differ? Here we shall emphasize one very major difference—the relative permanence of human boundaries. When we are dealing with unstable or short-lived territories like the market area of a seller or the domain of a small mammal, we can expect boundaries to be well adjusted to the forces that create them. Thus in

Figure 19-3 if seller 1 reduces its production costs more than other sellers, we would expect its territory to increase. But boundaries that are legally defined may persist long after the forces which created them have changed. We have already noted in Section 5-1 the gap between the legal limits of cities and the actual limits of built-up areas or commuting zones. International boundaries also reflect the political balance of forces at the time of their creation. The present boundaries between East Germany and West Germany and North Korea and South Korea relate to the military situations that existed in 1945 and 1953, respectively. Different segments of a country's boundaries may date from different periods. In the United States the boundary of Maine with Canada has a 1782 vintage, whereas that of Arizona with Mexico dates from 1853. Figure 19-4 shows the extent to which the boundaries of nation-states in tropical Africa are legacies of European colonial expansion and bear little relation to present cultural and economic realities.

Let us summarize as we conclude the first part of this chapter. We have seen that territoriality is strongly developed in certain animals other than human beings and that it is important in population control and selective breeding. Territories with a similar spatial form arise in human communities, although whether they serve a similar purpose remains a matter of debate. Finally, we have noted the distinctive institutionalized quality of many human territories that sets them firmly aside from their biological counterparts. It is this third type of territory that we shall be concerned with in the remainder of this chapter.

Figure 19-4. State boundaries and ethnic groups.
The decolonization of tropical Africa in the decade 1955 to 1965 saw the emergence of many new independent states. These states have, however, retained the boundaries of the original British, French, and Belgian (formerly German) colonies, boundaries which were drawn where opposing forces met during "the scramble for Africa" in the last two decades of the nineteenth century. As the map shows, the boundaries of the new states run across major ethnic boundaries rather than enclosing homogeneous ethnic groups. [From H. R. J. Davies, *Tropical Africa: An Atlas for Rural Development,* University of Wales Press, Cardiff, 1970, p. 23, Plate 10.]

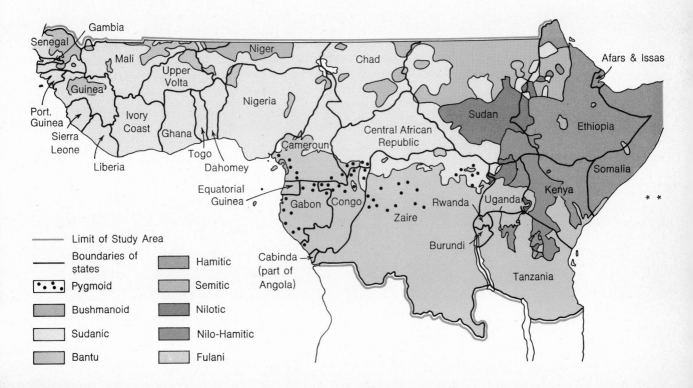

19-2
States as Political Regions

From the geographer's viewpoint the most evident territorial unit is the modern nation-state. There are about 200 nation-states in the world today, and they range in size and importance from the Soviet Union (with one-sixth of the world's land surface) to units of less than a square kilometer. Such states cover about 80 percent of the world's land surface (and the adjacent territorial waters) and divide it into a set of discrete bounded cells. (See Figure 19-5.) The areas remaining outside national sovereignty are controlled through colonial or joint-trusteeship arrangements.

The State as a Geographic Unit If we were searching for a single organizational unit in our organization of the world today, there would seem to be simple and persuasive reasons for using the nation-state as this basic unit. Nation-states are the principal "accounting units" for which comparative statistical data are regularly collected. States are decision-making units in that their central governments can affect the relationships between the population within their borders and the environment. They are clearly defined by boundaries that separate them from their neighbors, and these boundaries form noteworthy discontinuities in the pattern of human organization, sometimes evident in the landscape itself. Nation-states are increasingly be-

(a)

Figure 19-5. Boundaries between nation states.
(a) The unmarked boundary between Canada and the United States is crossed here by the Alaska highway in the unpopulated northlands of sub-Arctic Canada. (b) The heavily protected 20 km (12 mi) Berlin Wall constructed in 1961 to regulate migration from the German Democratic Republic (East Germany) to the Federal Republic (West Germany). (c) Part of the 2400 km (1500 mi) Great Wall of China built between B.C. 200 and A.D. 600 protect the northern boundaries of the Ch'in and Han empires against nomadic raids. [Photo (a) from Rotkin, P. F. I., (b) from United Press International, and (c) from the New York Public Library.]

(b)

(c)

ing organized as integrated economic units, and they collect and publish the data we use in building up a picture of the globe. Geographers often regard such states as the individual tiles out of which the world mosaic of regions is formed simply because they are the easiest units to compare.

Despite these advantages there are drawbacks, too. First, there is the difference in size and population. (See Table 19-1.) How can we usefully compare a country like the Soviet Union with countries like San Marino or Andorra? Second, large states have immense internal contrasts; in Canada the population forms a thin ribbon of settlement along the southern border—most of the country is virtually unsettled. Third, we cannot be always sure that we are comparing like with like, certainly a basic principle of analysis. In France we find a state where the influence of the central government is rather uniform everywhere. In Indonesia the government has much less control over the outlying, peripheral areas.

The boundaries of states are also a problem. They are often arbitrary geometric lines wholly unrelated to either the natural environment or population characteristics. (Recall Figure 19-4 showing how the boundaries of countries in tropical Africa cut across tribal divisions.) The borders may not enclose contiguous spatial areas. (Consider West and East Pakistan before the creation of Bangladesh in 1971 or the United States and Alaska.) And they may shift violently and abruptly from time to time.

The importance of central authority within a country is related historically to the amount of decision-making power the central goverment has acquired. To an increasing extent, decisions on resource exploitation, patterns of settlement, population growth, regional development, and pollution levels are being made at the central government, the most vital, level. As this trend continues, boundaries between states become more important to geographers in interpreting the environmental and spatial patterning of the earth's surface.

Centrifugal Versus Centripetal Forces From the geographic point of view we can think of a state as being something like a cell in biology. It has a

Atlas cross-check.

At this point in your reading, you may find it useful to look at the World Political map in the atlas section in the middle of this book.

Table 19-1 The world's largest and smallest countries[a]

Largest countries					Smallest countries				
	In area (km^2)	(mi^2)	In population (millions)			In area (km^2)	(mi^2)	In population (thousands)	
Soviet Union	22,402,000	(8,650,000)	China	825	Vatican City	0.4	(0.15)	Vatican City	1
Canada	9,976,000	(3,852,000)	India	586	Monaco	1.5	(0.58)	Nauru	7
China	9,561,000	(3,692,000)	Russia	252	Nauru	20	(8)	Andorra	16
U.S.	9,520,000	(3,676,000)	U.S.	212	San Marino	62	(24)	San Marino	19
Brazil	8,512,000	(3,287,000)	Indonesia	129	Liechtenstein	157	(61)	Liechtenstein	22
Australia	7,687,000	(2,968,000)	Japan	110	Barbados	430	(166)	Monaco	26
India	3,271,000	(1,263,000)	Brazil	104	Andorra	453	(175)	Qatar	95
Argentina	2,777,000	(1,072,000)	Bangladesh	75	Singapore	580	(224)	Maldives	109

[a]Data refer to the latest estimates for independent countries of the world.

distinctive shape enclosed by a boundary. It has a nucleus, the capital city. It has internal structure related to differences in population and development. It exports and imports across its boundary. These views of the state in biological terms were current in the last century, but later dropped since they could be used to support the aggressive expansion of a state.

Modern views see the state as a section of the earth's surface whose territorial boundaries are held in place in a state of tension. Figure 19-6 attempts to sketch some of the ideas inherent in the state idea in terms of the kind of human-environment systems discussed throughout this book. It is based on the ideas of Wisconsin geographer Richard Hartshorne (see Figure 25-7) who sees the state's existence depending on a dynamic equilibrium between centripetal and centrifugal forces. *Centripetal forces* act to bind a state together and cause it to survive: They include a common language and culture, a long common history, good boundaries. *Centrifugal forces* act to tear a state apart: They include internal divisions in culture and language, a short common history, and disputed boundaries. Continued existence of the state depends on the centripetal forces exceeding the centrifugal. Perhaps the most important but elusive of the binding forces is what Oxford geographer Jean Gottmann has termed a *national iconography* (Greek *eikon* = portrait or image). By this he means the psychological attitude of a people, drawn from a combination of past events and deeply rooted beliefs.

Boundaries of the State   Boundaries can be usefully divided into three categories on the basis of when they originated in comparison with settlement. (See Figure 19-7.) *Subsequent boundaries* are those that are drawn after a population has become well established in an area, and the basic map of social and economic differences has been formed. Thus in the case of an old

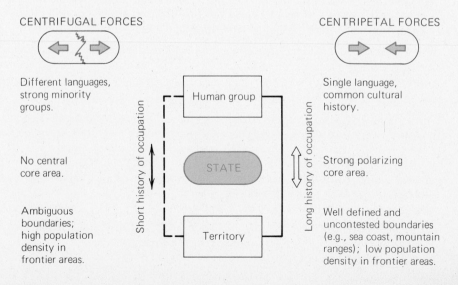

Figure 19-6.   Centrifugal and centripetal forces in state identity. The Hartshorne model of the state held in a dynamic balance between destructive (centrifugal) and constructive (centripetal) forces.

CENTRIFUGAL FORCES

Different languages, strong minority groups.

No central core area.

Ambiguous boundaries; high population density in frontier areas.

CENTRIPETAL FORCES

Single language, common cultural history.

Strong polarizing core area.

Well defined and uncontested boundaries (e.g., sea coast, mountain ranges); low population density in frontier areas.

Human group

STATE

Territory

Short history of occupation

Long history of occupation

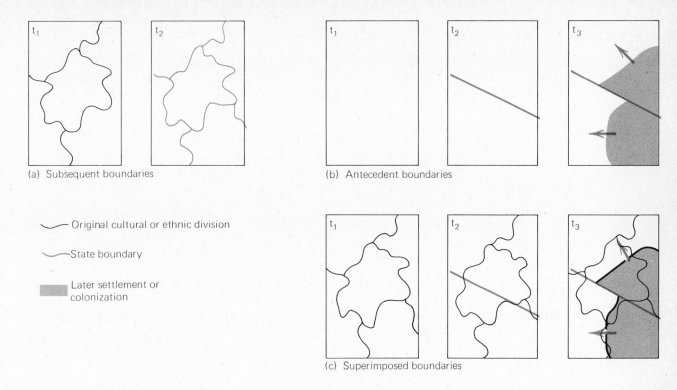

(a) Subsequent boundaries

(b) Antecedent boundaries

Original cultural or ethnic division

State boundary

Later settlement or colonization

(c) Superimposed boundaries

nation-state like France we may think of the present national boundaries as a rough approximation to the *limites naturelles* of the French nation. The idea of fitting the boundary around an existing ethnic group dominated the thinking of members of the Versailles Peace Conference after World War I; it played a significant part in drawing the line within the Indian subcontinent between India and Pakistan in 1948; it runs through the demands for a reunited Ireland or Germany. Yet if we consider the boundaries of most of today's nation-states, we discover that the boundaries represent arbitrary cutoff points. Many of the characteristics that give identity to a nation—language, ethnicity, a common history, and cultural traditions—do not end abruptly at its boundaries. The two German states do not include all German-speaking peoples; millions of Chinese live outside Asia (let alone a Chinese state); Israel contains only a fraction of world Jewry; and so on. Conversely, one very stable national unit (Switzerland) includes four distinct language areas.

By contrast, *antecedent boundaries* precede the close settlement and development of the region they encompass. Groups occupying the area later must acknowledge the existing boundary. The boundary separating the relatively uninhabited land between the United States and Canada, extablished and modified by treaties between 1782 and 1846, is an example of an antecedent boundary. Antarctica has also been divided up by a cartwheel-like system of meridians running north from the South Pole.

Figure 19-7. Three types of boundaries.
The maps are arranged in a time sequence t_1, t_2, t_3.

The third type, *superimposed boundaries,* is the converse of antecedent boundaries, in that they are established after an area has been closely settled. This type of boundary normally reflects existing social and economic patterns. The boundary between India and Pakistan, drawn at the division of British India in 1948, is an example of this type of boundary. Many new nation-states in Africa are emerging within a framework of international boundaries laid down by colonial poachers from Europe in an earlier day (Figure 19-4). Nigeria, Tanzania, and Zambia each inherited land areas that make sense in terms of colonial spheres of influence but have little overlap with either environmental or cultural discontinuities.

Drawing Boundary Lines   To settle a political dispute with a legal treaty is one thing, to translate that settlement into an identifiable line on the ground is another. The difficulty is most acute on the international level but is also encountered regarding boundaries within the state.

To clarify the problem of dividing land areas, suppose we consider the internal and external boundaries of the conterminous United States. In Figure 19-8 we see that over 80 percent of the international and interstate boundaries are *geometric.* These geometric boundaries are mostly lines of latitude like 49°N, which separates the United States from Canada, and north–south meridians such as sections of the Oklahoma–Texas boundary at 100°W. The ease with which these lines can be drawn by standard surveying methods makes them valuable when sparsely settled or empty areas are being divided up.

Other geometric boundaries are sometimes used, too. Part of the boundary between Delaware and Pennsylvania is an arc of a circle centered on the town

Figure 19-8.   Types of boundary lines. Different types of lines have been used for the international and interstate boundaries of the conterminous United States, The high proportion of geometric lines are typical of a situation where political decision predated close settlement. [From N. J. G. Pounds, *Political Geography* (McGraw-Hill, New York, 1963), p. 89, Fig. 32. Copyright © 1963 by McGraw-Hill, Inc. Used with permission of McGraw-Hill Book Company.]

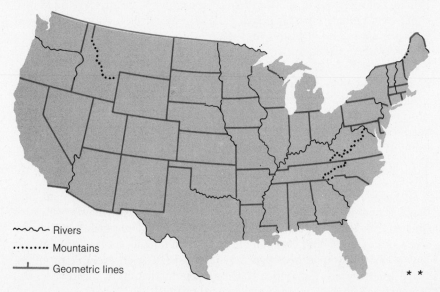

~~~~~ Rivers

········· Mountains

⊥ Geometric lines

* *

of New Castle. Still other geometric boundaries consist of straight lines between points where locations are known. The 1853 treaty defining the boundary between Arizona and Mexico describes a straight line joining a point 31° 21′N and 111°W with another point on the Colorado River. Figure 19-8 also shows some *nongeometric* boundaries; these boundaries generally follow the irregular course of natural surface features. Rivers are the most common natural feature used in drawing boundaries because they are self-evident dividing lines. As we shall see in the next section, they may nonetheless lead to disputes.

States exist in a permanent condition of inter-nation tension. Since the supply of territory is finite, pursuit of separate interests by each state must occasionally result in conflict. About seven percent of the world's gross product is spent on armaments, and in individual consumer countries arms spending may take up to one-half of its GNP. For the United States about one in nine jobs is created by the needs of the military-industrial complex.

19-3
Conflict Between States

It is clear that the world's political map is kept in place at very high cost. But even the stability of that map is perceived as being greater than it actually is. For example, although we think of Europe as having two main wars this century (1914-1918 and 1939-1945), an exact count shows a total of seventeen separate conflicts this century, including five since 1950. Over the years since 1815 Europe has seen 67 boundary changes (Figure 19-9) while a European state has appeared or disappeared on average once every three years. At the world level the late 1970s sees around 70 of its states in boundary disputes with one or more of their neighbors. Most are likely to be settled by agreement.

Pressure Points Within the State   Why do conflicts arise? An army of historians over the ages has probed the causes of war, and it would be impossible to try and summarize their complex and contradictory findings. We can, however, try to identify those *geographic* considerations which enter into the historians' models. Figure 19-10 is an attempt to do so. It shows a hypothetical state (termed here 'Hypothetica') beset by a series of conditions which could give rise to conflicts with its neighbors. Each condition is indexed by a number, and we shall discuss each briefly in turn and give an illustration from the actual record of international conflict. Note that the list we have stays within geographic bounds. The actual source of conflicts goes beyond this to include many nonspatial and nonenvironmental reasons outside our immediate concern. Conflicts that arise in the bordering offshore areas are discussed in the next chapter. (See Section 20-2.)

Landlocked States and Corridors   The first problem (1) lies in Hypothetica's landlocked position. It has no direct access to the world's oceans and

Figure 19-9. The stability of international boundaries.
The maps show contrasts in the permanence of land boundaries among European countries. Some land boundaries have changed little since the fifteenth century. Sea boundaries, like that between England and France, are not shown. The four oldest land boundaries, shown in (a), are those of Spain with Portugal, of Spain with France in the western Pyrennees, of Switzerland, and of the Low Countries. Note the large number of short-lived boundaries in Eastern Europe. What kind of factors make for very stable boundaries? [From N. J. G. Pounds, *Political Geography* (McGraw-Hill, New York, 1963), p. 29, Fig. 7. Copyright © 1963 by McGraw-Hill, Inc. Used with permission of McGraw-Hill Book Company.]

(a)   Over 400 years old   * *

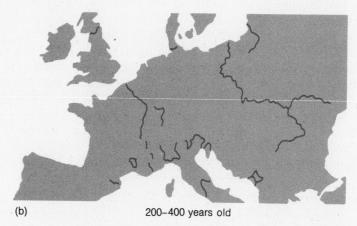

(b)   200–400 years old

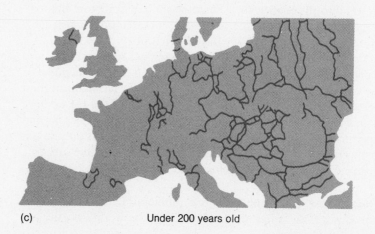

(c)   Under 200 years old

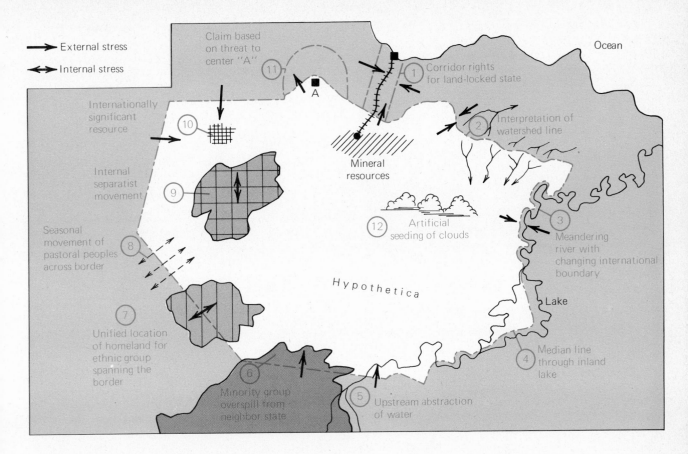

Legend:
→ External stress
←→ Internal stress

Ocean

Claim based on threat to center "A"

① Corridor rights for land-locked state

Internationally significant resource

A

② Interpretation of watershed line

Mineral resources

Internal separatist movement

⑫ Artificial seeding of clouds

③ Meandering river with changing international boundary

Seasonal movement of pastoral peoples across border

H y p o t h e t i c a

Lake

④ Median line through inland lake

Unified location of homeland for ethnic group spanning the border

⑥ Minority group overspill from neighbor state

⑤ Upstream abstraction of water

Figure 19-10. Geographic sources of international stress.

A landlocked country, Hypothetica, with some of the potential trouble spots (1) through (12) identified. Actual examples of the disputes illustrated here are discussed in the text.

yet needs to ship its raw materials from a tidewater port to its overseas markets. About one-fifth of the world's states are landlocked, and special transit arrangements have to be made for their goods to cross a foreign country. Figure 19-11 shows the present position of boundaries in Africa. Note the critical importance of the few railroads, and the special transit problems for the breakaway state of Rhodesia (Zimbabwe) in southern Africa which now has hostile neighbors along most of her borders.

Land corridors have provided one solution. The Polish Danzig Corridor to the Baltic (enshrined in Woodrow Wilson's proposals for European postwar reconstruction in 1918) was described by the French leader Marshal Foch, as the "root of the next war": events in 1939 were to prove him right. It illustrates an attempt by a state in Eastern Europe to secure a narrow strip of land linking its national territory to an adjacent sea. Outside Europe, the Eilat Corridor in Israel and the Antofagasta Corridor in Bolivia serve a similar purpose. Note that states may not need to be *completely* landlocked in order to need corridors.

Although most corridors have represented attempts to gain direct access to the sea, some have been aimed at indirect access by way of navigable rivers.

(a)

(b)

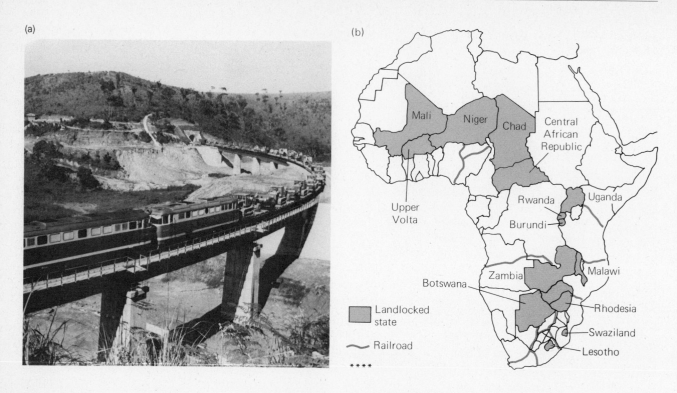

****

Figure 19-11. Communication problems for the landlocked state. (a) A freight train loaded with tractors makes its way over the Tanzania-Zambia Railway. (b) The 14 landlocked states of Africa and their relation to relevant railroads through neighboring states. [From R. Muir, *Modern Political Geography* (Macmillan, London, 1975), p. 62, Fig. 3.4. Photo by UPI.]

Ecuador has a long-standing claim to a strip of northern Peru that gives access to the navigable upper reaches of the Amazon; in 1922 Colombia obtained from Peru the narrow Leticia Corridor, allowing a 120-km (75-mi) frontage on the same river. In general, most landlocked states have secured access to the sea by international conventions that allow the movement of goods across intervening territories without discriminatory tolls or taxes. Switzerland, Czechoslovakia, Austria, and Hungary are European examples of states that use such agreements in their trade.

Conflict Over Natural Boundaries The next four trigger points are all related to rivers and lakes. Problem (2) relates to the use of a watershed boundary in defining an international dividing line. The classic case of this type of dispute occurred between Chile and Argentina. The original 1871 boundary assumed a line joining the mountain peaks would give watershed, whereas erosion had shifted it well to the east. A mutually acceptable boundary was finally worked out in 1966 and turned on detailed definitions of a master stream. (See the discussion of stream order in Section 3-2.)

Mountain barriers frequently pose such problems in deciding on the exact location of a divide or watershed. The 1782 treaty defining the northern boundary of Maine in terms of highlands divided rivers running to the Atlantic Ocean from those running north to the St. Lawrence River. But because of the complexity of the hydrology of the area, the boundary line was not finally established until the Webster–Ashburton Treaty was made in 1843.

Hypothetica also conflicts with her neighbors on two other river systems. The detailed demarcation of rivers (3) causes problems for two reasons: The course of the lower portion of a river changes continually, and the river has width and may have several channels. Figure 19-12 illustrates the rapidity of changes in the course of the Rio Grande separating Texas and Mexico. Clearly, the boundaries fixed at one time may produce anomalies a decade later. Boundaries may follow the navigable channel of waterways or, in the case of wide bodies of water (4), some median line between the two shores. A line equidistant from both shores was constructed through Lake Erie by the International Waterways Commission for the Canada–United States border.

The fifth problem relates to river flow coming downstream into Hypothetica from the territory of another state (5). Water may be drawn off or ponded back upstream, with severe consequences for irrigation places downstream. Tension between Israel and her Arab neighbors over the waters of the river Jordan and by India and Pakistan over the Indus waters relate to this hydrologic type of problem, while Welsh nationalists resent the export of water to English cities. So far most disputes have been settled by negotiation. For example, agreement between Canada and the United States has been reached to limit pollution of the Red River before it flows north across the forty-ninth parallel.

Conflict Over Minority Groups   Perhaps the most explosive and emotive problems in international tension relate to the location of minority groups. Hypothetica exhibits four such problems. Trigger (6) relates to a linguistic minority group lying along its borders. German occupation of the Sudeten parts of Czechoslovakia in 1938 had as its object the union of those German-speaking areas with the rest of Germany. Pressures to redraw the 1926 line dividing Northern Ireland from the rest of Eire to place Roman Catholic populations in the southern state exist today. Trigger (7) relates to a distinct ethnic group posed halfway between Hypothetica and a neighboring state. Both the West African states of Ghana and Togo have put forward claims to create a unified homeland for the Ewe people, a distinct African tribe arbitrarily divided by a colonial line dividing British and French spheres of influence in the late-nineteenth century. Where pastoralists cross boundaries in search of seasonal pastures for their flocks (8), small-scale friction can occur. Small changes in the French–Italian boundary in the Alpes Maritimes have eased this problem, but it remains unresolved between the states of Somalia and Ethiopia in East Africa.

Distinct minorities within a state (9) form an internal rather than an international problem. The claim for separate identity for the Indian tribal areas within the United States or Canada, the separatist movements of Quebec within Canada, Scotland within the United Kingdom, or the Basques

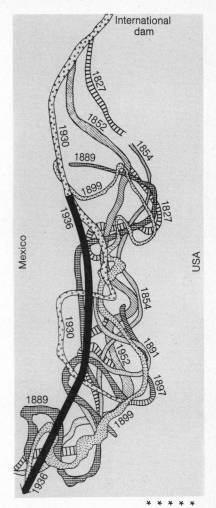

Figure 19-12.   River boundaries.
Local changes may occur in international boundaries because of geomorphic changes. The map shows shifts in the main channel of the Rio Grande near El Paso from 1827 to 1936. The international boundary between the United States and Mexico runs along the main channel and has therefore changed as the river changed. [After S. W. Boggs, *International Boundaries* (Columbia University Press, New York, 1940).]

within Spain all illustrate different levels of internal tension. Such problems become international when they lead to the creation of a breakaway state with its own international status or when external powers are attracted into the arena on behalf of the separatist area.

Other Sources of Conflict   The last three conflict triggers in our, by now, trouble-ridden Hypothetica are miscellaneous. Condition (10) relates to where our state contains a resource of international significance. This might be possession of a strategic resource (say, uranium or chrome) in critically short supply or a distinct cultural resource. Thus Jerusalem, currently part of Israel, contains in the Old City sites of central religious significance to the faiths of Jews, Moslems, and Christians (Roman Catholic, Greek Orthodox, and Protestant). From the eleventh-century Crusades onward, states have fought over such holy places. Point (11) is the reverse of (10). In this case, our state believes it has a location so vital to its own identity and survival that it is prepared to go to exceptional lengths to protect it. Thus the Soviet Russian claim in 1939 to Finnish territory in the neighborhood of Russia's Leningrad, on the grounds that it needed to give its second-ranking city a greater measure of security from possible artillery attack. French claims for enlarged ocean space around the Pacific island of Muraroa for nuclear tests is discussed in the next chapter. (See Section 20-2.)

A final trigger point (12) is included to remind ourselves that the sources of international conflict are evolving. So far legal disputes over cloud seeding for rain-making have involved only a few Great Plains communities in the United States, but—should technology progress—the consequences could spread upward to the international level.

The Richardson Model: Conflicts and Spatial Arrangement   One source of conflict not shown on the map is the number of neighbors a state has. Mathematician Lewis Richardson, in a book with the intriguing title *The Statistics of Deadly Quarrels*, thought this might be vital. Written before the intercontinental missile age, the book contends that the potential for any interregional relationship (including conflict) is a function of number of neighbors. Richardson argued that a state like Germany in 1936, which had nine other nation-states abutting its own territory, had a much higher chance of conflict, according to this line of thinking, than a nation-state like Portugal with only one neighbor. If territories were closely packed in an unbounded land area, the average number of neighbors could reach as high as six. Such areas might approximate the form of Christaller's hexagonal central-place territories that we met in Chapter 15. The spherical shape of the planet means that if states continue to grow larger, the number of neighbors must decrease until the world is dominated by only two states; then the number of neighbors drops to only one. Although according to Richardson's thesis this arrangement would decrease the potential for conflict, the scale of conflict in a two-state world would be immense.

Although such spatial speculations may play a part in international stress, they are overriden by far more powerful ones that lie outside the scope of this book. You may, however, care to browse through some of the writings in political geography in which these problems are explored in depth. (See "One Step Further . . . ," p. 474.) It is particularly worthwhile to look at a book written by geographer Isaiah Bowman nearly a half-century ago. In his *New World* he reviewed, continent by continent, the stress areas of a world just recovering from the horrors of the 1914-1918 war. Current problems like those of the Middle East (see Figure 19-13) can be seen in a context of deep-seated difficulties, and some conflicts can be seen to have a spatial persistence that is both intriguing in retrospect and sobering in prospect.

If spatial factors play an integral part in war strategies, geographers would contend that they should have an equal weight in strategies of peace. Certainly Richardson's conclusions on conflict were aimed at the reduction and resolution of wars. In the early 1960s the founding of the Peace Research Society by a group of behavioral scientists, including geographers, showed a similar sentiment. If good fences do indeed make good neighbors, then geographers need to determine from their research just where fences might best be built.

Figure 19-13. Critical areas of international stress.
The opening of the Suez Canal (center) through the narrow Afro-Asian isthmus in 1869 reduced direct sea-route distances between Western Europe and India by about one-half. Arab-Israeli conflicts over the last 25 years culminated in its closure after the Six-Day War in 1967. Renewed hostilities between Egypt and Israel in 1973 have led to a buffer zone being set up on the east bank of the Canal. The future of Sinai itself is likely to remain a source of conflict. This Gemini photograph looks southeast over the Nile delta and Sinai desert toward central Arabia. [Photo courtesy of NASA.]

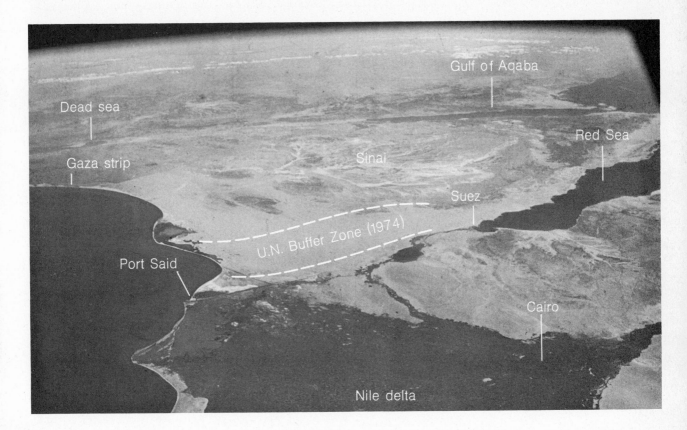

Summary

1. The concept of territoriality is a familiar one to students of animal behavior. Some similarities exist between the territorial behavior of humans and of animals.

2. Territorial divisions called Dirichlet polygons are generated by a model of equal competition for space in an area of agricultural colonization. These polygons may be distorted by the numerous variables which cause inequalities in competition for space. The location of territorial boundaries for city regions may be estimated by using modified gravity models.

3. The modern nation-state represents the most significant political region for geographic analysis and is a useful basis for global comparative studies. Its advantages include the direct links with decision-making, availability of data, and clear-cut boundaries. Its disadvantages include the extreme size and population differences between countries, large internal contrasts, differences in form of government, and the arbitrariness of country unit boundaries. The continued existence of such states is dependent on a balance being maintained between centrifugal and centripetal forces.

4. Spatial boundaries between states are of three types —subsequent, antecedent, and superimposed. The classification relates to the relative timing of settlement of an area on the one hand and boundary-drawing on the other. Lines dividing land areas can be geometric or nongeometric, the former based primarily on meridians and parallels, the latter on natural landscape features.

5. States exist in a permanent condition of international tension. Geographic analysis of pressure points within the states show a wide variety in the causes of conflict. For landlocked states the most critical stress comes from problems of access. Disputes over natural boundaries, minority groups, and the possession of strategic resources form another set.

6. The Richardson model argues that the potential for conflict is a function of the number of neighbors possessed by a state. Trends toward large "super-states" may reduce the number of conflicts, but increase their size.

Reflections

1. Debate the value of ethological evidence in attempts to understand human territoriality. Do you think that regarding man as a "naked ape" provides (a) insights into or (b) misleading analogies about our spatial behavior?

2. What is going on in Figure 19-3? Try introducing a third center, and plot the boundaries that might result.

3. Look at the variability in the age of the international boundaries in Figure 19-9. List factors which make for (a) stability and (b) impermanence in boundaries.

4. Select *three* areas of the world where you might expect international disputes over territory to occur in the next decade. Compare your selection with that of others in your class. Do any common patterns emerge?

5. Review your understanding of the following concepts:

ethology   antecedent boundaries
median lines   subsequent boundaries
Dirichlet polygons

One Step Further . . .

Many of the ideas presented in this and the next chapter are part of the legacy of political geography. Two recent books summarizing geographers' work in this area are

Muir, R., *Modern Political Geography* (Macmillan, London, 1975) and

Kasperson, R. E., and Minghi, J. V., Eds., *The Structure of Political Geography* (Aldine, Chicago, 1969).

The critical role of the state in shaping the world's political region is argued in a classic contribution,

Hartshorne, R., in James, P. E., and C. F. Jones, Eds., *American Geography: Inventory and Prospect* (Syracuse University Press, Syracuse, N.Y., 1954), Chap. 7.

Territoriality in nonhuman animal populations is widely established. For a controversial introduction to the biological literature on this subject, read

Ardrey, R., *The Territorial Imperative: A Personal Inquiry into the Animal Origins of Property and Nations* (Atheneum, New York, 1966).

For a view of market areas and the economist's views of spatial partitions, read

Richardson, H. W., *Regional Economics* (Praeger, New York, and Weidenfeld & Nicolson, London, 1969), Chap. 2.

Two classical studies by geographers who were intimately involved in the boundary problems that followed World War I are

Bowman, I., *The New World: Problems in Political Geography* 4th ed. (World Book Company, New York, 1928) and

Boggs, S. W., *International Boundaries: A Study of Boundary Functions and Problems* (Columbia University Press, New York, 1940).

It is worth browsing through a historical atlas to compare the kaleidoscopic change in some parts of the earth's surface with the relative stability in others. A highly recommended atlas is

Darby, H. C., and H. Fullard, *Cambridge Modern History Atlas* (Cambridge University Press, London, 1971).

There are no special geographic journals devoted to the topics treated in this chapter, and research is published in the general geographic serials. Journals like International Affairs *(a quarterly) and* World Politics *(also a quarterly) often carry interesting papers. The* Journal of Peace Research *(an annual) is devoted to applying academic ideas to the resolution of conflicts.*

In the same year that the inhabitants of Britain were voting about the European Common Market, the inhabitants of Boston had a different problem. The federal district court in June 1974 ruled that the Boston School Committee had fostered and maintained a segregated public school system. Revised plans called for a redistricting of the school system and transporting by bus more than 18,000 students in the 94,000 student system to achieve racial balance. The turmoil that ensued in Boston, like the furor in Britain, arose from boundary changes—from one geographic area being linked in a new way to another, while others found themselves on different sides of a line on the map. So *coalition* (the joining together of areas) and *partition* (the separation of areas) are two sides of the same coin.

Recall that in the previous chapter we concentrated on a primary territorial division, the state. In this closely linked chapter we go on to consider how states order their external and internal territorial organization through coalition and partition. First, we look at associations between states in the form of international coalitions. Why are these formed, and why do they have particular geographic structures? Second, we consider the international task faced in dealing with that 70 percent of the earth's surface which is sea-covered. As the potential of the offshore and ocean waters begins to be realized, so there is a tightening up of territorial rights. More countries want a piece of the cake, or want the size of their existing piece extended. How are the oceans to be divided? Finally we turn to the local scale. How do states organize their internal space? Are there efficient and inefficient ways of doing this? Are some ways fairer than others?

chapter 20

Coalition and Partition

So far in our discussion of the state we have dealt with a single unit in conflict with its neighbors. But states also need to be looked at collectively. So here we consider how states may group together to form coalitions. The word coalition comes from two Latin words meaning to grow up together. We look first at the various forms of coalitions. What brings the countries together? In particular what benefits come from removing or reducing boundaries? Last, we consider enforced coalitions in the form of empires.

Types of Coalition Self-interest brings individuals together into groups: the gang in a city, or the trade union in an industrial plant. Likewise a state may join together with other states to pursue some common gain. Table 20-1 shows some of the groupings at the present time. How can we make sense of this bewildering array?

One approach is to try to distinguish between the purpose of such associations. Let us take the case of OPEC (the Organization of Petroleum Exporting Countries). This group was formed in 1960 by five leading oil-exporting states. Its purpose essentially was to bring pressure on the major oil companies to raise their prices to the consumer countries and give back a greater share of profits to the producer countries. (We shall see more of the oil companies when we discuss the multinationals in Section 22-4.) OPEC's membership now includes the producers of half the world's oil, and it has been strikingly successful in gaining its economic objectives.

A number of the coalitions in Table 20-1 are, like OPEC, essentially *economic* in function. Thus EEC, EFTA, and CARIFTA are all regional trading blocks, the first two for Western Europe and the third for the Caribbean. They are paralleled by groupings for other purposes. NATO, SEATO, and the Warsaw Pact countries are all *defense* associations. Defensive alliances have probably existed at all levels of human organization, from the tribe to super-state as far back as historical records go. It is one of the oldest and most persuasive reasons for union. Third, there are *political* associations of countries wishing to strengthen a common political purpose between member states. The Arab League countries and the Organization of African Unity illustrate this purpose.

This threefold division into economic, defensive, and political groups is a useful simplification. Not all organizations fit into the scheme, however. The United Nations is a global association of over 140 states and has economic and policing functions as well as a primary political role. The British Commonwealth is the legacy of a once-powerful economic block, retaining surprising vitality as a cultural grouping with some political aspects. If we plot the states that make up the organizations shown in Table 20-1, then some overlap in membership is shown up. For example, Nigeria is a member of both OPEC and OAU, as well as of the British Commonwealth. Other countries, like Switzerland or Finland, go to some lengths to avoid membership in defensive or political blocks.

Geographers are interested in these block arrangements because they modify the dominance of the single state. For example, the growing cen-

Table 20-1 International organizations

Type of organization			
Political		**Economic**	**Defensive**
Global	Regional	Regional	Regional
United Nations,	Arab League, Organization of African Unity (OAU)	European Economic Community (EEC), Caribbean Free Trade Area (CARIFTA), Organization of Petroleum Exporting Countries (OPEC)	North Atlantic Treaty Organization (NATO), Warsaw Pact
British Commonwealth			

tralization of function in the nine countries of the EEC means that spatially significant decisions (e.g., over resource use or regional policy) are made in the community headquarters at Brussels, rather than in state capitals such as Bonn, London, Paris, or Rome.

Coalition Benefits: Boundary Reduction One of the benefits expected to flow from economic linkage between countries is improved trade. This comes from lowering the taxes (tariffs) on goods that cross the international boundary plus the enlarged market. But to understand the importance of removing or reducing a boundary we must first see how it blocks flows in the first place.

Geographers have developed a general graphic model of the blocking action of boundaries (Figure 20-1). When the barrier is a political one, marked by tariff differences, the potential field for trading is restricted in varying ways. Figure 20-1(b) shows the probable form if the political boundary can be crossed at all points along its length; Figure 20-1(c) gives the probable form if the boundary can be crossed only at a customs point. If the

Figure 20-1. Boundaries and flows of trade.
A general model of the impact of boundaries on spatial interaction. (a) A cross section of the market area around a trade center. If you follow the vertical dashed lines, you can see how this cross section is projected down to (b) to show the same situation in map form. Note how the shaded area indicates that part of the market that is lost by the effect of the border tax. Map (c) shows the still greater loss if the border can only be crossed at a single customs post. The effect of a natural boundary (hence, no border tax) with a constricted crossing point is shown in map (d).

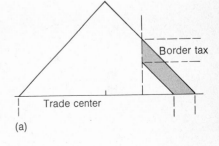

(a)

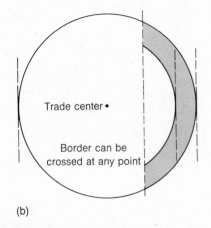

(b)

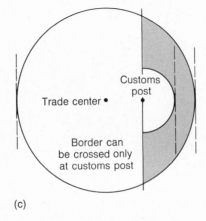

(c)

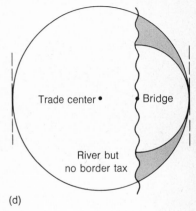

(d)

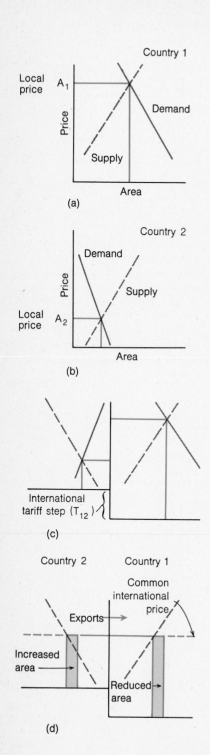

(a)

(b)

(c)

International tariff step (T_{12})

(d)

Country 2   Country 1

Common international price

Exports

Increased area

Reduced area

boundary is not a political one but a natural feature such as a river, with a single crossing point, the field will probably look like that in (d).

We can throw further light on the effect of tariff walls between countries by adapting some of the conventional economic models of trade. Let us assume that we have two adjacent countries, each of which produces a similar crop (e.g., wheat), separated by a tariff. In each country, the land area devoted to the crop is directly related to its price. As the price of the crop increases, the amount of land under cultivation increases, and vice versa.

Let us do as economists do and plot the relationship of the crop area to price in terms of supply and demand curves. (Readers who have already taken classes in economics will see that we are greatly simplifying the situation; those who have not need note only the conclusions we draw from the graph.) Figure 20-2 shows the area–price relationship as an upward-sloping supply curve. Price also affects the volume of demand for the crop, but in an inverse way: The demand for a high-priced good is assumed to be less than that for a low-priced good, and vice versa. We can represent this second relationship by a downward-sloping demand curve. The local price for the crop in each country is established when the demand for it and the supply of it are in balance.

If we regard the two countries as completely isolated, the demand and supply situation in each will be determined by the internal conditions in each. Let us suppose that in Country 1 farmers get a high price for the crop and a large area is devoted to it [Figure 20-2(a)]. In Country 2, conditions are different. The crop sells for a low price and the area devoted to it is smaller [Figure 20-2(b)]. What will happen if we drop our assumption of isolation? Logically, we should expect the low-priced crop from Country 2 to find its way into Country 1.

A tariff can prevent this flow of exports, however. We can see this from our model if we place our two diagrams back to back and displace them vertically

Figure 20-2. Tariffs and international trade.
A simplified model of the effect of international tariff between two countries on the supply of and demand for a common agricultural product. Charts (a) and (b) show the local conditions in each country in which the local price (vertical axis) and the area of the crop under cultivation (horizontal axis) are related through the conventional economic mechanism of demand and supply. The international tariff between the two countries is shown as a "step" (T_{12}) in chart (c). In this case the step is not high enough to keep production from being expanded in Country 1 (the low-cost producer) and the extra output exported to Country 2 (the high-cost producer). How high does the step have to be to keep out cheap imports from Country 2? This depends on the relative difference between the local price in Country 1 (A_1 in the diagram) and the local price in Country 2 (A_2). When the tariff (T_{12}) is *less* than the difference in price (i.e., $A_2 - A_1$), as it is in the diagram, then flows will occur as in (a). If, however, Country 2 raises its tariff so that T_{12} is greater than $A_2 - A_1$, no trade will take place.

[Figure 20-2(c)]. The amount of the displacement represents the size of the tariff imposed by Country 1. A flow of export occurs only if the difference in the local prices in the two countries is greater than the tariff. If the tariff is low enough for a flow to take place, a general international price will be established. The area devoted to the crop in Country 2 will expand, and the area devoted to it in Country 1 will contract [Figure 20-2(d)].

The analysis here is clearly oversimplified. The model refers to a highly simplified situation in which there is a single product and there are only two countries. Economists have developed a complex theory of international trade to explain the flow of many commodities among many nations. Nevertheless, the simple graphical example given here gives us some insight into the impact of tarrifs on trade between countries. You might like to ponder the effects of tariff reductions and the establishment of common international prices (as in the European Economic Community) on the crop areas of the countries affected. Does this have implications for farm production in your own country?

Enforced Coalition: Spatial Imperialism   Coalitions are not only voluntary. One state may be coerced into political and economic associations with another by conquest or colonization.

In geographic terms, important distinctions can be made between a state's expansion into (a) adjacent or nonadjacent territories and (b) occupied or unoccupied space. Occupied space is where another state already exists; in contrast, unoccupied space is not organized into political units at the state level and may be largely empty of population. Let us take some examples. Peripheral expansion into an already occupied area is illustrated by the absorption by the German state of the Austrian state in 1938. Expansion of Tsarist Russia into the adjacent but sparsely populated lands of central Asia is an example of the second case. Nonadjacent expansion by the European states into overseas lands provides the classic case of imperialism. The British Empire included occupation of the Indian peninsula (occupied space) and much of Australasia (largely unoccupied space).

Imperialism is geographically significant in that different colonial powers have left different regional imprints. For example, among the European imperial powers there were consistent differences in the attitudes taken to intermarriage of the colonists with non-European populations of the occupied lands. The racial cohesion among Brazil's 105 million people today stands in contrast to the apartheid splitting South Africa's 25 million people. This contrast has part of its roots in differences between Portuguese and Dutch colonial policy three centuries ago. Settlement patterns in French Canada or Australia reflect attitudes of imperial powers toward the occupation of their overseas lands. Trade flows, language and cultural links of the independent African states still strongly reflect the European states which were involved in their foundation. (Look back at Figure 19-4.)

Imperialism has attracted considerable interest from Marxist writers because of its role in world economic development. Lenin, writing in 1916, saw imperialism as one of the reasons for the persistence of the capitalist economic system in Western Europe—an economic system whose demise Karl Marx had predicted a half-century earlier. Overseas colonies were seen by Lenin to be excellent outlets for capital investment by industrial countries, providing not only sources of cheap raw materials but also captive markets for manufactured goods. Profits from imperialism would allow higher wages to be paid to workers in the imperial homelands (so reducing the likelihood of revolution), but at the expense of lower wages for colonial workers (so increasing the likelihood of revolution). Thus according to Lenin's thesis the impact of imperialism was to spatially transfer the class struggle from a within-state basis to the international level: i.e., developed western countries (the "haves") versus developing Third World countries (the "have nots"). We shall expand this theme in the next chapter, but we should note here that not all imperial powers are states. The emerging power of major international corporations (the multinationals) represents a new form of imperialism which may prove considerably more powerful than the older form of coalition discussed above.

Lenin's thesis was not the only model of a global conflict between coalitions. A confrontation between a land-base empire dominating the world's central Eurasian landmass and a sea empire controlling peninsular extremities together with the Americas, Australasia, and Africa was considered by an English geographer, Halford Mackinder, in the early years of this century. Mackinder's recognition of the strategic significance of Eastern Europe resulted in his 1904 dictum: "Who rules East Europe commands the Heartland, who rules the Heartland commands the World Island, who commands the World Island commands the World." (See Figure 20-3.) His *heartland* included much of European Russia, Siberia, and Soviet Central Asia, together with what is now northern Iraq and Iran. Mackinder's thinking was adopted by a few German political geographers such as K. Haushofer; how big a role it played in Hitler's global strategy for World War II is uncertain.

To sum up: Coalitions of states have come together for the purposes of economic union, defense, and political association. By reducing the importance of the international boundaries between members of a coalition, considerable economic benefits can be gained, though arguably at the expense of nonmember countries. Historically, not all coalitions have been voluntary. Empires represent enforced coalitions in which a single state has dominated other areas. Imperialism has left a geographically significant legacy in terms of population distribuiton and economic development and is still evolving in different forms.

20-2
Offshore and Ocean Waters

Seventy percent of the world's surface is covered by water. (Check back to Section 4-3.) For most of our history the high seas were either a barrier to expansion or, for the last few centuries, corridor space for intercontinental

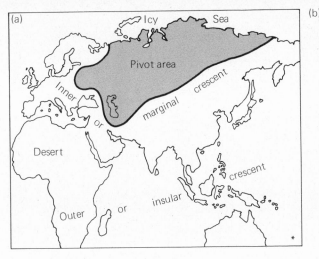

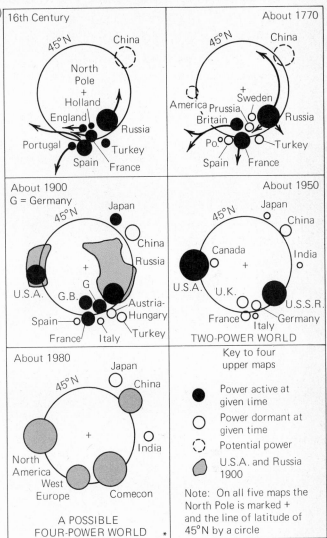

Figure 20-3. Mackinder's heartland model.
(a) English political geographer Sir Halford Mackinder saw control of the "pivot" area of Russia as being vital to the control of the "world island" (Asia, Europe and Africa) and hence of the world. (b) The shifting balance of political power in the world since the sixteenth century. Comecon is a general alliance of Soviet Russia and the countries of Eastern Europe. [From J. P. Cole, *Geography of World Affairs,* 4th ed. (Penguin Books, Harmondsworth, 1972), p. 109, Fig. 4.7. © J. P. Cole, 1959, 1963, 1964, 1965, 1972.

movements. Despite the importance of fishing banks, whaling resources, and strategic islands the oceans themselves were rarely a source of dispute.

Now all this is changing. England and Iceland fight "cod wars" over fishing grounds; Greece and Turkey quarrel over the ownership of parts of the potentially oil-rich seabed of the Aegean; Canada extends pollution control over areas not recognized by the United States. Suddenly ownership of ocean resources has become a hot issue, and finding ways of peaceful partition becomes urgent.

Territorial Status of Water-Covered Areas   Geographers usually divide the ocean into three zones in terms of territorial rights. First there are rights over the immediate *offshore areas* around a country. Second, there are claims for mineral rights in the *continental shelf* around a country. The continental shelf is an area of very smooth, gently sloping ocean floor which fringes all continents. Its limit is conventionally given as a depth of 100 fathoms (183 m or 600 ft) and its width varies from only a few kilometers to 350 km (about 200 mi). Third, there are the *ocean floors* themselves. Ocean floors make up over 90 percent of the world's ocean-covered areas and vary greatly in depth and smoothness. [Check back to Table 3-2.] We will look at these zones in turn.

Offshore Areas   Figure 20-4 shows the conventional legal divisions of water-covered areas in a simplified manner. The starting point for determining a state's control over adjacent waters is the *coastal baseline*. All water landward of this baseline is termed *internal waters* legally treated as part of the land surface of the state. This generally runs along the coastline itself but may skip across indentations (such as bays and estuaries) and include offshore islands. Iceland (Figure 20-5) provides a good example of this. Note how the coastal baseline sometimes hugs the coast but more frequently runs some miles out from it. In practice, therefore, there is considerable disagreement as to how the baseline shall be drawn. For example, Canada draws a baseline to include the whole of Hudson Bay as internal waters.

Seaward of the coastal baseline lies a country's *offshore territorial waters*. How far offshore such waters should stretch remains unresolved in interna-

Figure 20-4.   Dividing the offshore zone.
The extension of a state's sovereign territory outward from the coastline. Note the division of the water into four territorial zones—the internal waters, the offshore territorial waters, the outer contiguous zone, and the high seas.

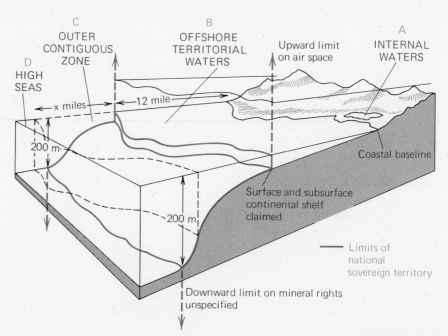

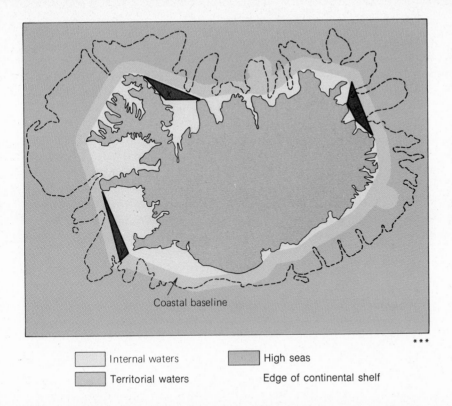

Figure 20-5. Offshore limits.
This map of Iceland shows the effect of changes in baselines on the limits of internal waters. The 12-mile (19.3-km) limit shown has subsequently been extended to 200 miles (320 km). This unilateral extension by Iceland has caused disputes with Western European countries (notably Great Britain and West Germany) who used to fish on the continental shelf outside the 12-mile limit. [From L. M. Alexander, *Offshore Geography of Northwestern Europe* (Rand McNally, Skokie, Ill., and Murray, London, 1963), p. 109, map 12.]

Coastal baseline

***

Internal waters	High seas
Territorial waters	Edge of continental shelf

tional law. Early claims of offshore waters were for the limited areas that could be controlled. These ad hoc "cannon shot" distances have led to a diversity of claims. A survey of the more than 100 maritime states in the late 1970s showed a huge range from Australia, which is bounded by territorial waters only 5.6 km (3.5 mi) in breadth, to El Salvador, which claims 370 km (230 mi). The Philippines claim as territorial waters all the immense sea areas between the islands of the archipelago. A preponderance of states claim a 19 km (12 mi) limit. It was not until 1958 that the United Nations brought together 86 states for the first Law of the Sea Conference. Only slow moves toward the standardization of territorial claims have been made since that date.

Continental Shelves   Beyond the limit of territorial waters, two further claims may be pushed. First, special rights may be claimed in a further offshore zone, termed the *outer contiguous zone*. Since the distance limit is nowhere defined in international law this is marked as width × in Figure 20-4. The main use of the outer zone is in the control of customs, protection of fishing grounds, national defense, and pollution control. An example of such outer zones is the 100-mile wide zone claimed by Canada off its Arctic seaboard for pollution protection. A second claim outside territorial waters is to surface and subsurface rights on the continental shelf. Under the Continental Shelf Convention of 1958 claims for control of the seabed and geologic resources below (but *not* the overlying waters) for shelf depths out to 200 m (650 ft) were recognized. The physical configuration of these shelf areas is extremely complex, and exploration permits have been granted by the United States government in water depths up to 1500 m (5000 ft).

Ocean Floors   Beyond the territorial and contiguous waters lie the *high seas*. These are the areas left over outside any current claims and make up the great part of the world's surface. They are areas for free international movement of shipping and for fishing fleets. In 1973 the French government broke international law and provoked strong international condemnation (notably from Australia and New Zealand) by imposing a 116 km (72 mi) ban on foreign shipping around Mururoa Atoll in the Pacific during nuclear tests.

Conflicting Offshore Claims   So far we have assumed each state can push out its claims seaward in isolation. But what happens if it has neighbors also claiming the same sea area? Who then owns it?

The question first became important where rivers and lakes run between countries. There the principle adopted was for each state to extend its sovereignty at an equal rate until it met the lake territory of adjacent and opposite states. Thus the method used for Lake Erie was an adaptation of the median line principle used in devising Dirichlet polygons. (See the polygons in Figure 19-2).

Although simple in theory, the application of this method to seas and oceans is complicated by ambiguous definitions of a country's coastline. Consider again the map of Iceland's territorial waters in Figure 20-5, and note how the irregular shape of the coastline is approximated by a polygon. This polygon, the coastal baseline, is the starting line for all territorial claims. But the baseline can be drawn in different ways (see the two areas marked x), by adopting a small offshore island or rock as one of the bases for drawing the polygon (see area y). Whether or not a bay is regarded as part of a country's internal waters, or an offshore island or sandbar is regarded as national territory changes the baseline from which the median line with a neighboring state is determined. It was for this reason that it wasn't until 1977 that France and England could decide where the median line down the English channel really ran (Figure 20-6).

Shelf Division   International boundaries in the continental shelf areas around Western Europe were of little more than academic importance until the middle 1960s. Then natural gas and oil discoveries highlighted the advantages that countries like the United Kingdom and the Netherlands had gained from what had previously been only paper titles. Figure 20-6(d) shows the effect of median line territorial boundaries in the North Sea on ownership of natural gas and oil sources. Natural gas is found mainly in the southern half

(a)

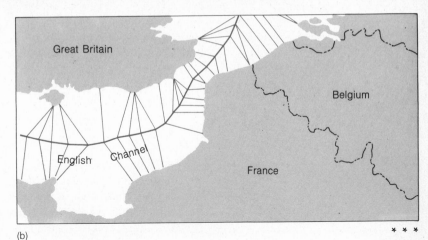

(b)

* * *

(c)

*

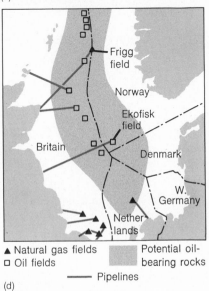

▲ Natural gas fields   Potential oil-
□ Oil fields   bearing rocks
── Pipelines

(d)

Figure 20-6. The principle of median boundaries applied to offshore areas.
(a) Offshore oil drilling. (b) International boundaries in the English Channel. (c) An outward extension of national sovereignty into the oceans, with each nation extending outward until it meets its adjacent and opposite neighbors. Note that while the boundaries in (b) have been ratified by the countries bordering the English Channel and are internationally recognized, those in (c) are merely hypothetical. (d) Impact of median boundaries on share of oil and gas in the North Sea. [Photo (a) by Martin Weaver, Woodfin Camp & Associates. From (b) L. M. Alexander, *Offshore Geography of Northwestern Europe* (Rand McNally, Skokie, Ill., and Murray, London, 1966), p. 58, Fig. 4; (c) T. F. Christy and H. Herfindahl, *Hypothetical Division of the Sea Floor,* a map published by the Law of the Sea Institute, Washington, D.C., 1968.]

of the North Sea; the earliest strike was in the Netherlands' waters, but larger sources were later found in the British area. Later oil strikes have come in the northern half of the North Sea, many close to international boundaries. The Ekofisk oil field is Norwegian but is connected by pipeline to Britain because a deep trough in the seabed off the Norwegian coast would have made pipeline construction very costly. Clearly some Western European countries have benefited more than others from the way the boundary lines have been drawn.

Ocean Floors   Although in the short term the greatest interest attaches to control of the immediate offshore waters, the long-term implications of ownership of the continental shelves and ocean floors may eventually be more important. For example, if we allow the present principle of the median line [Figure 20-6(b)] to be extended to the open oceans [Figure 20-6(c)], we find some extraordinary distributions of territory. Portugal, by virtue of its ownership of the Azores and Cape Verde Islands, would stand to gain the largest share of the North Atlantic seabed.

The mineral riches of the ocean floors are only just beginning to be realized. For whereas offshore drilling for oil and gas on the continental shelf had been established from the 1920s, the present decade is seeing an increased interest in the surface deposits of the deep oceans. Minerals such as manganese may accumulate on the ocean floor in the form of rich nodules. Preliminary estimates for the Pacific suggest that such nodules may contain enough manganese to meet world demand for the next 400,000 years (at present rates of consumption) and that the nodules may be accumulating faster than they could be used. But few hard facts are known and the exploration of the ocean's mineral resources (including dissolved minerals in seawater) promises to be one of the great frontier areas of research in the remainder of this century.

If the ocean floors are going to be very important mineral resources in the future, should they be divided in the way shown in Figure 20-6? What about the problem of a landlocked country with no sea boundary and thus no base for a claim? Suggestions at recent United Nations conferences state that the ocean floors out beyond the continental shelves should be divided up into a checkerboard of exploration areas. Then sections of the ocean could be allocated, perhaps on a random basis, between all countries. Although this is only a tentative idea, it will probably be made again as economic and technological problems in using ocean resources are solved. Ownership problems, if not solved by agreement will be resolved on a power basis.

Mineral resources are not the only reason for heightened international interest in the world's oceans. We have already seen in Part Two of this book the food productivity of the oceans and their sensitivity to pollution. There is therefore increasing concern by United Nations agencies over adequate control of pollution by oil, mineral elements like mercury, or nuclear waste, and the excessive reduction of the population of certain marine animals like whales.

To sum up: The world's oceans represent the last major frontier area left on this planet. With proper international control their vast potential resources, both renewable and nonrenewable, could be harnessed for immense good. Without such control the oceans threaten to be as great a scene of conflict in the future as were the land areas in the past.

20-3
Partitions Within States

Our third area of concern in this chapter presents a shift of scale: from the international political scene to local, grass-roots politics. For each state is made up of a hierarchy of local areas through which central governments keep in touch with grass-roots needs. Such areas include the state, county, and township in the United States; the province, county, and township in Canada; the county, district and parish in England; the département, commune, and arrondissement in France; and so on.

Most countries, whether western or nonwestern, developed or underdeveloped, have intermediate and local authorities with which they share the central authority of the nation-state. The existing sequence is frequently a historical patchwork, and from time to time central governments decide to reform the hierarchy along more relevant lines. The Napoleonic reform of the ancient French system of provinces, the sequence of Soviet reforms since the early 1920s, Salazar's reforms of the Portuguese system in the 1930s, the Swedish reforms of the 1950s, and the British reorganization of the 1970s are each part of a recurring cycle of revision, a cycle in which geographers are increasingly being called upon to help redraw maps.

Partition at the Regional Level   But what maps should be substituted for existing maps? On the basis of what criteria should new areas be drawn? How do we weigh a central government's demand for a few large efficient units against a local outcry for a small unit tailored to the needs of a particular community? The solution to these problems involves not just deciding on new units but arranging them into levels, or tiers. With a *single-tier system*, each local area has direct access to the central government; with a *multiple-tier system*, the local areas must work through an intervening *bureaucracy* on one or more regional levels.

The British Problem   We can examine the situation in a specific context by presenting a recent case of boundary reform in Britain. Until 1974 the country was split into over 1000 local districts, each with a variety of powers and each directly responsible to the central government in the performance of certain administrative functions. The system had been only slightly altered since the nineteenth century, so its lack of relation to the distribution of population and the needs of efficient management was becoming evident.

A Royal Commission was established to prepare recommendations. They proposed three criteria for the new system. First, they wanted the new units to be large enough to provide local government services at a low cost. "Large enough" depended on the service provided. While 500,000 was the minimal population for an efficient police force, the threshold for education services was 300,000. Lower figures were given by local health and welfare authorities (200,000 people) and by child-care services (250,000 people).

A second criterion for the new distribution was that the new units should be cohesive enough to reflect local community interests and to allow a local voice in political issues. In practice, cohesion and self-containment are measured by journey-to-work patterns, the range of public transport, newspaper circulation, and the organizational pattern of professional, government, and business organizations. This approach yields urban regions connected to rural areas by ties between workplaces and residences.

The third criterion for the delimitation of new units was the present pattern. Whenever possible, existing units and their boundaries, even on the county level, were used as building blocks in order to retain common interests and traditional loyalties, to preserve the skill and momentum of existing local authorities, and to minimize changeover problems.

The British Solution   Given the three basic criteria described above, we might assume that it would be relatively simple to find a common spatial solution. As geographers would have predicted, this is not the case (cf. Figure 11-9). The Royal Commission could not reach an agreement and proposed two alternative systems. The majority proposal was to divide the country into a few rather large local government areas (61 for England outside London itself) responsible for all local government services. In contrast, the minority proposed a two-tier system with 35 upper-level and 148 lower-level city regions. In this case the upper-level authorities would look after planning and transportation, while the lower-level authorities would control services like education, social welfare, and housing.

We can appreciate the differences between the two proposals by looking at some maps of southwest England (Figure 20-7). The first area shown [Figure 20-7(a)] is similar in size and population to the state of Massachusetts. The existing 12 units vary considerably in size and area. If the majority proposal were adopted, a slightly enlarged southwestern province would be replaced by eight new areas [Figure 20-7(b)] larger and more uniform in size and with less extreme differences in local revenue-raising ability. If the minority proposal were adopted, a two-tier system, set within a redrawn southwestern province [Figure 20-7(c)] would result.

Despite their differences, we can see that certain recurrent geographic features dominate both the traditional map and the two proposed revisions. First, we can identify eight core areas. Five of these are largely rural and center on an existing county town; the remaining three have metropolitan centers in the leading cities of Bristol, Plymouth, and Bournemouth. Second, there are certain persistent boundaries that recur in all three maps and may represent rather fundamental discontinuities in the socioeconomic structuring of the area.

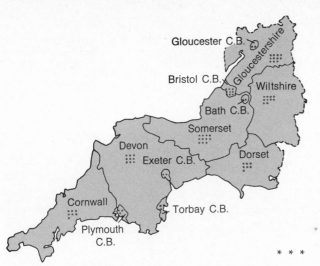

(a) The southwest economic planning region with existing county and county–borough (C.B.) boundaries

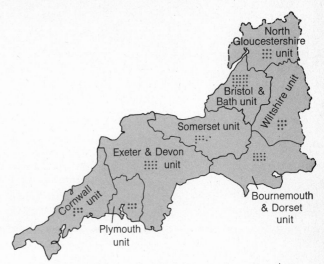

(b) The Redcliffe–Maud proposals for the southwest province, with unitary areas

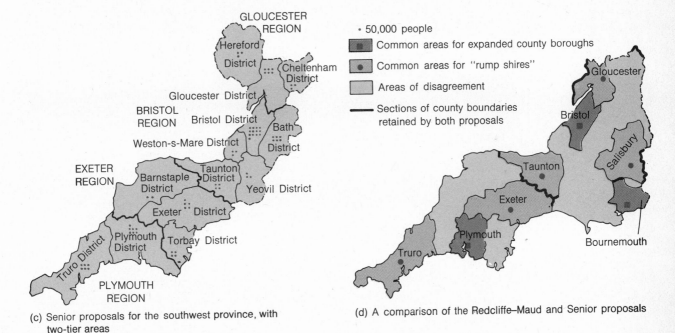

(c) Senior proposals for the southwest province, with two-tier areas

(d) A comparison of the Redcliffe–Maud and Senior proposals

Figure 20-7. Reform of administrative areas.
The maps show proposed reform of boundaries in southwest England. The first three maps show alternative proposals, and the fourth, the areas of agreement and disagreement. "Rump shires" are the core areas of the old countries or shires (a) left undivided by both sets of proposals (b) and (c). [From P. Haggett, *Geographical Magazine* **52** (1969), p. 215. Reproduced by courtesy of The Geographical Magazine, London.]

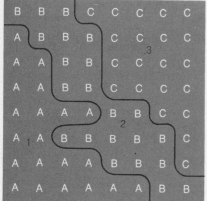

(a) Maximal segregation

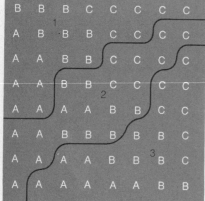

(b) Maximal desegregation

Figure 20-8. Alternative region-building strategies.
In (a) the boundaries create the greatest possible segregation of three hypothetical groups (A, B, and C). In (b) each area has a fair percentage of people from each group. Choices between these two types of zoning often face administrators drawing up school district lines in ethnically diverse areas. In (c) and (d) the problem faced by a school desegregation ruling in Boston is illustrated. [From J. D. Lord, *Spatial Perspectives on School Desegregation and Busing*. Association of American Geographers, Washington, D.C., 1977, p. 15, Fig. 13. Photo by United Press International.]

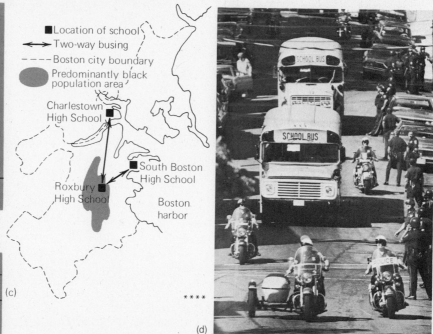

By recognizing recurrent features in proposed regional systems, geographers can isolate the broad features in any set of solutions and narrow the areas of search. In the case of southwest England the main disputes over boundaries occur in three critical zones, as shown in Figure 20-7(d). By concentrating on these zones of disagreement and accepting that substantial areas of agreement are possible, we can reconcile a head-on clash between the proponents of apparently conflicting regional systems. The proposals now adopted for the area do retain many of the old county boundaries.

Partition at the Local Level On the lowest spatial level geographers are concerned with how boundaries of local districts are arranged. Figure 20-8 shows the choices faced by legislators in drawing up boundaries. Let us assume that we must establish three new school districts with roughly equal numbers of school children in three ethnic groups (A, B, and C) in a city. Two alternative policies, maximal segregation and maximal integration, will produce two unlike spatial arrangements of boundaries. Moreover, these two solutions are only a fraction of the literally billions of alternative ways of partitioning the area. Can we ensure that the solution chosen is a fair one?

Gerrymanders: The Problem In 1812 Governor Elbridge Gerry of Massachusetts established a curiously bow-shaped electoral district north of Boston in order to favor his own party. In his memory the term "gerrymander" is used to describe any method of arranging electoral districts in such a way that one political party can elect more representatives than they could if the district boundaries were fairly drawn. Since 1812 the number of examples of gerrymandering has grown rapidly, and Figure 20-9 presents two of the most extreme ones.

If we wish to rig boundaries in an unfair way, we have two possible strategies. The first is to *contain* our opponents, to lump all the electors in the opposing party within a single boundary so that a representative receives an unnecessarily large majority of votes. The second possible strategy is to disperse our opponents, to split the electors of an opposing party between many districts so that nowhere are they numerous enough to elect one of their own representatives.

Within the United States the location of equitable political boundaries has been a subject of greater interest since the 1962 *Baker vs. Carr* decision of the United States Supreme Court. In this case the court ruled that legislative seats must be apportioned and states divided so the number of inhabitants per legislator in one district is approximately equal to the number of inhabitants per legislator in another district. The case was triggered by a group of voters in Tennessee who claimed that their votes were debased by inequities in the state's electoral districts. The situation arose because Tennessee had continued to elect its state legislators on the basis of an apportionment adopted in 1901. In the intervening 60 years, however, there had been radical shifts in the distribution of the population from rural areas to cities and suburbs. This led to a situation in which one vote in rural Moore County was equal to nineteen votes in urban Hamilton County (containing the city of Chattanooga). Parallel situations existed in other American states. In Vermont the most populous district had 987 times more voters than the least populous district. Many states had to redraw their district boundaries to comply with the Supreme Court decision.

Gerrymanders: A Solution? It is one thing to identify the problem of unfair electoral boundaries but quite another to discover a solution. Just how can we arrive at a "fair" spatial arrangement? The boundaries chosen for voting districts are likely to reflect a balance among three considerations. The first is that the voting population in each district should be *equal*. The ideal concept of "one person—one vote" must be modified to offer approximate equality because uncontrollable factors such as migration and natural change are continuously modifying the voting population within a district. The second consideration is that the voting district should be *contiguous*. Districts are usually in one conterminous unit and ideally should be compact within the sense that communication among the various parts of the district is easy. The third and more debatable consideration is that of *homogeneity*, or balance. Some may think it is desirable for the parts of a district to have a common social, political, or economic characteristic. Others may argue that the district should be a balanced mix representing a wide range of communities rather than a single one.

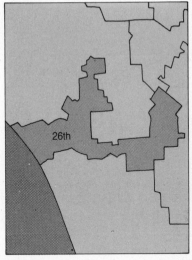

(a) Los Angeles, Calif. ✶ ✶ ✶ ✶

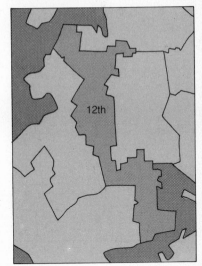

(b) Brooklyn, N.Y. ✶ ✶ ✶ ✶

Figure 20-9. Gerrymandering in the United States.

(a) California's twenty-sixth electoral district in 1960. (b) Brooklyn's twelfth electoral district in 1960. In both cases the boundaries were drawn to give a major electoral advantage to a political party. Following Supreme Court decisions such local gerrymanders have now had to be eliminated.

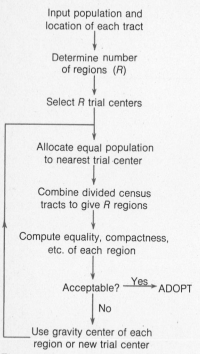

Input population and
location of each tract

↓

Determine number
of regions (R)

↓

Select R trial centers

↓

Allocate equal population
to nearest trial center

↓

Combine divided census
tracts to give R regions

↓

Compute equality, compactness,
etc. of each region

↓

Acceptable? —Yes→ ADOPT

↓ No

Use gravity center of each
region or new trial center

Figure 20-10.   Electoral districting.
This flow chart shows the main elements in a
computer program for allocating population
in census tracts to a set of compact electoral
regions with a similar number of electors in
each area. "Gravity center" is at the central
point within a region which is most
accessible for all its population.

Several computer programs have been designed in an attempt to achieve
equality, contiguity, and homogeneity (or balance) by a nonpartisan method.
One such method is outlined by the flow chart in Figure 20-10. It takes an
initial set of small tracts whose voting population is known and allocates them
to electoral regions. The original allocation is successively adjusted to make
the regions more equal in population while keeping them compact. (See the
discussion of a program for creating electoral districts on page 495.) Figure
20-11 presents the results of using such a method to redistrict a sample part of
New Jersey. Six districts were to be formed from over 50 tracts [Figure
20-11(a)]. In the first allocation the largest district, with a 12,500-person elec-
torate, was 5 percent greater than the average for all districts. In the second
allocation this discrepancy was reduced to only 1 percent.

In evaluating this approach to the electoral areas we should recall that
numerous district boundaries can be drawn with approximately the same
concern for equality and contiguity, and yet some will favor one party at one
time rather than another. To resolve the imbalance we need to inspect
various boundary maps and adopt the one most in accord with the prevailing
concept of political equity.

Apartheid   Contact between different ethnic groups has led historically to a
wide range of demographic situations. Interbreeding and blending with the
original groups into a single hybrid population is one outcome; annihilation,
expulsion, and stratification are others. South Africa today provides examples
of a further outcome, spatial separation of ethnic groups through a formal
partition procedure.

Figure 20-11.   The search for improved electoral districts.
These maps of Sussex County, New Jersey, in the United States, show the first stages
in the allocation of census tracts to electoral regions, using the kind of program shown
in Figure 20-10. The electoral districts in maps (b) and (c) are made up by combining the
original tracts (a). Figures indicate the population in each electoral district in thousands.
[From J. B. Weaver and S. W. Hess, *Yale Law Journal* **73** (1963), Fig. 1.]

Trial centers

(a) Original tracts

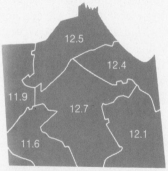

(b) First allocation

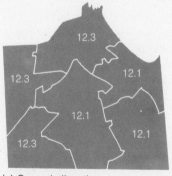

(c) Second allocation   ＊ ＊ ＊ ＊

A COMPUTER PROGRAM FOR SETTING ELECTORAL-DISTRICT BOUNDARIES

Consider the problem of dividing an area into a number of electoral districts (e.g., six) in an unbiased way. Let us assume that we know the population data of several small tracts. (It is 100.) These data allow us to estimate the probable number of voters in each tract. Our task is to assign tracts to districts in such a way that we create six districts with more or less the same number of voters and as spatially

compact as possible. The flow diagram in Figure 20-10 indicates the basic steps. First, choose arbitrary or reasonable trial centers for each district and mark their geographic coordinates. Second, compute the matrix of distances between the centers of each tract and trial centers. Third, assign tract populations to the nearest trial centers until each district has the same number of voters. (Split tracts are assigned to the district to which the majority of voters were previously assigned.) Fourth, compute the center of gravity for each district on the basis of all tracts assigned to that district in step 3. Fifth, inspect the districts. If the

new centers differ from the trial centers, then go back to step 2 and repeat the assignment process. The cycling process ends when the equality of the voting population and the compactness of the districts reaches an acceptable level or when there are no further shifts possible in the district boundaries. If an acceptable solution is not obtained, new trial centers should be adopted and the program begun again. See P. Haggett and R. J. Chorley, *Network Analysis in Geography* (Arnold, London, 1969), Chap. 4, ll.

While the past quarter of a century has seen almost all African peoples gain independence from white rule, events in South Africa have followed a different pattern. There the 18 million Africans who form 71 percent of the total population have been offered a prospect of "separate development" under the *apartheid* policy of the national government. The areas designated as African homelands are shown in Figure 20-12(a). Rather than forming a contiguous block, the homelands are split up into 20 separate sections related in part to the 9 main tribal divisions of the Bantu.

The African homelands appear unsatisfactory for a number of reasons. First, they represent only 14 percent of South Africa's surface area even though 6 out of 7 in the population are African. Second, the agricultural potential of much of the homelands is very poor. Third, only slightly less than half the Africans live in the homelands. The remainder live elsewhere in South Africa, mainly in the separate townships surrounding the major cities of the Witwatersrand conurbation, Durban and Cape Town. Since the urban-industrial economy is dependent on black labor, it is unrealistic to expect more than a fraction of the 8 million "outsiders" to move back to the homelands where industrialization has hardly begun.

The geography of apartheid, whether at the national level of separate homelands or at the local level of separate residential areas around major cities [Figures 20-12(b) and (c)], is clearly unstable. At present rates of growth the African population is expected to double by the end of the century, and to further increase its share of the country's total population. It is hard from the outside to discern any factors which are not likely to increase rather than decrease pressures on a change in the apartheid policy toward some more equitable system of ethnic coexistence.

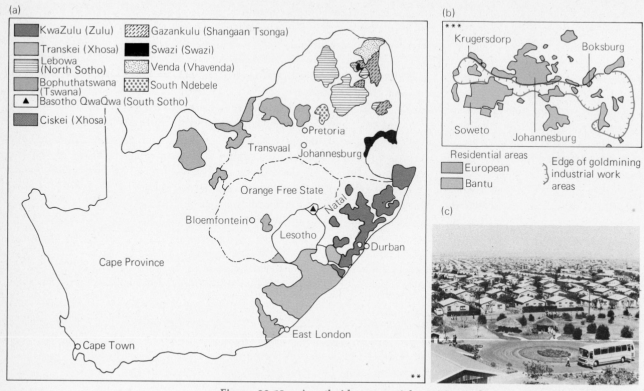

Figure 20-12. Apartheid as a spatial partition policy.
(a) Apartheid at the national level within South Africa with proposals for African (Bantu) homelands. The main tribal divisions of the Bantu are shown. (b) Apartheid at the urban level with separate residential areas for the Bantu on the edge of South Africa's major industrial area, the Witwatersrand conurbation (the "Rand") in the northern Transvaal. (c) The major African township on the Rand is Soweto, some 16 km (10 mi) southwest of Johannesburg. This has a population of around half a million Africans. [From D. M. Smith, *Human Geography: A Welfare Approach* (Arnold, London, 1977), p. 243, Fig. 9.1. Photo by Wide World Photos.]

The microgeography of separate divisions within a state represents the other end of a size spectrum which began with international politics. As we saw in our opening chapter (On the Beach), geographers are continuously engaged in fitting together jigsaws of different sizes. The state provides one important level in that hierarchy of jigsaws, and in the next chapter we look at its economic importance.

Summary

1. Coalitions are groups of states joining together for some common economic, political, or defense purpose. Some are geographically contiguous and regional, others discontinuous and global.

2. Economic benefits from reducing tariffs between member states in a coalition may be shown through barrier and multiregional trading models.

3. Coalitions may not always be voluntary. Enforced coalitions are demonstrated by the empires which have grown up in past periods. The Mackinder model

saw a continuing stress between a land-based empire, such as that on the area now occupied by European Russia and central Asia, and the outer rimland states.

4. International boundaries in offshore and ocean waters are being more specifically defined as their resource potential is recognized. Territorial status of water-covered areas includes internal waters, offshore territorial waters, an outer contiguous zone, and the high seas. The median line principle is commonly used in resolving overlapping claims between two or more states.

5. Partition within a nation state demands the iden-

tification of criteria for drawing sets of regional and local boundaries. Even where criteria exist in abstract, various levels of government may find it difficult to agree on their applications on the ground. At the local level many boundary problems result from the unfair gerrymander system, which had been used in the United States in the past to secure electoral advantage. "Fair" spatial arrangements should include equality of voting population, and homogeneity of balance in social, political, and economic characteristics. Such arrangements may be aided by the adoption of nonpartisan computer districting methods.

Reflections

1. Take one of the regional international organizations listed in Table 20-1 and find out which countries now belong to it. Look at the distribution of member countries on a world map. Does this provide cues to the factors which control membership?

2. Use the median line principle in Figure 20-6 and a map to divide a small enclosed sea like the Mediterranean or Baltic. Would it be fair to use this principle in dividing the world's oceans?

3. Gather data for your local area on the boundaries of electoral districts (e.g., senatorial, parliamentary, or city council districts) and on how many votes were cast for various

parties at a recent election. Was the result affected by the electoral boundaries? How would you redraw these?

4. Look at the distribution of school districts in your city. How far do these try to (a) follow or (b) cut across different social, income, or ethnic differences?

5. Review your understanding of the following concepts:

Mackinder's heartland	computer districting
coastal baseline	apartheid
offshore territorial waters	
continental shelf	
gerrymandering	

One Step Further . . .

Much of the literature leads directly on from that in political geography recommended for the previous chapter (Chapter 19), so we do not repeat it here. There remain, however, a number of areas of special interest. For a view of market areas and the economist's views of spatial partitions, read

Richardson, H. W., *Regional Economics* (Praeger, New York, and Weidenfeld & Nicolson, London, 1969), Chap. 2.

On the division of ocean waters and the control of offshore resources, see

Alexander, L. M., *Offshore Geography of Northwestern Europe* (Rand McNally, Skokie, Ill., and Murray, London, 1963).

Problems of spatial partition within the city and the public issues to which it gives rise are being studied increasingly by geographers. See

Cox, K. R., *Conflict, Power, and Politics in the City: A Geographic View* (McGraw-Hill, New York, 1973), and

Cox, K. R., D. R. Reynolds, and S. Rokkan, Eds., *Locational Approaches to Power and Conflict* (Halsted, New York, 1974).

The general issues of separation with a particularly full account of apartheid is given in

Smith, D. M., *Human Geography: A Welfare Approach* (Arnold, London, 1977). See esp. Chap. 9.

For whosoever hath, to him shall be given, and he shall have more abundance: but whosoever hath not, from him shall be taken away even that which he hath.

The Gospel According to St. Matthew, XIII, 12

Humorists throughout the ages have warned us to choose our parents with care. Geographic humorists might remind us to choose our birthplaces with equal caution! Because it sums up so many economic and cultural considerations, our location in terms of nationality continues to be one of the prime determinants of our life—and indeed has a bearing on whether we will even survive the trauma of birth itself.

In this chapter we try to answer three questions about the geographic importance of country units. First, we look at the main inequalities that exist between states at present. What is the spatial pattern of rich and poor countries? Second, we turn to the contributions of geographers to studies of the economic-development process. Does development follow a particular geographic pattern and show a specific spatial form? Finally, we look at trends in patterns of inequality. Are the different parts of the world converging? Are countries getting more like one another? Or do the rich get richer and the poor get poorer, as our opening quotation suggests?

chapter 21
Inequality Between States

21-1
Global Patterns of Inequality

You need hardly take a geography course to learn that some nation-states have populations more prosperous than others. *If* we make the naive assumption that richness means material wealth, even a layman will have no difficulty in separating Sweden from Senegal and placing it in the appropriate group in Table 21-1. But these are extremes. If we want to distinguish between geographic areas and see how the gaps between the rich and the poor are changing, then we need a reliable ruler that can measure finer differences. What rulers have been developed, and what patterns do they reveal?

The Search for a Yardstick of Development Geographers would ideally like to compute and map some quantitative index that serves as an unambiguous ruler for measuring the economic or social performance of a region. For instance, we might reasonably regard a country as poor in which the low level of health services and nutrition causes many infants to die and a country as rich in which the high level of health services and nutrition prevents infant deaths. If we take available figures for infant deaths per 1000 live births in the 1950s, we find Sweden (with 17) and the Netherlands (with 20) at one end of the scale and Tanganyika (with 170) and Burma (with 198) at the other. [See Figure 21-1(a).] But how valuable this single index is in separating poorer from richer countries is open to debate. Taiwan (with 34 infant deaths per 1000 live births) and Iraq (with 35) scores surprisingly high, while Yugoslavia (with 112) appears to be a very poor country. This difference in scores is due partly to differences in the way each country collects population data and partly to how the available wealth is distributed.

A single index of wealth, then, has evident disadvantages. The United Nations has suggested that standards of living can be properly defined only by using many indexes: indexes of health, food, education, working conditions, employment, consumption and savings, transportation, housing, clothing, recreation, social security, and human freedoms. [See, for example, Figure 21-1(b).] Although the United Nations has admitted that information is not available on many of these items, at least the principle of a multiple index has been firmly established.

If we were to measure all these indicators of wealth, they would not necessarily tell the same story. Figure 21-2 shows how New Zealand and Ceylon vary on ten different measures of development. Ways of combining multiple measurements into a few general indexes are now available. They involve collapsing several measurements into a smaller number of components. (See the discussion of principal-components analysis on page 504.) Brian Berry at the University of Chicago compressed 43 different indexes of economic development into a single diagram (Figure 21-3). Note that this figure has two axes. The first and most important is the longer, vertical axis which measures differences in *technical development*. This single index accounts for about 84 percent of the information in the many original measures. The second and less important axis is the short, horizontal one. This

Table 21-1 Countries of the world grouped by level of economic development[a]

Less developed countries (LDCs)		Moderately developed countries (MDCs)	Highly developed countries (HDCs)
AFRICA	El Salvador 4	AFRICA	AFRICA
Algeria 16	Guatemala 6	Libya 2	—
Angola 6	Guyana 1	South Africa 25	
Benin 3	Haiti 5		AMERICAS
Cameroon 6	Honduras 3	AMERICAS	Canada 23
Chad 4	Nicaragua 2	Argentina 25	Puerto Rico 3
Congo 1	Paraguay 3	Brazil 104	United States 212
Ethiopia 27	Peru 15	Chile 10	
Ghana 10	Trinidad 1	Costa Rica 2	ASIA
Guinea 4		Cuba 9	Israel 3
Ivory Coast 5	ASIA	Jamaica 2	Japan 110
Kenya 13	Afghanistan 19	Mexico 58	Kuwait 1
Liberia 2	Bangladesh 75	Panama 2	USSR 252
Madagascar 7	Burma 30	Uruguay 3	
Malawi 5	Cambodia 8	Venezuela 12	AUSTRALASIA
Mali 6	China 825		Australia 13
Mauritania 1	India 586	ASIA	New Zealand 3
Morocco 17	Indonesia 129	Cyprus 1	
Mozambique 9	Iraq 11	Hong Kong 4	EUROPE
Niger 4	Jordan 3	Iran 32	Austria 8
Nigeria 61	Korea (North) 15	Korea (South) 33	Belgium 10
Rhodesia 6	Laos 3	Lebanon 3	Czechoslovakia 15
Rwanda 4	Malaysia 12	Saudi Arabia 9	Denmark 5
Senegal 4	Mongolia 1	Singapore 2	Finland 5
Sierra Leone 3	Nepal 12	Taiwan 15	France 53
Somalia 3	Pakistan 68		Germany (East) 17
Sudan 17	Philippines 41	AUSTRALASIA	Germany (West) 62
Tanzania 15	Sri Lanka 14	—	Iceland 0
Togo 2	Syria 7		Italy 55
Tunisia 6	Thailand 41	EUROPE	Luxembourg 0
Uganda 11	Turkey 38	Bulgaria 9	Netherlands 14
United Arab Republic (Egypt) 36	Vietnam 43	Greece 9	Norway 4
Upper Volta 6	Yemen (North) 6	Hungary 10	Sweden 8
Zaire 24	Yemen (South) 2	Ireland 3	Switzerland 6
Zambia 5		Malta 0	United Kingdom 56
	AUSTRALASIA	Poland 34	
AMERICAS	Fiji 1	Portugal 9	
Bolivia 5		Romania 21	
Colombia 24	EUROPE	Spain 35	
Dominican Republic 5	Albania 2	Yugoslavia 21	
Ecuador 7			

[a]Latest population estimate rounded to nearest million.

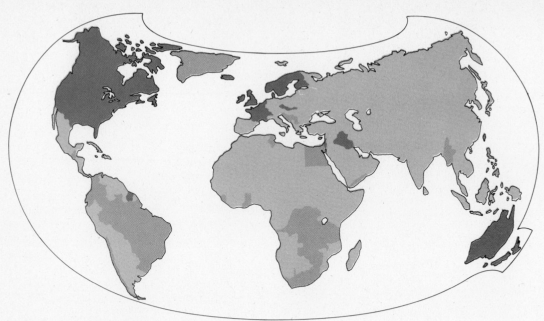

(a) Single index: infant deaths per 1000 live births

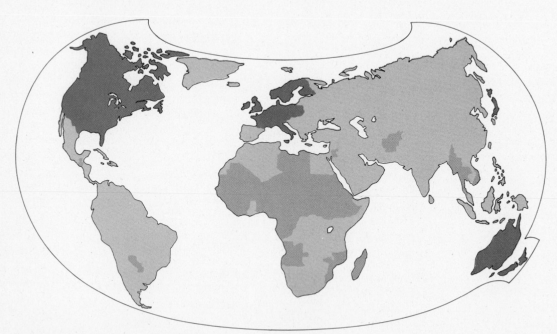

(b) Multiple index: Technological index based on 43 individual indexes

Most developed 20 countries   Most retarded 20 countries

Figure 21-1. Development patterns.
The maps show country-by-country variations in (a) infant mortality as measured by
infant deaths per 1000 live births and (b) a combined index computed from 43 variables.
On both maps only the countries with the 20 highest and 20 lowest scores are shown.
[Data from United Nations and other sources, for late 1950s. After N. Ginsburg, *Atlas of
Economic Development* (University of Chicago Press, Chicago, 1961), pp. 25, 111.
Copyright © 1961 by the University of Chicago.]

index measures contrasts in the *demographic stage* a country has reached.
Together, the two indexes account for 88 percent of the original contrasts
between the 95 countries.

Since Berry's analysis provides such a compact description of differences
between states, we shall look at each of his two indexes in more detail.

Contrasts in Technical Development Figure 21-1(b) shows the world distribution of technically advanced and technically backward countries, according to Berry's first index. The term "backward" causes natural irritation
to readers in those countries it is used to describe and is logically unsatisfying
since the situation in such countries is a dynamic one and most of them are
moving forward quickly. Let us, therefore, use the terms highly developed
countries (*HDCs*), moderately developed countries (*MDCs*), and less
developed countries (*LDCs*), which are more acceptable and accurate terms.
Table 21-1 shows how the world's leading countries appear to fit into these
three categories in the early 1970s.

**Figure 21-2. Development profiles
typical of HDCs and LDCs.**
Contrasts in the relative ranking of New
Zealand and Ceylon on ten measures of
socioeconomic development. These two
countries are selected at random from Table
19-2 to illustrate the profiles of highly
developed and less developed countries.
Note the consistently higher performance of
New Zealand on all measures compared to
the more variable, but generally medium to
low, ranking of Ceylon. All measures are
calculated on a percentage of
population-weighted basis to make the
scores of different countries comparable.

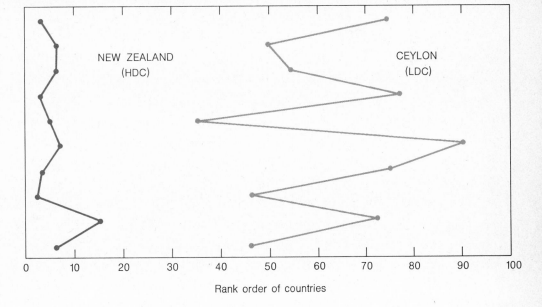

PRINCIPAL-COMPONENTS ANALYSIS

When multiple measurements are made of the same set of individuals (e.g., countries), we can usually transform the original set of variables into a new set of variables that are independent and account in turn for as much of the original variation as possible. To illustrate this concept, consider points A and B as describing the characteristics of two countries in terms of two original variables (*x* and *y*). The shaded elliptical area indicates the full set of observations, that is, the cluster of points representing all the countries examined. Using standard statistical techniques, we can identify the long axis of the ellipse (axis l, or the *principal axis*). Diagram (b) shows how this "synthetic" axis can be used as a ruler on which points A and B can

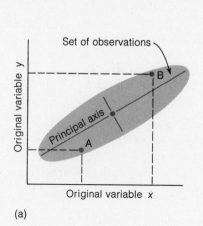

(a)

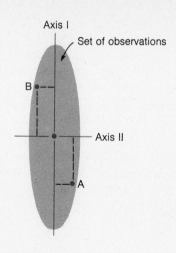

(b)

be described in terms of *both* the original variables (*x* and *y*) combined. Note that this principal axis accounts for much more of the original variation within the ellipse than the secondary axis (axis ll) drawn at right angles to it. This idea of "collapsing" a large set of original variables into a small number of basic dimensions or

composite variables underlies a large and expanding area of mathematics termed *principal components and factor analysis.* For a valuable nontechnical introduction to this subject, see P. R. Gould, *Transactions of the Institute for British Geography,* 42 (1967), pp. 53–86.

Superficial Theories of Development   A look at the map in Figure 21-1(b), the diagram in Figure 21-3, and Table 21-1 suggests that the levels of development and location on the globe are directly connected. Broadly speaking, all the LDCs have tropical locations, and all the HDCs have midlatitude locations. Such a correspondence between variables might lead us to suppose that development is a matter of natural environmental resources —and of climate in particular. Certainly the kind of conditions that we found in the climatic zones E and F in Chapter 4 are major bars to agricultural production. Natural resources do play a key role in development, but we can think of countries (Denmark, Japan, and Israel, for example) which are definitely HDCs but which have a rather limited resource base.

Two other superficial explanations of differences in development relate these to differences in race and culture. If we look at Table 21-1 it is possible to see some apparent link between the distribution of people of Western European (Caucasoid) stock and the distribution of HDCs and between people of Negroid stock and the LDCs. But again, the link is illusory, for example, Japan fails to fit into this picture. More important, people of the same racial stock have occupied quite different positions in the developmental "pecking-order" at different periods of their history and in the same locations. The economic vigor of migrant Chinese in Malaysia contrasts with the conservatism of folk of exactly the same stock in China itself.

Cultural differences also fail to provide a workable explanation of differences in development. As we saw in Chapter 11, religious beliefs do directly influence attitudes toward development. A culture which is preoccupied with the hereafter or despises material prosperity is unlikely to show the same concern with development as a Rockefeller or a J. P. Morgan. But Max Weber's emphasis on the link between capitalism and the Protestant ethic looks somewhat threadbare today. The most rapidly developing countries in the last two decades (e.g., Japan, France, and the Soviet Union) have been non-Protestant.

Interlocking Factors in Development   It is easier to destroy inadequate explanations of the contrasts in development shown in Figure 21-1(b) than it is to replace them. Climate and environment and race and culture are insufficient explanations not because they play no role in development, but because their effect is not simple. Nor is it always the same.

Figure 21-4 attempts to show how these factors may affect levels of development through their interaction with the four factors which economist Paul Samuelson sees as the "four fundamental factors" in understanding development—population, natural resources, capital formation (domestic or imported capital), and technology. In interpreting the diagram, it is important to note that each factor interlocks with the others. Countries showing sustained economic growth over this century (e.g.; Sweden) have tended to score high on all four counts. However, countries showing slow growth may well have been held back by any one of the critical ingredients.

Contrasts in Demographic Stages   The second axis in Berry's development diagram (Figure 21-3) showed demographic contrasts between countries. To understand the links between the population structure of a country and its level of development, you will need to refresh your understanding of the concept of a *demographic transition*, which we first met in Chapter 7. You might like to turn back and look again at Figure 7-13.

Broadly speaking, most of the world's HDCs are in Stages III or IV of the demographic transition. They have slowly expanding or stationary populations. Conversely, most LDCs are Stage I, and most MDCs in Stage II or III. Compare a map of the world birth and death rates (Table 7-4) with the map of technical development in Figure 21-1(b). One of the major effects of a country's position in the demographic transition on development is related to its age distribution, which affects the size of the active labor force in relation to the total size of the population that must be supported. Consider Figure 21-5, which compares the age distribution of three countries in different demographic phases. Mexico is a country in the second phase, with a

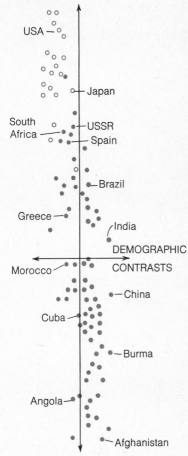

MORE DEVELOPED COUNTRIES

USA

Japan

South Africa — USSR
Spain

Brazil

Greece

India

DEMOGRAPHIC CONTRASTS

Morocco

China

Cuba

Burma

Angola

Afghanistan

LESS DEVELOPED COUNTRIES

● Tropical countries
○ Developed temperate countries

Figure 21-3.   Generalized development yardsticks: economic and demographic scales.

The vertical economic scale accounts for substantially more of the local variation than the shorter demographic scale. [After N. Ginsburg, *Atlas of Economic Development* (University of Chicago Press, Chicago, 1961), p. 113, Fig. 3. Copyright © 1961 by The University of Chicago.]

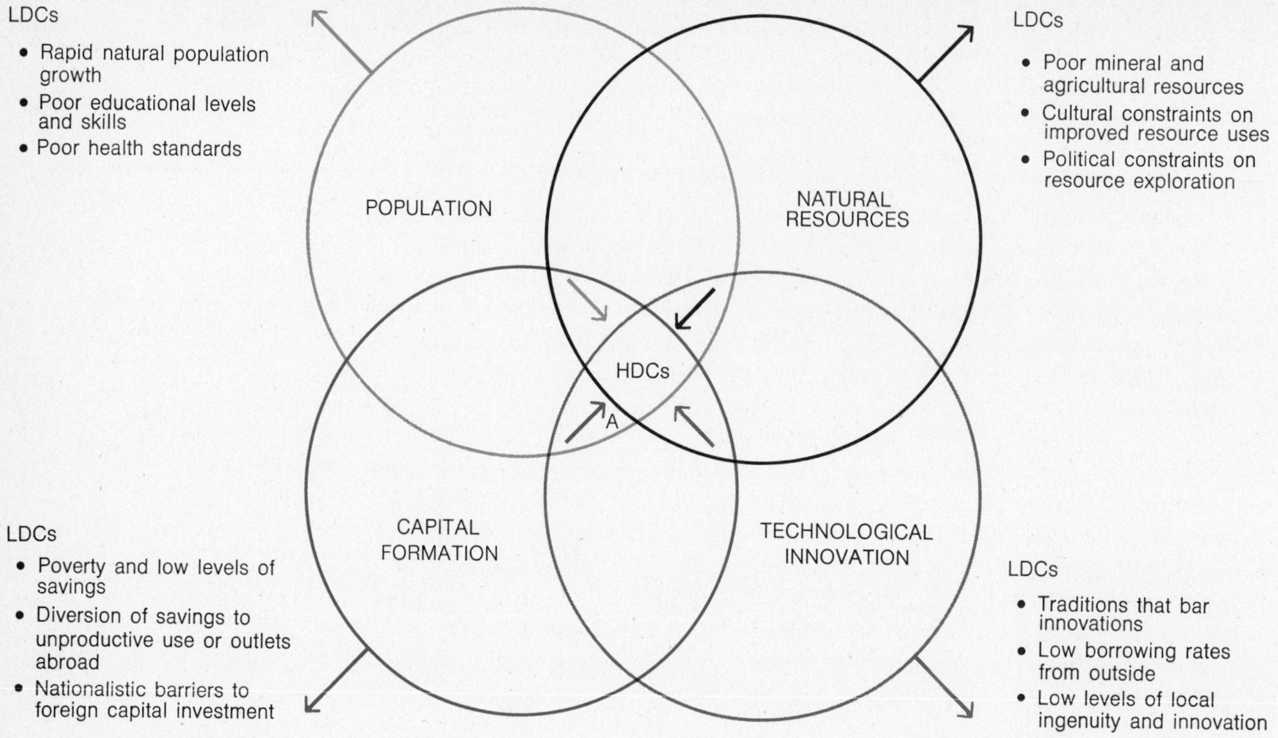

LDCs

- Rapid natural population growth
- Poor educational levels and skills
- Poor health standards

LDCs

- Poor mineral and agricultural resources
- Cultural constraints on improved resource uses
- Political constraints on resource exploration

POPULATION

NATURAL RESOURCES

HDCs

A

LDCs

- Poverty and low levels of savings
- Diversion of savings to unproductive use or outlets abroad
- Nationalistic barriers to foreign capital investment

CAPITAL FORMATION

TECHNOLOGICAL INNOVATION

LDCs

- Traditions that bar innovations
- Low borrowing rates from outside
- Low levels of local ingenuity and innovation

Figure 21-4. Constraints on economic development.

Four major factors in economic development are shown by a set of interlocking circles. HDCs tend to lie near the central overlap area, although some may attain high levels of development with an overlap of only three of the factors. Japan and Denmark are HDCs with rather limited natural resources and therefore lie in Sector A. What about the LDCs? If you followed the four outside arrows and mentally stretched the circles into ellipses, they could each meet on the reverse side of a sphere. This would place the most developed HDC (e.g., Sweden or the United States) on one pole and the lowliest LDC on the other! Three examples of the ways in which each of four constraints may operate on an LDC are indicated.

population increasing by more than 3 percent; Japan can be put in the third phase because its population increases by about 1 percent; Sweden is now increasing very slowly at a rate of less than 1 percent and is in the fourth phase. Note the contrast in the number of children (aged 0 to 14 years) in the three countries: 44 percent of the population in Mexico, 30 percent in Japan, but only 23 percent in Sweden. Likewise, the number of older folk (aged 65 years and over) is only 3 percent of the total population in Mexico, but twice this level in Japan, and three times this level in Sweden.

One of the critical questions, therefore, is whether a fall in birth rates will accompany the urbanization of the LDCs. United Nations surveys measuring the number of children 0–4 years of age in proportion to the number of women aged 15 to 44 years (i.e., potentially reproductive females) reveal a general correlation between this index of fertility and the stage of development of a country. HDCs have generally low index levels, and LDCs generally higher levels. Some demographers have claimed that it is possible to trace the movement of countries with long demographic records, like Sweden, through the four phases. Others have suggested that birth-rate increases may follow a growing abundance of food before the death rate starts to fall. Will countries presently in the second phase move necessarily and progressively to the third and fourth phases? If so, at what rates?

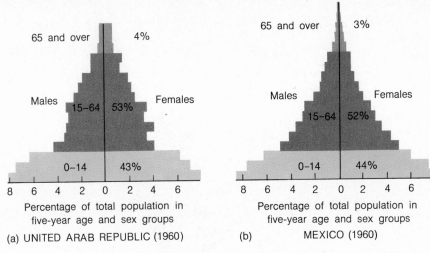

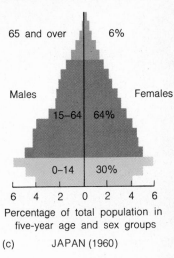

(a) UNITED ARAB REPUBLIC (1960)

(b) MEXICO (1960)

(c) JAPAN (1960)

Figure 21-5. Demographic contrasts and development stages.
Population pyramids show the age and sex distribution of the population of four countries at different stages in the demographic transition. Note that country (a) is an LDC, country (b) is an MDC, country (c) a "recent" HDC, and country (d) a "long-standing" HDC. [Data from United Nations, Demographic Yearbooks, 1957 to 1960. Adapted from J. O. M. Broek and J. W. Webb, *A Geography of Mankind* (McGraw-Hill, New York, 1968), pp. 447–456. Copyright © 1968 by McGraw-Hill, Inc. Used with permission of McGraw-Hill Book Company.]

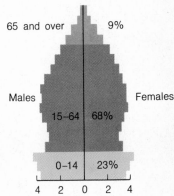

(d) SWEDEN (1957)

We could pattern the changes in birth and death rates in Western Europe since 1700 into a general model. These changes reveal a rather consistent S-shaped fall in death rates (from about 3.3 percent to around 1.5 percent) followed by an S-shaped decline in birth rates (from approximately 3.5 percent to around 1.7 percent). The greatest rate of increase in Europe was in the mid-nineteenth century, when the lag between the two rate curves was at a maximum. But post-World War II fluctuations in the birth rate indicate that the demographic transition model is an oversimplification of the demographic situation in advanced countries.

To sum up: The pattern of states around the world shows great contrasts in the levels of material prosperity. Many measures of that contrast are possible, but one useful summary is that of technical development and demographic stage. We now turn to look more closely at that pattern over time and space.

Many writers have tried to see the facts of world economic development as a linear progression through inevitable stages. For Adam Smith in the *Wealth of Nations* (1776) it was the calculus of fixed land and growing population that provided the key to a golden age in which all a country's product accrued to labor. For Karl Marx in *Das Capital* (1867) it was a one-way evolution from

21-2
Spatial Aspects of Economic Development

primitive culture, through feudalism and capitalism, to the end state of socialism and communism. For Walter Rostow in *The Stages of Economic Growth* (1960) it was again a multistage progression from a primitive society to an age of high mass consumption.

In the historical succession of economic theories, each of the major factors in Figure 21-4 has at one time or another been singled out for the dominant role. Earliest models of development stressed natural resources: Smith stressed the importance of labor, Marx of capital, and Rostow of technical innovation. We also noted in Section 14-2 the role of urbanization and changing job sectors.

For the economic historians and development economists the world story is a frustrating one. The facts of growth have rarely stuck to the predetermined timetables of theory. But geographers have not been wholly immune to the fascination of theory building. What kind of geographic models of growth have they produced? Have they been any more successful than their colleagues in economics?

The Rostow–Taaffe "Stages of Growth" Model   We have already seen in this book the kind of development models a geographer builds. In Chapter 13 we reviewed the work of the Swedish geographer Hägerstrand on *spatial diffusion* models, and in Chapter 14, the parallel work on *urbanization* models. Both these sets of ideas underline the approach we shall follow, emphasizing variations in *where* development takes place and the effect it has on the changing spatial organization of the world economy.

The Spatial Growth Model   Figure 21-6 shows a four-stage model of the spatial pattern of development of an idealized island country. It is based on the work of a group of geographers led by Edward Taaffe at Northwestern University in the early 1960s and draws heavily on Peter Gould's work on the modernization of West African countries, notably Ghana. It has close parallels with Rostow's division of economic development into four phases: a "traditional society," a "take-off" phase, a "drive to maturity," and a movement "toward high mass consumption."

In *Stage I* there is a scatter of small ports and trading posts on the coast. Each small port has a small inland trading field, but most of the interior villages are untouched by the coastal development. Subsistence agriculture dominates the island, apart from the few coastal pockets with trading links to the outside world.

Stage II is the critical stage. It is roughly analogous to Rostow's "take-off" phase, which is clearly based on the analogy of an aircraft which can fly only after attaining some critical speed. Other economists have termed this stage "the spurt" or "the big push." It is marked by two geographic characteristics: first, new transport links to the interior to tap new areas of natural resource for export; and second, differential growth of the coastal centers as some expand (centers a and b in Figure 21-6), some maintain their position (c, d, and e),

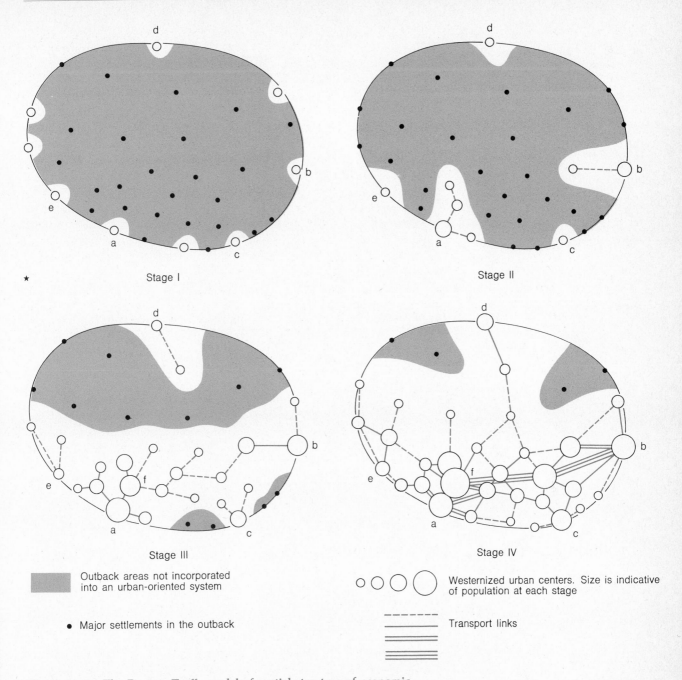

Stage I

Stage II

Stage III

Stage IV

Outback areas not incorporated into an urban-oriented system

Major settlements in the outback

Westernized urban centers. Size is indicative of population at each stage

Transport links

Figure 21-6. The Rostow–Taaffe model of spatial structure of economic development.
An idealized sequence of stages in the economic development of a hypothetical island. Note that Stage II is the critical "takeoff" period in which major transport links are first driven into the interior. The four maps should be viewed as representing the processes of growth discussed in the text rather than as an exact spatial reconstruction of actual events.

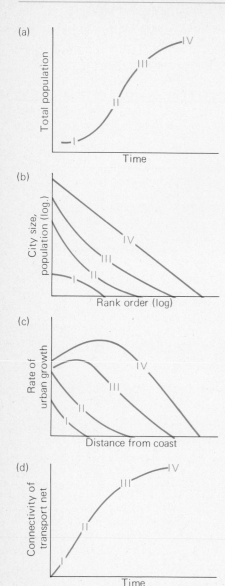

Figure 21-7. Spatial processes in the Rostow-Taaffe model.
Summary of the four basic growth processes underlying the geographic changes shown in Figure 21-6. Note that numbers I through IV relate to the four phases of growth shown in that figure.

and the remainder are pinched out. Studies of African colonial areas suggest that the tapping of mineral resources and the need for political and military control are key factors in the expansion of transport links to the interior of a developing country.

Stage III is marked by the rapid growth of the transport system about each of the major ports and the emergence of important new inland centers at transport junctions (e.g., center f). Note also the beginnings of lateral interconnections between a and b. The northern half of the island remains isolated, but the southern half shows rapid growth and urbanization.

In *Stage IV* the development of transport links continues. Note the development of high-priority shuttle lines between center b and center f. Center f has now taken over a's role as a primate city, marking the shift from an external, export-oriented phase to one in which the island country has major internal markets of its own. The north–south transport link is now complete, and the few remaining "primitive areas" take on a new role as heavily protected wilderness areas for the overurbanized folk of the south part of the island.

Evaluation of the Rostow–Taaffe Model Two questions are raised by this sequence. First, what processes are shaping the pattern? Second, does the sequence described match the events we actually observe?

An attempt to answer the first question is presented in Figure 21-7. This summarizes the four basic processes that have been built into our four-stage model. None of them is new to you. Each has been described in earlier parts in this book (see especially Sections 14-1 and 15-1), and you may wish to use this as an opportunity to review them. The four processes are (1) an S-shaped increase in population, reflecting the demographic transition; (2) the emergence of a family of rank–size rules with an early primate pattern in Stage II, and a more regular form in Stage IV; (3) the launching of a series of diffusion waves for rates of urban growth; and (4) an increase in the connectivity of the transport network as development proceeds. In Figure 21-7(c), the peak of the urbanization wave moves inland in sympathy with the faster growth of the inland center at location f.

How far does the historical pattern of development support our idealized model? We shall mention here two pieces of evidence from the many studies geographers have done on this topic. Figure 21-8 shows the ways in which the importance of New Zealand ports has changed over a 100-year period. Note the pinching out of small ports, particularly on the west coast of the South Island, and the increasing dominance of the Christchurch area.

The second piece of evidence is more general. You will recall from the discussion of transport networks in Chapter 18 (Section 18-2) the ways in which geographers use graph theory. One simple measure of increasing connectivity is the ratio between the number of links in a system and the number of nodes. This is termed the *Beta index*. Thus if we have a railroad system with 12 links and 8 nodes, its Beta index is $^{12}/_8$, or 1.50.

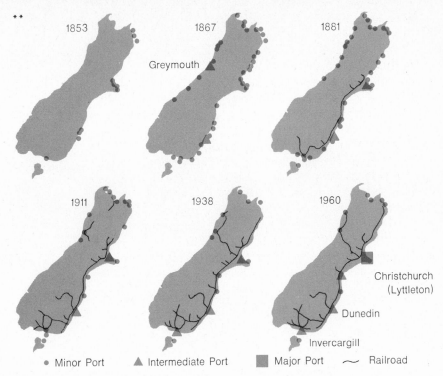

Figure 21-8. New Zealand spatial development.
A century of development on New Zealand's South Island shows some of the integration of transport links and the concentration of port facilities predicted by the model in Figure 21-6. Note the "weeding out" process by which the trade of the many small ports in 1867 was progressively captured by the few larger ports. In the case of the South Island there was a strong bias toward overseas export rather than internal exchange of trade.

● Minor Port   ▲ Intermediate Port   ■ Major Port   ～ Railroad

Figure 21-9 presents Beta indexes for the railway systems of several countries. The values range from around 1.33 to 0.50; when the index is less than 1.00, it indicates that the network is split into several separate subsections as in Stage II (Figure 21-9). Highly developed countries like France have high Beta indexes, whereas poorly developed countries like Ghana have low Beta indexes. The relation between economic development and the connectivity of transport networks is also shown by the changing position of a single country (French Indochina) over time. In the case of developed countries in the last few decades we should expect the relationship of the railway network to growth to be weaker. The closure of some rail links (and therefore a decreased Beta index) would be compensated for, in such cases, by the increased connectivity of other transport links (highways, air routes, etc.).

So, the Rostow–Taaffe model forms a useful, qualitative description of the spatial development process. As a pattern it allows geographers to see more clearly when an individual state or region is deviating from the normal course.

Friedmann's Core–Periphery Model   An alternative approach to modeling the spatial pattern of economic development has been proposed by planner John Friedmann of UCLA. He maintains that we can divide the global

Figure 21-9. Connectivity and economic development.
Connectivity values (as measured by the Beta index) are shown for the railway systems of countries with different levels of economic development. The maps on the left indicate the evolution of a simple transport system and the resulting connectivity values. N = the number of nodes, L = the number of transport links, and B = the ratio of links to nodes (i.e., the Beta index). [After K. J. Kansky, *Structure of Transportation* (Department of Geography, Research Paper 84, University of Chicago, Chicago, Ill., 1963), p. 99, Fig. 25. Photo by Ted Grant, DPI.]

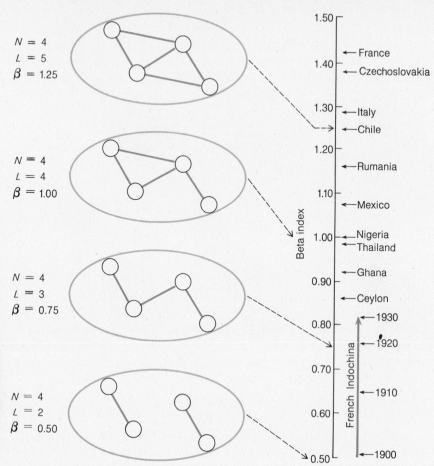

economy into a dynamic, rapidly growing central region and a slower-growing or stagnating periphery. There are four main regions in Friedmann's scheme.

Core Regions First, Friedmann describes *core regions*, which are concentrated metropolitan economies with a high potential for innovation and growth. They exist as part of a city hierarchy and can be distinguished on several levels: the national metropolis, the regional core, the subregional center, and the local service center. On the international level, the North Atlantic community comprising the metropolitan clusters of both eastern North America and Western Europe may be regarded as a core region for development in the western world.

Upward-Transition Regions   Friedmann's second and third regional elements are growth regions. *Upward-transition regions* are peripheral areas whose location is relative to core areas, or whose natural resources lead to a greatly intensified use of resources. They are typically areas of immigration, but this is spread over numerous smaller centers rather than being concentrated at the core itself. *Development corridors* are a special case of upward-transition regions that lie between two core cities. A typical example of an expanding corridor region is the Rio de Janiero–São Paulo corridor in Brazil.

Resource-Frontier Regions   *Resource-frontier regions* are peripheral zones of new settlement where virgin territory is occupied and made productive. In the nineteenth century the midcontinental grasslands of the world provided such a frontier region for grain and livestock production. In the present century agricultural colonization on this massive scale is not occurring. New agricultural zones are being opened up but through much effort (e.g., the Soviet occupation of the virgin lands of Siberia or the colonization of the Oriente, the trans-Andes lowland areas of Colombia, Ecuador, and Peru). Currently, resource-frontier areas are commonly associated with mineral exploitation (the North Slope of Alaska is a good example) and commercial forestry. The continental shelves are likely to be the important frontiers of exploitation by A.D. 2000. In a similar fashion, more intensive development of unused mountain, desert, and island areas for recreational resources is rapidly increasing their status and bringing them into this category.

Downward-Transition Regions   The fourth element in Friedmann's model is the *downward-transition region*. These regions are peripheral areas of old, established settlements characterized by stagnant or declining rural economies with low agricultural productivity, by the loss of a primary resource base as minerals are depleted, or by aging industrial complexes. The common problems of such regions are low rates of innovation, low productivity, and an inability to adapt to new circumstances and to improve their own economies.

Outside the four main types stand a few zones with special characteristics. Regions along national political borders or watershed regions fall into this group. Friedmann suggests that the four main regional types exist on various spatial scales. For instance, downward-transition areas exist on the global level (the rural part of much of the "underdeveloped world" of Latin America and Afro-Asia) and within the cities themselves (blighted and ghetto areas), as well as on the national level (the Italian south, the Mezzogiorno). They may also vary with respect to the status of the general spatial economy of which they are a part. Thus the problems of Appalachia, a depressed area within a core region, must be distinguished from those of depressed areas *within* downward-transition regions.

Friedmann's core–periphery model is linked directly to Thünen's zoning. (See Section 17-1.) Thünen himself thought of old Western European cities and the growing centers of eastern North America as proving a North Atlantic "world city" about which global land-use zones developed. By the 1860s the fall in transport costs—both ocean rates and rates for overland travel by rail—had made the sheep and wheat lands of central North America, the Pampas, and Australasia equivalent to the outer rings in the Thünen's model. In this historical context Friedmann's scheme fits firmly within the evolving sequence of ideas about the impact of changing accessibility of places on patterns of world development.

21-3
Convergence or Divergence?

Most countries are richer now than they were at the beginning of the century. The real pattern of development is thus a dynamic one, and the question of the direction of change is important. Are the rich countries growing richer —and the poor, poorer? We can use two lines of evidence in trying to decide whether international contrasts in wealth between countries are deepening or lessening: the historical evidence of statistical trends and the arguments of theoretical growth models.

Evidence From Historical Trends   One major block to using historical evidence is its highly variable quality. Estimates for current levels of income and the gross national product for most of the less-developed countries are crude, and reconstructions for earlier time periods vary with the assumptions made. Although the present differences between advanced western countries and the Third World are clear, the historical trends in those differences are obscure. Even when information is available on income or production, we lack the data on comparative costs needed to translate income or production figures into meaningful comparisons of regional welfare. There is simply not enough quantitative evidence to confirm the impression of an increasing difference between the "haves" and "have nots."

In the case of continental blocks and individual countries the situation is more hopeful. In the United States regional incomes for the nine main census divisions have converged since 1880. The trend was not, however, a steady one: In the 1920s, the regional contrasts in income actually increased rather than decreased. This was, however, an isolated phase linked to the different ways in which parts of the United States withstood the Great Depression. In the case of a smaller spatial economy, Great Britain, during the last 25 years there were rather weak tendencies toward convergence despite a very active regional equalization policy. The gap between the poorest and richest regions narrowed a bit, but the actual magnitude of this gap in Britain, as in most

Western European countries and particularly Scandinavia, was already narrow. In developing Afro-Asian and Latin American countries, where regional differences in income are greater, the data available are not sufficiently accurate to make firm estimates. The impact of equalization policies on long-term trends within the Soviet Union is not really known either.

Historical and empirical studies provide no strong convergent or divergent trends in regional income. What evidence there is points to rather weak and unsteady changes rather than strong and headlong processes. Regional changes have a complex spatial pattern with different trends operating in various ways on various scales; thus it is probable that divergence and convergence are occurring simultaneously on different spatial levels. Which appears to be taking place may be a function of the levels for which data are available.

Evidence From Theoretical Models　Theoretical models of regional growth have been largely produced as byproducts of general economic theory. They cover only certain parts of the regional growth and have little geographic detail. Here we look at the conditions of a few of the economic models.

Swedish economist Gunnar Myrdal has stressed that economic market forces tend to increase, rather than decrease, regional differentiation. The buildup of activities in prosperous, growing regions influences the less prosperous, lagging regions through two types of induced effects: spread effects and backwash effects. (See Figure 21-10.)

The positive impacts on all other regions of growth in a thriving region are called, by Myrdal, *spread effects*. This impact comes from the stimulation of increased demand for raw materials and agricultural products and the diffusion of advanced technology. Thus, to give a simple example of a spread effect, the medical services in a poor country may gain from the advances in drug therapy conducted in an advanced country without itself having to meet the high costs of the initial research.

Backwash effects of agglomerated growth are net movements of population, capital, and goods that favor the development of the growing area. One classic example of the backwash effect is the "brain drain," typified by movements of medical doctors to the United States from poorer countries. In this and similar selective migration flows the poorer region loses its most highly skilled workers. In a more extreme form, it may also lose its most active population (say, aged 20 to 40 years), leaving only the young and the old behind.

These two opposing forces do not imply the existence of an equilibrium situation. Indeed, Myrdal maintains that the two effects balance each other only rarely. What is more likely is a cumulative upward or downward movement over a considerable time that leads to long periods of increasing regional contrasts (Figure 21-11).

Although Mydral's model of economic growth has been criticized for its

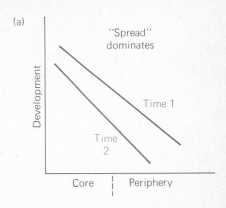

(a)

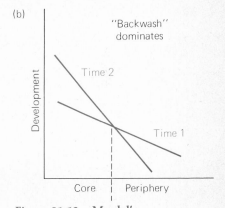

(b)

Figure 21-10.　Myrdal's spread–backwash model.

Changes in the balance between the two sets of forces. In (a) spread dominates so that all areas develop over time with a slight trend toward equalization. In (b) backwash dominates and there is an increasing gap between the core and the periphery.

Figure 21-11. Backwash effects in the Myrdal model.

Three examples of the circular, cumulative, causation process by which the gap between core and periphery areas is widened. The joint impact of (a), (b), and (c) is to give the downward deprivation spiral in (d). Note that the three cycles need not interlink. For example, migrants moving to the core may send back a share of their earnings to their family in the periphery. Also, governments may try to keep up the level of services in peripheral areas to counteract outward movement of capital and labor.

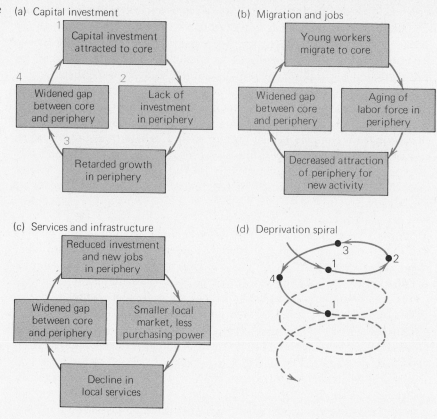

(a) Capital investment

1. Capital investment attracted to core
2. Lack of investment in periphery
3. Retarded growth in periphery
4. Widened gap between core and periphery

(b) Migration and jobs

Young workers migrate to core
Aging of labor force in periphery
Decreased attraction of periphery for new activity
Widened gap between core and periphery

(c) Services and infrastructure

Reduced investment and new jobs in periphery
Smaller local market, less purchasing power
Decline in local services
Widened gap between core and periphery

(d) Deprivation spiral

qualitative nature and lack of econometric substance, the more formal models of regional economic growth *also* fail to demonstrate conclusively the direction of movement. One modern model of the regional system (the Harrod–Domar model) maintains that interregional growth leads to divergence. Rapidly growing regions have high levels of income and net inward movements of labor and capital. In contrast, more traditional models of regional growth indicate that, though fast-growing regions have a net inward movement of capital, income levels in these regions are low and the net movement of labor is outward.

Future Trends   Despite the scarcity of evidence and the lack of agreement between economists, there is still an abundance of projections of future trends. Most of these start with trends in population. By A.D. 2000 it seems likely that the world population will have nearly doubled and that there will be 6.4 billion of us. The average annual rate of growth assumed in this estimate is a little below 2 percent. According to current United Nations estimates Africa and Latin America will be the two fastest-growing continents (both with a 2.7-percent annual increase). Yet the bulk of world population will continue to be in Asia, which has 58 percent of the world total. If we use the definitions in Table 21-1, then LDCs have about two-thirds of world population today. By A.D. 2000 this share will have risen to three-quarters.

What makes these figures disturbing is that they are not matched by equivalent projected changes in the gross national products of the countries in the two groups. (*Gross national product* [GNP] is the economist's term for the total value of goods and services produced by a country during a given time period, usually a year.) The present ratio between the GNP of the LDCs and the developed world (HDCs and MDCs) is 15 percent to 85 percent, and this seems unlikely to change much by the end of the century. For both worlds the per capita GNP is likely to rise—by about 3 percent a year—so standards of living will more than double by the end of the century. The rise will probably be faster in developed countries. The present gap in living standards between the two worlds, now standing at 12:1, will widen in favor of the developed world to around 18:1.

Zero Economic Growth?   In this chapter we have looked at some of the current patterns of rich and poor countries in economic terms. Most of the world's countries are very poor compared to the United States. Only about ten of the world's countries have gross national products greater than that of the state of California. Given their high standard of living and the concern over population pressures, resources, and pollution that we described in Part Two of this book, it is wholly understandable that a zero economic growth (ZEG) movement should have arisen in the western world. The idea that we must return to a self-sustaining economic system to prevent an ecological disaster has been strongly argued by figures like Rachel Carson, Paul Ehrlich, and Barry Commoner. We shall look at some computer models of the future growth dilemma in Chapter 24.

Whatever the appeals of ecological arguments for ZEG, a geographer must note that they are not equally convincing all over the globe. Readers of this book from LDCs (and such readers are likely to be very few) will be unimpressed by antipollution arguments when they still want to develop the industry that will pollute and use resources. Concern about pollution is understandably low in countries where famine and disease are the front-line problems. And do the readers from the fortunate lands of North America, Western Europe, and Australasia really want their incomes to be frozen for the next few decades—or want half their own country's GNP diverted to international aid?

In the longer run many changes are possible. Once we start measuring growth in terms of net social welfare (NSW) rather than GNP, the developed lands may look considerably less affluent than we now believe them to be

The possibility of regional revenue sharing and equalization appears, currently, to be greater at the more restricted spatial scale of the region than on the global scale. It is to this within-country level of spatial organization that we turn again in the next chapter.

Summary

1. Measuring the prosperity of a nation state or region requires more than a single index. Generalized indices combining multiple criteria are now used in an attempt to measure development. Among the most useful indices at the global level are those of technical development and demographic stage. Berry's model of economic development combines these factors with Samuelson's four factors for understanding development. The position of a country in the demographic stage strongly affects the age distribution of its population which is, in its turn, important in understanding the size and composition of the labor force.

2. Many models of regional economic development have been formulated, each with a different dominant factor. Geographers have assembled a four-stage model of the spatial pattern of development for a country based on an increasingly close network of transportation links. The degree of integration can

be measured by the graph-theory indices studied in Chapter 18.

3. The Friedmann core-periphery model postulates four main regions: the core; the upward-transition region; the resource frontier region; and the downward-transition region. This forms a dynamic version of the Thünen model studied in Chapter 17.

4. Both empirical evidence and theoretical models do not show that the economic gap between the more and less developed countries of the world is widening. Rather, the trends show a confused pattern of convergence and divergence with little overall movement in one direction rather than another.

5. Like zero population growth the idea of zero economic growth (ZEG) is attractive only in terms of some theoretical ecological balance. In practice it would seem to set up intolerable interregional strains particularly for presently less developed countries.

Reflections

1. Why are so many of the world's less developed countries located in the tropics? Do you rate the climate of tropical countries (a) a major factor or (b) an irrelevant point in your explanation?

2. What effect does a country's population pyramid have on its level of economic development and vice versa? Draw population pyramids to illustrate your points.

3. Consider the changing connectivity of the networks in Figure 21-9. Increase the number of towns (nodes) from 4 to 6, and draw links to illustrate a range of Beta index values from 0.5 to 1.5.

4. Write down your own views on zero economic growth (ZEG). Debate within the class whether this is (a) feasible and (b) desirable. Do you think you would hold the same views if you lived in a less developed country?

5. How do you expect your own country's share of the world's wealth to change between now and the end of the century? Why? Compare your views with those of your classmates.

6. Select any one LDC from Table 21-1 that interests you. Use encyclopedias and reference books to look into its background. Which factors appear to be the most critical in accounting for its low level of development?

7. Review your understanding of the following concepts:

technical development
   indexes
demographic indexes
population pyramids
the Beta index
core–periphery models
spread effects

backwash effects
gross national product
   (GNP)
zero economic growth
   (ZEG)

One Step Further . . .

One of the liveliest and most perceptive analyses of the development process is given by an Australian geographer,
> Brookfield, H., *Interdependent Development* (Methuen, London, 1975).

A world view of regional inequalities and their spatial distribution is given in
> Berry, B. J. L., E. C. Conkling, and D. M. Ray, *The Geography of Economic Systems* (Prentice-Hall, Englewood Cliffs, N.J., 1976), and
> Ginsburg, N., Ed., *Atlas of Economic Development* (University of Chicago Press, Chicago, 1961).

A variety of possible approaches to regional planning problems is provided in

> Friedmann, J., and W. Alonso, Eds., *Regional Development and Planning: A Reader*, 2nd ed. (M.I.T. Press, Cambridge, Mass., 1974).

Economists'views of the economic-development process are given in
> Rostow. W. W., *The World Economy* (Macmillan, New York, 1978), and
> Myrdal, G., *Rich Lands and Poor* (Harper & Row, New York, 1957).

The case against "growth" is presented in
> Mishan, E. J., *Costs of Economic Growth* (Penguin Books, Harmondsworth, 1967).

"I have a dream. . . . I've been to the mountain top . . . I've seen the promised land."

MARTIN LUTHER KING, JR.
Speech from the Lincoln Memorial, Washington, D.C., August 28, 1963

chapter 22
Inequality Within a State
Welfare Issues in Regional Planning

At the time of his death, Martin Luther King was preparing for a "poor people's march" on Washington, D.C., in early 1968. This march was to focus public attention on the deeply complex problems of poverty in America. Readers of the previous chapter, with its graphs and charts showing the United States heading the table of highly developed nations, must find it paradoxical that we begin a chapter on regional inequalities by focusing on the richest country in the world. When we talk about a country as being an HDC, MDC, or LDC, we are referring to the *average* conditions of the country taken as a whole. This ignores the strong regional variation within the country. To find pockets of poverty within the United States (say, parts of rural Appalachia) is no more noteworthy than to find pockets of affluence within an MDC (say, the rich Copacabana Beach area within Brazil). Variation in development occurs at all spatial levels, both within and between countries.

In this chapter we concentrate on differences *within* countries, focusing especially on the spatial inequalities in welfare within a state. But what is an inequality—and what is welfare? And once we define these things, how do we define a just distribution of welfare? These issues dominate the first part of this chapter and form a backdrop to the study of regional-planning practices in the second part. There we look at how Western European countries tackle the problem of poor regions. Third, we illustrate the kind of issues raised at the local level within the broader framework of regional planning. Finally, we look back at the whole of Part Five of the book and reconsider the importance of the state in a world where multinational corporations are increasingly important. To what extent are the corporations challenging the state? Do the multinationals have a distinctive geography of their own?

22-1
Spatial Inequalities and Welfare Issues

Before we look at the details of regional-planning practices, we must investigate three more general issues. What is spatial inequality? What is social welfare? And what is a geographically just distribution? None of these questions can be answered precisely, since the answer to each depends on the reader's view of society itself. This first section may perhaps serve as a framework for a debate which runs far outside the bounds of our particular focus of inquiry.

Lorenz Curves: Measuring Inequalities in Welfare Our first question is a technical one, and produces much less emotional reactions than the others. Given that differences between regions exist, how should we measure them?

One of the most useful measures of inequality is the *Lorenze curve* (Figure 22-1). This is a graphic representation of the distribution of any measure of welfare (e.g., income). If it is perfectly straight, the distribution is perfect. The more bowed a Lorenz curve is, the more unequally the proxy for welfare is distributed. The difference between an actual Lorenz curve and a straight line is called an *inequality gap*.

Figure 22-1. Measures of inequality. (a) The distribution of population and income among the four main ethnic groups in South Africa. (b) A Lorenz curve showing the inequalities between the four groups. The shaded area represents an "inequality gap." [South African data for 1967 from D. M. Smith, *An Introduction to Welfare Geography,* University of Witwatersrand, Johannesburg, Department of Geography, Occasional Paper No. 11, p. 95, Table 5.]

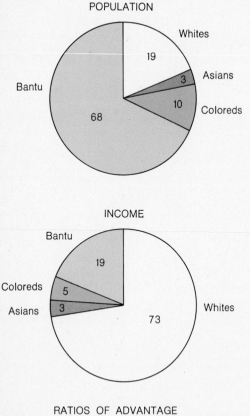

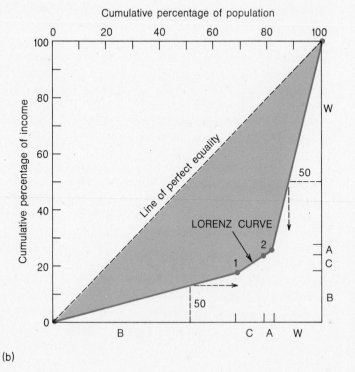

(a)
(b)

RATIOS OF ADVANTAGE

White (W) 73/19 = 3.84   Coloreds (C) 5/10 = 0.50
Asians (A) 3/3 = 1.00   Bantu (B) 19/68 = 0.28

The construction of Lorenz curves is illustrated in Figure 22-1, using the distribution of the income in South Africa as a measure of the distribution of welfare there. If we divide up the South African population (some 21.5 million in 1970) into major ethnic groups, there are strong contrasts in income per capita. The Bantu people make up about two-thirds of the population and receive about one-fifth of the income. Conversely, the 3.7 million white South Africans make up about one-fifth of the country's population but control nearly three-quarters of its national income.

By dividing their share of the country's income by the percentage of the population they include, we can compute a *ratio of advantage* for each ethnic group. Thus for the Bantu this ratio is 19 (their share of income) divided by 68 (their share of the population), or 0.28. Ratios above 1 indicate that a group is better off than the average group in the nation, and ratios below 1 that it is worse off. To draw a Lorenze curve, we take the group with the lowest ratio, in this case, the Bantu, and plot its position on a graph of population and income (point 1 in Figure 22-1). We then take the group with the next-lowest ratio, the "colored" population (about 2 million strong) and add its shares of population and income to those of the Bantu. We then plot the position of the two groups together (their cumulative position). This is point 2. The Bantu and the "colored" together make up 78 (68 + 10) percent of the population and share 24 (19 + 5) percent of the country's income. The shares of remaining groups are added in the same way, and further cumulating percentages of population and income are plotted.

In addition to being simple to construct, Lorenz curves have a number of properties that are useful in studying inequalities. If we look at the two 50-percent marks in Figure 22-1, we can see that the lower half of the population receives only around 13 percent of the country's income; conversely, half the income goes to only around 15 percent of the population. South Africa is used here only as an example. Lorenz curves for all the world's countries have this characteristic convexity. Inequality exists everywhere, though the degree of inequality varies. This variation can be shown by comparing countries with very different inequality gaps. An LDC like Thailand has a much greater gap than the HDCs. Even within the advanced countries, there are strong differences in the size of the gap. Sweden, with its progressive taxation system, has a Lorenz curve much closer to the line of perfect equality than the United States.

Alternative Indicators of Welfare So far we have used income to measure welfare. It is, however, only one index, and a very crude one, of the social welfare of an area. A *social welfare indicator* is simply a term for a statistic which shows the change from "worse-off" to "better-off" areas. As you might guess, it has proved extremely difficult to get people to agree on such measures, since they must first agree on what social welfare is.

From the geographic viewpoint, there are two aspects of social indicators we should note. First, no two indicators tell exactly the same spatial story. Consider Figure 22-2, which shows contrasts in the ten provinces of Canada in

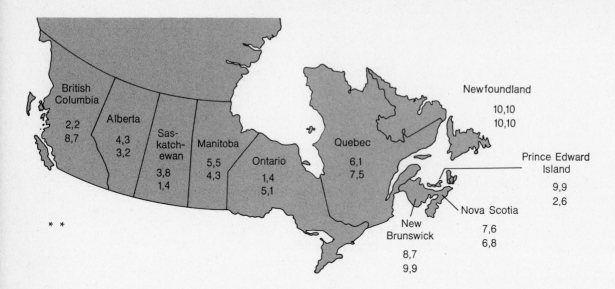

Figure 22-2. Spatial variations in socioeconomic health.

Each of the ten Canadian provinces is ranked using four alternative measures of income and job opportunities. "Healthy" provinces have low rank scores, and "unhealthy" provinces have high ones. Only Newfoundland, which ranks tenth every time, has wholly consistent scores. [Data from P. E. Lloyd and P. Dicken, *Location in Space* (Harper & Row, New York, 1972), p. 207, Table 10-5.]

the 1960s. These contrasts have been measured by four different indicators of social welfare: (a) per-capita income; (b) higher education, as measured by the number of college enrollments; (c) average unemployment rates; and (d) average job-participation rates for males. Each of these indicators tells us something about the "health" of the provinces in terms of, say, jobs and the potential skills of the population. In Figure 22-2 we have superimposed on each province its position in the marked list of Canadian provinces with respect to each indicator: A 1 indicates a position at the top of the table (the "best" score), and a 10 indicates a position at the bottom of the table (the "worst" score). Of the ten provinces, only Newfoundland has a consistent set of scores. The problems of the Maritime provinces show up, as does the relative prosperity of Ontario and Alberta. British Columbia is strikingly well off in some ways but badly off in others.

Ranking is of course a rough measure of contrasts, since small differences may force a province into a low position. "Roughness" is also a problem in Figure 22-3, which moves south of the border and looks at interstate contrasts in the United States. Here the nine measures are much more complex. They were produced by Kansas economist J. O. Wilson from the "domestic goals" proposed by Eisenhower's Commission on National Goals. Each index was created by collapsing a large number of raw measures of welfare in much the same way measures of wealth were collapsed in Chapter 21.

The profile of each state shows its rank position on each index. Profiles are plotted for three states with different average performances. Minnesota is in the upper half of the ranked list of all states, with particularly high rankings on indicators II and IX. North Carolina is in the lower half and Missouri is consistently in the middle of the table. *No* state has an entirely consistent

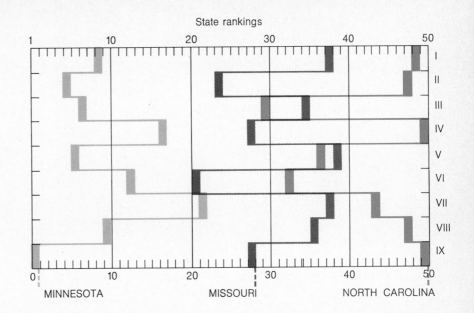

SOCIAL WELFARE INDICATORS

I Status of the individual

II Racial equality

III Democratic process

IV Education

V Economic growth

VI Technological change

VII Agriculture

VIII Living conditions

IX Health and welfare

pattern of rankings. For example, California, whose position on the first eight indicators ranges from first to fourth, plunges to fourteenth on the health and welfare index. Hawaii ranks first (ties with Utah) with respect to racial equality but fortieth with respect to technological change.

The second general question raised by social welfare indicators is one of stability. Do we wish to measure inequality in terms of an area's position in any one year or decade? Or should we be more interested in rates of change? Figure 22-4 illustrates the contrast between the answers to the two questions in a study of the quality of life in 18 metropolitan areas of the United States. Urbanists Jones and Flax, on whose data the figure is based, wanted to measure how Washington, D.C., stood in relation to other major cities and to see whether living conditions there were getting relatively better or worse. Using seven measures of the quality of life (ranging from housing costs through robbery rates to pollution levels), they came up with the rankings shown on the longer, horizontal axis of the chart. Minneapolis headed the list of high-ranking cities, and Los Angeles trailed the list of low-ranking ones. Washington, D.C., did not do too badly, but San Francisco—a shock to Bay Area fans (including the author)—was next to the last.

How were conditions in these centers changing? The results of a second study on changes during the 1960s are shown on the shorter, vertical axis of the chart. While conditions in a number of low-scoring cities like New York and Chicago were getting better, the quality of life in Washington, D.C., was getting relatively worse. In San Francisco, according to the study, conditions were deteriorating more rapidly than anywhere else. Of course, measuring changes over such a short time period may be misleading, since the results may reflect short-term influences such as a particular mayoral program.

Figure 22-3. Interstate differences in the quality of life.
The graph shows the relative ranking of three states (out of the 50) in terms of the nine social welfare indicators. Low rankings indicate a high quality of life; high rankings indicate poor conditions. The nine indicators were derived by statistical procedures from 85 different variables. [Data from J. O. Wilson, *Quality of Life in the United States*, Midwest Research Institute, Kansas City, Missouri, 1969, p. 13.]

Figure 22-4. Intercity comparisons of social welfare.

An attempt to see how Washington, D.C., ranked in relation to other metropolitan areas with respect to living conditions produced the results shown here. Each city is ranked on both the quality of life there (on the horizontal axis) and on changes in that quality in the 1960s (on the vertical axis). The high scores of Minneapolis and Boston were no surprise, but the poor showing of San Francisco was wholly at variance with its image. Note that all urban data are sensitive to the exact city boundaries used and how much of the suburban area is included. [Data from M. V. Jones and M. J. Flax, *The Quality of Life in Metropolitan Washington, D.C.: Some Statistical Benchmarks* (The Urban Institute, Washington, D.C., 1970).]

Spatial Aspects of Social Justice Let us assume that we have agreed on a proper measure of welfare and that we have found marked inequalities between provinces, states, metropolitan areas, or other regions. What then? A concern with spatial inequalities in welfare must raise, for geographers, ethical issues which have troubled philosophers from Aristotle to Marcuse. As Johns Hopkins geographer David Harvey puts it, how *do* we achieve "a just spatial distribution, justly arrived at?"

What kind of claims can the people of any disadvantaged region make on the larger, national community? From the welter of conflicting suggestions three major ideas stand out. First, they can make claims based on *need*. We may argue that all parts of a country should have a basic right to a certain standard of education or medical care, regardless of spatial differences in the cost of providing those services. Postal costs in remote rural areas in most countries are well above the national norm, but postal charges are usually common all across the country. Second, they may make claims based on their *contribution to common good*. Areas which contribute greatly to the good of the whole nation might be expected to receive a greater-than-average payment. A special subsidy to a city faced with the problem of preserving costly old buildings of historical value (e.g., as Amsterdam is in Holland or Venice is in Italy) would be merited because of the city's contribution to the heritage of the whole country. Third, the people of a region may make claims on the larger community on the basis of *merit*. The environmental challenge to human life varies from region to region. The allocation of extra resources to a region might be justified to protect areas of great potential stress from natural hazards (e.g., to protect marginal agriculture in semiarid lands) or from social hazards (e.g., to reduce crime in high-risk areas in inner cities and ghettos). Of these three bases for the distribution of social resources, need is arguably the primary one, with contributions to the common good and merit running second and third, respectively. Readers interested in following up these lines of thinking should definitely browse through David Harvey's *Social Justice and the City*. (See "One Step Further..." at the end of the chapter for bibliographic details.)

Let us stay with the simple concept of need and look at its spatial implications. Figure 22-5 shows how different philosophies might affect a spatial redistribution of wealth in an idealized five-region country. Four situations are described. In the first the government's policy is a "laissez faire" one of nonintervention, with no revenue sharing [Figure 22-5(a)]. In the second, problem areas (the two poorest areas) are designated as needing special help [Figure 22-5(b)]. This kind of approach—having a special strategy for special areas—poses problems we shall consider later in the chapter. In the third situation, there is a sliding-scale approach to regional needs [Figure 22-5(c)]. This is rather like the negative-income-tax approach, in which rich areas subsidize poor areas on the basis of need. In the fourth, a complete-equalization approach is taken and taxes are adjusted so that all areas are brought to the same income-level [Figure 22-5(d)]. In each case, a different interpretation of social justice leads to a different spatial pattern of levels of welfare. These cases are, however, highly simplified, and deal with a hypothetical country. We turn now to realistic examples of current regional-intervention policies in Western Europe and North America.

Figure 22-5. Enigmas in spatial justice.
The charts reveal the spatial implications of four alternative approaches to revenue sharing in a simplified five-region country. The arrows indicate the directions of transfers to the subsidized areas (shaded). Rearranging the diagrams (putting the poorer areas at the center and the richer ones on the periphery) would give us a picture of problems in big cities today.

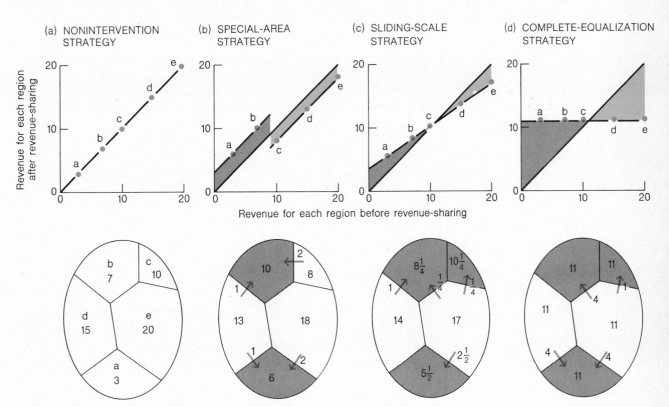

22-2
Intervention at the Regional
Level

Should central governments intervene to adjust regional inequities—or should they allow the normal equilibrium-seeking forces like migration to operate? If the locational theory stressed in Part Four of this book tells us anything, it is this: First, spatial specialization occurs because it is an efficient way to use immobile resources; and second, the basis of specialization is continually changing. That is, the spatial economy of a country appears to be both specialized and dynamic. Historically the emergence of new centers has been accompanied by the obsolescence of earlier, and less efficient, centers of regional production.

What appears to be at issue is not the necessity of spatial change but who should pay the costs involved. On the local scale the blight and decay in our city centers are part of the price paid for the benefits the automobile has brought to suburban areas. On the regional scale the limited opportunities of some old coalfield areas like those of Appalachia are a legacy of their specialized role in preceding decades. Few people would support a complete equalization policy like that in Figure 22-5(d) which, by moving resources to the population, effectively froze the present spatial pattern; but more observers are asking whether the costs of spatial changes should fall solely on a local part of a city any more than they should fall on a local part of a country. The logic of the argument does not stop at national boundaries; it has implications for future revenue sharing on the international level.

Tools of Regional Policy   If central governments wish to intervene in the regional growth process, what tools can they use? Three main strategies are currently in use.

The first is investment in the *public sector*. Regional investment of this type spans the construction of entire new cities in underdeveloped regions, like Brasilia in west-central Brazil [Figure 22-6(a)], to the building of new schools in a city ghetto. It is commonly aimed at improving the basic infrastructure of a region. Transport and power facilities are generally the prime targets for improvement. The dam-building by the Tennessee Valley Authority [Figure 22-6(b)] and the road-building in the program for developing Appalachia exemplify the emphasis in regional-support legislation.

Inducements to business in the *private sector* to invest in a region are a second strategy. These inducements may be positive, such as capital grants or

Figure 22-6.   Tools of regional policy.
(a) A large-scale national reorientation of resources underlay the building of a new federal capital at Brasilia by the Brazilian government. The new capital lies on the central plateau of Goias, about 900 km (560 mi) inland from the old coastal capital of Rio de Janeiro. Begun in the late 1950s, the new capital city has a population now approaching 500,000. (b) Public power and transportation program sectors are typified by the Tennessee Valley Authority program in the 1930s. (c) New agricultural settlements in Israel's southern arid zones are financed by the government. [Photographs (a) by Erwitt, Magnum, (b) courtesy of the United Nations, and (c) courtesy of Israel Information Services.]

(a)

(b)

(c)

tax concessions to industries already operating or willing to operate in undesirable areas, or negative, such as controls and penalties for companies in fast-growing areas. Companies operating in rapidly developing areas may have proportionately higher taxes and face legal restrictions on their expansion.

Inducements to *individuals and households* to locate in or leave a region are a third strategy. Migration from a declining area may be hindered by the inability of would-be migrants to sell their houses or land. Compensation to those farmers willing to move (plus aid to enlarge the farms of those who stay) is offered by the governments of both Ireland and Sweden to make migration from agricultural areas with low and declining prospects easier. When the need is to attract population, similar monetary and other inducements are used.

The choice of regional policy tools depends largely on the resources available in the country as a whole and, more importantly, on its sociopolitical system. For instance, Britain has tried to resolve the problem of unemployment in its peripheral coalfield areas by loans for new industry (initiated in 1934), building factories and industrial estates (in 1936), providing tax incentives for new industry, including depreciation allowances (in 1937), controlling industrial buildings outside the problem areas (in 1945), and awarding standard grants for new industrial buildings and new machinery (in 1960). The measure of central government control of industry's location in Britain is representative of that in a politically middle-of-the-road state with a mixed economy and a strong commitment to equalizing job opportunities between regions. States following a less socialist policy tend to allow industry to decide its location on commercial grounds; those following a more socialist policy favor a greater degree of direct control by central government.

Defining Areas of Need To allow regional policies to be introduced, we must draw a clear distinction between problem and nonproblem areas. The boundary chosen should clearly reflect the character of the inequality. To take a simple example, in the 1930s the Brazilian government defined the boundaries of the Nordeste (a problem region in the northeast "bulge" of that country) by rainfall figures. There the key problem was recurrent drought, leading to crop failure and famine, so it made sense to draw a boundary around the aided area in rainfall terms. Regions within the drought boundary received special government help, while regions outside did not.

We encounter much greater difficulties when a problem region is defined by population characteristics and levels of distress. In some countries the basis for preferential regional treatment is unemployment. But unemployment rates are very unstable over time. They tend to underestimate job opportunities; that is, areas of high unemployment also have outmigration and a low proportion of folk in jobs (e.g., there may be no jobs for women in the labor market). This means that the official unemployment figures are too optimistic in problem areas. Thus, before we can establish any realistic index of relative welfare, we need to modify our definition of unemployment to include other variables related to migration or wage levels.

No matter what index of regional need we adopt, we are left with the question of where to draw the line. For example, if we adopt a 4.5 percent unemployment rate to define the regions which will receive aid, then we invite protests from regions just below the threshold. An extreme if simplified illustration is shown in Figure 22-5(b) where unaided region *c* is made worse-off than aided region *b*. The limit may conceal sharp contrasts within the distressed region itself. The difficulty is made more acute by the fact that the same index may yield different value if different regional subdivisions are chosen. Generally, the smaller the system of subdivisions, the greater is the spread of index values, and vice versa. This gives great scope for gerrymandering the boundaries of a distressed region to make it qualify for government aid.

One reaction to this problem has been the replacement of a twofold division (between regions that receive aid and those that do not) by three or more levels of aid. Such a system has the theoretical advantage of matching the amount of aid to the degree of distress, but is objected to on the grounds of increasing administrative costs. Figure 22-7 shows that in England and Wales there is now virtually a four-level system of regions: (1) areas of rapid growth and high prosperity subject to negative controls on further expansion (e.g., Greater London), (2) "normal" areas which require neither positive nor negative intervention (e.g., most of southern England), (3) areas with moderate economic difficulties (e.g., Plymouth), and (4) depressed areas (e.g., South Wales) receiving the full range of government assistance. France has a similar five-stage series of zones, with metropolitan Paris and rural Brittany at opposite ends of the scale.

The difficulty of having a fully differentiated scheme of regional assistance is that, at one extreme, it heavily penalizes the very productive regions, while at the other it grants massive help to the least productive ones. If it was successful, it would freeze the existing geographic pattern of production. Such a freezing is, of course, wholly at variance with the change and diffusion already discussed. Indeed, it would be disturbing to think what effect a fully successful regional aid policy would have had on the United States if it had been in operation in 1920, or in 1820! There are clearly some regions (e.g., worked-out mining areas) where the levels of unemployment are so high and the prospects so poor that it is hard to justify a policy other than strategic withdrawal. However, *within* problem areas discriminatory policies should favor those parts of the problem regions that have the greatest growth potential. It is to this idea that we now turn.

Growth Poles as a Regional Policy   The notion of productive points within, or as near as possible to, distressed areas as centers of new investment is an attractive one. It has been strongly urged by French regional economist Francis Perroux in his concept of the growth pole (*pole de croissance*). Basically, a *growth pole* consists of a cluster of expanding industries which are spatially concentrated (usually within a major city) and which set off a chain reaction of minor expansions throughout a hinterland.

Figure 22-7. Zones of government aid.
The United Kingdom is divided into four types of areas which receive varying degrees of incentives to industrial development. Northern Ireland has its own system of incentives, broadly similar to those of the British developmental areas. Over the last half-century, a succession of British governments have juggled the boundaries and played with the particular range of policies followed. This map shows the situation in the early 1970s; some boundaries have been changed since then.

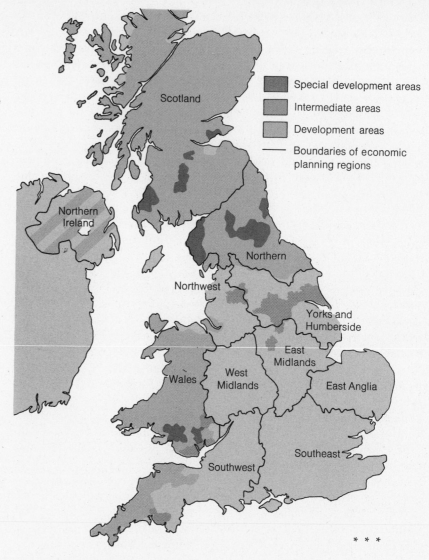

* * *

A growth-pole policy in regional development means the deliberate selection of one or a few potential poles in a problem area. New investment is concentrated in these areas rather than being spread thinly "in penny packets" over the whole area. The arguments in favor of this policy are that public expenditures are more effective when they are concentrated in a few clearly defined areas and that new industries there will stand a better chance of building up enough agglomeration economies to achieve some degree of self-generating growth. Agglomeration economies are the benefits that come from the sharing of common infrastructure (roads, power supplies, water, etc.), the increased size of the labor market, and reduced distribution costs.

Although the evidence is conflicting, the necessary population of a growth pole may lie between 150,000 to 250,000 people. Only in cities of this size will there be the basic ingredients of large-scale diversified growth. Whether spatial concentration is an *essential* part of growth pole theory is open to doubt. Some plants—petrochemical or metallurgical smelting plants, for example—may need specific locations away from urban concentrations of population. They bring benefits to the poorer region in terms of increased income (particularly through contributions to local taxation) rather than through more jobs.

In practice a growth-pole policy is likely to run into two kinds of difficulties: the technical ones of selecting the best potential pole, and the political ones of convincing the unsuccessful poles of the wisdom of the policy.

Strategic Withdrawal as a Regional Policy An inverse growth-pole policy may be used when the general economy of an area is experiencing a long-term structural decline. For example, Figure 22-8 presents the British government policy toward the mining villages in the Durham coalfield in the northeastern part of England. This area had its heyday during the last century when Durham coal was in demand in a country that was a world leader in the Industrial Revolution. Today most of the mines are closed or closing, made uneconomic by the competition from more efficient fuels and from better

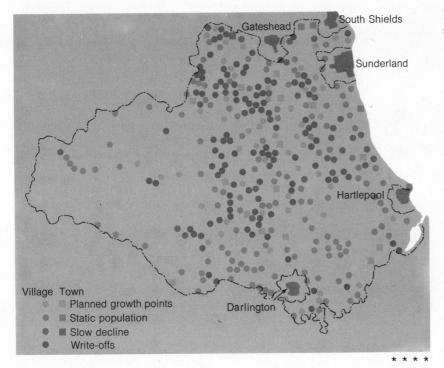

Figure 22-8. A local strategy of spatial concentration and withdrawal. This map of Durham County in northeast England is based on a development plan adopted in the 1950s. Since then, some settlements have been reclassified. The fourth group of villages, termed ''writeoffs'' by the plan's critics, are mainly small mining villages in which the colliery has closed. In these settlements new capital spending was to be limited to the social and other facilities needed for the life of the existing property. [Data courtesy of County Planning Department, Durham County Council, England.]

Village Town
- Planned growth points
- Static population
- Slow decline
- Write-offs

* * * *

coalfields in other parts of England. Unemployment rates are high, job opportunities low, and there is a steady outmigration from the mining villages. An extreme interventionist might argue that the economic and social waste of unused manpower and an underused infrastructure (roads, railways, schools) should be prevented by subsidizing new employment opportunities and by fixing high prices for coal. Conversely, we could argue that migration is an effective solution to the problems of the area, that subsidies can only be granted at the expense of the more productive sections of the country, and that the new generation (and particularly their children) will lead happier and fuller lives once they have established themselves in the more prosperous and job-rich parts of the country.

Government policy, as Figure 22-8 shows, has reflected a mixture of both viewpoints. On the regional level the response has been interventionist. A new town has been built, new industries have been attracted to it, new roads constructed, and so on. On the local level, however, the situation is different. The smaller and more remote villages are being progressively abandoned, and people are being encouraged to move to the larger and better endowed areas. What we are seeing is, in effect, a policy of strategic withdrawal that balances the social costs of nonintervention (borne locally) against the economic costs of intervention (paid largely by the rest of the country).

Regional Planning in the United States   The notion of regional planning in the context of the five-year plans of communist states like the Soviet Union or the social-democratic traditions of Sweden or Britain is a familiar one. The political institutions of a country like the United States make it harder for the central government to intervene at the regional level. What kind of regional planning occurs within a more capitalistic state like the United States?

In the 1930s regional planning in the United States was rather narrowly confined to the development of water resources, with special attention to river-basin planning. President Franklin Roosevelt's establishment of the Tennessee Valley Authority was an outstanding example of this kind of intervention, which was widely copied by other countries. Federal intervention in the economic life of the nation was, however, seen as the thin end of the wedge of a socialist-style intervention in some quarters. Thus it was not until 1961 and the Kennedy Administration that the Area Redevelopment Act was passed. Under this act areas with unemployment rates greater than 6 percent or running at specified levels above the national average were eligible for assistance. Although over 1000 counties were designated to receive aid (see Figure 22-9), the total impact of the act was rather small. Some industrial parks were established, but the major share of the act's resources went for recreation and tourist projects.

The major contribution of the Johnson Administration was the passage of the Appalachian Regional Development Act in 1965, which involved federal aid being coordinated on an interstate basis. The main areas of public investment were transportation (particularly through extensions of the inter-

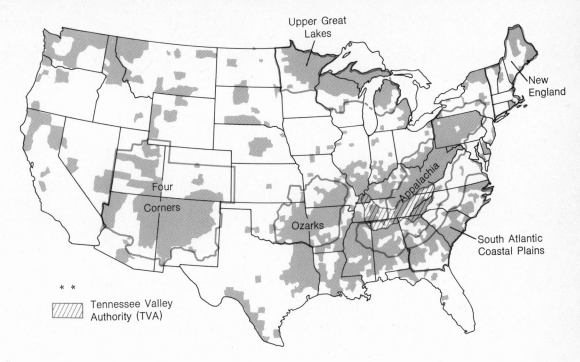

Figure 22-9. U.S. regional planning. The map shows two examples of federal programs in the United States. The shaded areas indicate counties designated to receive aid under the Kennedy Administration's Area Redevelopment Act. Heavy boundaries enclose the six interstate planning regions set up by the Johnson Administration's Economic Development Act. The main federal effort has been in the depressed Appalachian area, where it overlaps with part of the TVA area set up in the early 1930s.

state highway program), the development of natural resources, water-control schemes, and social and educational programs. By 1968 five further interstate areas had been set up under the Economic Development Act.

Aid appears to have been spread somewhat thinly, and the total funds available were, by Western European standards, small in relation to the United States' huge GNP. Later administrations (Nixon and Carter) have given less emphasis to regional policy. By the 1970s public concern had shifted somewhat from the plight of the less prosperous rural regions to the more concentrated problems of the inner cities. Currently it is the plight of the cities and the problems of metropolitan revenue-sharing that dominate the scene. As the suburbs have sprawled outward and industry has decentralized, more American cities have been left to cope with a severe housing blight, high crime rates and other inner city ills, and a dwindling tax base.

Intervention in regional problems may range all the way from international aid programs affecting half a continent down to decision-making on specific issues of local significance. Here we look at three examples of intervention on a limited spatial level: One is concerned with land use, one with location; and one with an interlinked ecological problem.

22-3
Local Planning Problems

Problems of Land Use   As population density on the fixed area of the earth has increased, so has the pressure to use land in intensive ways. In Chapter 8 we saw something of the effect of this increasingly intensive land use on the level of environmental change. Nevertheless, the idea of using the same piece of land for a variety of purposes is very attractive.

Such ideas are not new. Indeed, many so-called "primitive" ways of using land, such as the shifting cultivation systems in the humid tropics, represent an integration of crop production, grazing, and timber use within a limited area. In the context of contemporary western society, however, the term multiple land use refers to the integration of major uses like forestry and recreation. Often, but not always, it implies that the lands being used are publicly owned. The subject is a hotbed of controversy. What kind of land uses can be mixed? Which mixtures are long-term and stable ways of using the land, and which are dangerous and short-sighted? And who decides what is an optimal land-use strategy?

At the outset we should separate two different types of multiple land uses. First, there is the *common use* of the same tract of land for two or more purposes. For example, the same river basin might be used for recreation, forestry, and grazing. The physical possibilities and limitations of such common uses depend on their *compatibility*. Table 22-1 indicates the physical compatibility of nine land uses with one another. For example, there is high

Table 22-1   Compatibility of major land uses

Primary land use	Urban purposes	Recreation	Agriculture	Forestry	Grazing	Transport	Reservoirs and water management	Wildlife	Mineral production
							Physical compatibility with secondary use for		
Urban purposes	Complete	High for city parks; zero for others	None	None	None	Very poor, except city streets	None	Very poor	Very poor
Recreation	None	Complete	None	Poor to moderate	Very poor to none	Very poor	Poor	Fairly high	Very poor
Agriculture	None	Very poor	Complete	Zero	Zero	Zero	Very poor	Poor to moderate	Poor
Forestry	None	High	None	Complete	Variable— none to fairly high	Zero	Zero	High	Poor to moderate
Grazing	None	High	None	Usually very poor	Complete	Zero	Poor to fairly high	High	Poor to high
Transport	None	None directly; incidental on rights of way	None	None	None	Complete	None	None	None
Reservoirs and water management	None	Poor to high	Very poor	Very poor	Poor to moderate	None	Complete	Poor to high	Very poor
Wildlife sanctuaries	None	High	Very poor	Moderate	Moderate	None	Poor	Complete	Very poor
Mineral production	None	Poor	Poor	Fair	Fair to moderate	Fair	Poor	Poor to fair	Complete

SOURCE: M. Clawson *et al.*, *Land for the Future* (Johns Hopkins University Press, Baltimore, Md., 1960).

compatibility between forestry as a primary use and wildlife refuges as a secondary use for land primarily devoted to agriculture. About some combinations no generalizations can be made at all. For instance, some urban areas (city parks) very definitely have secondary recreational uses, while others have not. The compatibility of reservoirs and water management with other secondary uses of land may vary considerably from place to place.

What the table makes clear is that certain land uses are highly intolerant of others. When land is used primarily for urban and transport purposes, usually secondary use for the other seven types is ruled out. In contrast, grazing and forestry are tolerant both of each other and of numerous secondary uses. Of course conditions vary within each category; for example, agriculture may be very intensive (and intolerant of other uses) or very extensive (and tolerant). Note also that the table refers only to physical compatibility of different land uses. Some of the economic, social, and legal problems of common use are formidable.

A second main type of multiple land use is *parallel use* of the same tract of land for two or more primary purposes. In this case individual land uses are kept spatially separate but are intermingled within a given tract and administered as a single unit. National forests, for example, contain limited recreational strips (e.g., along highway margins or lakeshores), and commercial forestry is confined to the back country. In practice the parallel use of intermingled tracts has often involved publicly owned land, usually in areas with relatively low land values. Privately owned land in areas with higher land values is usually divided into small parcels; parallel land use in these areas poses more difficult economic and legal issues.

How do we decide whether a particular strategy of multiple land use is efficient? Ideally, we should survey the benefits and costs that would arise from a given land use combination. However, the process of defining, let alone measuring, the costs and benefits is complicated by several practical difficulties. Some land uses produce their benefits in spatially distant areas but bring few local benefits. Thus watersheds create benefits like water not only for local residents but also for remote cities. In a similar fashion recreation and wildlife benefits can accrue to the whole population and not only to local residents. In the same way some types of land use may bring penalties for the local population. A watershed-control program that would bring great benefits to the water supply of a distant city might sterilize land for local agriculture. Wildlife reserve areas in East Africa restrict the herding and hunting possibilities for the local population; urban freeways bringing suburban commuters to a CBD increase the noise and pollution levels and decrease the visual amenities for the population in the inner areas of a city. Finally, some benefits from multiple land uses may be so intangible or long term that they cannot be enjoyed by the local population.

Problems of Location   The decision to locate any public facility is usually a contentious matter. Public facilities come in three forms. First, "desirable" facilities such as local public library or health center which have no objec-

Figure 22-10. Location of "desirable" public facilities in Sweden.
Work by Swedish geographers on determining the optimum location of major regional hospital centers. The boundary lines enclose areas that are closest to each of the six main medical centers for persons traveling by car. [From S. Godlund, *Lund Studies in Geography B,* No. 21 (1961), Fig. 8.]

tional attributes and would be welcomed by most communities. Figure 22-10 shows the locations decided for regional hospital centers in Sweden, an example of desirable facilities. Second, "undesirable" facilities such as airports or garbage incinerators which are needed by the population as a whole but no one wants as their immediate neighbor. Clearly everyone may gain the economic benefit from a new airport, but the costs in noise or congestion is borne by the nearby residents. Third, "hybrid" facilities which confer a mixture of benefits and costs on local groups. The mixture may depend on both location and social group. A noisy discotheque may disturb nearby residents, but it would probably be regarded as a valuable amenity by local teenagers.

Because they represent the most acute problems we shall take as our example here the second of our three cases. Foremost among these unwanted facilities are international airports like Kennedy Airport in New York. We can use the conflict surrounding these facilities as an example of more general locational conflicts.

Consider the frantic efforts to find a suitable location for a third London airport within the crowded southeastern part of England (Figure 22-11). An initial decision to locate a new airport at Stansted, north of London, was abandoned in 1967 after loud local protests. In 1968 a government commission was set up to advise on the site. The commission examined 78 potential sites and rejected most on grounds that ranged from meteorological conditions to accessibility to London to local construction costs. No ideal site was found, but four sites—Cublington, Thurleigh, Nuthampstead, and Foulness—were considered reasonable possibilities. Estimates of the total costs and benefits associated with each of these sites were made. As Table 22-2 shows, two kinds of cost were dominant: the costs to the airlines of aircraft movements and the cost to the passenger of movements from home to the new airport. Note that the first type of cost showed much less variation from site to site than did the second. Costs falling on the local community are relatively small in terms of the total (i.e., unless you happen to live in that community), but vary greatly from one site to another. After many calculations of this kind the commission recommended in 1971 that the third London airport be located at Cublington.

The selection of the inland site at Cublington raised a storm of protest. In a minority report one member of the commission described selecting an inland site in a rural area with a high level of amenities as "an environmental disaster." Citizens questioned the utility of applying cost–benefit analysis, an economic tool, to an essentially political situation, and they challenged the detailed calculations of the costs falling on the local areas. (For example a Norman church that would have to be bulldozed despite its considerable historical and architectural interest was costed at its fire-insurance value.) In response to this protest, the government decided to adopt the coastal Foulness site, despite higher construction costs there. Even this location has met a storm of protests and plans have now been put on the shelf.

The sheer scale of an international airport lies at the heart of the locational problem. In this case, although the area within the airport perimeter would be only a few square kilometers (about the size of central London itself), the direct and indirect effects of the facility would be felt over a far larger area. Some 65,000 people would be directly employed at the airport, and a new city of around 275,000 (or an equivalent expansion of nearby urban centers) would be needed to cope with the population directly or indirectly generated by the airport. The new highways and railways linking the airport with London would take up thousands of acres of land well outside the immediate neighborhood of the airport. Above all, a noise umbrella would extend over an area of 750 km² (290 mi²), and an inner zone of 125 km² (48 mi²) of land would be virtually uninhabitable as a normal residential area. (See Figure 22-12.)

The conflict over all unwanted facilities—be they on the massive scale of an airport or the local scale of a penitentiary—is that while the benefits arising from them would be *regional* in scale and affect a large population, the costs in disruption and a loss of amenities would be *local* in scale and would fall on a rather small proportion of the population. Thus in England there is a general agreement that the London region needs expanded airport facilities, but no particular part of the region wants the airport. In this situation no amount of detailed locational analysis will resolve the problem. Rather, the whole community must reevaluate its system of calculating costs and benefits so that those who suffer from the creation of the facility will be appropriately and realistically compensated. (See the discussion of alternative ways of measuring costs on the next page.)

Interlinked Ecological Problems   Although in textbooks we can separate one planning issue, like multiple land uses, from another, like airport sites, in

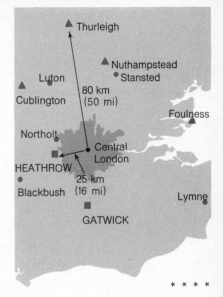

*   *   *   *

■ Main London airports
● Small airports
▲ Potential sites for new airport

Figure 22-11. The search for airport sites.

The map shows the short list of four alternative sites for a third London airport put forward by the Roskill Commission. The Commission's choice, on the grounds of costs, was an inland site, Cublington; the final government choice, on environmental grounds, was a coastal site, Foulness.

Table 22-2   Costs of alternative airport sites in southeast England[a]

Type of cost	Size of cost (in millions of pounds)			
	Cublington	Thurleigh	Nuthampsted	Foulness
Movement of aircraft from new airport to their destinations	960 [$2304]	972 [$2333]	987 [$2369]	973 [$2335]
Travel of passengers from their homes to new airport	887 [$2129]	889 [$2134]	868 [$2083]	1041 [$2498]
Disruption of the local community to be displaced by the new airport	55 [$132]	55 [$132]	66 [$158]	45 [$108]
Noise costs inflicted on the residents of the immediate airport vicinity	14 [$34]	14 [$34]	24 [$58]	11 [$26]
Capital costs of building new airport	289 [$694]	288 [$691]	285 [$684]	252 [$605]
Total net costs	2265 [$5436]	2266 [$5438]	2274 [$5458]	2385 [$5724]

[a]All costs are given in millions discounted to a common base year (1975). Bracketed numbers indicate millions of dollars.
SOURCE: Roskill Commission, *Report on the Third London Airport* (H.M.S.O., London, 1970).

ALTERNATIVE WAYS OF MEASURING COSTS

In calculating the cost of ecological or spatial alternatives, geographers should be aware of the different types of costs accountants measure. These include:

1.   *Opportunity costs.* These measure the loss or sacrifice involved in using a resource or location for one purpose rather than another. Opportunity costs are the best measure of the cost of using a resource or space for one purpose when there are competing alternatives and the resource or space is being *fully used.*

2.   *Marginal costs.* These measure the extra costs incurred in using spare resources or space for a particular purpose. Marginal costs are important when an area is not fully used and the issue is whether or not to utilize *spare capacity.*

3.   *Full costs.* This is an accounting concept for comparing the full costs (both fixed and variable) of using a particular resource or space with the full benefits (both long term and short term). Full costs have the advantage of being relatively easy to calculate in a standard way, thereby allowing the interregional comparison of alternatives. The concept is common in *cost-benefit* analysis.

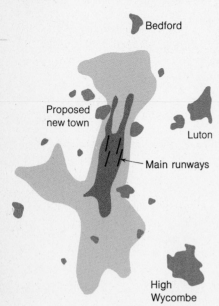

Figure 22-12.   Environmental hazards around major airports.

The map shows the noise umbrella (lightly shaded) that would surround the projected London airport site at Cublington. Normal residential life in the inner zone (heavily shaded) would be severely disrupted by the noise.

Bedford

Proposed new town

Luton

Main runways

High Wycombe

practice we find that problems of location and land use are closely intertwined. Let us consider a small-scale English example, one that could be repeated many hundreds of times in other countries. Figure 22-13 shows a small man-made lake, the Chew Valley Reservoir. Despite its pastoral and apparently peaceful quality this 12-km² stretch of water has been the center of three conflicts, each of which involved ecologic and spatial issues so critical that it is worth examining them as a microcosmic example of conflicts on all spatial levels.

The first conflict broke out in the late 1950s, when the lake was being planned. The city of Bristol, about 15 km (9 mi) distant, was demanding ever more water, and a new storage reservoir was urgently needed. But where should it be located? The Chew Valley was ideal because of its hydrology, its unpolluted catchment, and its nearby location. However, the area to be flooded was officially designated as first-class agricultural land. The conflict of interests between the city (containing around 400,000 people) and the area to be flooded (containing only a few farms) raised significant questions about the location of unwanted facilities, the permanent elimination of food-producing land, and individual-versus-state rights. The final decision to build the reservoir in the Chew Valley hinged on opportunity costs; similar reservoirs in other sites would have caused still more difficulties.

The second conflict, in the late 1960s, concerned the ecology of the newly formed lake. To increase their agricultural yield, farmers in the watersheds around the lake were using larger amounts of artificial fertilizers. Creeks draining the watersheds carried downstream to the lake water that had absorbed small quantities of fertilizers. The newly formed lake acted as a trap in which the nutrient levels in the lake built up. (Nutrients are the chemical elements essential for the growth of plants.) Water lost by evaporation, overflow, or abstracted as a water supply was replaced by inflowing enriched waters from the catchment. In the enriched waters new organic food chains (see Section 3-2) were created. These chains produced spectacular color changes in the lake as large colonies of algae turned the water green. Oxygen

levels in the deeper lake waters decreased, and the fish population fell. This process in which lake water is enriched by nutrients which cause excess algae growth is called *eutrophic*. Although the long-term process of eutrophication is difficult to stop or reverse, cooperation between the water authorities and local landowners has reduced the rate at which it is occurring.

The third conflict in the early 1970s centered on the use of the lake environment. Its original and primary purpose was simply to provide an urban water supply. Fishing was permitted and the lake stocked, and ornithologists were allowed to observe the wildfowl population. Pressure to use the lake for other purposes has been successful. Sailing was permitted in 1968, and other recreational uses will be allowed in the near future. Not all the uses proposed for the lake are compatible with each other, or with the primary use of the

Figure 22-13. Problems in land-use planning.
The photo shows the man-made Chew Valley Lake created in 1952 to meet increasing demands for water from cities in southwest England. Its location, its pollution, and its multiple uses have made it a source of considerable conflict among different land and water users. [Ordance Survey. Crown copyright reserved.]

EUTROPHICATION AND "LAKE DEATH"

A *eutrophic* lake like that of the Chew Valley is simply a nutrient-rich lake. As the nutrient supply (particularly nitrogen and phosphorus) of a lake builds up from the inflow of fertilizer-enriched water from the surrounding fields, some major changes in the lake's biology and chemistry take place. Water weeds become very abundant. Later, thick mats of blue-green algae (called blooms) also become evident. As the algae die, they sink to the lower layers of water, decaying and consuming the oxygen dissolved in the waters.

Eutrophic lakes are undesirable for man in several ways. The surface mats of algae interfere with recreational use, choke out game fish (trout are particularly affected in the Chew Lake), and affect the water's taste. Decayed algae clog water treatment filters and slowly make the lake more shallow. Thus, in the very long run, the lake literally dies.

Eutrophication is not a new phenomenon. Blue-green algal blooms were reported from Switzerland's Lake Zurich in 1896. There is no doubt that eutrophic processes have been greatly speeded-up in the last few decades, due to increasing use of fertilizers in agriculture and the introduction of synthetic detergents (containing phosphorus) into domestic use from the 1950s. Phosphorus levels in Lake Erie have increased threefold since the 1940s. Eutrophication may also occur in the estuaries (mouths) of major rivers. Algal blooms were first observed on the Potomac in 1925 and have been massive and persistent since 1962.

Attempts to control eutrophication and stop "lake death" are difficult. Nitrogen has so many natural and artificial sources and is so critical in agricultural production that its control is not generally feasible. Attempts to reduce phosphorus, and particularly the tight control of synthetic detergents, is more promising. For a good nontechnical account of eutrophication, see Donald E. Carr, *Death of the Sweet Waters* (Norton, New York, 1966).

---

area for a domestic water supply. Figure 22-14 presents some zones of activity that now allow parallel use of the lake for different purposes.

The three conflicts over this small English lake raise a number of questions about conflicts among interest groups. The location of the reservoir created a controversy over the importance of the city's need for water compared to the economic and social costs that would be inflicted on a small farming community by the creation of the reservoir. The eutrophication problem stirred up the conservationists against farmers who wanted to improve the productivity of their land. The lake's potential uses created cross conflicts among groups wishing to use the lake primarily for their own purposes. Each conflict involved spatial considerations at different geographic scales and pointed up the need for a long-term ecological viewpoint. And each conflict was exacerbated by the crowding caused by a high population in a limited and intensively used land area.

22-4
Multinationals and the Corporate State

In Part Five of this book we have argued that it is the *state* which, either singly or in coalition, is the strategic political unit of our time. We have seen in the previous four chapters how differences between and within states leave an indelible stamp on the world geography. But is the state the only decision-making unit which geographers need to know about?

Galbraith's New Industrial State Economist John Kenneth Galbraith would say "no" to that question. In his *New Industrial State* (1967), Galbraith argues that the large corporation is fast becoming the strategic economic unit of greatest significance. He was able to show that whether we measure at the level of an industry, a country, or the whole world, business activity is becoming more concentrated. For the United States the hundred largest manufacturing corporations now control just over one-half of the total assets of *all* business corporations: Fifty years ago it was little more than one-third. In the United Kingdom two out of five employees work for the hundred largest manufacturing firms—in the early 1930s it was one out of five.

Corporations have grown both by expanding the size of individual plants, but more significantly by expanding the number of their plants and the range of their products. Typically the large business corporation is both multiplant and multiproduct, and so has a highly diverse structure in both product and space.

The Size of an Industrial Corporation Measurement of the size of a business corporation may be conducted in different ways. *Fortune*, a magazine which publishes regular surveys on the size of industrial corporations both inside and outside the United States, uses three measures: These are the total sales of a corporation in a year, its corporate assets, and the numbers of workers it employs. Figure 22-15 shows the leading 25 corporations. This combined list was made up by taking the 15 top-ranking corporations on each of the three measures. Note that over half the corporations are American with British, Dutch, German, Japanese, Iranian and Italian making up the remainder. (The state-owned corporations of Communist bloc countries are not included.)

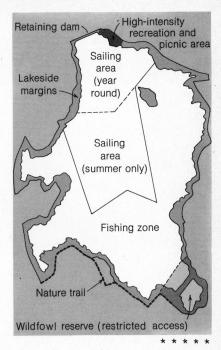

* * * * *

Figure 22-14. Zoning solutions for multiple land uses.
The map shows the proposed zoning of Chew Valley Lake for competing uses. [Data from Bristol Waterworks Company. From C. Hartley, *Countryside Community Recreation News Supplement,* No. 3 (1971), p. 9, Fig. 1.]

Figure 22-15. World's largest industrial corporations.
The 25 largest industrial corporations were defined in terms of sales, employment, and assets. Note that they do not include banking and insurance combines or institutions within communist bloc countries. Since data refer to the middle 1970s, the sales position of oil corporations may be slightly inflated. [From data given in *Fortune Magazine*, May 1975, © 1975 Time Inc.]

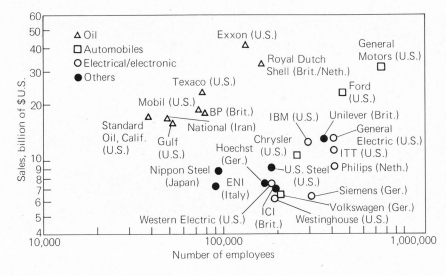

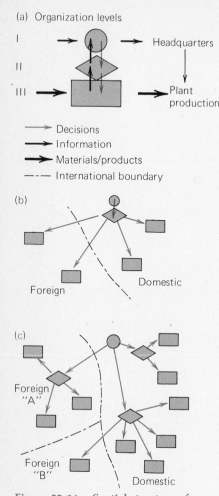

(a) Organization levels

I → ◯ → Headquarters

II ◇ ↓

III → ▭ → Plant production

→ Decisions
→ Information
→ Materials/products
--- International boundary

(b)

Foreign   Domestic

(c)

Foreign "A"

Foreign "B"   Domestic

Figure 22-16. Spatial structure of multinational corporate activity.

(a) Tornqvist's concepts of a three-level decision-making hierarchy within a single plant. (b) – (c) Different spatial arrangements for a multiplant corporation with foreign plants.

In terms of each corporation's products the graph shows a clear distinction between the eight oil companies in the list, which have a high ratio of sales to employees, and the others. These others are companies in electronics, electrical machinery, and automobiles with steel, chemicals, and foodstuffs also represented. Note however that the categorization of manufacturing corporations in terms of a dominant product is slightly misleading since corporations are tending to become more diversified.

Let us take as an example General Motors, the corporation that has occupied first or second place on all three lists for the last half-century. This corporation has over 120 plants within the United States producing nearly 100 different major products ranging through the familiar automobile and its accessories to domestic washing machines and military hardware. About half the plants concentrated on a single product, but the remainder produced from two to twelve different lines. ITT produces not only a wide range of electronics, but also food, car parts, insurance, and cosmetics.

Spatial Ecology of the Corporation   As the scale and diversity of corporate organization has grown, so too has the spatial diversity. We need here to check back on the work of Swedish geographer Gunnar Tornqvist described in Section 14-3. Recall that he analyzed the spatial structure of the corporation in terms of three levels of organization. At the highest level, *Level 1*, the head office is concerned with the general long-term planning of the corporation's activities. Here the main functions are nonprogrammed in that chief executives are involved in highly flexible negotiation with government and public officials and with similar executives in other corporations. The primary locational need is for an information-rich site with easy opportunities for face-to-face contacts and facilities for rapid travel to other major centers. These needs will usually be met only in a national capital city or large metropolitan center. At the second level, *Level II*, are the more routine administrative funtions. There is still a need for good location, but contacts can be carried out indirectly via telephone, telex, or written communication. *Level III*, the lowest level, is concerned with the routine day-to-day organization of work at the plant level. Locations here are controlled by the manufacturing needs of the product and fall in line with arguments on industrial location presented in Section 17-2. Products dependent on particularly low labor costs may be located in areas where production costs are low, and so on.

In the simplest corporate unit, the single one-product plant, the three levels recognized by Tornqvist may be combined in a single location as shown in Figure 22-16(a). Here the flows of information and decisions take place within the plant, while the input of raw materials and output of finished products reflect the geographic distribution of resources on the one hand and of markets on the other. With corporate growth the situation becomes spatially more complex. Figure 22-16 shows just two of the possible arrangements of Levels I and II activities as the number of plants and the range of products increases. In the first [Figure 22-16(b)] there is a twofold geographic structure with the head office and administration in one location and the operating units in another. In the second [Figure 22-16(c)] the structure is threefold with complete spatial separation of the three levels. The larger and more diversified the corporation, the stronger the arguments for a threefold structure become.

Multinationals and the Problem of Foreign Control   The large corporation clearly poses a control problem for central or regional government within a particular state. In communist countries any mismatch between the goals of the corporation and the goals of the state may be reduced by taking over the corporation and converting it into a state enterprise. In western capitalist countries, the restrictions may come in the form of legislative action controlling monopoly or prices. The mixed economies of Western Europe show various intermediate solutions such as government purchase of stock, full nationalization, and so on.

But these measures apply only to the home-based corporation. How do governments control the large multinational corporation whose headquarters lie wholly outside its territory? It was the French politician Jean-Jacques Servan-Schreiber who in *Le Défi américain* (1967) showed the size of the problem for Western Europe. He predicted that in the next two decades, the world's third greatest industrial power after the United States and Russia, would be not Europe but American industry in Europe. A good deal of Servan-Schreiber's predictions have come to pass. Over two-thirds of all the Western European computer industry is American-owned (55 percent by IBM). In employment terms, the leaders are ITT, Ford, and General Motors, all with over 100,000 employees in Europe.

Although the American-based multinationals are the largest, the problem is a much broader one. For multinational corporations are of increasing importance in shaping development levels all around the world. If we exclude communist bloc countries, then multinationals probably account for approaching one-quarter of the world's total industrial output. For the last decade their foreign output has been increasing at something like twice the world growth in GNP. They are much more dominant in technologically advanced industries, just those with the highest present and future growth potential (e.g., electronic engineering, chemical engineering, automobiles). About half the multinational assets are American-owned, with European countries (notably Britain and West Germany) and Japan controlling most of the remainder.

Foreign multinationals are of concern to state governments in that each has legitimate interests quite different from the other. Plants within a country are, however large, still essentially branches whose ultimate control is in an overseas capital. (See Figure 22-17.) Such corporations have, at least in theory, the option of cutting their losses in a particular country and investing elsewhere. But there are substantial counterbalancing benefits. Arguably, the foreign multinational represents the most efficient sector in many of the

Figure 22-17. Americanization of the Canadian economy.
These two graphs show aspects of the important role of United States corporations in Canada. (a) Historical trends in the share of United States investment in the total of foreign investment in Canada. Note the surge of investment after each of two world wars as the British share contracted sharply. (b) Spatial shares of Canadian and United States-controlled jobs in Canadian manufacturing. The horizontal axis of the graph measures distance from the economic hub of Toronto. Notice the heavy dominance of the United States in Ontario province; both Quebec and British Columbia have predominantly Canadian control. [Data from B. J. L. Berry, E. C. Conkling, and D. M. Ray, *The Geography of Economic Systems* (Prentice-Hall, Englewood Cliffs, N.J., 1976), p. 284, Fig. 15.6.]

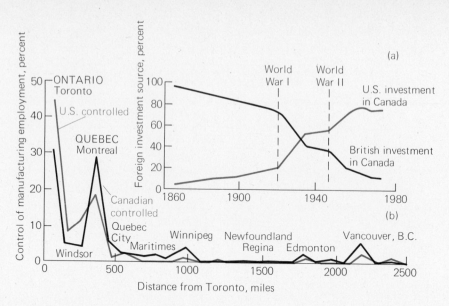

world's countries, and contributes an increasing amount to their growth. Increasingly, Level II activities (to use Tornqvist's term) are being decentralized to leading cities of those countries in which the multinationals operate.

It would be wrong, therefore, to conclude this section by leaving the impression that multinations represent some cancerous growths on the world economic scene. Large multiplant, multiproduct, multinature corporations have arisen in the middle of the twentieth century as a logical response to the technologies and economics of the manufacturing process. There remain, however, the inevitable problems of regulating and controlling a twentieth-century economic institution in terms of a unit of human political organization (the state) which has different historical origins and a different geographic scale of operation. Both institutions have legitimately different areas of interest, and geographers will watch with concern which plays the greater role in setting the regional patterns of a changing economic world.

The State as a Corporation   Although the state and the corporation are seen as different institutions, they share some features of geographic interest in common. The role of state intervention in economic affairs is well known in the communist countries of the world, but it is less frequently realized how large is the state's role in western economies. Let us take the United States as an example. From 1902 to 1970 total government expenditude in the United States increased from 7 percent of the gross national product to over 32 percent. Not only did expenditures increase greatly, so also did government jobs. These now account for about a third of the total workforce in the United States. In one sense, then, the federal government within the United States' economy is acting as the largest single corporation! Trends which are true for this country are also true for a number of European countries and are even more accentuated in the case of Great Britain, France, and Italy.

The significance of these broad swings toward additional government employment on the spatial organization of the economy is not yet clearly understood by geographers. Some work has been done on the contentious issue of the impact of defense expenditures. The main defense industries consist of aircraft, steel, electronics, and related services. Studies show that a number of states within the United States have benefited greatly from government expenditure on arms and equipment. California, New Mexico, and Colorado in the west and Washington, D.C., and Maryland in the coast areas of the east have been the main beneficiary regions. A second area of study has been the impact of government expenditures department by department. In Canada study of expenditure by the Ontario Provincial Government Departments showed severe spatial bias in favor of the Toronto region. In this sense, therefore, the government can be seen as an agent within the spatial economy tending to reinforce the advantages of the core region.

There is evidence for a fundamental paradox in government policy. As the state becomes a larger and larger employer of labor, so its own policies play a more important part in determining the spatial structure of the economy. To some extent the growth of government and its centralization may be exacerbating the difficulties of the distressed and marginal areas. In that sense it has contributed to the very problems which it is trying to solve through the kind of regional policies discussed in Section 22-2.

In this chapter we have discussed how central governments intervene to adjust regional differences within a country. But the issues which intervention raises have as much to do with ethics and politics as with economics. Insofar as these matters of capital and conscience have a spatial framework, they are of concern to geographers because their resolution affects world distributions. Communities are showing increasing concern with the consequences, both ecological and socioeconomic, of locational decisions and spatial inequalities. Geographers view this concern with mixed reactions, since they are keenly aware of the spatial nature of the inequalities, yet equally skeptical of any naive argument for blanket uniformity. Regional policies which aim either at achieving greater uniformity or at avoiding unpalatable locational decisions (be they the choice of growth centers at a regional level or of sites for reservoirs at a local level) may well be justified on social grounds in the short term; but they may prove very expensive for society as a whole in the long run. A balanced assessment of the costs and benefits involved demands that we try and see the future spatial and ecological consequences of present decisions. In the Epilogue, we shall take a look at some first faltering steps in this direction.

Summary

1. Consideration of spatial variations in welfare is an important aspect of regional planning. Research into the definitions and geographic aspects of social welfare has raised a number of questions regarding the best way to measure differences between regions. Some regional inequalities may be measured graphically by use of a Lorenz curve. All countries of the world display Lorenz curves which indicate imperfect welfare distribution between their regions.

2. Even when spatial inequalities in social welfare have been defined, we still face the problem of solving the related but more important question, what a just spatial distribution of social benefits is. Three criteria are seen by Harvey as important in satisfying claims of disadvantaged regions: need, contribution to the common good, and merit. Each leads to different spatial patterns of help.

3. The problem of government intervention in the adjustment of regional inequalities may be helped by use of investment in the public sector, inducements to companies in the private sector, and inducements to individuals and households. Drawing the boundary between problem and nonproblem areas is difficult as productive regions may be expected to contribute resources, while the nonproductive regions receive large benefits. Investment is often restricted to those parts of problem regions with the greatest growth potential. Under the growth-pole concept, there is a spatial concentration of investment in selected nodes in order to bring about secondary growth in their hinterlands. Outmigration from problem areas has also been used as a solution in areas where persistent economic decline is seen as inevitable.

4. Local planning problems often develop out of needs for more intensive land use. Multiple land use is one solution to increasing intensity. There can be common use of the same land for two or more primary purposes, limited by compatibility of use; or parallel use through spatial separation of uses. The location of public facilities raises difficult problems of trading off local costs versus general benefits.

5. The position of the nation state as the most significant political unit for geographic study is being challenged by the growing importance of major multinational corporations. Industrial corporations are increasingly important agencies in shaping both between-state and within-state economic growth. Paradoxically, the government sector within the modern nation-state is itself taking on many of the characteristics of an industrial corporation. This is leading to a conflict of roles for government at both the regional and local scales of planning.

Reflections

1. How do geographers measure spatial inequality? Assume that a country is divided into five regions, each with 20 percent of the country's population but with unequal shares of its total income (say, 50, 25, 15, 8, and 4 percent, respectively). Plot this variation in regional incomes as a Lorenz curve. (Go on to plot actual curves for your own province or state, if you have the time.)

2. Gather data on the different levels of welfare in counties in your home state. Plot the values on a map and study the resulting distributions. Identify high and low areas, and attempt to explain their distribution.

3. Examine Figure 22-5. What kind of regional strategy would you favor for your own country? On what basis would you give some areas special aid? Which areas would receive this aid? Compare your list of needy areas with the areas suggested by others in your class.

4. List the arguments you would use to (a) support and (b) oppose a policy of strategic withdrawal of population from an area.

5. Examine the factors that determine the location of a major international airport. Should the local population of an area be able to veto the construction of an airport there?

6. What do we mean by "a just spatial distribution"? Can the geographer contribute to this debate?

7. Review your understanding of the following concepts:

Lorenz curves	revenue sharing	compatible land uses
inequality gaps	programs	benefit-cost analyses
social welfare	growth poles	eutrophication
indicators	strategic withdrawal	multinationals

One Step Further . . .

The general philosophical issues in regional-help programs are discussed in
Harvey, D. W., *Social Justice and the City* (Arnold, London, 1973), esp. Chap. 3.

Problems and issues in the United States are discussed in
Morrill, R. L., and E. H. Wohlenberg, *The Geography of Poverty in the United States* (McGraw-Hill, New York, 1971), and
Smith, D. M., *The Geography of Social Well-being in the United States* (McGraw-Hill, New York, 1973).

Many texts review national styles of regional economic planning. For a discussion of French and British styles, see
Boudeville, J. R., *Problems of Regional Economics Planning* (Aldine, Chicago, and Edinburgh University Press, Edinburgh, 1966) and
McCrone, G., *Regional Policy in Britain* (Verry, Lawrence, Mystic, Conn., and Allen & Unwin, London, 1969), Chap. 3.

For a discussion of the issues raised by planning with respect to land and water resources and biological resources in America, look at

White, G. F., *Strategies of American Water Management* (University of Michigan Press, Ann Arbor, Mich., 1969),
Tunnard, C., and B. Pushkarev, *Man-Made America: Chaos or Control?* (Yale University Press, New Haven, Conn., 1963).

The problem of the multinational corporations and their role in regional economic development is reviewed in
Lloyd, P. E. and P. Dicken, *Location in Space*, 2nd ed. (Harper & Row, New York, 1977), Chap. 9.

Regional problems are well covered in the regular geographic periodicals, but Antipode *(a quarterly) is the forum where the issues of welfare geography and spatial justice are argued out in the most challenging manner. You might also like to look at a few of the wide variety of specialized journals:* Regional Studies *(a quarterly), the* Journal of Regional Economics *(another quarterly), and* Papers of the Regional Science Association *(a biannual publication) are all interesting.*

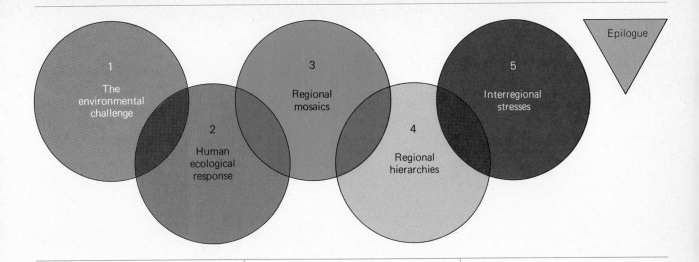

The Epilogue is concerned with the future in three senses. *Frontiers in Space* (Chapter 23) describes important and rapidly developing areas of geographic research. The discussion ranges from the use of remote-sensing devices in earth-orbiting satellites at one end to new techniques of map making at the other. *Frontiers in Time* (Chapter 24) describes the increasing concern of geographers with the multiple worlds of the future. We look at the techniques geographers can use to project the future spatial patterns of people on the planet on all scales from the global to the local, and we draw together some of the speculations on our future tenure and organization that have come up in earlier chapters. *On Going Further in Geography* (Chapter 25) is concerned with the future in a more personal sense. It has been written specifically for the student considering whether to pursue further courses in geography and tries to outline the main structure of the field, its development, its internal controversies, its job possibilities, and so on. It returns, after one complete whorl of the helix, to some of the issues raised in *On the Beach* (Chapter 1) and serves as a beginning to the more advanced courses that lie on the next and higher circuit.

epilogue

The Future Task

Huck Finn to Tom Sawyer in their flying boat: "We're right over Illinois yet. And you can see for yourself that Indiana ain't in sight. . . . Illinois is green, Indiana is pink. You show me any pink down there, if you can. No sir; it's green." "Indiana pink? Why, what a lie!" "It ain't no lie; I've seen it on the map, and it's pink."

MARK TWAIN
Tom Sawyer Abroad (1896)

chapter 23

Frontiers in Space

In the last four sections of the book we have been looking at the environmental challenge the earth poses for human beings. We have asked four questions. What is our ecological response? What is our cultural response, and how has it affected the mosaic of world regions? How have these world regions been shaped into a hierarchy of city regions? What conflicts and stresses have been set up between these regions? All four questions underline the critical need for more and more geographically relevant information.

Here we are faced with a paradox: growing information, yet growing uncertainty. Let us take the first point. In the 1980s the level of existing information is, of course, higher by many orders of magnitude than it was a century ago. From wide-ranging surveys of the growth of scientific information we know that information grows in an exponential fashion; that is, the greater the amount of information that exists, the faster it grows. Depending on what we measure, it is possible to estimate, roughly, that the amount of environmental information tends to double within a period of 10 to 15 years or so. If one accepts the general form of this growth curve, the amount of information available to geographers such as Alexander von Humboldt and Carl Ritter in the early part of the nineteenth century was about 1000 times smaller than the amount available to the current generation of geographers. To create the images produced in a few seconds in our opening photograph would have required decades of survey and calculation by an earlier generation of mapmakers.

Despite this dramatic increase in data, the demand for geographic information is likely to be far higher in the last quarter of the twentieth century than at any previous time. For, despite its impressiveness, the increase in total geographic information has scarcely kept pace with the growing realization of our ignorance about the earth. Shifts in the geographic focus of human activities—to the jungles of Cambodia, the subarctic steppes of the Alaskan North Slope, or the inner blighted areas of American cities—may show how sparse is the stock of information available. Also, shifts in emphasis on different resources have swung the spotlight to some areas where there is critical lack of information.

In this chapter we focus on two views of the information problem. First, we look at the world from outer space and examine the contributions of remote

sensing to our understanding of the world. Second, we look at the problem of mapping information and ask how we can formulate a valid picture of the whole world from the limited evidence of a few sample investigations. Both questions link back to some of the basic questions we asked in the prologue of the book.

23-1
Environmental Remote Sensing

The first man to try to photograph the earth's surface from a balloon was probably a Parisian photographer, Gaspard Félix Tournachon. His attempts to capture "nothing less than the tracings of nature herself, reflected on the plate" date from 1858. The mania for balloon photography caught on (Figure 23-1), and by July 1863 even Oliver Wendell Holmes was conceding that Boston "as the eagle and wild goose see it" was a very different place from that seen by its solid citizens on the ground.

As we saw in Chapter 2, it was World War I, with the advent of aircraft and the widespread use of military intelligence, that converted aerial photography from a pastime into a program. The war led to a growing interest in aerial photography for resource surveys from the 1920s onward. And over the last 20 years the development of satellites has provided a new dimension of airborne surveillance. Artificial satellites circle the earth along orbital paths and stay in space for lengths of time dependent on their size and distance from the earth.

Figure 23-1. The beginnings of remote sensing.
Jacques Ducorn using a dryplate camera from a balloon in 1885. [From Gaston Tissandier, *La Photographie en Ballon* (Gauthier-Billars, Paris, 1885).]

The impact of this remote sensing on the provision of geographic information has come from two factors: first, the rapidly extending range of sensors, and second, improvements in the spacecraft themselves.

Improvements in Sensors

Photography   *Sensors* are instruments used to detect the electromagnetic energy associated with a particular object on the earth's surface. Black and white (panchromatic) aerial photography has for long been the principal sensing technique for the geographic study of the earth's surface. This uses the energy reflected by visible light falling on that surface. However, during the last 30 years there have been significant extensions in both the range and the capability of sensors. The array of remote-sensing systems now available has been expanded by the exploitation of different parts of the electromagnetic spectrum. (See Figure 23-2.)

Developments in photographic chemistry have vastly extended the amount of information that can be captured on photographs. For example, look at the series of photographs in Figure 23-3. The subject is the same in all three photos, but the film picks up and accentuates different aspects of the scene. Black-and-white panchromatic film (used in the second photo) is sensitive to all colors, whereas black-and-white orthochromatic film (used in the third photo) is sensitive to all colors except red. Thus the panchromatic film represents all colors as shades of gray, and the orthochromatic film records red objects as black. New types of film are sensitive to electromagnetic waves on the borders of visible light.

Geographers' interest in variations in land use makes them especially concerned with films that can differentiate between vegetation colors, notably shades of green. Infrared photography uses film that picks up varieties of tone and color that are not evident in either black-and-white or true-color photographs. Figure 23-4 shows how the reflective properties of forest-covered terrain vary for emissions of different wavelengths. Differentiation between the two types of forest is clearly much easier than in the visible sections, where the properties of the two types are similar and partly overlap. Most landscape features appear to have similar distinctive spectral signatures. These *signatures* are the unique pattern of wavebands emitted by a particular environmental object.

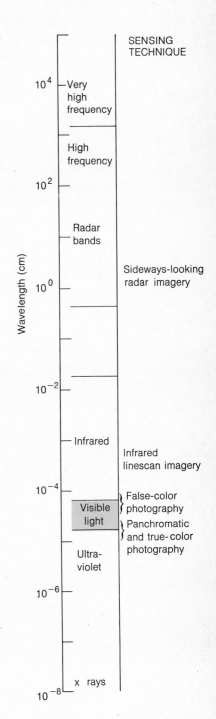

Figure 23-2.   Sensing techniques and the electromagnetic spectrum.
Remote sensing was at first confined to emissions of visible light, and conventional photographic techniques were used. Equipment in use today can also detect infrared and radar waves. As more sophisticated sensors become available, an ever-wider range of electromagnetic emissions from the earth's surface can be recorded and mapped. *False color* photographs are used to emphasize and separate important features by giving them distinctive and contrasting colors. In such photos, healthy vegetation, for example, may show up as bright red rather than green.

(a)

(b)

(c)

**Figure 23-3. Film types and
landscape images.**
Photographs of mixed woodland using (a)
infrared, (b) panchromatic, and (c)
orthochromatic film. Note the differing
capablility of the films to reduce haze and
differentiate broadleaf from needleleaf
species. Compare these photos with the
chart in Figure 23-4. [Photos by Tony
Philpott.]

Other Sensors Newer, nonphotographic sensing systems have advantages
and liabilities when it comes to the analysis of terrestrial phenomena. *Radar
sensors* direct energy at an object and record the rebounded energy as radio
waves. There are two advantages of radar images over conventional photographic images for mapping purposes. First, radar imagery is independent of
solar illumination and is unaffected by darkness, cloud cover, or rain. This
means that any part of the globe can be scanned on demand, including the
zones especially difficult or impossible to photograph by conventional methods—such as the cloud-covered humid tropics and regions of polar night. The
time available for radar scanning in the humid midlatitudes is five or ten
times greater than that available for taking aerial photographs of acceptable
quality. Second, radar imagery gives greater detail of the terrain. For example,
radar images on a scale of about 1:200,000 provide information on drainage
patterns which is roughly equivalent to that derivable from a 1:62,500 topo-

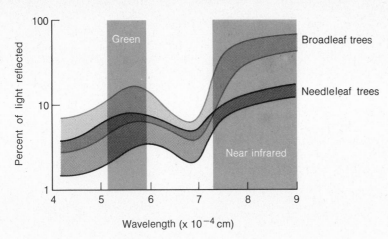

Figure 23-4. Image separation at different wavelengths.
The light reflected by the foliage of broadleaf and needleleaf trees is shown by the two broad curves. Note that the curves for the two types of foliage slightly overlap at the shorter wavelengths of visible light, making foliage difficult to separate on ordinary panchromatic or orthochromatic film. However, the two curves are widely separated in the longer wavelengths of near infrared light. It is therefore easier to map the different species using infrared film. [After R. N. Colwell, in R. U. Cooke and D. R. Harris, *Transactions of the Institute of British Geographers,* No. 50 (1970), p. 4, Fig. 3.]

graphic map. High-resolution radar can pick up very fine irregularities in the ground surface, and even show differences in subsurface conditions to a depth of a few meters. Figure 23-5 gives an example of a radar image of farmland in Kansas. The various crops appear as different shades of gray; the lightest shades indicate sugar beets.

Improvements in Satellites The first American satellite (Explorer I) was launched in January 1958. The first generation of observation satellites was formed by the eight members of the Tiros (Television and Infrared Observation Satellites) family launched between 1960 and 1963. As their name implied, the satellites carried two types of sensing devices. First, there were television cameras that transmitted pictures of the visible part of the spectrum back to earth. The first detailed weather pictures of cloud patterns were received as early as 1960, and Tiros III discovered its first hurricane (hurricane Esther) in July 1961. Second, Tiros carried infrared detectors that measured the nonvisible part of the spectrum and provided information on local and regional temperatures on the earth's surface.

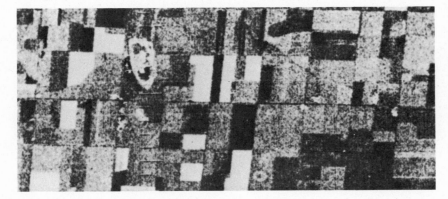

Figure 23-5. Radar images of landscapes.
Farmland in Kansas as seen by radar. Different crops are distinguished by various tones. The lightest areas are planted with sugar beets. [Photo by Westinghouse Electric Corporation.]

TERMS USED IN REMOTE-SENSING STUDIES

Bands are sections of the electromagnetic spectrum with a common characteristic, such as the visible band.

Enhancement refers to processes which increase or decrease contrasts on received images (e.g., photos) so as to make them easier to interpret.

Ground truth is information about the actual state of any environment at the time of a remote sensing flight overhead.

Imagery is the visual representation of energy received by remote sensing instruments.

Line scanning produces an image by viewing and recording a picture one line at a time, as on a cathode-ray tube (or TV set).

Multispectral sensing is the recording of different portions of the electromagnetic spectrum by one or more sensors.

Platforms are objects on which a remote sensor is mounted, usually an aircraft or satellite.

Radar is a sensor which directs energy at an object and records the rebounded energy as radio waves.

Resolution is the ability of a remote sensing system to distinguish signals which are close to each other, in time, space, or wavelength.

Sensors are instruments used to detect the electromagnetic energy associated with a particular object on the earth's surface.

Signatures are the unique pattern of wavebands emitted by a particular environmental object.

Synoptic images are those giving a general view of a part of the earth's surface, usually from high-altitude satellites.

Thermal infrared records the thermal energy *emitted* by objects of different temperatures on the earth's surface.

From a geographer's viewpoint, the most important development in global satellites was the ERTS. This satellite was launched from California in July 1972. It makes 14 revolutions a day around the earth, its sensors covering a series of 160-km (100-mi) wide strips. The strips overlap, so that the whole surface of the earth is covered once in every 18 days. Thus in the first year of the satellite's life each part of the planet came within range of its sensors 20 times. Of course, many areas were cloud-covered, but it is estimated that in the first year the ERTS provided cloud-free coverage of about three-quarters of the world's land masses. Pictures reaching the satellite are converted into electronic signals, stored on tape, and then broadcast back to three ground stations, at Fairbanks, Alaska; Goldstone, California; and Greenbelt near Washington, D.C.

Three years later ERTS-1 was joined by a second earth resources satellite. With the launching came a change in name and the two satellites are now named LANDSAT-1 and LANDSAT-2. Since each is following an orbit 800 km (500 mi) above the earth's surface the area covered on a single frame is correspondingly large—some 30,000 km² (11,600 mi²) or an area roughly the size of Maryland and Delaware combined. Despite this, the image can be "blown up" in photographic terms to show an object the size of the Washington Monument in Washington, D.C.

Unlike the images recorded by earlier satellites, the images received by LANDSAT are passed through a *multispectral scanner* which is sensitive to different parts of the electromagnetic waveband which we have already met (Figure 23-2).

Atlas cross-check.
At this point in your reading, you may find it useful to look at the LANDSAT color photos in the atlas section in the middle of this book.

Each spectral band picks up different information about the earth surface below. For example, water produces a distinctive spectral signature. It gives medium values on the green and red bands but low and very low on the two infrared bands. In musical notation, water would have a sound ♫ ! By contrast, pastureland has the sound ♫ ! Note that both look alike on the first two notes but differ on the third and fourth. While not all features have such clearly defined signature tunes, the principle of using information from all four bands remains. This means that the signals transmitted to earth can be recombined in many ways to bring out unsuspected features of the planet's surface. For example, when the signals are decoded and exposed on photographic film, they may produce false color images to emphasize and separate important terrain features. For example, vigorously growing vegetation may show up as bright red and diseased crops as pale yellow. Clear water may show up as black, while contaminated water carrying silt and sewage may appear bright blue.

The use of satellites for remote sensing is clearly still in its infancy. At the time of writing work is going ahead in the United States on both a space shuttle and spacelab program which will provide further opportunity for direct observation by scientists of the earth's surface. The NASA organization is paralleled by a major space-research program in the Soviet Union while the Western European countries have combined since 1975 to form ESA, the European Space Agency.

23-2 Monitoring Environmental Change

We have shown earlier how aware geographers must be of environmental change. (See Chapters 5 and 6.) Remote sensing from satellites provides unique opportunities for monitoring this change on a scale unthinkable to an earlier generation of researchers. Here we look at some potential areas of research and then examine more closely a regional example.

Potential Research Frontiers   From the geographic viewpoint, satellites are immensely useful because of their very wide coverage and the fact that difficult or inaccessible terrain presents no bar to data collection. Three areas of research applications seem exceptionally promising at this time.

The first is studies that take advantage of the potential *worldwide coverage* of satellite systems. Typical are proposals for a worldwide study of surface temperatures using infrared scanning systems (Figure 23-6). For despite a growing network of information on atmospheric temperatures, the data on temperatures of the ground surface itself are rather sparse and generally available only for developed areas. Records of daily and seasonal changes in temperature would tell us more about the heat and water balances between the earth and the atmosphere. Such data bear directly on human needs by helping to identify areas where the temperature is suitable for certain crops.

There is a similar lack of worldwide information on precipitation. Rainfall is normally recorded daily by meteorological bureaus' gauges scattered over the earth's surface. Satellites promise the rather exact location of rain and

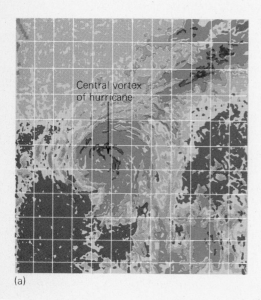

(a)

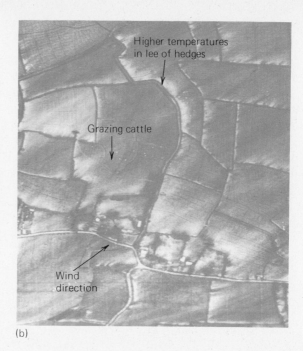

(b)

Figure 23-6.   Heat landscapes.
(a) Infrared data on Hurricane Camille (1969) with differences in temperature showing up as different shades. Note the characteristic vortex shape with violent winds moving around a calm central area. [Photographs courtesy of NASA.] (b) The image shown by an infrared linescan sensor of the river Axe lowlands, southwest England. Note shelter effect of hedges showing in light tones of those ground areas with higher temperature. Grazing animals are seen as light spots by reason of their high body temperature. A conventional air photograph of the same general area is shown in Figure 2-11(c). [Courtesy Dr. L. F. Curtis, Bristol University, and Royal Signal and Radar Establishment, Malvern.]

snow as they are actually occurring. Through links with conventional ground stations and surface radar stations, the type and intensity of precipitation can be measured, and improved forecasts can be given.

An ideal system of global watching would need a system something like that shown in Figure 23-7. This consists of three elements. First, geosynchronous satellites would continuously swing around in time with the earth's rotation and cover its low-latitude areas (effectively latitude 40°N to 40°S). Four are shown, but if this number could be increased, then their orbits could be nearer to the earth and their revolving power improved. Second, nongeosynchronous satellites would cover the middle and upper latitudes. Two are shown following a polar orbit at lower altitudes. Finally, a low level satellite in equatorial orbit would provide the fine detail not available from the more distant geosynchronous platforms. Provision to complete such a system by the 1980s has been made by United Nations organizations as part of its World Weather Watch program of global atmospheric research.

Another proposal for taking advantage of the global coverage of the satellite system relates to the completion of the World Land Use Survey and to cloud interpretation studies. The World Land Use Survey was initiated by British geographer Sir Dudley Stamp to provide a series of maps, on a scale of 1:1,000,000, of the human use of the earth's land surface. Examples of cloud interpretation studies are presented in Figure 23-8.

The second promising area of research is the study of phenomena that occur mainly in *inaccessible* and therefore *sparsely monitored* parts of the

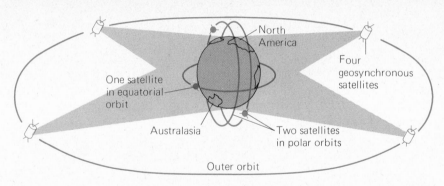

Figure 23-7. Worldwatch surveillance systems.
A proposed global observing system for atmospheric surveillance using seven satellites. These are then linked to conventional weather observations. [From work by V. E. Suomi reported in E. C. Barrett and L. F. Curtis, *Introduction to Environmental Remote Sensing* (Wiley, New York, 1976), p. 160, Fig. 9.3.]

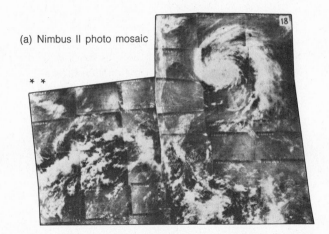

(a) Nimbus II photo mosaic

Figure 23-8. Cloud analysis using satellite photographs.
(a) Nimbus II photographs of the central American region, June 11, 1966, compiled into a photo mosaic. (b) The identification of different cloud types, each characteristic of different meteorological conditions. This is termed nephline analysis. The line of crosses (+ + + + +) shows the position of the intertropical convergence on this day. (See Figure 4-5(c) for the average position of the convergence zone.) (c) The smoothing of nephlines to indicate the main patterns of air circulation. [From E. C. Barrett, *Progress in Geography* **2** (1970), pp. 192–193, Figs. 19, 20.]

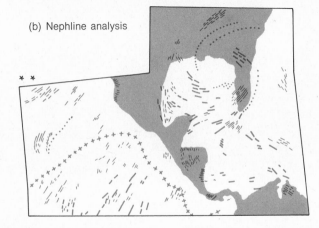

(b) Nephline analysis

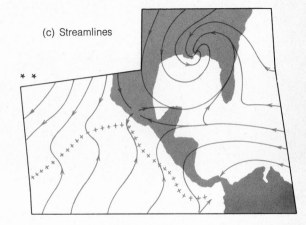

(c) Streamlines

earth's surface. Most of the ice masses of the world, for instance, are located either in polar or high mountain areas where both ground and airborne surveys pose logistical problems. With satellite surveillance we can regularly record changes in the extent of ice masses, their surface characteristics, or the presence of new snow deposits. Such studies are of value because glaciers contain three-quarters of the world's fresh water and are critically connected to short-term changes in the earth's hydrology and climatology, as well as to long-term changes in sea levels. Studies of glacial budgets (that is, gains or losses in the volume of ice in a glacier or ice sheet), and rates of iceberg measurement at the seaward margins are all feasible with existing satellite and sensor technology. Work is going ahead in this field.

The third promising area of satellite research is the study of *ephemeral* or *highly mobile* phenomena that cannot be recorded with present survey techniques. We can include in this area proposals to monitor the distribution of bush fires and their effect on natural vegetation. High-resolution images with sharp detail also can record the occasional, but ecologically critical, human use of wildlife and recreational areas. Photographs from aircraft are already helping to determine the location and intensity of vehicle parking in national forests and wildlife areas. Satellite photographs also can check on the increasing pressures on these areas and give us early warning of their overuse.

More modest projects use spacecraft for recording changes in the distribution of airborne sediments (e.g., dust clouds and pollution), movements of coastal sediment, movements of traffic in metropolitan areas, and the distribution of seagoing craft. One particularly intriguing suggestion in the realm of historical geography is that sensing devices could help us reconstruct caravan networks in the western Sahara and Takla Makan deserts. Trails frequented by animals have a surface composition and chemical content unlike the untrodden and unfertilized terrain around them, and it seems possible that such differences are detectable by appropriate sensors.

Early-Warning Systems   Keeping the earth under routine surveillance from satellites may allow the development of early-warning systems. Let us look at an example of a United Nations Food and Agriculture Organization (FAO) program. Since the earliest Egyptian records, locusts (Figure 23-9) have been known to erupt in massive swarms, devastating crops over hundreds of square miles. The Old Testament describes them as one of the plagues which descended on Pharoah's Egypt. Although this member of the grasshopper family lives in the arid and semiarid lands of the world, its breeding cycle is critically dependent on rain. Rains supply the moist soil conditions which allow the insect to get through the 40- to 100-day gap between egg-laying and the adult stage when locusts then emerge in massive swarms. These are both hungry and highly mobile, having been known to move distances as far as 3,500 km (2,200 mi) in as little as three weeks.

Information about rain in the potential breeding grounds in the desert margins of the Sahara is of vital interest to the FAO locust control network

(a)

(b)

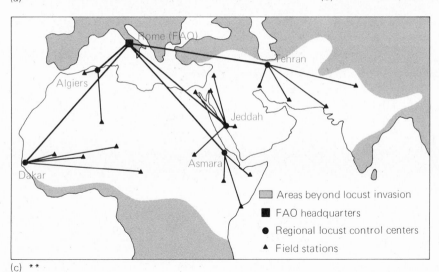

(c) ******

Figure 23-9. Use of remote sensing in pest control.
(a) Locust swarm. (b) NASA photo of conditions over the Persian Gulf (black areas) and adjacent land areas (gray). On this day, dry weather dominated with a band of cirrus clouds (white) not leading to rainstorms. (c) Network of locust control centers in North Africa and the Middle East based on the United Nations' Food and Agriculture Organization (FAO) head-quarters in Rome. [Photo (a) by Wide World Photos and photo (b) from E. C. Barrett, Consultant's Report to the Food & Agriculture Organization of the United Nations on Assessment of Rainfall in North Eastern Oman, Bristol, January, 1977.]

(Figure 23-9). But information is hard to gain since these areas are very sparsely peopled, there are few meteorological stations, and when rains do come they fall as a random pattern of intense but highly localized storms. So studies have been started to try and use environmental remote sensing. In southern Algeria and central Arabia LANDSAT records have been used to identify areas where *rains are probable* (based on the cloud characteristics), and where *rains have recently fallen* (based on the green "flush" of plant growth which comes after the rains). Once key areas have been identified from the photos, then the remote sensing team at Rome can alert the regional headquarters of the locust control program. Control officers then fly over the areas with light aircraft, or conduct ground surveys in the more accessible areas. If the presence of locusts in their larval stages is confirmed, then control measures can be put under way.

The early-warning program is still in its infancy but, if successful, this application of remote sensing may help to save millions of dollars in crop loss.

23-3
Mapping Geographic Space

Since the Greek geographers first conceived of the planet Earth as a sphere and set about measuring its dimensions, one geographic goal has been to fill in the world map as accurately as possible. As we have seen in this chapter, the advent of remote sensing has allowed that goal to be achieved. One might imagine that, as the last blank areas in the world map are filled in, all other geographic puzzles will be solved too. No longer can we debate, as did our forebears, the location of the source of the Nile or whether a northwest passage between the Atlantic and Pacific Oceans really existed.

But even though one set of spatial puzzles has been solved, new ones have been uncovered. As we have just noted in earlier chapters, geographers are becoming increasingly aware of mental maps. We have also seen the ways in which travel times and transport costs can crumple and distort the familiar world map. The conventional world map describes a space that is continuous, isotropic (i.e., movement is equally possible in all directions), and three-dimensional. Nevertheless, the real space in which people move is discontinuous, anisotropic (i.e., the costs of movement vary markedly over the map), and change rapidly over time—which means it has four dimensions rather than three. Mapping this real space poses fundamental questions for map-makers and requires a reassessment of conventional Euclidean geometry. (See the discussion of non-Euclidean space on the next page.)

Spatial Transformations   Some analysts view this break from traditional mapping as paralleling the changes in theoretical physics when concepts of *absolute space* (in which the spatial coordinates have a fixed structure) were replaced by *relative space* (in which the coordinates reflect the structure of what is being described). The distinction between the two is shown by the maps of Sweden in Figure 23-10. Here we see a transformation from a rectangular coordinate system in absolute space into one in which locations are plotted relative to their direction and distance from a given center. Swedish geographer Torsten Hägerstrand used this transformation to describe the migration field of a small parish in the south central part of his country. Most of the migration movements were over short distances, but a few migrants traveled large distances, such as 5000 km (3000 mi) to the United States. Local movements could be clearly distinguished only if they were plotted on a large-scale map, but movements to the main Swedish cities and to overseas countries needed to be plotted on small-scale maps. The problem of matching the different scales was overcome by using a special projection centered on the village from which migration was occurring. In this projection all radial distances from the center are transformed to logarithms,

NON-EUCLIDEAN SPACE

The geometric concepts synthesized in Euclid's *Elements* (written about 300 B.C.) form the basis of geographical measurement of the globe. Consider, for example, the distance between locations A and B on a plane.

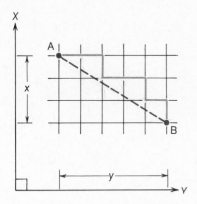

In Euclidean space the distance between these two points (d_{AB}) is given by the Pythagorean theorem as

$$d_{AB} = \sqrt[2]{x^2 + y^2} = \sqrt[2]{3^2 + 5^2} = 5.8 \text{ units.}$$

The variables x and y measure the differences between the two locations. If we superimpose on our continuous plane a Manhattanlike grid of streets, the distance from A to B becomes

$$d_{AB} = \sqrt[1]{x^1 + y^1} \quad \text{or} \quad x + y = 8 \text{ units,}$$

as we can no longer walk directly from A to B.

When we compare the formulas for estimating the distance between the same two points in Euclidean space and "Manhattan" space, we see that the differences lies in the exponents;

these have a value of 2 in the first case but 1 in the second. Formal geometries have been developed to handle spaces where distances are both greater and less than those given by the Pythagorean theorem, but so far these non-Euclidean geometries have been little explored by geographers. See D. W. Harvey, *Explanation in Geography* (Arnold, London, 1969), Chap. 14.

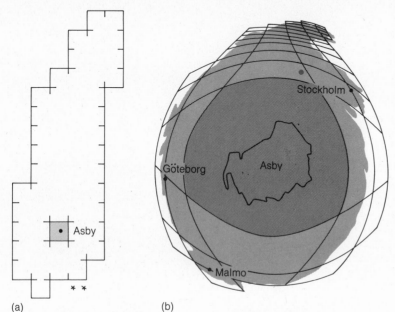

(a)   (b)

Figure 23-10. Sweden: a migrant's view.

A conventional map of Sweden (a) differs greatly from the same space as viewed by migrants from a single Swedish parish. The map of Sweden centered on the parish of Asby (b) can show population migration over both short and long distances. Note the correspondence between the central cell with curved boundaries and the corresponding square cell on the conventional map. [From T. Hägerstrand, *Lund Studies in Geography B*, No. 13 (1957), p. 73, Fig. 38.]

while all directions remain true. The result is a map on which local and international movements can be plotted on a single sheet. Although the transformed map of Sweden is unfamiliar and distorts its familiar outline, it reflects a limited field of space rather accurately.

This simple illustration shows just one of the ways maps can be transformed for different geographic uses. Considerable geographic research effort is going into ways of plotting far more complex relations (e.g., the "time maps" of New Zealand and the South Pacific shown in Figure 14-9). Since this research area is rather mathematical, we must leave its consideration to more advanced courses.

Computer Mapping   The growing need for environmental information in new map forms has thrown special emphasis on the role of the electronic computer in map production. Although digital computers were available around 1945, only in the last decade or so have geographers exploited their enormous capability for the storage, transformation, and retrieval of vast amounts of environmental data.

Figure 23-11 presents a typical computer map. Many are produced according to a system pioneered by a group at the Harvard Laboratory for Computer Graphics. The laboratory's work has been built largely on the Synagraphic Mapping Technique (SYMAP), a set of programs for creating a diversity of maps by combining alphabetical and numerical keys in fast-line

Figure 23-11. Computer cartography. This three-dimensional map, produced on the high-speed graph plot of a computer, shows the percentage of blacks in the population of different areas of New Haven, Connecticut, in 1967. By examining different views of the map and changing the vertical scale, the image best suited to emphasize the spatial distribution can be selected. Usually alternative maps are examined on a cathode-ray tube display screen before the required map is chosen for printing. [From D. L. Birch, *Journal of the American Institute of Planners* **37** (1971), p. 85, Fig. 5. Reprinted with permission.]

printers. By superimposing combinations of numbers and letters on the printer, one can develop a series of "tones" that simulate the gray scales of conventional isarithmic and choropleth maps. (For example, if you print a single character, say, an "O," this gives a very light tone. If you overprint "X" onto "Z" and then onto "H," you get something which gives a much darker tone.) Maps can be produced quite rapidly. If fact, geographers call for maps of environmental conditions from a stored set of data (a data bank) and have them within a few seconds.

The fastest and inherently most flexible output is provided by cathode-ray tube (CRT) displays. Maps are displayed on a television tube and can be photographed to provide permanent copies. In addition to the line printer and the CRT display are numerous incremental graph plotters that can quickly produce maps and graphs from data stored in the computer.

In addition to being used for direct mapping, digital computers are increasingly being relied upon to prepare spatial data in map form. For example, the transformation of map coordinates from a system of Cartesian coordinates to one of polar coordinates is a trivial task even for mechanical calculators. The digital computer allows much more difficult transformations of geographic data from one map projection system to another. This means special projections can be tailored to particular jobs; we can draw a map for New Yorkers on a projection system centered on New York, a map for Muscovites centered on Moscow, and so on.

Yet another research byway that the computer has opened is in historical geography. By reading in the coordinates of places given on old maps and charts and comparing them with the true coordinates, we can re-create the original projection system used on the old maps. This allows us to estimate the probable location of settlements long since abandoned and provides some clues for archeological research. The use of this method for locating the position of old shipwrecks, however, has yet to be reported!

Mapping From Samples   In their drive to make valid worldwide statements about the earth as the home of the human species, geographers often assume the need to inspect all corners of the world and to make a complete environmental inventory. When we compare our situation with that of other investigators, we see that this assumption can be relaxed. The geologist makes inferences about rock strata from the records of a few bore holes; the pollster makes predictions about our voting behavior by interviewing a few thousand citizens, not millions.

In a way the earth's surface is analogous to a population. To be sure, the population is an unusual one because it does not consist of a finite number of individuals (for instance, the 212 million individuals who make up the U.S. population) but is continuous. Therefore, when we want to designate an individual within this spatial population, we have to do so in an arbitrary way, by specifying a point location like 68°30′N 27°07′E, an areal unit like 1 km $\times$ 1 km, or a census division like a tract or county. These individuals, however defined, form *sampling units,* from which we can construct a picture of the population as a whole.

Making inferences about a population from a small part of it is a dangerous procedure. We may select samples that are biased in some way or the sample may be too small for a reliable estimate. But how do we decide what is "too small"? Fortunately, many of the rules governing the relationships between a sample and a population were worked out by mathematicians like Karl Pearson and R. A. Fisher in the first half of this century. Today there is a large body of well-substantiated sampling theory to which geographers can turn in setting up their investigations. Sampling theory helps us to decide how much error is likely to occur with sample designs of different kinds. For example, in simple *random sampling* (Figure 23-12), the accuracy or sampling error is proportional to the square root of the number of observations. This means that if we were to increase the number of sampling points in Figure 23-12 from 25 to 100, we could expect to improve our accuracy not by four times as much, but by only $\sqrt{4}$, or 2.

Geographers design their sample surveys of environmental characteristics in close cooperation with their colleagues in statistics, and most of the

Figure 23-12. Simple random sampling.

It is frequently too costly to inspect every part of a region before making generalizations about it, so geographers rely increasingly on a series of sample observations. But how should such observation points be selected? To avoid bias, the location of each sampling point within the watershed shown here is given a pair of random numbers on the two axes of the coordinate system.

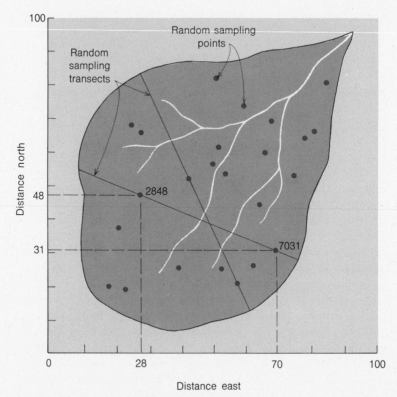

SAMPLING DESIGNS

Geographers use different sampling designs to investigate particular spatial distributions. Each involves randomization procedures.

Stratified sampling divides the study area into separate strata, and individual sampling points are drawn randomly from each strata. The number of sampling points is made proportional to the area of each strata. Stratification can be used when major divisions within the area are already known to the investigator.

Systematic sampling employs a grid of equally spaced locations to define the sampling points. The origin of the grid is decided by the random location of the first sampling point. The grid also can be randomly oriented. Systematic location gives an even spatial coverage for mapping purposes.

Nested, hierarchic, or *multistage* sampling divides the study area

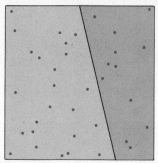

(a) Stratified sampling

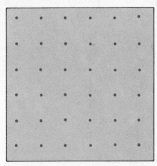

(b) Systematic sampling

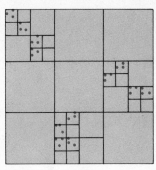

(c) Nested sampling

into a hierarchy of sampling units that nest within one another. Random processes select the large first-stage units, the smaller second-order units within these, and so on. The final location of points within each small unit can be randomly determined. Nested sampling is useful in reducing field costs and in investigating variations at different spatial levels within the same area but it entails higher sampling errors. For a discussion of the applications and limitations of these and other designs see B. J. L. Berry and D. F. Marble , Eds., *Spatial Analysis* (Prentice-Hall, Englewood Cliffs, N.J., 1968). Chap. 3.

---

problems they encounter are general statistical ones we need not be concerned with here. Geographic research has been confined mainly to working out the efficiency of different kinds of spatial arrays of sampling points. Each spatial sampling design has particular disadvantages and advantages and is used for specific types of environmental investigations. (See the above discussion of sampling designs.)

Conclusion: The Unfinished Map　Cartography was one of the earliest areas of research by geographers and played a key part in the geographic training of the Greeks. In recent years its study had declined and it has become something of a Cinderella subject in geography departments. The world map appeared to be nearly complete, and there were few discoveries to be made: *Terra Australis Incognita* (the "unknown southern land") had been added to the map in the eighteenth century, darkest Africa had been opened up in the late nineteenth century, and the polar areas and ocean floors have been explored in our own century. Today the situation has changed. Just as one mapping task appears to have been successfully completed, another and more difficult one of mapping in new kinds of economic and social space has appeared. The research frontier of inner space has succeeded the closed frontier of outer space. Geographers find that they are back with the Greeks, pondering just what kind of spatial world they are really living in.

Summary

1. The total amount of geographic information has increased enormously over the past century, doubling every decade or so. Paradoxically, greater knowledge of the earth has led to a keener appreciation of the gaps that exist in this information.

2. One of the major new ways of plugging this information gap is through the use of remote sensing. Various types of sensors have been developed for remote sensing from both aircraft and satellites. Photographic sensors include the use of panchromatic, orthochromatic, and infrared films. Nonphotographic sensing systems such as high-resolution radar are used to solve the visibility problems posed by cloud cover and darkness.

3. Satellites in orbit around the earth enable photographs to be taken of the entire earth's surface once every day. Of particular interest to geographers is the earth resources technology satellite (ERTS-LAND-SAT) whose signals are passed through a multispectral scanner to produce special-use photographs of earth features. Spacecraft enable us to increase our worldwide coverage, to study inaccessible or sparsely monitored areas, and to monitor ephemeral or highly mobile phenomena.

4. Remote sensing has helped solve the need for some kinds of world maps only: the need for up-to-date economic and cultural maps remains as acute as ever. Attempts by geographers to map economic change in space–time dimensions has created new problems not solvable by ordinary Euclidean geometry. In this new kind of mapping the older concepts of absolute space are abandoned and the concept of relative space is used in which locations are plotted relative to their direction and distance from a given center.

5. New map forms and the large quantities of data on the environment have led to the development of computer mapping programs such as the Synagraphic Mapping Technique (SYMAP).

6. Sampling techniques are widely used as a basis for mapping and for collecting environmental information, thereby overcoming the need for complete surveys.

Reflections

1. Try to obtain some satellite photos of your own area on different scales.* What kinds of information do these photos show that cannot be obtained from more conventional maps?

*Satellite photos are available from the EROS Data Center (Sioux Falls, South Dakota 57198, USA).

2. Radiation-sensitive sensors now enable satellites to capture differences in *heat* on the earth's surface on film. List ways in which this capacity provides valuable information for the geographer.

3. There has been much criticism of the amount of federal money invested by NASA in the space program. Set up a debate in class on the value of this investment. On which side would you wish to speak? Why?

4. Consider the view that, now most of the world is mapped, the geographer can hang up his or her boots. What is the fallacy in this viewpoint?

5. Review your understanding of the following concepts:

remote sensing	isotropic and
infrared sensors	anisotropic space
radar images	Euclidean space
false color photographs	random sampling

One Step Further . . .

An introduction to remote sensing, stressing its geographic applications, is given in
Barrett, E. C., and L. F. Curtis, *Introduction to Environmental Remote Sensing* (Wiley, New York, 1976).

Excellent examples of earth photographs from spacecraft are available in a number of NASA publications, for example,
National Aeronautics and Space Administration, *Earth Photographs from Gemini III, IV and V* (Government Printing Office, Washington, D.C., 1967), and
Short, N. M., and others, *Mission to Earth: Landsat Views the World* (Government Printing Office, Washington, D.C., 1977).

Some of the problems involved in handling the more complex kinds of geographic space are reviewed in
Harvey, D. W., *Explanation in Geography* (St. Martin's, New York, 1970), Chap. 14.

Although the results of current research are given in the regular geographic journals, you will need to look at specialized technical journals such as Photogrammetric Engineering *and* Remote Sensing of Environment *(both quarterlies) in order to keep abreast of the most recent developments in remote sensing.*

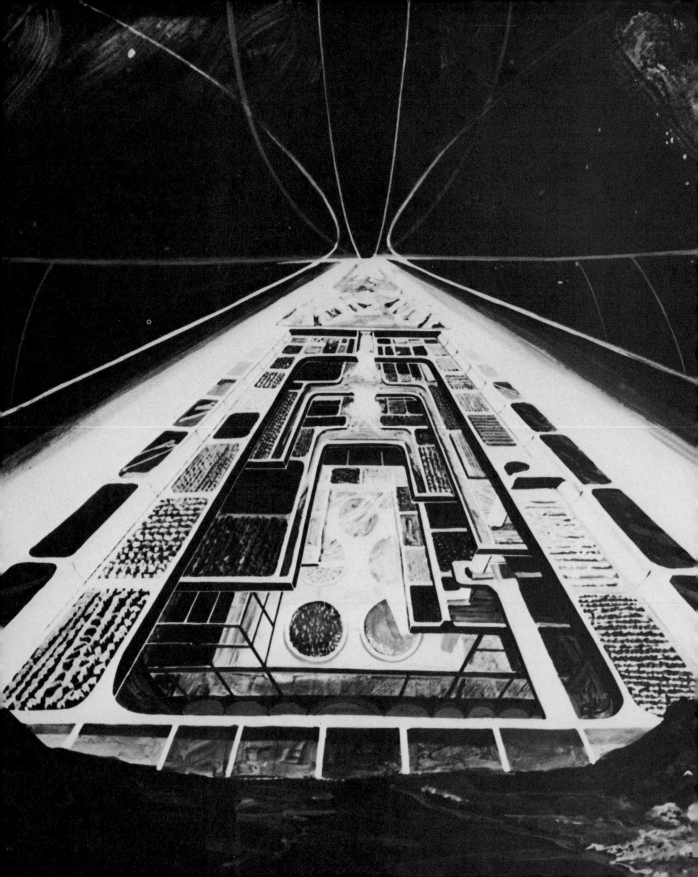

After performing the most exact calculation possible in this sort of matter, I have found that there is scarcely one tenth as many people on the earth as in ancient times. What is surprising is that the population of the earth decreases every day, and if this continues, in another ten centuries the earth will be nothing but a desert.

MONTESQUIEU
Lettres persanes (1721)

Just over 100 years ago the purchase of Alaska by the United States government was a hot issue in Congress. Representative Cadwalader C. Washburn had no doubts that it was a bad buy! The *Congressional Globe* for 1868 reports his statement that "the possession of this Russian territory can give us neither honor, wealth, or power." (North Slope oil resources were still a century away.) Senator Samuel White had said much the same in 1803 about the Louisiana Purchase ("the greatest curse that could ever befall us"). Looking back over the track record of statements on the future potential seen in different parts of the earth does not fill the reader with confidence. On the whole forecasters have consistently failed to get the balance right. Edens have turned out to be deserts, and deserts, Edens. How far do you consider the NASA artist's conception of an agricultural area of a space colony (opposite) is or is not an accurate forecast?

Despite the difficulties, geographers continue to be fascinated by the task of charting the future potential of this planet. We have already seen some of the questions they ask. How much more food can particular regions and biomes produce? How much larger will world population become? Will the cities of America's eastern seaboard link up into a megalopolis? Will states be successful in reducing inequalities among themselves and within their own territories? These questions have no precise answers at the present time. For while the geography of the past is accomplished and unchangeable, future geography is still fluid.

But it is precisely in this uncertain area that many of our most pressing spatial and ecological problems lie. So in this chapter we will see how geographers set about trying to reduce that uncertainty by narrowing the zone within which the most probable futures lie. We begin first with some general characteristics of forecasts and then look at very short-term forecasts. Then we consider what chance have we of look-

chapter 24

Frontiers in Time

ing still further ahead toward the end of the century? Finally, we ask what effect the growing interest in forecasting has on the way geographers go about their work. We have already met some aspects of forecasting—for example, in our study of population growth in Chapter 7 and resource uses in Chapter 9. Here we try and draw some of these threads together in looking at the way geographers are responding to the challenge of the future task.

24-1
Some Principles of Forecasting

Forecasting is a risky business, and it gets riskier the further ahead we try to look. Let us take the example of the spatial distribution of the United States population. We know what it looked like in 1970 because of the census taken that year. Given information on trends in births, deaths, and migration, we could predict with some accuracy the 1980 map. But the 1990 map poses more problems and is likely to be less accurate. Beyond that date our forecasts get progressively riskier. Will many more of us be living in Colorado or Florida in 2030? How important will New York City be as a population center in 2080? The only honest answer is that nobody now knows.

Conditional Assumptions   In the absence of certain knowledge we have to make what appear at the time of the forecast to be reasonable assumptions. In the case of population we begin with the advantage of a long history of population records—since 1790 in the case of the United States. As the human being is an animal, we also know something about her breeding patterns and his biology—for instance, we know that 206 babies are needed on average to produce 100 girls since there are slightly more male than female births. But these areas of certainty are swamped by what we don't know. What will happen to preferred family size? Will marriage and childbearing be postponed until later in life? Will the present dearth of children in western countries be followed by a baby boom in the last decades of this century? Will nonwestern countries follow western demographic trends?

Since we don't know the answers, there is a temptation to throw in the towel at this point. But that isn't possible. Government (both federal and local) and business are hungry for information about future numbers. Decisions on school building or power plants may hinge on numbers expected to enter first grade or to switch on their air conditioning. No forecasts would mean we were back to tossing coins about the future, or going as the Greeks did to the oracle at Delphi.

So forecasters do the best they can. They begin by looking at the "don't know" areas and reducing them to a set of *conditional assumptions*. For example, they may assume an increase in the proportion of married women in the labor force. This *conditional assumption* may have a range of values from a very conservative, small increase to a more massive swing towards parity. Figure 24-1 shows one way in which two conditional assumptions may be

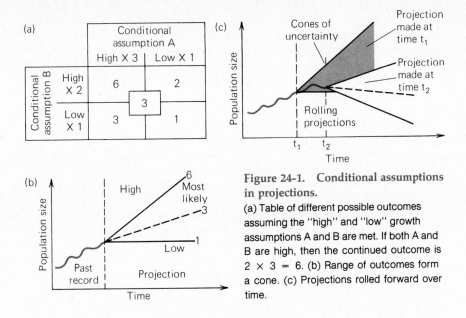

Figure 24-1. **Conditional assumptions in projections.**

(a) Table of different possible outcomes assuming the "high" and "low" growth assumptions A and B are met. If both A and B are high, then the continued outcome is $2 \times 3 = 6$. (b) Range of outcomes form a cone. (c) Projections rolled forward over time.

combined to project a series of possible increases. In the simple example shown assumptions are multiplied. But in other models they might be averaged.

Note the contrast between the most likely trend (the midpoint curve) and the curve of uncertainty, which gets wider as we look further into the future. Turn back now to Figure 7-14 earlier in the book to see how an actual forecast (world population, 1965-2000) behaves. Clearly the further ahead we try to look, the less sure we are of what the future holds. In practice forecasts are rolled forward to take account of changes, as in Figure 24-1(c).

So all forecasts are conditional. They project what is expected to happen if, and only if, the basic assumptions turn out to be correct. Better forecasts are not as often a product of better mathematics as of firmer foundations for making assumptions.

Regional and Local Modifications   So far we have concerned ourselves with global forecasts. But for geographers this isn't enough. They are concerned with the fine detail of local and regional contrasts, i.e., with producing forecast maps. Although the principles remain the same, the task is more difficult and the practices more complex.

Let us look at one procedure. First, separate forecasts are made for small geographic areas—say, the counties within a state or province— on the basis of their past records. Second, each country is compared with its neighbors. The argument here is as follows: A fast-growing area may spill over some of its growth to the others in its region. So, to know what is likely to happen in one area, we must also know where its neighbors are heading. Third, each county is compared with the state or province of which it forms a part. Clearly the sum of the local forecasts must fit with that of the regional forecast as a whole. Fast growth in some counties may be at the expense of slower growth elsewhere, that is, the region taken as a whole may be rather static even though small parts of it are booming.

The final forecasts for each local area are therefore checked and rechecked both sideways (with neighbors) and upward (with the regional forecast) until a consistent figure is obtained. Figure 24-2 shows population forecasts for the Windsor–Quebec City axis of Canada. When the map for 2001 was put through the checking process described above, a substantial downward revision followed. The map shown is therefore an overestimate of likely population growth.

Now that we've seen something of the general problem, we turn to some regional examples. We begin with short-term and then move on to long-term forecasts.

24-2
Short-Term Forecasting

The simplest forecasting models are those in which we already have a long set of records from the past and we wish to look only a short way into the future. Such *forecasting models* are simplified representations of how the real world works, used in attempts to predict the future. Such models may be expressed by mathematical equations, graphs, stated in words, or drawn on a map. We have already seen how geographers use past climatic records (in Chapter 6) and past population censuses (in Chapter 7) in this way. Now we shall look at this kind of trend-projection method more closely.

Projecting Past Trends Consider Figure 24-3, which shows the record of unemployment over 9 years for a small English town (Bridgwater) of around 25,000 people. Data are available for the percentage of the local labor force out of work for each of the 108 months in this period, and they form an erratic picture [Figure 24-3(a)]. The forecaster's problem is this: Given this information, can we make any useful predictions about the likely trends in jobs in the tenth year?

One way is to look for any apparent patterns or trends in the past record. We find, for example, that unemployment in this town has been getting worse, and we can generalize this rise as a long-term trend. This trend is represented in Figure 24-3(b) by a straight line. The line was fitted to the past values for unemployment and is the best estimate we can make on the linear element in the rise in joblessness. By continuing this line forward into 1969, we can see how long-term trends will affect the job situation in the town if, and only if, the 1969 trend is generally the same as in the rest of the decade.

Although the trend line gives a useful general indication of change, it ignores shorter term *cyclic* variations. For example, there are also important seasonal variations. More people are unemployed in the winter, which is a slack period for the construction industry and the tourist industry. February

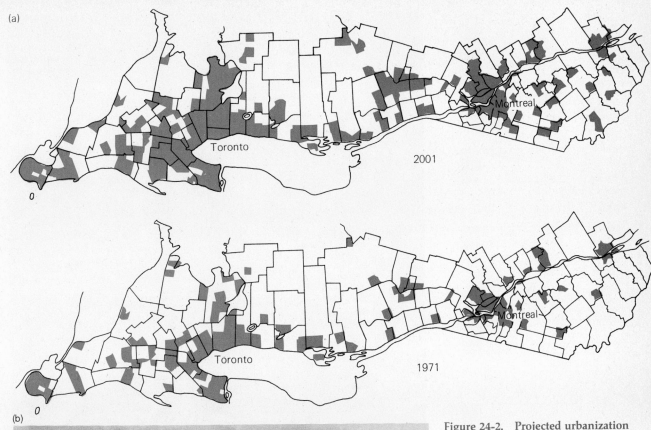

Figure 24-2. Projected urbanization of Canada's main axis.

(a) One of a set of forecasts of population density in the Windsor–Quebec city area by the year 2001 compared to the 1971 situation. In both maps census subdivisions with densities of over per km² (i.e., 120 persons per mi²) are shaded. (b) Most of the projected growth is expected in the western half of the area where Toronto forms the main generator of economic growth. [From M. Yeates, *Main Street* (Macmillan of Canada, Toronto, 1975), p. 315, Fig. 9.4. Photo courtesy of the Canadian Consulate General.]

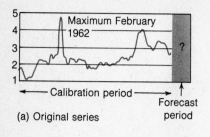

(a) Original series

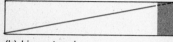

(b) Linear trend

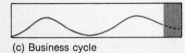

(c) Business cycle

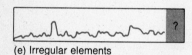

(d) Seasonal cycle

(e) Irregular elements

Figure 24-3. Projected time series.
Graphs (b) through (e) show how the time
series in (a) can be broken down into regular
and irregular components. The regular
components may be either linear trends or
recurrent cycles (e.g., business cycles or
seasonal cycles). Regular components can
be projected forward as elements in a
forecast. [Data are monthly unemployment
figures for Bridgwater, England (shown at
right) for 1960-1969, supplied by the
Department of Employment. Photo by
Aerofilms.]

tends to be the worst month of unemployment, and July the best. With
other statistical devices we can pick out these seasonal swings and, again,
project them as in Figure 24-3(d). In addition to this seasonal cycle of work
there are other recurrent changes that have a longer duration. In Bridgwater,
business-cycle variations had a duration of about five years, with peaks in the
summer of 1961 and the winter of 1964. Again, this fact can be recognized by
statistical methods and its effect projected forward as in Figure 24-3(c). If we
now combine the long-term trend, the business cycle, and the seasonal cycle,
we have a better chance of projecting the 1969 values accurately. However, we
are still left with irregular elements, like the closure of a single plant, whose
recurrence cannot be predicted from past records. [Figure 24-3(e)].

The Question of Accuracy   How can we be sure that our projections are
accurate ones? The only reliable way is to wait to see what happens and
compare the actual values with the projected ones. If we do this for Bridgwater, as in Figure 24-4, we find that although the projections are reasonably

good for the whole of 1969, they become progressively worse in the following years.

In this particular instance a comparison of actual values with forecast values was possible. But this would not be so for long-range forecasts. If we are forecasting the population of New York State in 1990, we want to judge its probable accuracy *now*, not in 1991 or whenever the relevant evidence is in. Comparing real values with forecast values is therefore a poor way of evaluating the utility of a forecasting model since we can arrive at a verdict only *after* the results have happened.

For these reasons we need tests for projections that are applicable *before* and not *after* a forecast is made. Such tests depend partly on common sense and partly on statistics. We look first at the internal logic of a forecasting model, that is, whether it seems to make reasonable sense in terms of what we know about a situation. Alternatively, we can use the past record itself as a test by arbitrarily separating it into two halves and using one half to project the other. Tests with split series are named *Janus tests*, after the Roman god of doorways, who had two faces, one looking backward and one forward.

Forecasting Contagious Diffusion Let us return for a bit to the models of spatial diffusion discussed in Chapter 13. We noted there that computer simulation plays a vital part in the geographer's study of diffusion processes. For example, Washington geographer Richard Morrill used computers to simulate the block-by-block northward expansion of the black ghetto in the city of Seattle. The basic Hägerstrand model of diffusion fitted to the actual pattern of expansion from 1940 to 1960 by using ten 2-year cycles. Rules were created to model the inmigration of a black population from outside Seattle, to calculate the critical pressures within the ghetto, and to trace the direction and distance of spread of families from the ghetto into surrounding city blocks. The rules were modified to allow rapid absorption into the ghetto of middle-class, single-family houses to the north, but only slow inroads by blacks to the west and along the lake to the east (where there was rather expensive housing and apartment blocks). Figure 24-5 presents a typical 2-year cycle and shows the direction of family movements. Having built the model to fit past processes accurately, Morrill could run the simulation for 2-year cycles into the future, tracking the expansion of the ghetto. Computer-generated maps will, like trend projections, contain increasing errors over time.

In other cases we may be interested in slowing down a diffusion process and limiting its spatial spread. For example, Canadian geographer Roland R. Tinline devised optional strategies for the control of foot-and-mouth disease in England. In 1967 England suffered an epidemic that forced the slaughter of about 443,000 animals and cost millions of dollars. The control strategy involved slaughtering infected animals and banning the movement of all animals in a cordon around the outbreak. Tinline devised a mathematical model of the spread of the actual outbreak. This model can be used to create

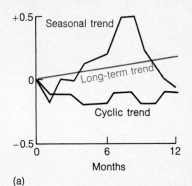

(a)

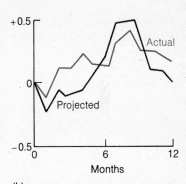

(b)

Figure 24-4. Projected and actual trends.

(a) Here, the shaded "forecast period" of Figure 24-4 is enlarged to show the three regular trends being projected forward over one year. (b) These three trends are summed to give a projection that can be compared with actual values for unemployment.

THE LANGUAGE OF FORECASTING

Bellwethers are areas within a country that show trends earlier than the rest of the country. Literally the term means a sheep (wether) wearing a bell that walks ahead of the flock.

Business cycles are recurrent fluctuations in general economic activity. They may be observed in fluctuations in the production or employment within a country or region.

Delphi forecasts are qualitative statements about the future based on the averaged view of experts in a particular field. They are usually confined to forecasting the timing of technical "breakthroughs."

Forecasting models are simplified representations of how the real world works, used in attempts to predict the future. Such models may be expressed by mathematical equations, graphs, stated in words, or drawn on a map.

Naive models are forecasting models that do not involve any theoretical basis. For example, we may use a straight line projection where one looks at how a region's economy has grown in the past and projects that it will grow at the same rate in the future.

Postdictions are the opposite of predictions. They forecast "in reverse" by using present and past records to project a situation at a still earlier period.

Projections are the forward extension of trends recognized from past and present records.

Scenarios are informal forecasts of future situations based on a sequence of arguments rather than on mathematical equations.

Simulation models are models which reproduce complex, real-world situations usually in terms of a computer program. Literally, simulation is the art or science of "pretending."

Time series are data gathered over a period of time, usually at regular time intervals.

Turning points are points in time when economic activity in a region or country changes direction. When activity is falling during a recession and then swings upward again, the precise time at which it begins to swing is the turning point.

Figure 24-5. Simulation of ghetto expansion.

The map shows one two-year period from a computer simulation of the expansion of the northern part of Seattle's black ghetto. The arrows show simulated movements of families within the ghetto. Shading indicates city blocks around the edge of the ghetto where housing is desired by black families. A distinction is made between an unsuccessful attempt to move (light shading) and a successful attempt (dark shading, with an arrow entering the block). [From R. L. Morrill, *Geographical Review* **55** (1965), p. 357, Fig. 13. Reprinted with permission.]

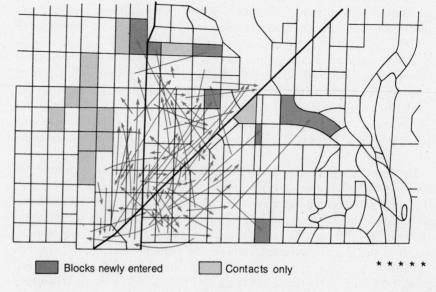

■ Blocks newly entered   ▢ Contacts only

* * * * *

control strategies for further outbreaks by simulating a mathematical outbreak and by testing its spread over different parts of England given various kinds of control strategies.

Bellwethers and "Early-Warning" Regions   Stock-market analysts keep a watchful eye on certain shares that tend to move a little ahead of the rest of the market. Similarly, electoral pollsters tend to watch closely the returns from particular primary elections in the United States (e.g., the slogan "As Vermont votes, so votes the nation"). Geographers use similar bellwethers, looking for regions that are consistently ahead of others in terms of spatial diffusion. So *bellwethers* are areas within a country that show trends earlier than the rest of the country. It is hoped that the changes in such regions will like the barometer provide some early warning of storms ahead. We can illustrate this idea with a map of southwest England [Figure 24-6(a)] which shows the timing of upturns in two business cycles for each of 70 local areas. A ring of small towns 20–50 km (12–31 mi) from the regional capital (Bristol) tends to respond several months earlier than the capital city itself, and nearly half a year ahead of some of its more remote westerly brethren.

Figure 24-6.   Lead and lag areas.
(a) These maps of southwest England indicate areas that reached low points on the business cycle three or more months after Bristol (lag areas). Bristol, the regional capital and major employment center, reached its own trough in the months indicated. (b) Leads and lags between 25 cities in the American Midwest. Note the way the steel cities or the Pittsburgh group show fluctuations in their employment trends 3 to 5 months between the car-making cities of the Detroit group and farm-machinery cities of the Indianapolis. The five independents reflect national rather than regional swings. [(a) From K. Bassett and P. Haggett in M. D. I. Chisholm, A. E. Frey, and P. Haggett, Eds., *Regional Forecasting* (Butterworth, London, 1971), p. 398, Fig. 6. (b) Data from L. J. King *et al.*, *Regional Studies,* **3** (1969), Table 2.]

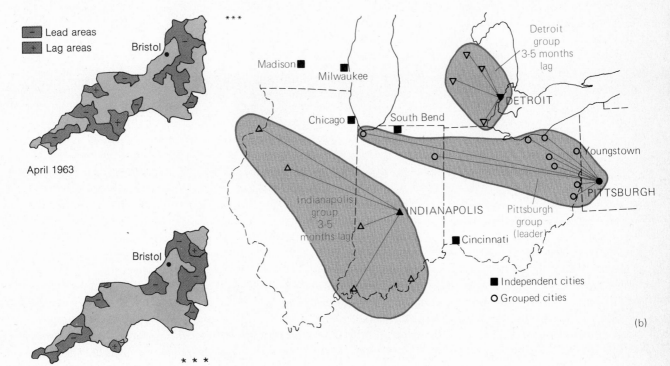

For geographers, as for stock-market analysts, the key issue is to judge how consistent these early-warning signs are. In the case of regional business cycles attempts to isolate lead areas have produced intriguing results. A study of unemployment in midwestern cities in the early 1960s shows that a group of cities around Pittsburgh regularly led the Detroit and Indianapolis areas by 3 to 5 months [Figure 24-6(b)]. Regional unemployment data for 10 British regions shows the Midland region leading most other regions by 3 months and Scotland and the North leading by 6 months. A lot more work needs to be done before we can be sure, however.

How can geographers handle this multilevel, multinational, multiindustry diffusion? Figure 24-7 shows, in a simplified form, one of the possible ways to do it. We assume a three-level urban hierarchy in which changes in the upper level are passed on to the two lower levels. The changes are represented by a simple wave indicating varying levels of economic activity in the city. As each wave passes down to the next city it may be *lagged* (occur later in time) and *modulated* (be damped down or amplified). We know that small swings on the national level may be transformed into large swings in the local economy. Areas with large numbers employed in a heavy industry like steel may have economic cycles more pronounced than that of the nation as a whole; that is, the rises may be higher and the slumps lower. Conversely, areas with service industries like universities may experience milder than average swings in economic activity.

24-3
Long-Term Scenarios and Speculations

Thus far we have been considering short-term methods of forecasting that are essentially projections firmly based on past records. They are largely dependent on the numerical analysis of historical records. But what if these records are inadequate indicators of the future, or what if we wish to forecast changes that are difficult to quantify? Then we must proceed in a more speculative manner. We look here at two approaches at the world level. You may wish to compare these with the long-term forecasts of physical geographic phenomena we encountered in Chapters 5 and 6.

World Futures: A Pessimistic Model   We begin by looking at a model of the world economy built by an MIT team led by systems analyst Jay Forrester. Although the work of Forrester's team on urban systems was well known, it was the publication of *World Dynamics* in 1971—followed by an extended and popularized version, *The Limits to Growth*—that caught the public imagination. This book extended the same systems approach used in Forrester's city model to the global level. Special attention was paid to the interrelationship of five fundamental factors: global population, agriculture, resource use, industry, and pollution. Figure 24-8 shows the links between these elements as a series of positive and negative feedback loops. (We have already met and discussed these terms in our study of ecological systems in Chapter 3.) Even in this highly simplified diagram it is possible to see how population,

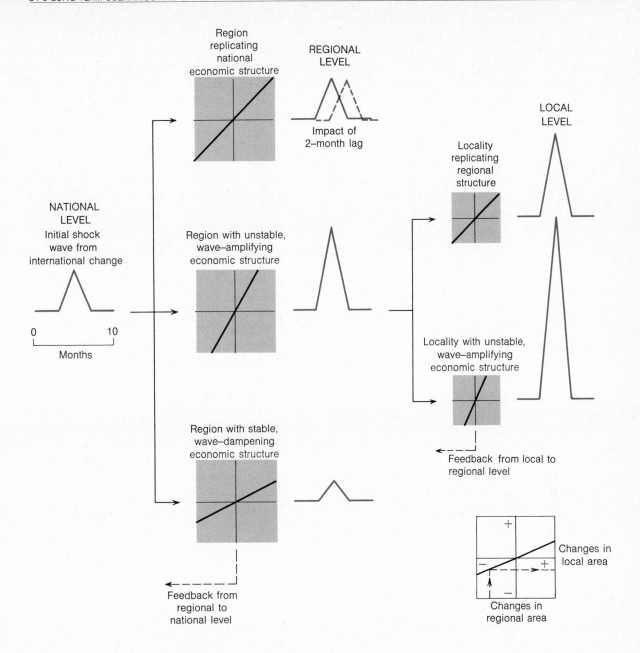

Figure 24-7. Hierarchic transmission of shocks.
This schematic model shows how changes in economic activity on one level may be transmitted to others. The *transformation boxes* (shaded) describe the degree of amplification or dampening caused by the economic structure at the regional or local level. The diagram is, of course, greatly simplified. In practice the waves are highly irregular and may be lagged in time, the shocks may start at levels other than the national one, and the effects of several shocks may intermingle and overlap. The box at the lower right shows how a major fall in economic activity in a regional area is reflected in a dampened form at the local level.

Figure 24-8. Main elements in the MIT model.

This chart gives a highly simplified view of the five main elements in Forrester's global model. Most of the links connecting population, agriculture, and industry are positive feedbacks (i.e., self-reinforcing over time), whereas both pollution and resource use create negative feedbacks (self-canceling over time). The actual computer model had around 140 elements, rather than the few shown here.

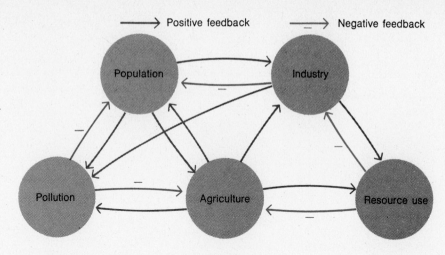

agriculture, and industry tend to be linked via positive loops, so that increases in one tend to drive up the level of the others. An exception is the link between industry and population. Industrialization is associated with lower rates of population increase. The addition of the resource and pollution factor brings major "governors" into the system, slowing down and even reversing growth through powerful negative feedback loops.

Figure 24-8 is a very simplified picture of the real MIT model, which linked the five basic elements through a complex network of over 130 rates, multipliers, and equations. These were linked in a computer program to project trends over a past 70-year period (1900 to 1970) over the next 130 years (i.e., 1970 to 2100).

Findings of the MIT Model   What findings did the MIT team come up with? Figure 24-9 attempts to summarize their results by reproducing some of the many computer runs of the model. In looking at the charts, remember that the vertical axis has deliberately been left unlabeled to underline the uncertainty attaching to the values.

The team studied first a "standard" world model. This model assumed "no major change in the physical, economic, or social relationships that have historically governed the development of the world system." As Figure 24-9(a) shows, population peaks out about the year 2050 as the rapidly dwindling resource base forces a slowdown of industrial growth. Industrial output (measured in per capita terms) peaks out around 2000, but the momentum of both population and pollution continues to drive them upward for another generation.

Given this first set of gloomy results, the MIT team then looked at responses to the crisis the model was predicting. Four strategies for overcoming the crisis were examined: increasing capital investment, reducing birth rates, cutting pollution, and stepping up agricultural production. Figure 24-9(b)

(a) WORLD MODEL, STANDARD RUN

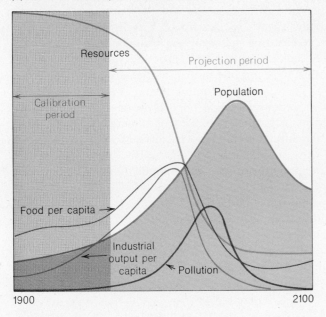

(b) "UNLIMITED RESOURCES" ASSUMPTION

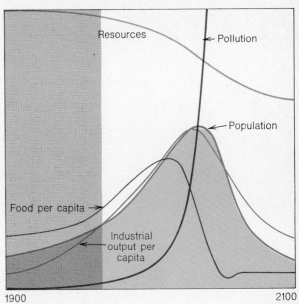

(c) FOUR-POLICY CONTROL PROGRAM

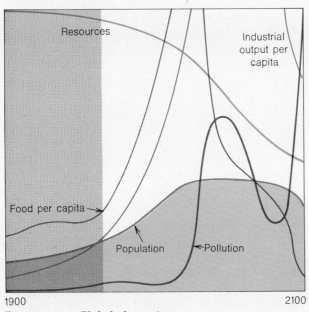

(d) STABILIZED WORLD MODEL

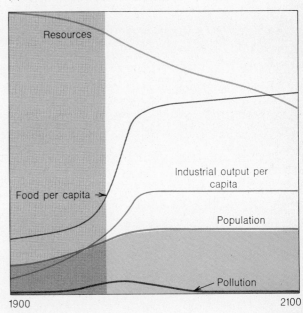

Figure 24-9.   Global alternatives.
The charts show simulated growth paths for world population and resources projected by the MIT model. The vertical scale in each diagram is left purposely blank to indicate the lack of numerical precision in the forecasts. For a discussion of each alternative and the basis of the projections, see the test. The shaded area at the left represents the relatively short calibration period (up to the mid 1960s) on which the long projections are based. [From J. Forrester, *World Dynamics* (Wright-Allen Press, Cambridge, Mass., 1971).]

shows the impact of the first strategy. It assumes that "unlimited" nuclear power will double the world's resource reserves and make extensive recycling possible. But population again tops out around the year 2050, its growth halted this time by accelerating pollution and dwindling per-capita food production. The other three strategies were found to be equally unrewarding. But suppose that, instead of trying one strategy at a time, we impose all four together. What then? The computer output in Figure 24-9(c) tells the story. The situation is much improved, in that population stays at a plateau for most of the next century. After that, however, pollution rates again begin to soar, industrial and agricultural production nosedives, and population again begins to fall.

This last result is disappointing since a stable population was held for only 80 years. What do we need to do to create long-term stability? Figure 24-9(d) shows an answer. If we raise food production by 20 percent and cut everything else (pollution by 50 percent, natural resource use by 75 percent, capital investment by 40 percent, and birth rates by 30 percent), the global population drops to slightly below the 1970s level. While resources are still declining, the effects of this decline are largely offset by recycling and the substitution of one resource for another.

Evaluation of the MIT Model   How much importance can we attach to these projections? There are two kinds of evaluations we can make. The first is a technical one. If we look carefully at the model, we can find a number of debatable assumptions. For example, we have already seen in this book that natural resources are *not* a stockpile wasting away (Section 9-1) and that world population will probably follow a logistic rather than an exponential path (Section 7-3). Also, the structure of the computer model has been widely criticized as being both too simplified and structured so that small errors can be compounded into major trends. In addition, economists have been particularly concerned by the absence of any recognition of the fact that scarcity of a resource brings a chain of counterbalancing reactions.

Second, we can evaluate the model in terms of its general message. It is successful insofar as it has drawn massive public attention to the long-term problems of sustaining life on this planet. More particularly, it has shown that attempts to intervene on one front (e.g., through population control) may be counterproductive and self-defeating. Stabilization as a goal is shown to be achievable only through the coordination of very many measures in a careful and integrated way. Even if we reject the specific projections of the MIT team, the model provides a benchmark from which further work can proceed. Other models, with greater accuracy and realism, are now being developed.

World Futures: An Optimistic Scenario?   The MIT team gives us one view of the future. But it is one of many views. Scholars in this last decade have been obsessed, as were their nineteenth-century forebears, with the end of their century. We have seen a succession of experiments in bringing together teams of individuals to think about the form of society, technology, the environment, or the world community in A.D. 2000. Groups like the Hudson institute in New York or the Futuribles group in Paris have been established to undertake future studies. One of their methods has been to design *scenarios,* that is, alternative pictures of the future, each with a supporting document stating how it was constructed—and whether it is likely to occur.

As an example, Figure 24-10 presents a world scenario of the probable economic ranking of nations in A.D. 2000. This scenario was developed by Herman Kahn, director of the Hudson Institute, and assumes a world population then of slightly less than 6.4 billion. The nations are divided into six categories, based on their gross national products per capita: postindustrial, early postindustrial, mass consumption, mature industrial, transitional, and preindustrial economies. Most of the world's population will live in countries in the fifth, or transitional, groups, which will include some of the world's largest countries—China, with an estimated 1.3 billion people; India, with 0.95; Pakistan, with 0.25; and Indonesia, with 0.24.

Kahn stresses that the scenario has three main elements. First, there are *fixed elements,* which include the fixed locations of large regional agglomerations of population. Second, there are what he terms *surprise-free projections* like decreasing communication costs. Third, there are *variable-choice elements* such as a growth-oriented society versus a stability-oriented one. The map contains significant long-term and surprise-free elements, and we would not expect this pattern to be totally reversed. But there is room for

Figure 24-10.   World scenario, A.D. 2000.

This projection of the state of the world in A.D. 2000 by the Hudson Institute is more optimistic than that of Forrester's MIT team. The map shows the economic ranking of the nations of the world at the end of the century, based on projected per capita earnings. Six types of economies are shown. [From H. Kahn, *Science Journal* **3** No. 10(1967), p. 120, Fig. 1.]

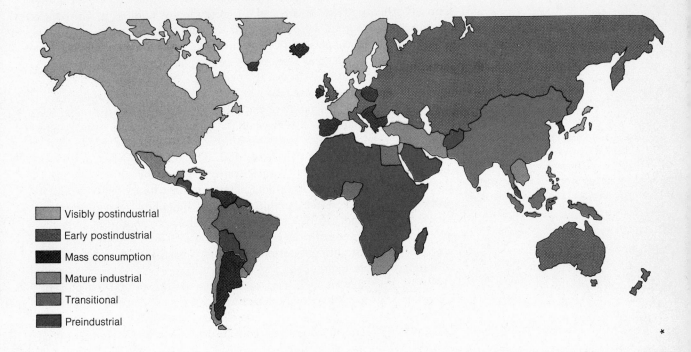

Visibly postindustrial

Early postindustrial

Mass consumption

Mature industrial

Transitional

Preindustrial

substantial changes in the individual ratings. The increase in China's industrial capacity may well be greater than Kahn suggests; whereas the current Japanese boom may be partly offset by stronger competition from other East Asian countries. This world picture is accompanied by other scenarios for individual resource elements that show sharper differences. For example, the scenario for the world's fuel resources is affected by whether the forecasters assume a primary role for nuclear power or whether they believe that we will continue to be heavily dependent on traditional fuels.

24-4
Forecasting and Geography

What effect does this growing interest in forecasting have on the way geographers go about their work? We shall look, in this section, at its effect on the study of the way geographers look at man's past as well as his future on the globe. We shall note also the emergence of new and important traditions in our ways of working.

One of the curious fallouts of research in forecasting has been to stimulate projections of past events. Archeologists and prehistorians have shown increasing interest in the locational models developed by geographers to describe the spatial distribution of the human population. For example, if we take archeological data on the size of Roman cities in Gaul as measured by the area enclosed by city walls we find that it shows the same kind of rank-size regularities we described in Chapter 15 in the discussion of modern city distributions. Similar work on Roman settlements in England and Mycenaean settlements in the Aegean shows some evidence that they conform to a Christaller-like pattern. What is the significance of findings of this kind? Broadly, they hold out the promise that we can predict which sites—from a series of archeological possibilities—are most likely to be worthwhile places to dig. New frontiers in joint research by geographers, and archeologists and other students of the past are being opened up. (See the discussion of forecasting the past on the next page.)

Concern with the geography of the future is a relatively recent phenomenon among geographers. A review of the geographic publications in any single year would reveal an overwhelming commitment on the part of geographers to the study of the recent past. The books, journals, and maps published in the late 1970s are largely concerned with the world of the early 1970s—its spatial patterns, its ecological relationships, its regional systems. A proportion of the research is concerned with describing and interpreting earlier decades, and with the geography of still earlier centuries. Very few indeed are concerned with the world beyond the 1970s. Thus we can represent the general traditional pattern of research by the kind of skewed distribution shown in Figure 24-11.

We do not have to look hard to find the reasons for such a pattern. Geographers have been heavily dependent, in their research, on empirical data published by official agencies. There is time lag, sometimes several years in length, between when an agency collects data and when it publishes it. If

Figure 24-11. The characteristic time distribution of geographic research. Most published geographic research deals with the recent past; very little is concerned with the future. The horizontal scale is in years.

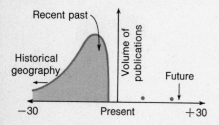

FORECASTING THE PAST: PUTTING GRAVITY MODELS IN REVERSE SHIFT

Some of the models we have discussed in this book are useful to archeologists and prehistorians (students of the past) as well as futurists. Let us take a hypothetical example of a Greek settlement dating from around B.C. 2000 that has been "lost" by a rise in the level of the Mediterranean Sea. Records of trade at excavated sites show it existed and describe its size, but don't say *exactly* where it was located. There may be several promising sites for the submarine archeologists and the problem is to know which is the most promising? One ingenious approach derives from work by Michigan geographer Waldo Tobler. Recall the gravity model discussed in Section 18-1. (See the discussion on page 439, "Gravity Models: Spatial Interaction".) It was shown there that a gravity model can describe flows (F) within a regional trading system in which the size of places (M) and their distance apart (D) was known. This was described by a simple model

$$F_{12} = a \frac{M_1 M_2}{D_{12} b}$$

The numbers 1 and 2 identify the two settlements between which flow is occurring, and a and b are con-

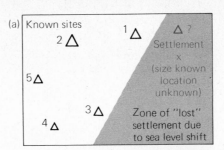

(a) Known sites

1 △   △ ?
2 △   Settlement
              x
5 △   (size known
            location
            unknown)
  3 △   Zone of "lost"
4 △   settlement due
          to sea level shift

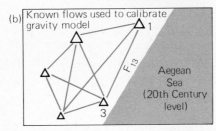

(b) Known flows used to calibrate gravity model

F_{13}

Aegean Sea (20th Century level)

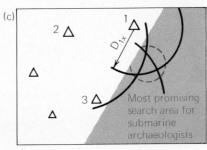

(c)

2 △   1
             D_{1x}
 △
      3 △   Most promising
 △   search area for
              submarine
              archaeologists

stants. You should check that you understand the formula and how it works by reading over again the fuller description and worked example on page 439.

By rearranging the terms in the equation we can bring distance (D) into the limelight

$$D_{1x} = b \sqrt{a \frac{M_1 M_x}{F_{1x}}}$$

Those of you who enjoy equations can work this out on the back of an envelope; those of you who don't may be happy to skip the algebra. What the equation is saying is that if we know the size of two places, the flow between them, and the two constants (a and b), then we can estimate how far they are apart. So if we assume that the two settlements (M_1 and M_x) each had 200 houses, that the flow between them (F_{1x}) was recorded as 100 jars of olive oil, and that the constants are conveniently a = 1 and b = 2. Then the distance between our known settlement and our "lost" settlement (D_{1x}) works out at

$$2 \sqrt{1 \frac{200 \times 200}{100}} = 20 \text{ km}$$

By repeating the calculation from other known settlements we can also estimate *their* distance from x. Combining the results of the various estimates (the circles in the diagram) suggests the most promising area within which a closer archaeological search for the site may be focused. Readers interested in following up the use of geographical methods in archeological research may like to look at I. Hodder and Orton, *Spatial Analysis in Archaeology* (Cambridge University Press, Cambridge, 1977).

---

we add the time needed for analysis and writing, this alone is almost enough to explain the pattern in Figure 24-11. But not quite. Econometricians face the same problem, but spend more of their time in forecasting. Part of the reason geographic research seems to favor analysis of the past might be the geographer's concern with descriptive rather than predictive or planning models. *Descriptive models* replicate selected features of an existing geographic system and attempt to show how it operates. *Predictive models*, on the other hand, rearrange the structure of descriptive models so that the value of variables of interest at the end of the causal sequence can be predicted from the value of variables earlier in the sequence. Finally, *planning models* incorporate alternative decisions into a predictive model so that we can evaluate their effect.

So, there are persuasive reasons why geographers should continue, and indeed intensify, their efforts to look both forwards (and backwards too). Forecasting is an integral part of human decision-making. As individuals, families, societies, and nations, we order our lives in terms of what we think will happen in the next day, month, year, or decade. We constantly make forecasts, so that the question we must really ask is not "Should geographers make forecasts?" but "How can geographers make their forecasts more accurate and therefore more useful in evaluating the spatial, ecological, and global futures that lie ahead?"

Summary

1. Although the track record of past forecasts of changing world geography is distinctly *not* encouraging, the need for geographic forecasts attracts increasing attention. Geographic forecasts involve (a) cross-checking between adjacent areas for spatial similarities in trends and (b) reconciling upper level (regional) with lower level (local) forecasts.

2. Short-term forecasting models are mainly confined to cyclic phenomena (business activity, unemployment, disease waves). The recurrent nature of such waves allows a more precise estimation of their likely accuracy.

3. Accuracy in forecasting can only rarely be achieved by waiting until actual values are available for comparison. Typically, models have to be calibrated and tested on a past record.

4. Several geographers have built models of spatial diffusion in an effort to forecast the spread of such phenomena as population and disease. Geographers often use bellwethers or leading regions that are foremost in terms of spatial diffusion. Others study movements of phenomena that diffuse through a hierarchy. Diffusions of this type vary from the relatively simple, as in the case of disease spread through a hierarchy of urban places, to the very intricate, involving many phenomena at several spatial levels.

5. Forecasting may be put into reverse (sometimes termed postdiction) by historical geographers interested to understand decisions already made in the past. Spatial reconstructing "lost" settlement sites illustrates this trend.

6. Geographers have done most of their research in terms of the geography of the recent past. This may result from the geographer's interest in descriptive rather than predictive or planning models. Moves toward an increased emphasis on the second and third type of model is likely to increase the work on forecasting.

Reflections

1. Most of you reading this book will still be alive in A.D. 2025. What changes do you expect to see in the environment and landscape of your own area by that date?

2. Gather data on the economic development of your own region or country over the last decade. (Figures for unemployment rates are useful, and they are often available on a

monthly or quarterly basis.) Plot the data on graph paper and look for a pattern. Is a long-term trend clear? Can you spot any seasonal or other cyclic influences?

3. Review the somewhat pessimistic graphs of the MIT model in Figure 24-9 and compare them with Figure 24-10. Are you (a) very worried or (b) complacent about these doomsday philosophies? Debate your views with the rest of the class.

4. Has geographers' concern for the future increased their interest in the past? Why?

5. Review your understanding of the following concepts:

conditional forecasts	bellwether regions
long-term trend	simulation models
components	scenarios
cyclic variation	postdiction

One Step Further . . .

Philosophical problems in forecasting and distinctions between long- and short-term projections are discussed in
De Jouvenal, B., *The Art of Conjecture* (Basic Books, New York, 1967).

A thoughtful applied study for central Canada is provided in
Bourne, L. S., R. D. MacKinnon, J. Siegel, and J. W. Simmons, *Urban Futures for Central Canada* (University of Toronto Press, Toronto, 1974).

Flexibly constructed, long-term scenarios are described in
Kahn, H., and A. J. Wiener, *The Year 2000: A Framework for Speculation on the Next Thirty Years* (Macmillan, New York, 1967) and
Hall, P., *London 2000*, 2nd ed. (Praeger, New York and Faber, London, 1970).

For more formal computer-dependent models, look critically at the MIT team's work, reported in
Forrester, Jay W., *World Dynamics* (Wright-Allen Press, Cambridge, Mass., 1971) and
Meadows, D. H. *et al.*, *The Limits to Growth* (Universe Books, New York, 1972), Chap. V.

A mixture of the two approaches, with a variety of geographic applications to regional planning problems, is given in
Chisholm, M. D. I., A. E. Frey, and P. Haggett, Eds., *Regional Forecasting* (Butterworth, London, 1971).

The growing volume of research in this area is beginning to appear in the regular geographic journals. There are also special journals devoted to forecasting, such as the Journal of Long Range Planning *(a quarterly), which has occasional papers of geographic interest.*

I do not know what I may appear to the world; but to myself I seem to have been only like a boy playing on the seashore, and diverting myself in now and then finding a smoother pebble or a prettier shell than ordinary, whilst the great ocean of truth lay all undiscovered before me.

SIR ISAAC NEWTON

(from Brewster, *Memoirs of Newton*, 1855, Vol. II, Chap. 27)

chapter 25
On Going Further in Geography

On the bachelor's level not even one graduate in every 150 in United States universities majors in geography. Even within the social sciences division, in which it is often placed, the numbers of geography majors are dwarfed by those in psychology or economics. On the graduate level, for every doctorate granted in geography there were 10 in physics, nearly 20 in chemistry and engineering, and 25 in education. Within the social sciences division, geographers were only one-seventh as numerous as their colleagues in economics or history.

This small population of newly hatched geographers should be seen in perspective, however. We need to see the situation in the United States' position in an international context. Geography as a modern university subject had its origins in the institutes of Germany and France during the early nineteenth century. The German contributions to the field, certainly until World War II, were dominant ones, and a majority of the classic works in geographic literature appeared first in the German language. In the universities of Western Europe, and those of Britain and the Commonwealth (to which geography spread, largely, in the first half of this century) the number of geography students is relatively large. Canada and Australia have strong research centers. The situation in the Soviet Union is also much more advanced; Soviet geographers outnumber their American colleagues by about three to one.

Also, the situation is changing. The number of geography students enrolled in U.S. colleges and universities more than doubled over the 1960s decade, and the membership of the leading professional society, the Association of American Geographers, increased by more than three times during the same period.

University geography is a curious mixture. In size it is perhaps comparable with subjects with small enrollments like anthropology or archeology rather than major fields like mathematics or history. In its recent spur of growth, however, it behaved more like a modest version of bio-chemistry or computer science. Today it follows, in common with most university subjects, a more steady-state pattern.

This final chapter is addressed to students who have found enough in the preceding 24 chapters to interest them in taking further courses in geography. We shall look first at the past growth of the field to try to understand the path followed in get-ting to the present position. From the present we shall look forward to the future; we shall project the present drift and contrast this projected future with other targets for the field. Then we shall ask four basic questions: How did geography evolve as a sepa-rate field of study? What is its present structure? What is its likely future—if present trends continue? What should its future role be?

25-1
The Legacy of the Past

We can understand the character of geography as an academic field in the 1980s only if we see it as one scene in a lengthy play. It is useful to divide the play into three acts. The first act is dominated by isolated research by individual scholars, the second by organized research by groups and societies, the third by the incorporation of research into national and international organizations. It is clear that the stages cannot be precisely fixed in time; each is continuing in different subdivisions of the field or in different countries in various stages of growth.

Act I: The Individual Scholar   The first growth period, from the beginning of formal geographic study in ancient Greece to the mid-nineteenth century, can be characterized by geographic studies sporadically distributed in time and space. The number of scholars who would have counted themselves as geographers was always small, and it was only occasionally that clusters of workers formed—in Alexandria in the second century B.C., in Portugal in the fifteenth century, or in the Low Countries in the sixteenth century. The patronage that encouraged these groups to join together usually came from an interest in practical problems: methods of surveying the earth, instruments for marine navigation, mapmaking, and the printing of atlases. In this early period people found the answer to many questions about the general shape of the earth and ways of putting spatial information on maps. The maps of the period include some of the most majestic products of Renaissance Europe. (See Figure 25-1.) Most of the geographic schools were short-lived, however, and had fluctuating fortunes.

Figure 25-2 shows some of the leading scholars in geography since 1775. Remember, in interpreting this diagram, that the number of names that might have been included was less in 1800 than it was in 1900. Indeed, as we shall see later, the general growth of geography over the two centuries spanned by this diagram has been logarithmic. Most of the geographers who have *ever* lived are alive today! Note too that boundaries between disciplines

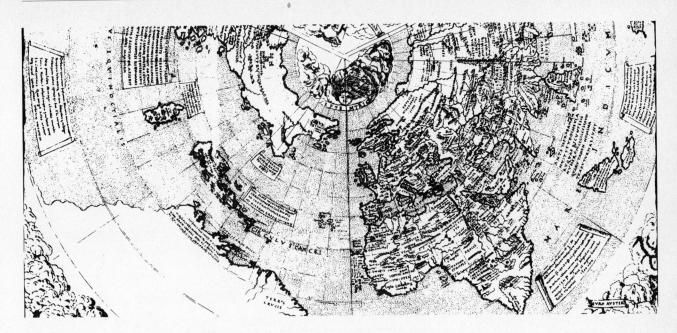

were loosely drawn. It is not surprising that individuals like Immanuel Kant, Alexander von Humboldt, and T. R. Malthus played notable roles in the growth of various sciences. The diagram also underlines the significant part played by Germany in the early part of the period (until 1929 nearly half of all geographic publications were in the German language). Inevitably, the diagram reflects the biases of the period. There are very few women geographers in the list, and certain countries isolated by language from the mainstream (notably China) are underrepresented.

The exact significance of individuals in an overall growth process is difficult to assess. Science often involves a snowball effect, because of which more than due emphasis is placed on the contributions of a few individuals, for example, on the work of the few physicists or chemists who win Nobel Prizes. This so-called "Matthew effect" (after the Gospel's remark that to him that hath more shall be given) also occurs in geography, and Figure 25-2 inevitably reflects it. Yet it would be impossible to think of German geography in the mid-nineteenth century without Carl Ritter and Friedrich Ratzel, France without Vidal de la Blache, the United States without W. M. Davis, or Britain without Halford J. Mackinder (Figure 25-3).

Our perspective on the present century is too short to allow the key figures to be discerned with certainty. In any case, team research in tending to blur the importance of the single scholar working in isolation.

Act II: Groups and Societies   The second period of growth, beginning in the early 1800s, was marked by the organized interlinking of research. One of the earliest methods of linkage was the foundation of societies to foster common

Figure 25-1.   Cartographic traditions in geography.
Mapmaking and exploration of the world played a key role in the early development of geography. This late-Renaissance world map was the first printed map to show America. It was made by an Italian, Contarini, in 1506 and illustrates the compromise between accuracy and adornment typical in maps of this period. For a description of modern developments in cartography, see Chapter 23. [Map courtesy of the Trustees of The British Museum.]

true

true

true

true

true

true

Figure 25-2. Geography, 1775–1975.
The chart shows some leading scholars in the most recent period of geographic research. It indicates the emergence of some major schools and the changing national balance of research. (Compare the German names in the nineteenth century with the American ones in the twentieth.) The geographers' names have been mainly taken from a list in Sir Dudley Stamp's *Longmans Dictionary of Geography* (Longmans, London, 1966), largely concerned with western scholarship. No comparative information was available on the growth of, say, Chinese geography. Some names have been included to illustrate the major contributions to geographic thought by scholars from other disciplines, particularly earlier in the period when the boundaries of fields were loosely drawn. Names have been chosen with a view to illustrating the emergence of significant research themes. Inevitably, other geographers would choose other names. Yet, while there might be only a modest overlap, the same general pattern of evolution of research themes over time would probably emerge. Very recent trends (within the last decade or two) are discussed in Section 25-3. Note that the country given after each geographer's name is sometimes nominal since he or she may have worked for long periods in two or more countries.

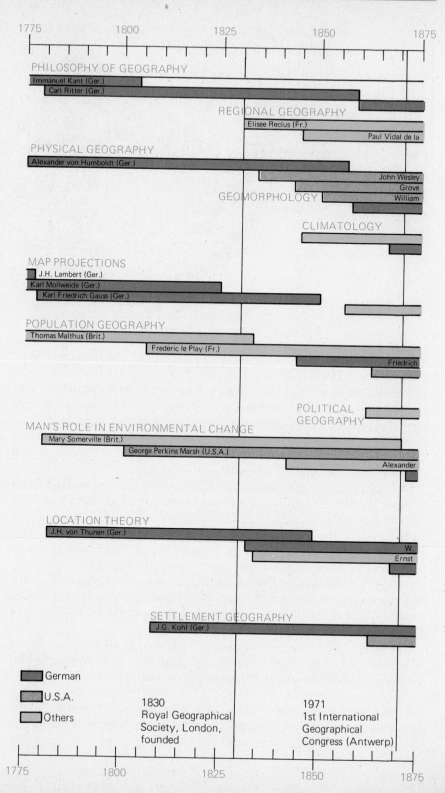

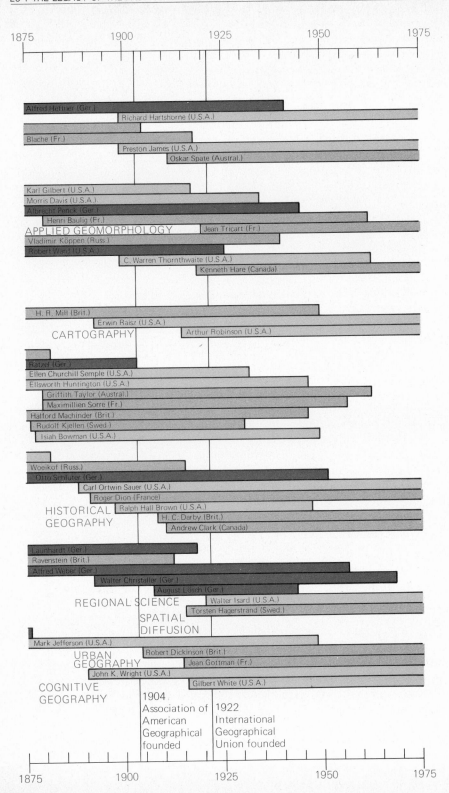

1875   1900   1925   1950   1975

Alfred Hettner (Ger.)
Richard Hartshorne (U.S.A.)
Blache (Fr.)
Preston James (U.S.A.)
Oskar Spate (Austral.)

Karl Gilbert (U.S.A.)
Morris Davis (U.S.A.)
Albrecht Penck (Ger.)
Henri Baulig (Fr.)
APPLIED GEOMORPHOLOGY   Jean Tricart (Fr.)
Vladimir Köppen (Russ.)
Robert Ward (U.S.A.)
C. Warren Thornthwaite (U.S.A.)
Kenneth Hare (Canada)

H. R. Mill (Brit.)
Erwin Raisz (U.S.A.)
CARTOGRAPHY   Arthur Robinson (U.S.A.)

Ratzel (Ger.)
Ellen Churchill Semple (U.S.A.)
Ellsworth Huntington (U.S.A.)
Griffith Taylor (Austral.)
Maximillien Sorre (Fr.)
Halford Mackinder (Brit.)
Rudolf Kjellen (Swed.)
Isiah Bowman (U.S.A.)

Woeikof (Russ.)
Otto Schluter (Ger.)
Carl Ortwin Sauer (U.S.A.)
Roger Dion (France)
HISTORICAL   Ralph Hall Brown (U.S.A.)
GEOGRAPHY   H. C. Darby (Brit.)
Andrew Clark (Canada)

Launhardt (Ger.)
Ravenstein (Brit.)
Alfred Weber (Ger.)
Walter Christaller (Ger.)
August Lösch (Ger.)
REGIONAL SCIENCE   Walter Isard (U.S.A.)
SPATIAL   Torsten Hagerstrand (Swed.)
DIFFUSION

Mark Jefferson (U.S.A.)
URBAN   Robert Dickinson (Brit.)
GEOGRAPHY   Jean Gottman (Fr.)
John K. Wright (U.S.A.)
COGNITIVE   Gilbert White (U.S.A.)
GEOGRAPHY

1904
Association of
American
Geographical
founded

1922
International
Geographical
Union founded

1875   1900   1925   1950   1975

(a)

(b)

(c)

Figure 25-3. "Modern" geographers.
Leading scholars from the first three
generations of modern geographers. (a)
Alexander von Humboldt, 1799–1859, a
German. (b) George Perkins Marsh,
1801–1882, an American. (c) Paul Vidal de
la Blanche, 1845–1918, a Frenchman.
[Photos courtesy of (a) Dr. Franz Termer,
(b) the Trustees of Dartmouth College, and
(c) the American Geographical Society of
New York.]

interests in geographic research. Such geographic societies generally fall into
four groups. The first group, the national societies, emerged in the early half
or middle years of the nineteenth century. These societies have a strong
interest in global exploration. For instance, the Royal Geographical Society,
in London (Figure 25-4), dates from 1830 and marks the merger of several
early exploring clubs like the Association for Promoting the Discovery of the
Interior Parts of Africa, founded in 1788. The American Geographical
Society of New York was founded in 1852 by a group of businessmen to
provide a center for accurate information on every part of the globe.

The second main group of societies is national professional groups, largely
dominated by university and research geographers. These groups are usually
later in date, smaller in membership, and less catholic in scope than the
national societies. The Association of American Geographers (1905), the
Institute of British Geographers (1933), and the Regional Science Association
(1954) are typical of this group. The third group consists of societies devoted
primarily to promoting geographic education in schools; the Geographical
Association in Britain and the National Council for Geographic Education in
the United States are examples. The fourth and most rapidly expanding set of
groups originated during the 1950s. These organizations are subgroups within
national professional organizations concerned with a particular aspect of
geography (e.g., cartography, geomorphology, or quantitative methods).
Geography appears to be following the pattern of other sciences in the rapid
growth of this fourth group.

The prime function of the societies was to foster common research interests through the reading of papers and the publication of journals. The establishment of journals like the *Annals of the Association of American Geographers* represented key breakthroughs in the circulation of research findings. Other journals were published by interested individuals, as was *Petermann's Geographische Mitteilungen* in 1855, or by small groups, as was Ohio State University's *Geographical Analysis* in 1969. The growth of journals provides a useful indicator of the increasing volume of geographic research. As Figure 25-5 indicates, the field has been rapidly expanding since the seventeenth century. The number of all scientific periodicals doubles about every 15 years, and the number of geographic periodicals increases at about half that rate. The slower rate of increase in geography is typical of all the older, established sciences like geology, botany, and astronomy because the total increase in scientific publications reflects the high birth rate of new scientific fields like computer science.

Figure 25-4. Geographic societies. The Royal Geographical Society typifies the groups formed in the middle of the nineteenth century to foster the research and exchange of geographic information. Its parkside headquarters in west London house a major library, map collection, and archives. Such societies predate the establishment of geography departments in universities, and many played a leading role in pressing for geography to be included in the school and college curriculum. The Royal Geographical Society was an important force in establishing geography at Oxford University and Cambridge University in England. [Aerofilms photo.]

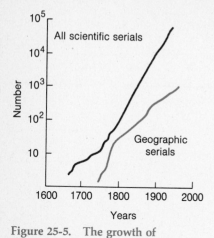

Figure 25-5. The growth of geographic research.
The chart shows the cumulative total of scientific and geographic periodicals founded since the middle of the seventeenth century. The vertical scale of the graph is logarithmic: The number of both scientific and geographic periodicals has been growing exponentially. [From D. R. Stoddart, *Transactions of the Institute for British Geographers,* No. 4 (1967), p. 3, Fig. 1-3.]

Act III: National and International Organizations   A critical role of geographic societies was to convey the importance of the problems they studied to the rest of the community. Their partial success was marked by the beginning of a third phase of geographic study, overlapping the second, in which geography departments were formally established in major universities and in which some countries set up government-sponsored research centers.

In the university sector, Germany again took the lead, with a considerable number of departments established by 1880. Developments in France were only slightly less rapid, but developments in the United States, Britain, and the Commonwealth lagged considerably. New geography departments often showed an irregular spatial diffusion pattern, marked by curious regional concentrations and sparse areas; Figure 25-6 shows the distribution of degree-giving departments in the United States, where there is presently a strong midwestern emphasis. Meanwhile, the need for national geographic research centers has led to the emergence of institutes like Brazil's *Instituto Brasileiro de Geografia e Estatistica* or the Soviet Union's *Akademiya Nauk SSSR*, each charged with the investigation and publication of regional data within their vast national territories. The latter has a staff of over 300 geographers working in ten divisions and a massive publishing program, including the bimonthly *Izvestiya, Seriya Geograficheskaya*. Even when separate geographic institutes have not been set up, geographic research has been increasingly incorporated into such organizations as Britain's Ministry of Town and Country Planning (now part of the Department of the Environment) and Australia's Commonwealth Scientific and Industrial and Research Organization (CSIRO).

Since 1922 the initiation and coordination of geographic research requiring international cooperation has been handled through the International Geographical Union (IGU). This organization holds meetings at intervals of four years. Between congresses it appoints commissions to study special subjects like arid zones, quantitative methods, or economic regionalization. The number of member countries has now swollen to over 60. Different member countries have a different interest in various areas of research; for example, problems in applied geography and regional planning dominate much Eastern European research. If we compare member countries, we can see that the size of their geographic research effort is broadly related to their overall scientific budget. However, some smaller countries play a role in research out of proportion to their size. One outstanding example is Sweden, which, with few universities and a small group of geographers, has led research in several important areas. The volume and quality of research from New Zealand is also remarkably high in relation to its small number of research centers.

25-2
The Present Structure

Geographers have repeatedly tried to define their field at each stage of its growth. For those who like formal definitions, Table 25-1 gives a variety of often-quoted ones. None of them will satisfy all geographers, but all geographers will recognize some common identifiable elements. (See Figure 25-7.)

Table 25-1 Some contemporary definitions of geography

Definition	Source
"Geography is concerned to provide an accurate, orderly, and rational description of the variable character of the earth's surface."	R. Hartshorne, *Perspectives on the Nature of Geography* (Murray, London, 1959), p. 21.
"Its goal is nothing less than an understanding of the vast, interacting system comprising all humanity and its natural environment on the surface of the earth."	E. A. Ackerman, *Annals of the Association of American Geographers* **53** (1963), p. 435.
"Geography seeks to explain how the subsystems of the physical environment are organized on the earth's surface, and how man distributes himself over the earth in relation to physical features and to other men."	Ad Hoc Committee on Geography, *The Science of Geography* (Academy of Sciences, Washington, D.C., 1965), p. 1.
"Geography is concerned with giving man an orderly description of his world . . .[however] the contemporary stress is on geography as the study of spatial organization expressed as patterns and processes."	E. J. Taaffe, Ed., *Geography* (Prentice-Hall, Englewood Cliffs, N.J., 1970), p. 1.
"Geography . . . a science concerned with the rational development, and testing, of theories that explain and predict the spatial distribution and location of various characteristics on the surface of the earth."	M. Yeates, *Introduction to Quantitative Analysis in Economic Geography* (Prentice-Hall, Englewood Cliffs, N.J., 1968), p. 1.

Unity and Diversity in Geography Let us try to summarize what these common elements in definitions of geography are.

First, we have seen that geographers share with other members of the earth sciences a concern with a common arena, the earth's surface, rather than abstract space, but that they look at that arena from the viewpoint of the social sciences. They are concerned with the earth as the environment of humanity, an environment that influences how people live and organize themselves and at the same time an environment that people helped to modify and build.

Second, geographers focus on human spatial organization and our ecological relationship to our environment. They seek ways of improving how space and resources are used, and emphasize the role of appropriate regional organization in reaching this end. Their work provides a perspective of our tenure on the earth and various forecasts—both optimistic and pessimistic—of our future on the planet.

Third, we have seen that geographers are sensitive to the richness and variety of the earth. They do not believe in blanket solutions to development problems; instead, they feel that policy should be carefully tuned to the spatial variety concealed by terms like "tropics," "Appalachia," and "ghetto." On each geographic scale, they seek always to disaggregate and dissect the uniform space of the legislator within the complex space of the real world.

Within this broadly defined area of agreement, different branches of geography have proliferated, each concerned with a limited research topic. Table 25-2(a) gives a summary of the orthodox division of geography into the study of regions (regional geography) and an analysis of their systematic characteristics (systematic geography). Each may be further subdivided into more specific fields like the regional geography of Latin America or urban geography. Some geographers recognize as a separate branch the study of the regional or systematic geography of past periods (historical geography), but others argue that time is an essential component in all geographic studies.

ALASKA [1]

Alberta

Saskatchewan

Simon Fraser

British Columbia

Calgary

Victoria

Washington

[4]

Man

Oregon

Oregon State

[3]

[3]

[5]

[1]

[3]

Humboldt SU

[1]

Utah

Northern Colorado

[2]

CSU Chico

Davis

Berkeley

[1]

[7]

Colorado   Denver

Nebraska

UC Davis   CSU Sacramento

CSC Sonoma

UC Berkeley

San Francisco
SU

CSU Hayward

San Jose SU

CSC Stanislaus

[25]

[4]

[7]

[5]

UCLA

Oklah
Sta

Riverside

[3]

CSU Fresno

Arizona State

[7]

[3]

Okla

CSC Bakersfield

Arizona

CSU Los Angeles   CSU Northridge

[12]

UC Santa Barbara

CS Polytechnical, Pomona

HAWAII   [2]

Texas A

UC Los Angeles (UCLA)

CSC, San Bernardino

UC Riverside

Hawaii

U Southern California

UC Irvine

CSU Long Beach

CSU Fullerton

Texas

CSC Dominguez Hills

San Diego SU

UC   University of California
CSC California State College
CSU California State University
SU   State Universtiy

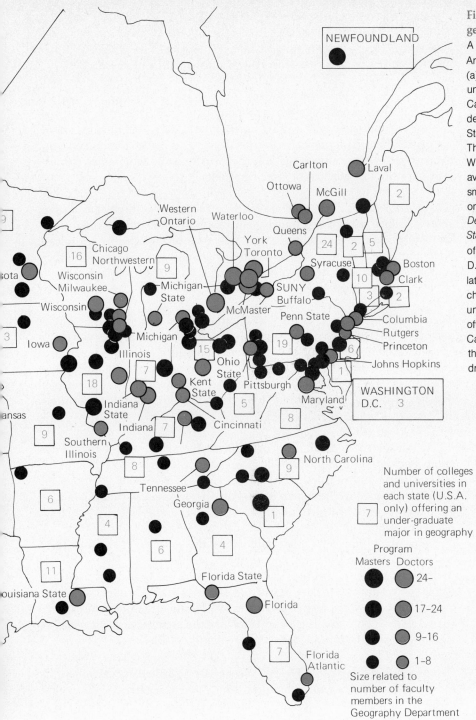

NEWFOUNDLAND

Carlton   Laval
Ottowa   McGill
Western   2
Ontario
Waterloo
Chicago   Queens
Northwestern   York   24   2   5
                  Toronto
Wisconsin   Michigan   Syracuse   Boston
Milwaukee   State   10   Clark
Wisconsin   SUNY   3   2
                  McMaster   Buffalo
Michigan   Penn State   Columbia
                  15   Rutgers
Illinois   Ohio   19   6   Princeton
          7   State   1
Iowa   Pittsburgh   Johns Hopkins
          18   Kent
Indiana   State   Cincinnati   WASHINGTON
State   5   D.C.   3
          9   7   8   Maryland
Indiana
Southern   North Carolina
Illinois   8   9
                  Tennessee
          6   Georgia   1

          4

          11   Florida State
                 Florida
Louisiana State
                 7   Florida
                      Atlantic

Number of colleges
and universities in
each state (U.S.A.
7   only) offering an
     under-graduate
     major in geography

Program
Masters Doctors
●   ● 24–

●   ● 17–24

●   ● 9–16

●   ● 1–8

Size related to
number of faculty
members in the
Geography Department

Figure 25-6.   A geography of geography departments.
A cross section of geography in North American universities in the late 1970s. (a) Geography departments offering an undergraduate major in the state of California. (b) Location of geography departments in Canada and the United States offering courses at the graduate level. Those with doctoral programs are named. Where the number of faculty was not available, the location is shown by the smallest of the four circles. Maps are based on data given in *Guide to Graduate Departments of Geography in the United States and Canada 1976–1977* (Association of American Geographers, Washington, D.C., 1976), and you will need to consult the latest version of this annual publication to check out more recent changes. Some universities not shown on this map may now offer geography programs, and the Canadian coverage is less complete than that for the United States. [Map from drawing by Andrew Haggett.]

Figure 25-7. *The Nature of Geography.*
Geographers are indebted to University of Wisconsin geographer Richard Hartshorne for much of the most thorough of modern analyses of the philosophy of their field. His *Nature of Geography* (1939) represents one of the fullest accounts of the historical evolution of its central ideas and has been followed by other major statements on the spirit and purpose of geography as an academic study. (See *One Step Further . . .* , p. 614.) [Photo by Tony Philpott.]

These conventional divisions are important, not least because most university catalogues describe courses in these terms. But perhaps a more helpful way of structuring geography is by how it approaches its problems [Table 25-2(b)]. In this book we have distinguished three different approaches.

Spatial Analysis   The first approach, termed *spatial analysis,* studies the locational variation of a significant property or series of properties. We have already encountered such variations in interpreting the distribution of population density or of rural poverty. Geographers ask what factors control the patterns of distributions and how these patterns can be modified to make distributions more efficient or more equitable.

Ecological Analysis   A second approach to geography is through *ecological analysis,* which interrelates human and environmental variables and interprets their links. We have already studied such linkages in the hydrologic cycle and land use cycles. In this type of analysis geographers shift their emphasis from spatial variation between areas to the relationships within a single, bounded, geographic area.

Regional Complex Analysis   The third approach to geography is by way of *regional complex analysis,* in which the results of spatial and ecological analysis are combined. Appropriate regional units are identified through areal differentiation, and then the flows and links between pairs of regions are established. We have already discussed some of the difficulties in regional complex analysis (Chapter 11) and looked at a few of its possible applications in regional planning (Chapter 22).

The advantage of looking at geographic problems in terms of these three approaches rather than the orthodox divisions is that they stress the unity of physical and nonphysical elements rather than their diversity. Geographers concerned with water resources or human settlement may find a common ground in the ways in which systems are studied or in their parallel search for efficient regional units.

The work of most geographers falls within the "triangle" formed by these three approaches to the field. Some may specialize, or, if you like to think of it this way, move toward one of three corners of the triangle. Indeed, the whole subject appears to have zigged and zagged over the decades, sometimes staying in the regional corner (as in the 1930s), sometimes lurching toward spatial analysis (as in the 1950s and 1960s). In the present decade it seems on the move again, now headed for the ecological corner.

Geography and Supporting Fields   Geography is particularly dependent on the flow of concepts and techniques from more specialized sciences. For example, in regional climatology we adapt models originally developed by meteorologists, who in turn draw their concepts from basic physics. Likewise, our models of regional growth borrow from the econometrician.

Table 25-2   The internal structure of geography

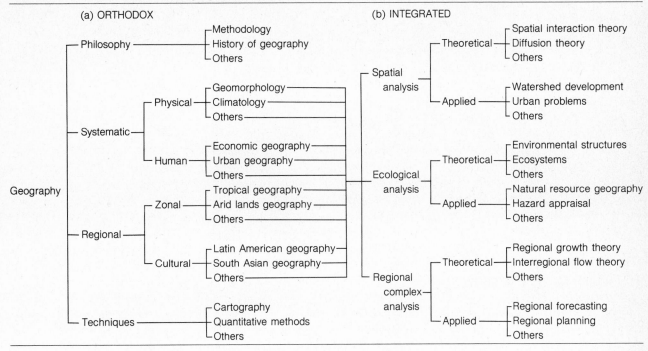

This dependence underlines the fact that sovereign subjects are about as irrelevant as sovereign states. Good botanists need to be reasonable biochemists, good engineers need to be fair mathematicians, and so on. Familiarity with these supporting fields is normally obtained by taking parallel courses in other departments. Some geographers argue that special importance attaches to mathematics because it provides a common language in which geographers can express spatial, ecological, and regional concepts in a concise and comparable way.

Let us assume that your particular interest is in the humanities and that you are especially attracted to the study of one of the major cultural areas like China. Then the appropriate courses to support regional courses in East Asian geography might include the history and economic structure of China; clearly, courses in Cantonese or Mandarin would be needed for more serious research. By contrast, if you wanted to specialize in environmental problems, you would need earth science courses like hydrology or oceanography, perhaps supported by a course in resource economics (Figure 25-8). Which courses you elect to take will depend on a combination of factors: your interest, your ability and previous training, and your long-term expectations. It is to the third of these we now turn.

Jobs in Geography   Although it is now less fashionable to earmark university courses as "career-oriented" or "general education," the distinction still lingers. Law, medicine, and geology typify fields where most graduates have a

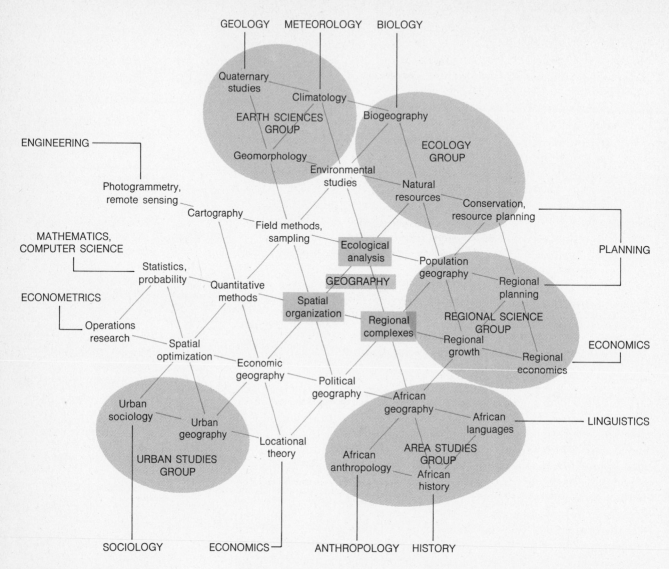

Figure 25-8. **Links between geography and supporting fields.** The courses indicated are intended to be representative rather than exhaustive. Departmental boundaries are drawn differently in some universities and colleges. For example, geomorphology tends to be taught within geography departments in British universities but within geology departments in America. African studies are used to illustrate area studies; a similar diagram could be constructed, for example, for Latin American or South Asian studies.

rather well-defined occupational outlet ahead of them. By contrast, fields like philosophy or political science have a high educational content but less obvious career prospects. Where does a degree in geography take us?

Jobs in geography can be broadly divided into two categories. First, one can find a job that is "field sustaining" in the sense that it is concerned directly with supporting the continuing study of the field itself. Thus many geographers find jobs in geographic teaching or research. Teaching opportunities exist on all levels from the elementary school to the postdoctoral level. In the countries of Western Europe geography is an important high school subject, and many graduates go back to teaching on this level. The position of geography in United States schools is less strong, though an important new curriculum for this level was initiated through the High School Geography

Project. The two-year college formed a significant and expanding job market for geography teachers in the 1960s and there was a recurrent shortage of PhDs for advanced teaching on the university level at that time. There is, however, a delicate feedback raelationship between the number of graduates produced and job opportunities (see Figure 25-9), and periods of abundant jobs and job scarcity tend to follow a cyclic pattern. The 1970s have shown a lower rate of growth. The 1980s pattern is still unclear.

Second, there are jobs outside the academic arena. Geographers have a long tradition of public service on all levels, from global agencies to city planning commissions. On the *international* level the tradition goes back at least to World War I, when geographer Isiah Bowman (later president of Johns Hopkins University) was prominent among the geographers advising the American delegation at the Versailles Peace Conference. This tradition continues today, with the British geographic team advising on the settlement of the Argentine–Chilean boundary dispute. Geographers are also represented in the specialist United Nations agencies.

On the *national* level geographers are well represented in federal departments of the United States; indeed, nearly one tenth of the American As-

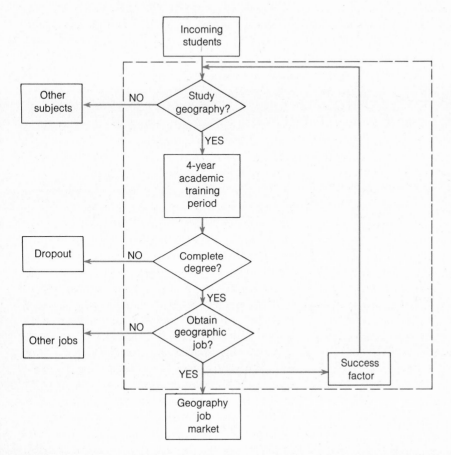

Figure 25-9. The geographic job market.
The chart shows the feedback between job opportunities and the number of graduate students. The negative feedback tends to introduce a short-term cyclic element into the general long-term expansion in numbers. Thus the 1960s with a rapidly expanding university sector attracted more graduate students into geography than the standstill late 1970s. A useful outline of opportunities in geography is given in *Careers in Geography,* 2nd edition (1976) published by the Association of American Geographers. [After J. W. Harbaugh and G. Bonham-Carter, *Computer Simulation in Geology* (Wiley-Interscience, New York, 1970), p. 15, Fig. 1-11.]

sociation of Geographers' membership works in federal establishments. This sort of work began in the 1920s, when geographers played an important part in the Soil Conservation Service. In the succeeding years their involvement has widened to include work for several agencies, such as the Geography Division of the Census Bureau and the Water Resources Board. Despite their growing role, however, American geographers are not as well represented in federal agencies as are their colleagues in countries like the Soviet Union, Sweden, and Britain. In Britain many geographers are working in the newly formed Department of the Environment, where they are engaged in research that ranges from regional planning to local land use. Below the national level extends a hierarchy of regional commissions and local planning agencies, each with geographers represented on them.

Besides working in the public sector, geographers work in the *private* sector as independent consultants or employees of large corporations. Much of their work is in locational consulting, environmental impact studies, and regional intelligence for investment or marketing. The optimal location of a new regional shopping center, new airstrips for an expanded flying doctor service, or the environmental impact of a new industrial plant all involve the kinds of problems discussed in this book and analyzed more deeply in more advanced geography courses. More geographers are establishing their own consulting agencies and so multiplying the opportunities for applied research.

But these broad categories hardly do justice to the range of opportunities that individual geographers—like individuals in any other field—create and develop. All university geographers' mail regularly includes letters (more often, postcards) from former students using their geographic training in careers as unlike as the administration of a regional hospital system and the planning of highways in New Guinea. Geography graduates are also giving more of their spare time to community counseling and are becoming increasingly interested in local planning issues involving ecological or locational decisions at the community level.

25-3
The Future Prospect

There are two elements in the future prospects of geography: first, the trends that can be foreseen from a continuation of movements already occurring; second, the goals toward which we wish to steer. Here we look at each element in turn.

Projections of Existing Trends   Perhaps the only sure forecast we can make about geography is that it will persist. The questions geographers ask are so basic that it is impossible to imagine a world without them, a world where regional differentiation is not significant or where general theories fit local circumstances with sheathlike precision. We certainly expect the causes of spatial differentiation to change, for as we gain uniformity in one sphere we lose it in another. However specialized science becomes, some scholars will still want to integrate and synthesize in the traditional geographic manner.

Beyond this, four more precise trends seem likely. The first projection concerns quantity. As we saw in Figure 25-5, the quantity of geographic research as measured by the number of specialized geographic serials has been doubling every 30 years since around 1780. For the increase in more recent years a variety of other measures is available, including membership in geographic societies, enrollments in classes, degrees granted, research published, and so on. These indicators all confirm that though individual decades are likely to have spurts or slowdowns above or below a long-term rate of increase, it would be unreasonable not to expect an increase in geographic research over the next decade.

A second projection relates to the fission of geography into specialized subdisciplines. Each individual scholar finds increasing economies of scale by specializing in a limited range of problems; limited time and resources (books, equipment, maps, computing time, or whatever) can be concentrated on in-depth study of a limited topic or region. As a result an increasing number of geographers tend to think of themselves as South Asian geographers, or diffusion specialists, or arid-zone geomorphologists. This trend is not confined to geography. The general botanist or zoologist, not to mention the still more general biologist, has long since been displaced in most universities by endocrinologists, conchologists, or other specialist.

A third projection relates to quantification. One of the most striking differences between the research papers in the current geographic journals and those of the 1950s is the greatly increased proportion of research using mathematical techniques. In the earlier period, the main mathematical applications were of spherical geometry in cartography and surveying, and of probability and statistics in climatology. Today the range of mathematical models has significantly expanded, and the applications now affect most branches of the field. We find historical geographers fitting polynomial surfaces to the spread of early settlement, or industrial location analysts modeling siting decisions as Markov chains. During the 1960s there was a skirmish between those geographers anxious to innovate with mathematical methods and those skeptical of their usefulness in solving orthodox problems. Today the general acceptance of such techniques, the more complete mathematical training of a new generation, and the widespread availability of standard computers on campuses make the conflict of a decade ago seem unreal. Mathematical methods are now seen as just one of many tools for approaching geographic problems; these methods are appropriate for some tasks, inappropriate for others.

Paradigm Shifts Within Geography   While the first three trends are linear or S-shaped, the fourth represents a more abrupt switch. Human geographers are getting increasingly uneasy about the "positivist" nature of much geographic research. *Positivism* is a philosophical approach which holds that our sensory experiences are the exclusive source of valid information about the world. This attitude developed in the natural sciences (like physics) but has

PHILOSOPHICAL SCHOOLS IN HUMAN GEOGRAPHY

Our relationship to geographic space and to the natural environment has been viewed differently in various periods of the growth of geography. Some of the "schools" of thought you may encounter in reading further are listed below.

Environmentalism is the view that natural environment plays the major role in determining the behavior patterns of man on the earth's surface.

Normative describes approaches to geography that establish a norm or standard. Thus, they are largely concerned with establishing some optimum condition (e.g., the "best" location or the "best" settlement pattern).

Phenomenology is an existential philosophical school which admits that introspective or intuitive attempts to gain geographic knowledge are valid.

Positivism is a philosophical school which holds that man's sensory experiences are the exclusive source of valid geographic information about the world.

Probabilism is a compromise position between environmentalism and possibilism that assigns different probabilities to alternative patterns of geographic behavior in a particular location or environment.

Methodology and *epistemology* are terms to describe the study of these different schools of thought and their contribution to the philosophy of geography.

been borrowed by geographers working in social-science areas. A positivistic approach leads to the discussion of human behavior in terms of analogies drawn from the natural sciences. Thus in Chapter 18 we discussed human migration in terms of Newton's laws of gravity. Much geographic effort of the 1960s went into trying to explain patterns of human behavior with neat, lawlike statements. Ultimate causes and the essential nature of phenomena like migration were put aside as being unknowable or inscrutable.

Given its historical pattern of evolution, it is understandable that positivism should have played a major part in geographic explanations of phenomena. Whatever its virtues in the more physical parts of the field, its effect on human geography was to lead to a stylized, sometimes overacademic kind of research. What then should we replace it with? The current decade favors renewed interest in a phenomenological approach. *Phenomenology* is an existential philosophical approach which admits that introspective or intuitive attempts to gain knowledge are valid. It accepts subjective categories as appear to be in the experience of the person behaving. The Great Plains are a "desert" if the person or group settling them believes them to be one and acts accordingly! Phenomenologists look with skepticism at the attempts at lawlike statements so characteristic of the previous decade.

This shift in philosophical position has also occurred in other fields like social anthropology and social psychology and has reinforced the links between geography and other social sciences. Of course no geographer's work is ever "purely" positivistic or "purely" phenomenological. Most geographers adopt a position between these two extremes, with systematic approaches to the field (like that in this book) tending to stress the positivistic side and works with a regional emphasis adopting a more phenomenological viewpoint.

These sudden changes of position or "flips" in a subject are different in character from the long-term swings. Recently, light has been thrown on our understanding of this subject by the French mathematician Rene Thom who sees many such changes in the natural world in terms of *catastrophe theory*. By "catastrophe," Thom does not mean "disaster," but simply a sudden change. For example, if we progressively lower the temperature of water there is a sudden switch in state from water to ice. If confidence slowly falls on Wall Street, there may be a sudden loss of nerve in which the bottom abruptly falls out of the stock market. Although catastrophe theory is very complex, it is possible to use its simpler models to throw light on such switches within geography. (See the discussion on the opposite page.)

New Goals for Geography?   Each new generation of geographers builds on earlier work but reinterprets its goals to match the prevailing scientific and social mores. Judging by the interests of the current generation of graduate students or current issues of radical journals such as *Antipode*, geography is likely to become more strongly oriented toward applied fields and practical problems. The questions debated over coffee cups are outward looking, socially relevant, and action-oriented.

PARADIGM SHIFTS

The last twenty years have seen two abrupt paradigm shifts in human geography: the move in the middle 1950s toward a logical positivist view and the move in the middle 1970s toward a more phenomenological approach to the field. Such moves appear to be relatively rapid in terms of the timespan of the subject.

Why do such "jumps" appear more characteristic than gradual, evolutionary change? Let us begin our answer by assuming that a geographer's decision to accept or reject a given paradigm is based on the relative balance of its perceived benefits and costs. (By benefits we mean the insights or analytic advantages that come from a particular philosophical approach.) Acceptance rests on a net balance of benefits over costs, and vice versa. The diagonal line in (a) therefore represents an indifference line where costs and benefits are exactly equal; it marks the boundary between acceptance and rejection. We may go on to represent the same line in a three-dimensional version [as in (b)].

But this simple accept/reject dichotomy as shown in (b) is a gross oversimplification of geographers' actual behavior. For, where both costs and benefits are very low, we may be indifferent to either outcome, i.e., the paradigm is too weak to be worth either supporting or opposing. Where costs and benefits are very high we

may tend to persist with a given decision once we have adopted it. Both these modifications are taken into account in (c) by showing the acceptance/rejection decision as a surface with a symmetrical fold lying along the axis of the indifference line shown in the two preceding figures.

Adoption of this revised version of the acceptance or rejection of paradigms within geography has some interesting implications. To show these, let us project the surface shown in three dimensions in (c) onto the two-dimension base delimited by the costs and benefits axes. The fold in the surface now forms a triangle shaded in (d). Note that the shaded triangle may now take on three *different* series of values since it is essentially a triple-sheeted fold in the original acceptance/rejection surface. Two important perspectives on our observed behavior follow:

(1) *Persistence of attitudes once a position has been adopted.* Consider the position of a geographer at location 1 in (d). If we assume benefits remain unchanged but evidence on the cost of maintaining an existing paradigm builds up, his position shifts horizontally across the benefit/cost space. Since he is on the upper part of the fold shown in (c) he changes to rejection *not* at point 2 when benefits are exactly equal to costs [i.e., the indifference line of (a)] but at point 3 where he suddenly falls from the

upper "accept" to the lower "reject" surface. Contrast this with the position of an initial rejector located at point 4 in (e). In this case the assumptions are reversed and his position moves over the lower surface of the fold as evidence on the benefits is accrued. Again attitude change occurs not at point 5 but at point 6 when he suddenly leaps onto the upper surface of the fold. Of course, the examples chosen are oversimplified in order to make a point; actual patterns might be expected to be much more complex.

(2) *Divergence of attitudes from initially similar positions.* The position of two geographers located originally at points 7 and 8 is shown in (f). Both the cost and benefits are small, and both people may be expected to be indifferent to the positions they hold. However, if the data on costs and benefits accrue over time in the way shown by the parallel paths, we can see how our two geographers diverge. One follows the upper surface of the fold, the other the lower. Note that although given exactly the same evidence, one geographer increasingly accepts the paradigm while the other increasingly rejects it. In other words we dig in our heels to hold our initial positions ever more strongly since there is now so much more intellectual capital at stake. See P. Haggett, *Midterm Futures for Geography* (Monash University Publications in Geography, No. 16, Melbourne, 1977).

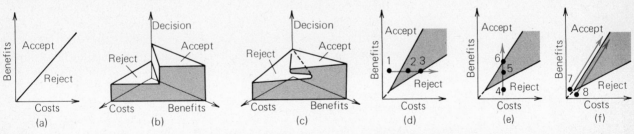

(a)   (b)   (c)   (d)   (e)   (f)

Can we help society to chart a middle way between the mindless exploitation and pollution of the natural world implicit in a short-run materialist economy and the unrealistic pipe dreams of a pastoral, protected, but unproductive world? Can we predict the welfare implications of different types of locational situations or spatial arrangements? Can we help to redraw political boundaries in order to equalize resource opportunities and also minimize the likelihood of future conflicts? What will the world geography or that of the United States look like by A.D. 1990 or A.D. 2030? These and many other questions are being asked today.

If these questions are typical of the broad, long-term issues young geographers wish to tackle, there will have to be some significant departures from the presently projected trends in the field. For example, instead of increasing specialization, there will have to be a greater emphasis on extending the ecosystems approach (see Chapter 3, Section 3-2), still largely confined to the physical and biological world, to include our own environment-modifying activities. Physical and human geographers will have to spend more time at each other's seminars to exploit their unique opportunities for cooperative research. A rapprochement will be needed between those interested in building quantitative models and those interested in the realities of individual regional complexes. Geographers will also need to stop acting as "terrographers" and devote more time to at least the continental margins of the ocean-covered 70 percent of the earth. Finally, more geographers will be needed outside the classroom, concerning themselves with the consequences of their learning.

That there are problems to be overcome is undeniable if geographers are to make their fullest contribution—improving, as far as they are capable, the relationship between us and our environment. Geographers have no special philosopher's stone that gives them instant insight or a divine right to be heard or consulted. The full credentials of painstaking research and tested theory are still being established. Their special curiosity is at once a strength and a weakness, and geographers are conscious that their own studies often lack the rigor of an econometrician's or the temporal perspective of an historian.

We have tried to show areas where these difficulties are being faced and overcome—and others where significant progress remains to be made. The book will have served its purpose if it attracts students from other disciplines into those areas that have, for too long, remained a narrow geographer's monopoly. These students may help to provide new insights and solutions to some of the puzzles that have, for too long, baffled the small group of geographic scholars. We hope that you will accept the invitation—and the implied challenge offered here—and go on from this brief and prefatory book to the rewarding geography courses that lie ahead.

Summary

1. Although geography is a small academic field, it has shown a rapid exponential rate of growth over the last 200 years. Some recent decades have shown above average growth (notably, the 1960s), while growth is slower in the present decade.

2. The history of the field shows three distinct phases. First, a period dominated by the ideas of individual scholars. Second, that of the geographic societies which began to develop in the late eighteenth century with the purpose of bringing together scholars with common research interests. Third, the foundation of university geography departments from the mid-nineteenth century and national geographic research centers. From 1922 the work of geographic research in individual countries has been coordinated by the International Geographical Union (IGU).

3. Although different definitions of geography abound, there are common elements in all such definitions. These include a focus of attention on the earth's surface, a concern with the earth as the home of humanity, with the spatial organization of the human population and with its ecological relationship to the global environment. All definitions stress geographers' awareness of the great spatial variety of the planet Earth, and the distinctiveness of its regions.

4. In this book three different approaches to geography have been illustrated: spatial analysis, ecological analysis, and regional complex analysis. These approaches are based on the essential unity of the physical and nonphysical elements within geography.

5. Jobs in geography are found both in college and university teaching and research, and outside academe in public service and private industry. Geographers in the public sector have worked largely in regional and local planning; geographers in the private sector have worked mainly in locational consulting, environmental impact studies, and regional intelligence.

6. Trends in geography during the 1960s were directed toward further specialization and quantification. The general emphasis was on a positivistic approach to geographic phenomena. Trends in the 1970s have been in an opposite direction, emphasizing the humanistic elements in the field and a concern with applied problems. These shifts within the philosophy of the field may be seen in terms of Kuhn's concept of paradigm shift.

Reflections

1. Look back at the notes you made in response to Question 1 at the end of the first chapter. How accurate were your predictions? Compare your own findings with those of the rest of the class.

2. "Adopt" any one of the individuals named in Figure 25-2. Browse through at least two of his or her works, and, if the person is now deceased, read through his or her obituary notice in one of the leading geographic journals. (The *Annals of the Association of American Geographers* is a particularly useful source of information on North American geographers.) List three major influences on the person's work.

3. Read through the definitions of geography in Table 25-1. Which is closest to your own definition of geography? Why?

4. Which subjects in other fields do you consider most useful for geography students? Justify your choice.

5. How should the research agenda of geography (or any academic field) be decided? Evaluate the relative importance of (a) experience (i.e., the past history of research endeavors, successful and unsuccessful), (b) the views of the present generation of geographers, and (c) the needs of society in determining this agenda.

6. Review your understanding of the following concepts:

the Matthew effect	quantification
spatial analysis	positivism
ecological analysis	phenomenology
regional complex analysis	paradigm shifts

One Step Further . . .

By the time you reach this point in the book, many of you will have had your fill of "further readings" in geography. But those few of you who are thinking of going further in the field might like to look at a publication that provides useful information on geography departments in North America:

Association of American Geographers, *Guide to Graduate Departments in the United States and Canada* (AAG Washington, D.C., annual).

For information on departments in other countries turn to The World of Learning *(Europa Publications, London, annual).*

The job market in geography and the career options open to graduating geographers are outlined in

Natoli, Salvatore J., Ed., *Careers in Geography* (Association of American Geographers, Washington, D.C., 1976).

Historical perspectives on the growth of geography are given in

James, Preston E., *All Possible Worlds: A History of Geographical Ideas* (Odyssey, New York, 1972).

While some of the radical directions being taken in the present decade are described in

Peet, R., Ed., *Radical Geography* (Methuen, London, 1978).

Serious scholars will want to delve into the two classic statements on the field by a leading geographic philosopher, Richard Hartshorne:

The Nature of Geography: A Survey of Current Thought in the Light of the Past *(Association of American Geographers, Lancaster, Pa., 1946) and*

Perspectives on the Nature of Geography *(Rand McNally, Skokie, Ill., 1959).*

Geographers' ideas about their subject are constantly changing and evolving. The Annals of the Association of American Geographers *(a quarterly) and the* Publications of the Institute of British Geographers *(a semiannual) carry reviews of current changes in the AAG or IBG presidents' "State of the Union" addresses to the annual conference. You might also look at* Progress in Physical Geography *and* Progress in Human Geography *(both quarterlies) which are international journals devoted exclusively to papers reviewing current developments in specialized fields within geography.*

Geographers are by definition internationally minded people. Most of the world's population measures its environment in terms of the metric system, and an increasing proportion of scientific research is now reported in metric terms. Measurements in this book are therefore given in the metric form. Nonmetric equivalents are given, in most cases, in parentheses. In a few cases where the historical contexts demand it—as in the discussion of the township and range system—the original nonmetric forms are retained in the discussion. Temperatures are given in degrees centigrade, with fahrenheit equivalents in parentheses.

appendix A
Conversion Constants

Measures of Distance

Kilometers to miles (1 km = 0.62137 mi)

1	2	3	4	5	6	7	8	9	10	km
0.62	1.24	1.86	2.49	3.11	3.73	4.35	4.97	5.59	6.21	mi

Meters to feet (1 m = 3.28084 ft)

1	2	3	4	5	6	7	8	9	10	m
3.28	6.56	9.84	13.12	16.40	19.69	22.97	26.25	29.53	32.81	ft

Centimeters to inches (1 cm = 0.393701 in.)

1	2	3	4	5	6	7	8	9	10	cm
0.39	0.79	1.18	1.57	1.97	2.36	2.76	3.15	3.54	3.94	in.

Square kilometers to square miles (1 km^2 = 0.386102 mi^2)

1	2	3	4	5	6	7	8	9	10	km^2
0.39	0.77	1.16	1.54	1.93	2.32	2.70	3.09	3.47	3.86	mi^2

Hectares to acres (1 hectare = 2.471054 acres)

1	2	3	4	5	6	7	8	9	10
2.47	4.94	7.41	9.88	12.36	14.83	17.30	19.77	22.24	24.71

Measures of Temperature

Degrees centigrade to degrees fahrenheit [$°C = \frac{5}{9}(°F - 32)$]

−80	−70	−60	−50	−40	−30	−20		°C
−112	−94	−76	−58	−40	−22	−4		°F
−10	0	10	20	30	40	50	60	°C
14	32	50	68	86	104	122	140	°F

Exponential Scales

Large numbers are frequently easier to describe in an exponential form. Thus scientists will often say 10^6 instead of 1,000,000, 10^7 instead of 10,000,000, 10^8 instead of 100,000,000, and so on. The number 10 is the most convenient base for an exponential scale, since conversions can be accomplished merely by moving decimals. Thus, typical conversions are

$$10,800,000 = 1.08 \times 10^7$$
$$2,370 = 2.37 \times 10^3$$
$$-481,000 = -4.81 \times 10^5$$
$$0.00171 = 1.71 \times 10^{-3}$$

Introductory courses in geography vary from country to country, between universities and colleges, and from department to department at the same level. The following chart outlines ways in which this book may be modified from its original purpose—to serve as the basis of a full one-semester introduction to geography—in order to meet the needs of shorter and more specialized introductory courses. Syllabuses and teaching strategies for such courses are discussed at length in the excellent series of publications of the Association of American Geographers Commission on College Geography. See especially *New Approaches in Introductory Courses* (#4, 1967), *Introductory Geography: Viewpoints and Themes* (#5, 1967), and *Geography in the Two-Year Colleges* (#10, 1970).

appendix B
Using the Book in Introductory Courses

In using the chart, instructors should note the following points:

1.   The sequences charted here are *only* suggestions. In any book, chapters have to be arranged in a linear (and hopefully logical) sequence. The material in this book has, however, a much more complex structure, and the sequence of chapters represents only one of many possible compromises. Indeed, the twenty-five chapters could be arranged in more than a billion different ways! The text has been redesigned into more modules than the earlier editions so that instructors can more easily rearrange and structure material to meet individual needs.

2.   Each sequence assumes that one week of each teaching period (either a semester or a quarter) will be used for reviews, tests, and the like.

3.   *Introduction to Geography.* The book is designed to provide material for a full 20-week, one-semester course introducing beginning students at the college level to the richness and range of geography. (The basic philosophy behind this approach is discussed in the *Preface.*) The chapter sequences in the chart suggest how the book can be adapted for use in shorter introductory courses.

4.   *Introduction to Cultural or Human Geography.* In many colleges the introductory geography courses are split into two halves—an environmental sciences course (an "Introduction to Physical Geography") and a social sciences course (an "Introduction to Human Geography" or an "Introduction to Cultural Geography"). The chart suggests ways in which this book might be used in such courses. How many of the chapters not directly included should be assigned for collateral reading or skipped entirely is a matter for the instructor to judge in the light of time constraints and the background of members of the class.

LENGTH OF TEACHING PERIOD	TITLE OF COURSE	Prologue 1 2	The Environmental Challenge 3 4 5 6	Human Ecological Response 7 8 9 10	Regional Mosaics 11 12 13	Regional Hierarchies 14 15 16 17 18	Interregional Stresses 19 20 21 22	Epilogue 23 24 25
One semester (20 weeks)	Introduction to Geography	◖ ◖	○ ○ ◐ ◖	○ ○ ○ ◖	○ ○ ◖	○ ○ ○ ◖ ◖	◖ ◖ ○ ◖	◖ ◖ ○
	Cultural/Human Geography	◖	◐ ◐ ○ ◖	● ● ○ ○	● ● ◖	○ ○ ○ ○ ○	○ ○ ○ ◖	◖
	Economic Geography	◖	◖ ◖	○ ○ ○ ◖	◖	● ● ● ● ●	○ ○ ○ ●	○ ◖
One semester (16 weeks)	Introduction to Geography	◖	○ ○	○ ○ ○ ◖	○ ○ ◖	○ ○ ○	◖ ◖ ○ ○	◖
	Cultural/Human Geography	◖		● ○ ◖ ○	● ● ○	○ ○ ○	○ ○ ◖ ◐	◖
	Economic Geography	◖		○ ○		● ● ○ ○ ○	○ ○ ○ ●	○ ◖
One quarter (10 weeks)	Introduction to Geography	◖	○ ○	○ ◖ ◖	○	○ ○	○	◖
	Elements of Physical Geography, Cartography, General techniques	○	● ● ○ ○	◖				○ ○ ◖
	Cultural/Human Geography			○ ○ ○ ◖	● ● ○	○	○	
	Economic Geography			○ ○		○ ○ ○ ○ ○	○	○ ○
	Senior Seminar/Colloquium		○ ○	○	○	○	○	○ ○ ○

○ One-week teaching period   ● One-and-a-half-week teaching period   ◖ Half-week teaching period

5. *Introduction to Economic Geography.* Another common variant is to introduce beginning students to geography through the medium of spatial analysis. Schemes for using this book in courses that take this approach—usually listed in the catalog as "Introduction to Economic Geography"—are outlined in the chart. Again, the emphasis is on introducing students to the most elementary principles of the discipline, and the course should be regarded as a broadly based forerunner of more detailed specialized courses.

6. *Other specialized one-quarter (10 week) introductory courses.* A fourth possible use for this book is in brief introductory courses in particular areas. Two examples of such usage are given in the chart. First, a course concentrating on "Elementary Physical Geography, Cartography, and Geographic Techniques" for students with little previous acquaintance with these topics. Second, more advanced topics in each chapter can be pursued in the "Senior Seminar/Colloquium."

Index